# INSTANT NOTES IN SPORT AND EXERCISE BIOMECHANICS

This is the clearest and most straightforward biomechanics textbook currently available. By breaking down the challenging subject of sport and exercise biomechanics into short thematic sections, it enables students to grasp each topic quickly and easily, and provides lecturers with a flexible resource that they can use to support any introductory course on biomechanics.

The book contains a wealth of useful features for teaching and learning, including clear definitions of key terms, lots of applied examples, guides to further reading, and revision questions with worked solutions. It has been significantly expanded to encompass rapidly developing areas, such as sports equipment design and modern optoelectronic motion analysis systems, and it includes a number of new sections that further develop the application of biomechanics in sports performance and injury prevention. A new companion website includes a test bank, downloadable illustrations and, where appropriate, suggestions for learning outcomes and/or lab-based sessions for lecturers.

*Instant Notes in Sport and Exercise Biomechanics* has been an invaluable course companion for thousands of students and lecturers over the last decade. Engaging, direct, and now fully refreshed, it is the only biomechanics textbook you'll ever need.


**Paul Grimshaw** is Associate Professor in Sports Engineering within the School of Mechanical Engineering at the University of Adelaide, Australia, where he is also Director of Engagement and Bachelor of Engineering (Mechanical and Sports) Program Coordinator. His main research interests are concerned with the prevention of injury in the knee and lower back, especially in golf.

**Michael Cole** is Senior Lecturer in Neuroscience and Motor Control within the School of Exercise Science at the Australian Catholic University, Australia. He has been involved in a number of multi-disciplinary research programmes that focus on injury prevention and span the areas of sports biomechanics, clinical biomechanics and neuroscience. His particular areas of expertise are in the assessment of postural instability, gait disability and falls in people with Parkinson's disease and the evaluation of invasive and non-invasive therapies for movement disorders.

**Adrian Burden** is Reader in Sport and Exercise Biomechanics in the Department of Sport and Exercise Sciences at Manchester Metropolitan University, UK, and is a Senior Fellow of the Higher Education Academy. He has held workshops on surface electromyography on behalf of the British Association of Sport and Exercise Sciences, and his research interests lie in the application of electromyography in clinical, sport and exercise settings.

**Neil Fowler** is Dean of Students at the University of Salford, UK. Before joining the University of Salford, Neil held the roles of Head of Exercise and Sport Science and Associate Dean of Learning Teaching and Quality Enhancement at Manchester Metropolitan University, UK. His research interests include spinal mechanics and pain management, the energetics of wheelchair propulsion and Paralympic sport, and the biomechanics of jumping.

# INSTANT NOTES IN SPORT AND EXERCISE BIOMECHANICS

SECOND EDITION

PAUL GRIMSHAW, MICHAEL COLE,
ADRIAN BURDEN AND NEIL FOWLER

Routledge
Taylor & Francis Group
LONDON AND NEW YORK

Second edition published 2019
by Routledge
2 Park Square, Milton Park, Abingdon, Oxon, OX14 4RN

and by Routledge
52 Vanderbilt Avenue, New York, NY 10017

*Routledge is an imprint of the Taylor & Francis Group, an informa business*



First edition published by Garland Science 2006

*British Library Cataloguing-in-Publication Data*
A catalogue record for this book is available from the British Library

*Library of Congress Cataloging-in-Publication Data*
A catalog record has been requested for this book

ISBN: 978-1-138-64023-8 (hbk)
ISBN: 978-1-138-64024-5 (pbk)
ISBN: 978-1-315-63676-4 (ebk)

Typeset in Sabon
by Swales & Willis Ltd, Exeter, Devon, UK

Visit the companion website: www.routledge.com/cw/grimshaw

The authors would like to dedicate this text to Professor James G. Hay (1936–2002) whose book *The Biomechanics of Sports Techniques*, first published in 1973, inspired us all to develop careers in the area of sport and exercise biomechanics.

# CONTENTS

## SECTION G
## Measurement and analysis techniques **315**



## SECTION H
## Appendices **383**

# PREFACE

This second edition of *Instant Notes in Sport and Exercise Biomechanics* follows the same successful format as the first edition by presenting a series of "Instant Notes" or "Lecture Topic Areas" that cover the fundamental principles of biomechanics in a thorough yet succinct way. This fully revised and updated version of the text contains a wealth of useful features for teaching and learning, including: i) more applied examples; ii) clearer definitions of key terms; iii) a greater number of problems and solutions; iv) revision questions with worked solutions; and v) further readings to guide extended learning. Furthermore, the book has been significantly expanded to encompass rapidly developing areas, such as sports equipment design and modern optoelectronic motion analysis systems, and it includes a number of new sections that further develop the application of biomechanics in sports performance and injury prevention. For the most part, the changes made to the second edition of the text have focused on ensuring its contemporaneity, by introducing and discussing concepts, technologies and applications that are relevant to current and future biomechanics students. To improve continuity amongst the many sections of this revision, the authors have endeavoured to use the *Système International* terminology (SI units) throughout. However, in some cases, it was necessary to use alternative (non-SI) units that are also commonly used in the field of biomechanics.

# ACKNOWLEDGEMENTS

As lead author of this book, I would like to take this opportunity to thank Michael Cole for coming on board with this project to replace Adrian Lees, who has since retired from the field of sports biomechanics. Michael's commitment to getting this second edition "over the line" has been the continuous source of motivation that we all needed. Finally, I am sure I speak for all the authors in acknowledging the support of our respective families and friends. In closing, I want to say thank you to my wife, Merilyn Lock, because without her encouragement during the planning of this second edition and her initial help with my sections, this revision, quite simply, would not have happened.

Paul Grimshaw

## SECTION A

# Kinematics of linear motion

# LINEAR MOTION

A1

*Paul Grimshaw*

**Biomechanics** can be defined as the study of forces and the effects of these forces on living things. In mechanics, further subdivision includes **kinematics** and **kinetics. Kinematics** is the study of displacement, velocity and acceleration (spatial and temporal) while **kinetics** examine forces, moments and torques. Biomechanics and mechanics are used extensively to study human motion. *Figure A1.1* defines biomechanics and kinematics in more detail.

Human movement can often be classified into components of **linear and angular motion.** This leads to a description that is termed general motion. **Linear motion** (or translation) is movement along a line which may be either straight or curved and where all the body parts are moving in the same direction at the same speed. This can be classified as either rectilinear motion (motion along a straight line) or curvilinear motion (motion along a curved line). Angular motion (discussed elsewhere in this text) involves movement around an axis (either imaginary or real) with body parts (or individual body parts) moving through angles (the same or different angles) in a certain time frame (*Figure A1.2*).

**Linear kinematics** deal with quantities that describe the motion of bodies such as **distance, displacement, speed, velocity** and **acceleration.** These can be classified as either scalar or vector quantities. **Scalar quantities** are represented by magnitude (size) only, whereas **vector quantities** possess both magnitude and direction. Vector quantities are represented mathematically, as a symbol with a direction sign or graphically by scaled straight lines or arrows. For example, **speed** is defined as the **distance travelled** per unit of **time (i.e. 12 m/s (metres per second) or 26.8 mph (miles per hour)**) and as such is a scalar quantity (it has magnitude but no direction).

$$\text{Speed} = \frac{\text{Distance travelled (miles)}}{\text{Time taken (hours)}}$$

## EXAMPLE

If an athlete ran a marathon 26 miles 385 yards (42 kilometres (km)) long in 2 hours and 28 minutes, what was the athlete's average speed in miles per hour (mph) and in kilometres per hour (km/h)?

Speed = distance/time

Distance = 26 miles 385 yards or 42 kilometres

Time = 2 hours 28 minutes or 2.47 hours

**Biomechanics**

The study of forces and the effects of forces on living things

**Linear motion** (or translation) is movement along a line that is either straight or curved – **Angular motion** (or rotation) is movement about an axis of rotation (either real or imaginary)

**Linear kinematics**
A sub-division of mechanics that is concerned with distance, displacement, speed, velocity and acceleration

**Angular kinematics**
A sub-division of mechanics that is concerned with angular displacement, angular velocity and angular acceleration

**Linear kinetics**
A sub-division of mechanics that is concerned with forces that cause or result from movement

**Angular kinetics**
A sub-division of mechanics that is concerned with torques and moments that cause or result from movement

**Figure A1.1** Biomechanics, kinematics and kinetics

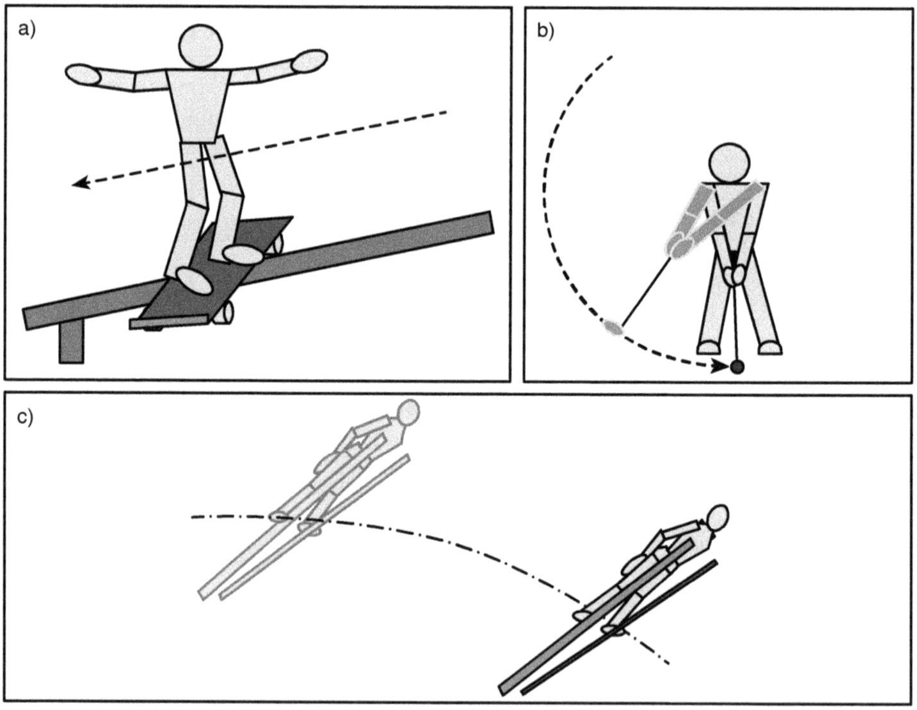


**Figure A1.2** Different types of motion: a) linear (translatory), b) angular and c) curvilinear

## Miles per hour (mph)

1 mile = 1760 yards or 1609.344 metres (or 1.609 km)

Marathon = 26.219 miles

Speed = 26.219/2.47

Speed = 10.615 mph

## Kilometres per hour (km/h)

26 miles 385 yards = 42 km

Speed = 42/2.47

Speed = 17.00 km/h

***Note: these values are the average speed in mph or km/h over the entire race.***

This would represent the **average speed** for this athlete over the whole 42 km marathon running activity. Speed in this case is a scalar quantity and is expressed in magnitude only (i.e. 10.615 mph or 17 km/h). In this example while the speed was presented in mph and km/h we could have expressed speed in many different units, for example **metres/second (m/s),** which is the standard SI (the international system of units from the French *Système International*) unit for speed or velocity. The athlete covered the distance of 26 miles 385 yards but we do not know whether this was in a straight line, in a series of curves, or indeed in a circle starting and finishing at the same point. In this example, the term speed is used because there is no directional component specified.

Also in this example, the term **distance** was used because there was no directional component to the race (or at least we did not know it). If a direction of the race was known (north-east, or a straight line between points A and B), the term **displacement** could be used. The result would have been the same (because the athlete covered the same distance/displacement in the same time) but the quantity would be a vector quantity because there would now be a directional component to the solution. This vector quantity could be expressed mathematically or graphically to scale by an arrow on a piece of paper (*Figure A1.3*).

Often within biomechanics it is useful to be able to express both **speed** and **velocity** components. Speed is a scalar quantity and velocity is a vector quantity but the units for each can be the same.

Speed = distance/time (mph, m/s, km/h)

Velocity = displacement/time (mph, m/s, km/h)

Sometimes it is only the average speed that is of interest (such as when an athlete runs a marathon race and the coach is interested in getting a quick and simple overall measure of how the race was performed). As this **average speed** would be presented over a 26.219-mile running distance it does not really describe the specific and important details of the race but it may be a useful basic measure for training.

Both **speed** and **velocity** can be uniform or non-uniform quantities. **Uniform** describes motion that is constant over a period of time (i.e. constant velocity or speed (no acceleration or deceleration)) and **non-uniform** describes varying or changing velocity or speed over time (i.e. with some acceleration or deceleration). In human motion, it is usually the knowledge of non-uniform motion that is more beneficial to the athlete, coach, scientist and student of

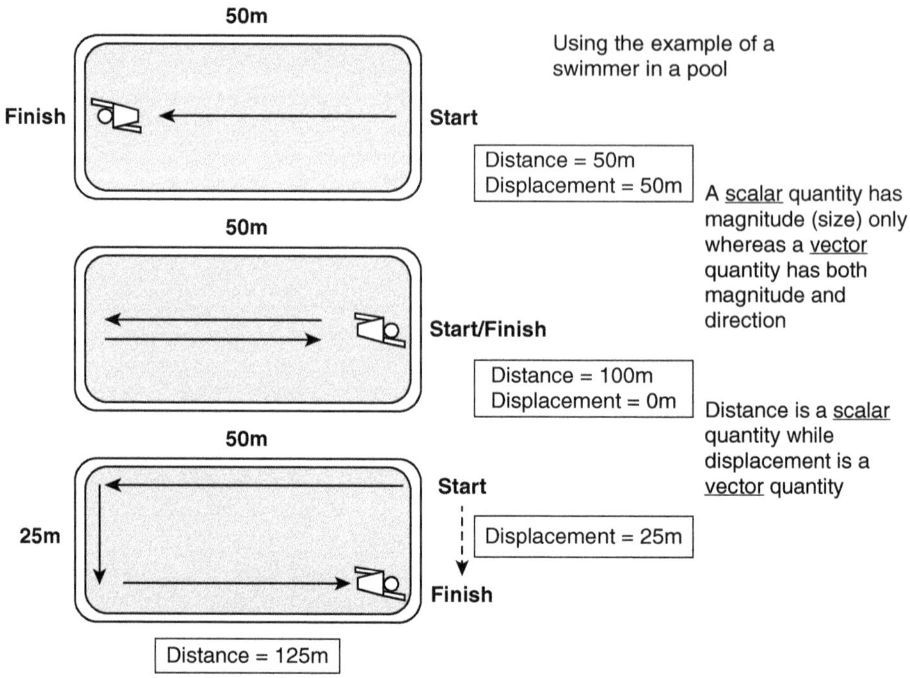


**Figure A1.3** Defining the terms distance and displacement, scalar and vector

biomechanics. For example, in the case of our marathon runner it would be more useful to know what changes in the runner's speed or velocity occurred throughout the activity. Such information would have important training and performance implications and would be as valuable in a sprint race lasting no more than 10 seconds (i.e. 100 m sprint) as it would be in a marathon event lasting several hours.

**Linear velocity** (vector) and linear **acceleration** (also a vector) are important measures within biomechanics that are used to describe and analyse the motion of human bodies. *Table A1.1*

**Table A1.1** Data for a 100 m sprint world record performance (RT = Reaction time)

| *Displacement (m)* | *Time (s) 10 m intervals* | *Cumulative time (s)* |
|---|---|---|
| **RT** | **0.165** | |
| 10 | 1.85 | 1.85 |
| 20 | 1.02 | 2.87 |
| 30 | 0.91 | 3.78 |
| 40 | 0.87 | 4.65 |
| 50 | 0.85 | 5.50 |
| 60 | 0.82 | 6.32 |
| 70 | 0.82 | 7.14 |
| 80 | 0.82 | 7.96 |
| 90 | 0.83 | 8.79 |
| 100 | 0.90 | 9.69 |

illustrates the 100 m sprint data from a world record performance (9.69 seconds) by Usain Bolt (Jamaica) set in 2008. Note: this is not the current world record (which at the time of this publication is 9.58 seconds, also set by Bolt in Berlin, Germany in 2009). Also indicated in this data is the reaction time (RT) that Usain Bolt had as he came out of the starting blocks. This time is inclusive in the 1.85 seconds to the first 10 m point.

From *Table A1.1* it is possible to see that the athlete covered the 100 m displacement (horizontal displacement in a straight line along a track) and that this 100 m displacement is divided into 10 m sections or intervals. For example, the first 10 m was covered in 1.85 seconds and the second 10 m in 1.02 seconds (or 20 m in 2.87 seconds (cumulative time)). It is possible to see from this data that the athlete covered the whole 100 m displacement in 9.69 seconds. We can now use this data to determine **average velocity** over smaller increments (such as every 10 m interval). Such information would provide us with a biomechanical description of the whole 100 m event in 10 m intervals.

## TASK

Use the data provided (*Table A1.1*) in Microsoft Excel to determine the average velocity over each 10 m interval and then plot the velocity (m/s) versus time (s) graph.
For each 10 m interval you should use the formula:

$$\text{Velocity} = \frac{\text{displacement}}{\text{time}}$$

For the first 10 m point this would be:

$$\text{Velocity} = \frac{10}{1.85} = 5.41 \text{ m/s}$$

The presentation and analysis of this velocity can be seen from the consideration of the calculations and data identified in *Table A1.2* (which was derived from a Microsoft Excel spreadsheet). Note: it is important to point out that this can be expressed as velocity (a vector quantity) because we have a directional component (i.e. horizontal displacement along a straight 100 m track) and even though we are considering the velocity (average) over much smaller increments (i.e. 10 m intervals) it is still an average velocity over that horizontal displacement interval or section. In this context taking even smaller time or displacement intervals will eventually lead to an "**instantaneous**" value for the calculation of speed or velocity. Such analysis provides a more detailed biomechanical breakdown of the event of the 100 m sprint race.

The data shown in *Table A1.2* such as the average velocity for the whole race (10.92 m/s), the specific velocity for each 10 m section of the race (which we calculated through Microsoft Excel) or "instantaneous" values for smaller displacement or time increments (if we had the data) could be compared with values for other Olympic and World athletic performances. Obviously, such knowledge of individual and comparative performances would have important training implications for both the athlete and the coach.

**Acceleration** is defined as the **change in velocity per unit of time** and it is usually measured in metres per second squared ($m/s^2$). This means the velocity of an object will increase/decrease by an amount for every second of its motion. For example, a constant (uniform) acceleration of 2.5 $m/s^2$ indicates that the body will increase its velocity by 2.5 m/s for every second of its motion (2.5 m/s for 1 second, 5.0 m/s for 2 seconds, 7.5 m/s for 3 seconds and so on). *Table A1.3* shows the Microsoft Excel calculations that now include the acceleration data for the 100 m sprint performance used in the previous example.

**Table A1.2** Microsoft Excel calculations (to two decimal places) of the velocity for each 10 m interval in the 100 m sprint data (d = displacement; dt = time for each 10 m; sum t = cumulative time; v = velocity; sd = standard deviation; max = maximum value; min = minimum value)

| *d (m)* | *dt (s)* | *sum t (s)* | *v (m/s)* |
|---|---|---|---|
| 10 | 1.85 | 1.85 | 5.41 |
| 20 | 1.02 | 2.87 | 9.80 |
| 30 | 0.91 | 3.78 | 10.99 |
| 40 | 0.87 | 4.65 | 11.49 |
| 50 | 0.85 | 5.50 | 11.76 |
| 60 | 0.82 | 6.32 | 12.20 |
| 70 | 0.82 | 7.14 | 12.20 |
| 80 | 0.82 | 7.96 | 12.20 |
| 90 | 0.83 | 8.79 | 12.05 |
| 100 | 0.90 | 9.69 | 11.11 |
| total | 9.69 | | |
| average | 0.97 | 5.86 | 10.92 |
| sd | 0.32 | 2.59 | 2.08 |
| max | 1.85 | 9.69 | 12.20 |
| min | 0.82 | 1.85 | 5.41 |

**Table A1.3** Microsoft Excel calculations (to two decimal places) of the acceleration for each 10 m interval in the 100 m sprint (d = displacement; dt = time for each 10 m; sum t = cumulative time; v = velocity; a = acceleration; sd = standard deviation; max = maximum value; min = minimum value)

| *d (m)* | *dt (s)* | *sum t (s)* | *v (m/s)* | *a (m/s²)* |
|---|---|---|---|---|
| 10 | 1.85 | 1.85 | 5.41 | 2.92 |
| 20 | 1.02 | 2.87 | 9.80 | 4.31 |
| 30 | 0.91 | 3.78 | 10.99 | 1.30 |
| 40 | 0.87 | 4.65 | 11.49 | 0.58 |
| 50 | 0.85 | 5.50 | 11.76 | 0.32 |
| 60 | 0.82 | 6.32 | 12.20 | 0.52 |
| 70 | 0.82 | 7.14 | 12.20 | 0.00 |
| 80 | 0.82 | 7.96 | 12.20 | 0.00 |
| 90 | 0.83 | 8.79 | 12.05 | –0.18 |
| 100 | 0.90 | 9.69 | 11.11 | –1.04 |
| total | 9.69 | | | |
| average | 0.97 | 5.86 | 10.92 | 0.87 |
| sd | 0.32 | 2.59 | 2.08 | 1.60 |
| max | 1.85 | 9.69 | 12.20 | 4.31 |
| min | 0.82 | 1.85 | 5.41 | –1.04 |

Acceleration can be represented by the equation:

$$\text{Acceleration} = \frac{v_2 - v_1}{t_2 - t_1}$$

where:

$v_1$ = initial velocity of the object at time $t_1$ (the beginning time point that we are interested in)

$v_2$ = final velocity of the object at time $t_2$ (the final time point that we are interested in)

*Positive acceleration* is when the velocity increases over a time period (speeding up) and *negative acceleration* is when the velocity decreases over a time period (slowing down).

Now that we have both the velocity and the acceleration data for each of the 10 m intervals it is possible, again using Microsoft Excel, to plot the velocity versus displacement interval (or time interval) and acceleration versus displacement interval (or time interval) so that we can analyse each section of this 100 m race.

From these graphs, it is possible to see that Usain Bolt is both accelerating and decelerating (speeding up and slowing down) throughout the whole activity. If we now look at the velocity versus time graph (shown in *Figure A1.4*) we can see that it is possible to read values directly from this graph for specific time points (e.g. 6.32 seconds into the race). Between 0 and 6.32 seconds we can see that there is a continuous positive acceleration (indicating the athlete is speeding up over this period of time). This positive acceleration continues at different rates up to the 6.32 s point. For example, the initial acceleration is high (2.92 and 4.31 m/s$^2$) as the athlete is sprinting rapidly out of the starting blocks and going from a position of no movement to moving very quickly. Then the athlete continues to accelerate (albeit at a much lesser rate +1.30, +0.58, +0.32 and +0.52 m/s$^2$ which are all positive values) to around the 60 m or 6.32 s mark. This effect can also be seen in the velocity curve in that it is seen to continually increase to this point. Then at 7.14 s the athlete is seen to have a zero value for acceleration,

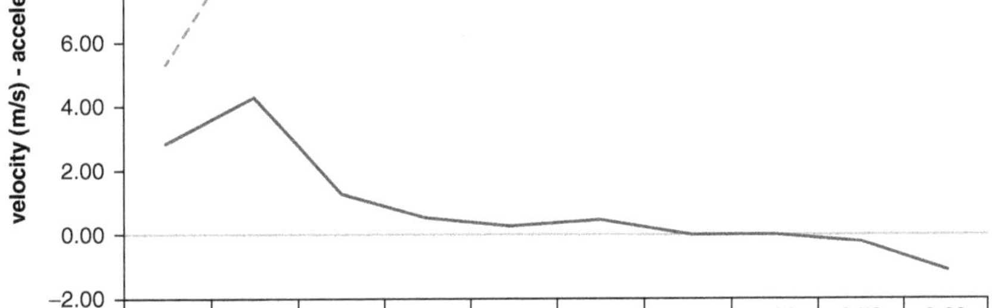


| | 1.85 | 2.87 | 3.78 | 4.65 | 5.5 | 6.32 | 7.14 | 7.96 | 8.79 | 9.69 | Time (s) |
|---|---|---|---|---|---|---|---|---|---|---|---|
| Velocity (m/s) | 5.41 | 9.80 | 10.99 | 11.49 | 11.76 | 12.20 | 12.20 | 12.20 | 12.05 | 11.11 | |
| Acceleration (m/s$^2$) | 2.92 | 4.31 | 1.30 | 0.58 | 0.32 | 0.52 | 0.00 | 0.00 | −0.18 | −1.04 | |

**Figure A1.4** Microsoft Excel graphs of velocity and acceleration against time for the 100 m sprint world record performance of 9.69 seconds set by Usain Bolt in 2008

which continues until 7.96 s. This means that during this time the athlete has **constant velocity** and is neither accelerating nor decelerating (neither speeding up nor slowing down). Again, this is seen in the velocity curve with a straight line or plateau at these points. Then from this last point onward there is a negative value seen in the acceleration, which means that the athlete is slowing down (at 8.79 seconds there is a value of $-0.18$ m/s$^2$ and at the finish line (9.69 s) a value of $-1.04$ m/s$^2$). This indicates the athlete is slowing down as they approach and cross the finish line (which is a clear characteristic we can often see in Usain Bolt's technique when watching him run). Again, this is seen in the velocity curve as the curve is decreasing during this time period.

In the context of these graphs the values that are plotted are between the points of displacement or time (i.e. indicating an average between two points that is actually expressed at the mid-point). In addition, considering that velocity is a vector quantity, the positive and negative sign would represent the directional component. A positive velocity value (which we have throughout the whole race) would indicate movement along the 100 m track towards the finish line, whereas a negative value for horizontal velocity would indicate movement back towards the start (which in a 100 m sprint race would not usually happen). However, in terms of acceleration, a positive value would indicate speeding up (accelerating) and a negative value slowing down (decelerating). For example, on the acceleration curve all values above the zero line are positive and mean that the athlete is speeding up, even though the curve is going down (they are just speeding up at a lesser rate). In this example, the velocity and acceleration signs (positive and negative) are independent of each other.

However, it is also possible to have a negative acceleration value when an object is actually speeding up (increasing velocity or accelerating). For example, in the case of acceleration due to the gravity of the Earth the acceleration is often expressed as $-9.81$ m/s$^2$ (or negative 9.81 m/s$^2$). This indicates a *downward* (towards the Earth) acceleration of 9.81 m/s$^2$ (i.e. an object will speed up (increase its velocity) as it falls towards the centre of the Earth (see the section on Gravity, Weight and Vertical Projection elsewhere in this text)). However, in the case of acceleration in the horizontal direction (as in the example of our 100 m sprinter) a negative acceleration value would indicate a deceleration (slowing down) of the athlete in that direction.

Finally, in terms of biomechanics it is useful to be able to present all of this data in a series of graphs (as we have done with Excel). In order to analyse performance, the coach and the athlete can use the graphs for velocity/time and acceleration/time.

From these graphs plotted for a world-record 100 m performance in 2008 by Usain Bolt (9.69 s), it is possible to see that the velocity data indicates the athlete increases velocity from the start and reaches a peak at around the 60 m point in the race (or at about 6.32 seconds). At this point the athlete manages to hold this peak velocity for about 1.5 seconds to 80 m before it then begins to fall towards the 100 m point in the race. This is confirmed by the acceleration/time graph, which shows positive (increasing velocity) values up to 60 m. Although it is obvious that the acceleration/time graph is decreasing during this section, the values are still positive and are hence indicating acceleration or speeding up. The acceleration/time graph then passes through (or holds its position at) zero (which at this point would indicate no acceleration), as the athlete would have constant horizontal velocity for this brief 1.5-second period. Next, the acceleration/time graph becomes negative, indicating a deceleration or slowing down (i.e. from about 80 to 100 m). Hence, the statement made by many athletics coaches and biomechanists of *"he/she who slows down the least wins the sprint race"* appears to be true of our 100 m world-record-level sprinter. This characteristic speeding up (increasing horizontal velocity) to a peak at around 60–70 m, holding this speed or velocity for about 1 to 1.5 seconds and then slowing down as they approach 100 m is typical of many

100 m performances at many different levels (from amateur to Olympic athlete). Hence, it is obvious that such biomechanical analysis may have important implications for both training and performance.

## TASK

The data in *Table A1.4* is taken from two different world-record 1500 m freestyle swimming performances (Kieren Perkins 1994 and Grant Hackett 2001). Use these data to calculate and plot the average horizontal velocity and acceleration over each 100 m displacement interval. Also, see if you can provide a brief analysis of each swimmer's race and suggest why Grant Hackett managed to break the long-held 1500 m World record set by Kieren Perkins in 1994.

Note: in this context it may be important to qualify that the displacement in a swimming event such as this is technically zero (i.e. the athlete starts, swims 50 m (down the pool length), turns, and then returns to the start again). Hence, technically the terms distance and speed are probably more appropriate in this application.

## TASK

From the different sets of 100 m sprint data shown in *Table A1.5*, select two athletes and use Microsoft Excel to determine and plot their velocity and acceleration data over each 10 m interval. From the graphs produced, describe the performance of the two athletes and explain why one of the athletes ran the faster race.

**Table A1.4** Two sets of world-record 1500 m freestyle swimming data shown over 100 m intervals

| *Distance (m)* | *Time (minutes : seconds . hundredths of a second)* | |
|---|---|---|
| | 1994 Perkins | 2001 Hackett |
| 100 | 54.81 | 54.19 |
| 200 | 1:52.91 | 1:52.45 |
| 300 | 2:51.48 | 2:51.29 |
| 400 | 3:50.37 | 3:50.18 |
| 500 | 4:49.04 | 4:48.82 |
| 600 | 5:48.51 | 5:47.45 |
| 700 | 6:47.72 | 6:45.96 |
| 800 | 7:46.00 | 7:44.47 |
| 900 | 8:45.28 | 8:43.05 |
| 1000 | 9:44.94 | 9:41.78 |
| 1100 | 10:44.63 | 10:40.56 |
| 1200 | 11:44.50 | 11:39.51 |
| 1300 | 12:44.70 | 12:38.51 |
| 1400 | 13:44.44 | 13:37.89 |
| 1500 | 14:41.66 WR | 14:34.56 WR |

**Table A1.5** Six sets of 100 m sprint data shown over 10 m intervals; athletes are Maurice Greene (Mo), Tim Montgomery (Tim), Asafa Powell (Asafa), Tyson Gay (Gay) and Usain Bolt (Bolt). RT = Reaction Time. OT = Olympic Trial. Dark shaded squares indicate the fastest possible split time (and final time) for the given data set. Combining these times (BEST column) would achieve a hypothetical world record time of 9.44 seconds. Table adapted from: http://speedendurance.com.

*MEN'S WORLD RECORD 100 METRE 10 m SPLITS*

| | *Mo '01* | *Tim '02* | *Asafa '05* | *Gay '08 OT* | *Bolt '08* | *Bolt '09* | *BEST* |
|---|---|---|---|---|---|---|---|
| RT | 0.132 | 0.104 | 0.150 | n/a | 0.165 | 0.146 | 0.104 |
| 0–10 m no RT | 1.70 | 1.79 | 1.74 | | 1.69 | 1.74 | 1.69 |
| 0–10 m | 1.83 | 1.89 | 1.89 | 1.95 | 1.85 | 1.89 | |
| 10–20 m | 1.00 | 1.03 | 1.02 | 1.03 | 1.02 | 0.99 | 0.99 |
| 20–30 m | 0.92 | 0.91 | 0.92 | 0.89 | 0.91 | 0.90 | 0.89 |
| 30–40 m | 0.89 | 0.87 | 0.86 | 0.87 | 0.87 | 0.86 | 0.86 |
| 40–50 m | 0.86 | 0.84 | 0.85 | 0.83 | 0.85 | 0.83 | 0.83 |
| 50–60 m | 0.83 | 0.83 | 0.85 | 0.80 | 0.82 | 0.82 | 0.80 |
| 60–70 m | 0.83 | 0.84 | 0.84 | 0.80 | 0.82 | 0.81 | 0.80 |
| 70–80 m | 0.86 | 0.84 | 0.84 | 0.82 | 0.82 | 0.82 | 0.82 |
| 80–90 m | 0.89 | 0.85 | 0.85 | 0.83 | 0.83 | 0.83 | 0.83 |
| 90–100 m | 0.91 | 0.88 | 0.85 | 0.86 | 0.86 | 0.83 | 0.83 |
| TIME (s) | 9.82 | 9.78 | 9.77 | 9.68* (wind +4.1 m/s) | 9.69 | 9.58 | 9.44 |

*Key notes*

| | |
|---|---|
| **Biomechanics Kinematics and kinetics** | The study of forces and the effects of forces on living things. Subdivisions of mechanics that are concerned with displacement, velocity and acceleration (kinematics) and forces that cause or result from motion (kinetics). |
| **Linear and angular motion** | Linear motion (or translatory motion) is movement along a line that is either straight or curved and where there is no rotation and all body parts move in the same direction at the same speed. Angular motion involves movement around an axis of rotation. |
| **Scalar quantity** | A quantity that is represented by magnitude (size) only. |
| **Vector quantity** | A quantity that is represented by both magnitude and direction. |
| **Distance and displacement** | The term distance is classified as a scalar quantity and is expressed with reference to magnitude only (e.g. 140 mi (miles)). Displacement is the vector quantity and is expressed with both magnitude and direction (e.g. 100 m along a track in a straight line from point A to point B). |

| | |
|---|---|
| **Speed and velocity** | Speed is the scalar quantity that is used to describe the motion of an object (e.g. 4 m/s). It is calculated as distance divided by time taken. Velocity is the vector quantity and it is also used to describe the motion of an object (e.g. 4.5 m/s from north to south). It is calculated as displacement divided by time taken. |
| **Acceleration** | Is defined as the change in velocity per unit of time and hence is a vector quantity. It is calculated as velocity divided by time taken. |
| **Average and instantaneous** | Average is the usual term for the arithmetic mean (in this context often over larger periods of time or displacement). Instantaneous (or tending to instantaneous (towards zero)) refers to smaller increments of time or displacement in which the velocity or acceleration calculations are made. The smaller the increments of time between successive data points the more the value tends towards an instantaneous value. |

## FURTHER READING

The following two resources provide additional reading on the concept of linear kinematics in sports and exercise.

1 McDonald, C., & Dapena, J. (1991). Linear kinematics of the men's 110-m and women's 100-m hurdles races. *Medicine and Science in Sports and Exercise*, *23*(12), 1382–1391.
2 Murphy, A. J., Lockie, R. G., & Coutts, A. J. (2003). Kinematic determinants of early acceleration in field sport athletes. *Journal of Sports Science & Medicine*, 2(4), 144–150.

A2

# CENTRE OF GRAVITY AND CENTRE OF MASS

*Paul Grimshaw*

The centre of gravity (COG) of an object is defined as the point at which the entire weight of an object is assumed to be concentrated. This can be further clarified to mean that the centre of gravity of an object is the point at which the force of gravity for the entire object can be placed so that the object will behave the same as when the force of gravity is distributed across the entire object. The term centre of mass (COM) is defined as the point that corresponds to the mean (average) position for the concentration of the entire matter in the body. Within biomechanics the two terms are often used synonymously (i.e. having the same meaning). The terms centre of gravity and centre of mass are used for imaginary points (i.e. they do not physically exist as a point that can be seen) that describe concentrations of weight or matter.

Every object has a centre of gravity, and for bodies of uniform density (such as regular shapes and where density is defined as the mass per unit volume) the COG is at the geometric centre of the object. However, for bodies with non-uniform density (such as the human body) the COG has to be calculated. As we know, the human body has non-uniform density, is an irregular shape and assumes many different positions during sport and movement. Hence, the COG may be constantly moving (although it can also be stationary) and it may also be located either within or outside of the body. For example, in the high jump athletes often allow the COG to move outside of the body so it can pass under the bar while the athlete travels over the bar. *Figure A2.1* shows some of the considerations for the COG in more detail.

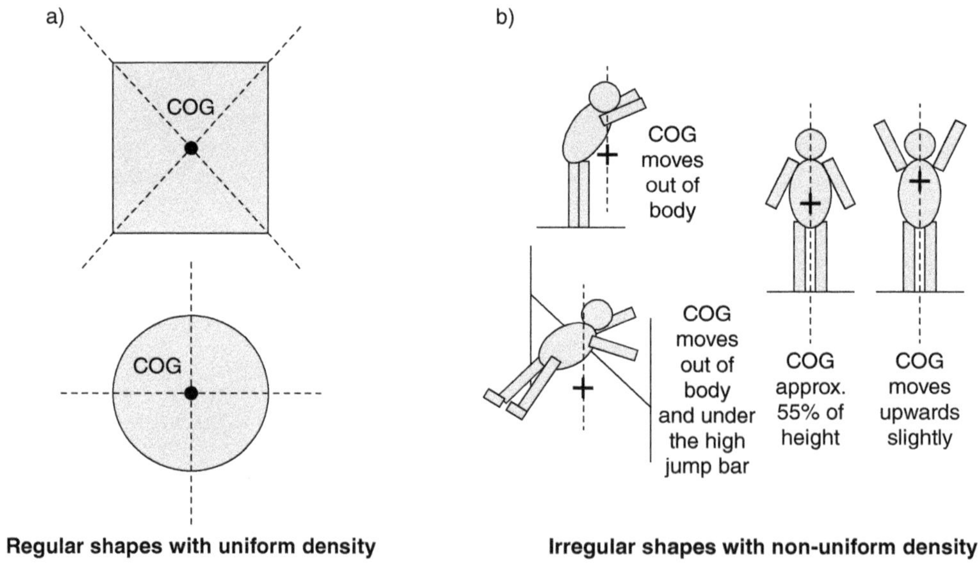


**Figure A2.1** Centre of gravity for a) regular shapes (uniform density assumed) and b) irregular shapes

The stability of an object is its ability to maintain its posture or position. This stability is affected by the position of the centre of gravity. For example, if a vertical line through the COG passes within the base of support then the object is more stable; conversely, if the COG falls outside the base of support it is considered more unstable. Stability is achieved when the COG lies near the lower part of the object. An increase in the area of the base of support will provide greater stability. Additionally, a heavier object is generally more stable than a lighter object because the force (and corresponding torque) needed to topple the heavier object over would be greater.

## TASK

Imagine the position of the COG while you are standing upright (i.e. it will be at approximately 55 to 57 per cent of your height (vertically) if you are standing symmetrically) and then imagine the position of the COG if you sit on the floor. The example where you are sitting will cause the COG to be closer to the ground (base of support) and this will offer a much greater degree of stability than when you are standing upright. For example, it will be more difficult to topple you over when you are sitting (i.e. an increased base of support and a lower COG position (closer to the base of the support)). Experiment with different positions of the body (preferably on a safe soft surface) or different objects to see how easy or difficult it is to topple the body or object over.

Within human movement many situations require the body to be able to move quickly from a stable to an unstable situation. For example, the 100 m sprinter is required to be in a stable position at the start of the race (particularly so they do not cause a false start) yet also to be in such a position that they can quickly move off into the sprint at the sound of the starting signal. *Figure A2.2* illustrates some examples of static and dynamic (moving) stability in relation to the COG of the human body during sport.

The vertical (and horizontal) position of the centre of gravity of the human body can be calculated by using the principle of moments. A moment is a force multiplied by a perpendicular distance from an axis of rotation and is discussed in more detail elsewhere in this text. Consider *Figure A2.3* where the human body is lying flat on what is termed a centre of gravity board.

Using the principle of moments (sum of the clockwise moments ($\Sigma$CWM) = sum of the anti-clockwise moments ($\Sigma$ACWM) – which is explained elsewhere in this text) the following equation can be developed (in this equation the full stop is used to represent a multiplication process). Remember that a moment is defined as a force (weight) multiplied by a perpendicular distance from an axis of rotation or, in this case, the fulcrum (*Figure A2.4* and *Figure A2.5*).

$$W_1.x_1 + W_2.x_2 = W_3.d$$

where:

$W_1$ = weight of human body acting at the COG position

$W_2$ = weight of board acting at the board's COG position

$x_1$ = horizontal distance from the fulcrum to human body COG (weight line)

$x_2$ = horizontal distance from the fulcrum to board's COG position (weight line which is probably half the board length)

$d$ = distance from the fulcrum to scales (centre)

$W_3$ = weight reading on the scales

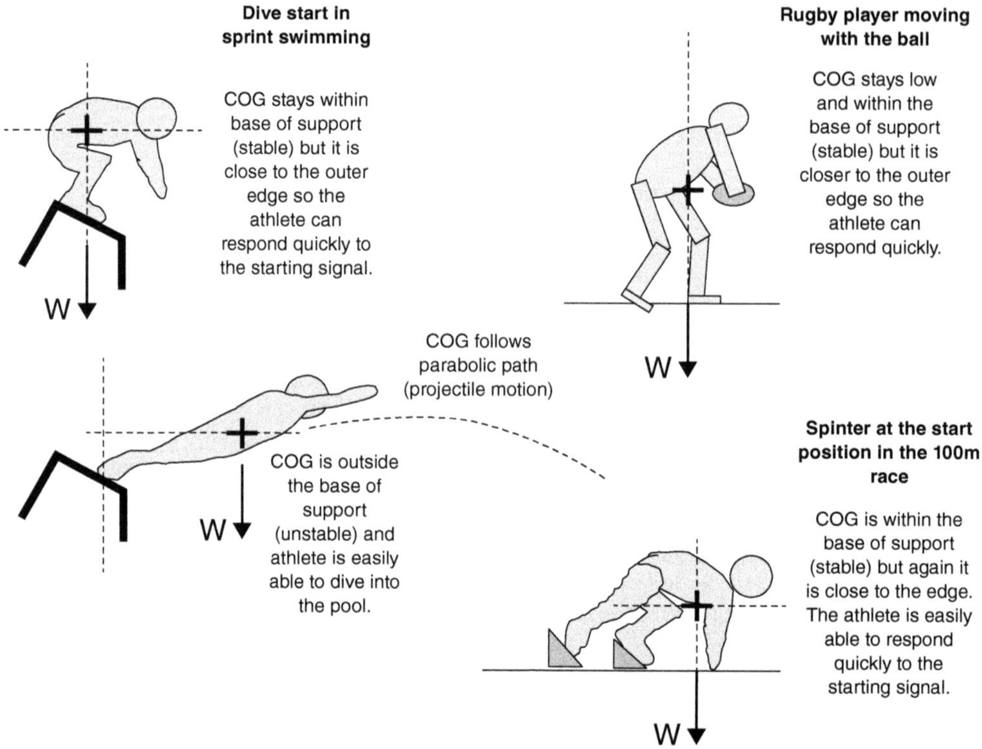


**Figure A2.2** Examples of static and dynamic stability in relation to the centre of gravity of humans during sport

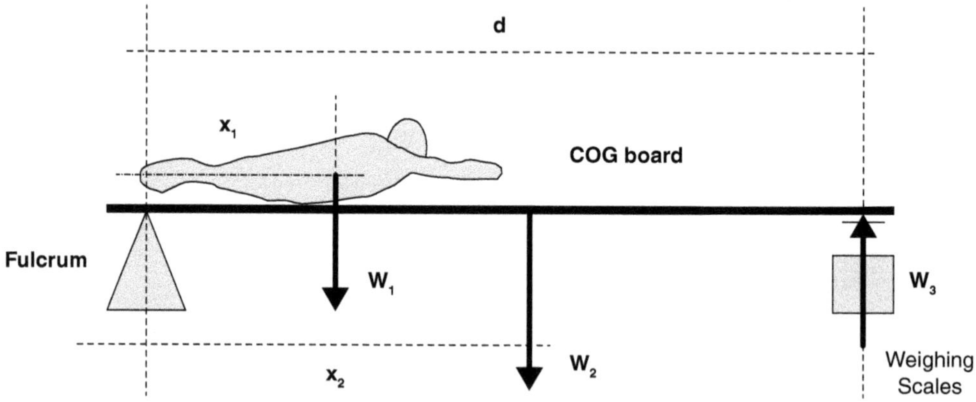


**Figure A2.3** The centre of gravity board

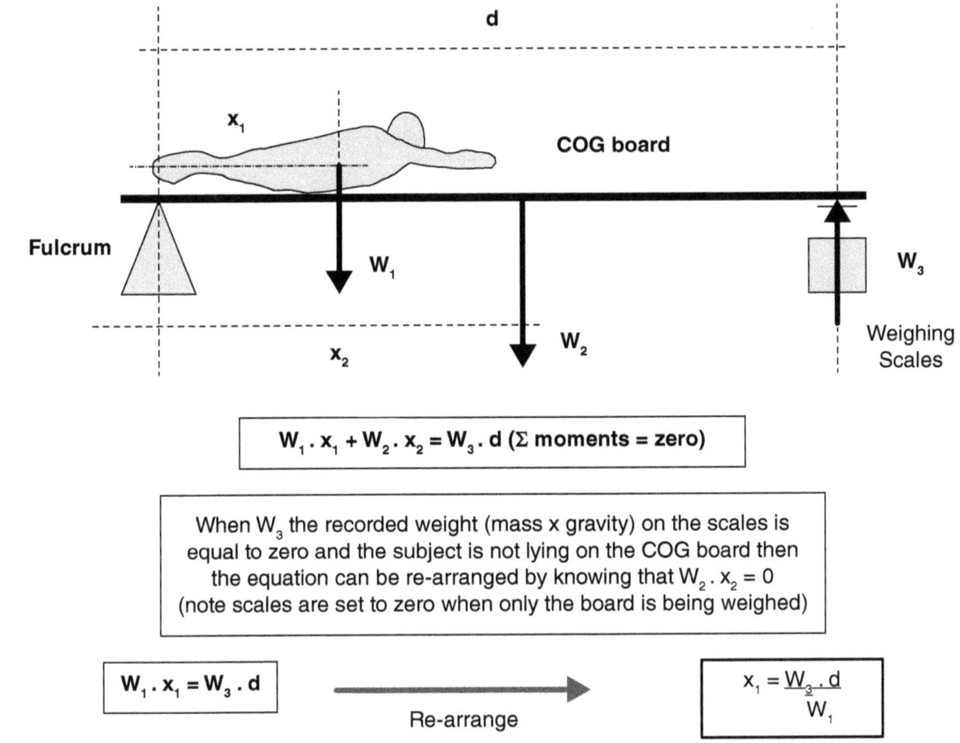


**Figure A2.4** Procedures for determining centre of gravity using the centre of gravity board

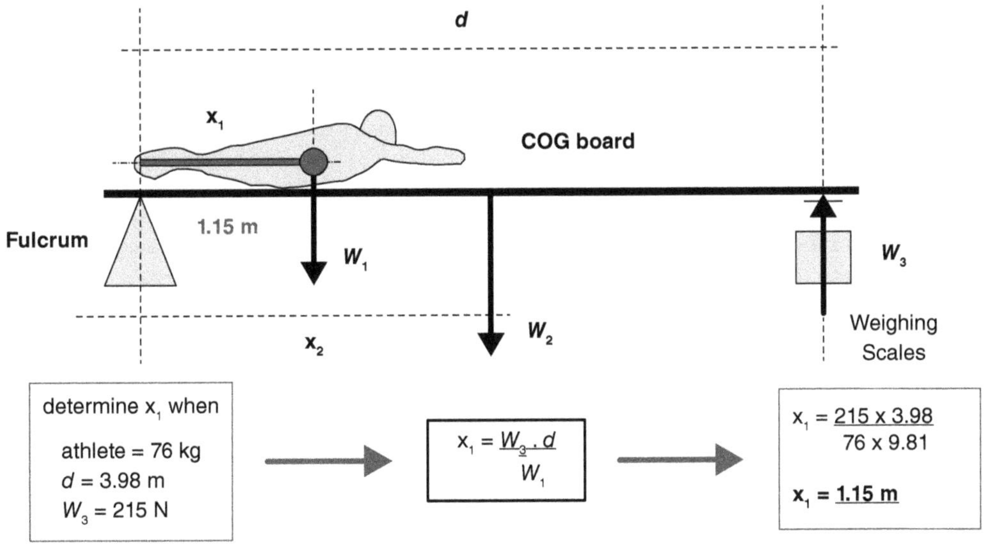


**Figure A2.5** Calculation of the centre of gravity from a centre of gravity board

When $W_3$ (the recorded weight (mass × gravity) on the scales) equals zero and the subject is not lying on the centre of gravity board, then the equation can be rearranged as follows. Note: it is important to clarify that the scales are actually set to zero (when the board alone is in place) as in this instance (and if not set to zero) they would actually record in part the weight of the board (i.e. the effect of gravity on the board acting at a position from the fulcrum because one end of the board is already supported).

$$W_2 \,.\, x_2 = 0$$

Hence, we can now eliminate ($W_2 \,.\, x_2$) from the previous equation (because it is zero).

$$W_1 . x_1 = W_3 \,.\, d$$

Rearrange this equation to produce the formula for $x_1$ (the position of the body's centre of gravity from the fulcrum) and we have the following.

$$x_1 = \frac{W_3 \,.\, d}{W_1}$$

In the above equation, $x_1$ represents the vertical height (actually expressed as horizontal position in our example because the athlete is lying flat on the board) of the subject's centre of gravity from the feet. Note: in this example we are only calculating the vertical position of the COG from the feet. The effect (i.e. the difference between one position and another) on the COG position by movement of the legs and arms can then be calculated by the same method but by taking the difference between the scale readings for the two relative positions. For example, if the arms are moved down to the side of the body we would expect the COG position to move lower down the body. Similarly, the athlete could stand on the board or lie in a different direction on the board to determine the COG position in other planes and axes of the body (but this method would still only give one position in one direction (i.e. in one plane and one axis) at a time).

Within human movement the centre of gravity has a location that will depend on the position of the body. The COG is constantly moving (although remember it can also be stationary) depending upon the body position and as we have seen it can often lie outside of the body. In this case we would need to modify the centre of gravity board experiment in order to account for this lateral or medial (in this example this may be left/right or forward/backward) deviation of the COG. To do this we would need to use a centre of gravity board that has both vertical and horizontal (x and y) axes.

In this case the same method of calculation applies but the athlete would assume a position on a board that had weighing scales at each of two positions on the board (i.e. a three-point reaction board – two scales and a fulcrum). Using the same principle of moments described previously, the two-dimensional (2D) position (x and y coordinates) of the COG for various body orientations on the board can now be calculated. This method is essentially how complex three-dimensional (3D) movement analysis systems determine the COG of the whole body.

Throughout human movement the centre of gravity location may be constantly changing, so in order to determine the COG in various body positions throughout human movement we must use an alternative method to the COG board. Usually this is achieved by a computer taking digital image sequences of the athlete performing an activity. The computer then digitises the precise location of body landmarks – shoulder, elbow, hip, knee, ankle, etc. – in each picture of this sequence in order to produce a two-dimensional or three-dimensional image or computer model of the human body. The human body is calculated as various body segments represented by lines (or "sticks"). The centre of mass (COM) for each of these segments is plotted along each "stick" or line length that represents a limb or segment. In order to do this, the segments need to be defined using points such as those presented in *Table A2.1*.

The centre of mass of each segment (*Table A2.2*) can now be expressed at a point that is a percentage of the length of the segment (from both a proximal and a distal location). These percentage lengths are derived from data presented in the text by David Winter. *Figure A2.6* illustrates the meaning of these two tables in more detail. It is also important to indicate that there are several other methods of locating COM positions, which are derived from techniques such as magnetic resonance imaging (MRI) and mathematical modelling.

**Table A2.1** Location of points to determine segment lengths within the body

| *Segment* | *Location of segment points* |
|---|---|
| Head and neck | C7–T1 (vertebrae) and 1st rib/ear canal |
| Trunk | Greater trochanter/glenohumeral joint |
| Upper arm | Glenohumeral axis/elbow axis |
| Forearm | Elbow axis/ulnar styloid |
| Hand | Wrist axis/knuckle II middle finger |
| Upper leg (thigh) | Greater trochanter/femoral condyles |
| Lower leg | Femoral condyles/medial malleolus |
| Foot | Lateral malleolus/head metatarsal II |

Adapted with permission from *Biomechanics and Motor Control of Human Movement* (3rd edition) by David Winter published by Wiley-Interscience publishers, New York, 2004

**Table A2.2** Location of centre of mass and proportion of total body weight for individual segments of the human body

| *Segment* | *Location of centre of mass (expressed as a % of the distance from each end)* | | *Proportion of total body weight (expressed as a % for each individual segment)* |
|---|---|---|---|
| | *Proximal (closest to trunk/ body mid-line)* | *Distal (furthest from trunk/ body mid-line)* | |
| Head and neck | 100 | 0 | 8.1 |
| Trunk | 50 | 50 | 49.7 |
| Total trunk, head and neck | 66 | 34 | (57.8) |
| Upper arm | 43.6 | 56.4 | 2.8 |
| Forearm | 43 | 57 | 1.6 |
| Hand | 50.6 | 49.4 | 0.6 |
| Total arm | 53 | 47 | (× 2 = 10.0) |
| Upper leg (thigh) | 43.3 | 56.7 | 10.0 |
| Lower leg | 43.3 | 56.7 | 4.65 |
| Foot | 50 | 50 | 1.45 |
| Total leg | 44.7 | 55.3 | (×2 = 32.2) |
| Total (all segments) | | | 100.00 |

Adapted with permission from *Biomechanics and Motor Control of Human Movement* (3rd edition) by David Winter published by Wiley-Interscience publishers, New York, 2004

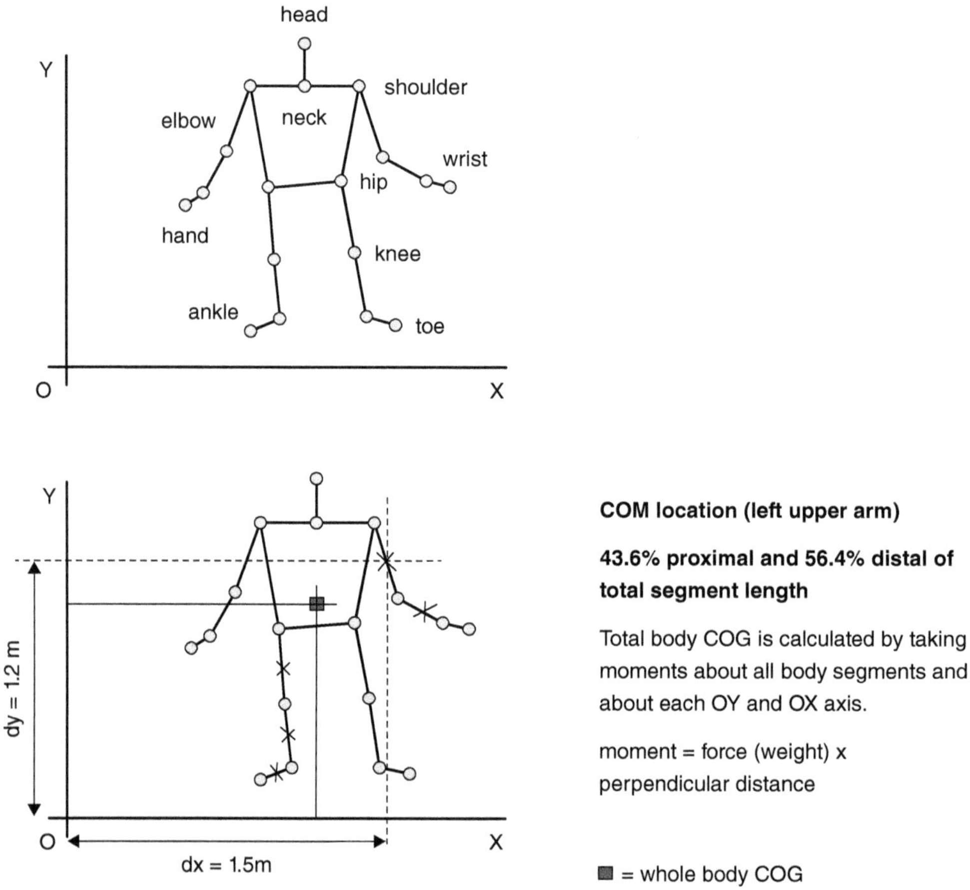


**Figure A2.6** An illustration of a) a "stick figure" human body representation and b) the centre of mass (COM) location for selected segments (frontal plane – anterior view of body)

In order to determine the centre of gravity position of the whole body in the example shown in *Figure A2.6,* moments about each axis for each body segment are taken. For example, the moment about the OX axis (the horizontal axis) for the left upper arm (because it is a frontal plane anterior view of the "stick" or line image of the body) would be the distance from this axis to the segment COM position multiplied by the weight (mass multiplied by acceleration due to gravity, 9.81 m/s$^2$) of this particular segment. For a 75 kg athlete and the left upper arm position shown in *Figure A2.6* this would be as follows.

## Taking moments about OX

Moment = force × perpendicular distance from axis of rotation (expressed in newton metres)

Moment (upper arm) $= ((0.028 \times 75) \times 9.81) \times 1.20$

*Note: it is the perpendicular distance along the Y axis (i.e. from the OX axis) that is used*

where:

0.028 = the percentage of mass of the whole body for this segment expressed as a decimal (i.e. a proportion of 1.0 or 2.8%)

75 = athlete's mass in kilograms (kg)

9.81 = acceleration due to gravity expressed in $m/s^2$

1.20 = perpendicular distance from OX expressed in metres (m) along Y

Moment about OX (left upper arm) = $((0.028 \times 75) \times 9.81) \times 1.20 =$ **24.72 Nm**

In order to calculate the centre of gravity of the whole body in this position all the moments for each individual segment about each axis of rotation are calculated and summated (i.e. sum the moments about OX and OY separately). Next, since we know that the sum of the centre of gravity of all the segments will equal the total centre of gravity of the whole body (i.e. 100 per cent of the weight) we can divide these respective summed moment values by the athlete's body weight ($75 \times 9.81 = 735.75$N in this case (expressed in newton units)) in order to find the respective OX and OY coordinate positions of the whole body centre of gravity. This can then be drawn on the diagram shown in *Figure A2.6* as the intersection of two lines from the respective OX and OY axes.

This method of determining the centre of gravity of the whole body is now used extensively in many computer software packages. The same principle is applied in the 3D analysis of human movement, but in this case moments are also taken about a third OZ axis. A recognition of the centre of gravity calculation and the movement patterns associated with the whole body during an activity is critical. Such applications range from the interpretation of how a child begins to walk to how an athlete can effectively clear a hurdle in athletics. The movement patterns of the whole body are often used in coaching and sports science and also in the study of pathological movement disorders.

*Key notes*

| | |
|---|---|
| **The centre of gravity** | This is the point in an object at which the entire weight of an object is assumed to act. More precisely it is the point at which the force of gravity for the entire object is considered to act such that it would behave in the same way as when the force of gravity is distributed across the entire object. |
| **The centre of mass** | This is defined as the point which represents the mean position for the concentration of the entire mass of the body. |
| **Imaginary points** | The centre of gravity and centre of mass are imaginary points, as they do not physically exist and hence they cannot be seen. |
| **Human body centre of gravity position** | The centre of gravity position in the human body may continually change (or it may also be stationary). It is a point that can be located within the body or it is a point that can be located outside of the body. The high jumper will allow the centre of gravity to pass under the bar during the high jump clearance phase. Stability is the ability of an object to maintain its posture or position. The stability of an object is affected by the position of the centre of gravity. Stability is achieved when the centre of gravity (COG) of the object lies nearer the lower part of the object. If a vertical line through the COG falls within the base of support the object is considered to be more stable. If the COG falls outside the base of support then the object is considered to be more unstable. |

*(continued)*

*(continued)*

| | |
|---|---|
| **Calculation of human body centre of gravity** | The centre of gravity of the human body can be calculated by various methods. The centre of gravity board is one method that can easily determine the position of the COG in static postures. Alternative methods involving 3D computation using video digitisation of body landmarks can determine the COG position used in the analysis of human movement. The COG position in either two or three dimensions or in static or dynamic conditions is determined from the principle of moments. Moments about different axes are taken in order to calculate whole body COG position/location. Individual segment centre of mass data is used in the determination of whole-body COG calculations. Different techniques exist that are used to determine individual segment centre of mass positions. |
| **Application** | A recognition of the centre of gravity position within the human body is critical for the application of biomechanics to the study of human movement. The centre of gravity movement patterns over the hurdle are an important consideration for the athletics coach. Similarly, the centre of gravity movement pattern in the child's pathological walking gait has important implications for clinical assessment. |

## FURTHER READING

The following two resources provide additional reading on the concepts of centre of mass and centre of gravity.

1 Drillis, R., Contini, R., & Bluestein, M. (1964). Body segment parameters. *Artificial Limbs*, *8*(1), 44–66.
2 Lee, C. R., & Farley, C. T. (1998). Determinants of the centre of mass trajectory in human walking and running. *The Journal of Experimental Biology*, *201*(21), 2935–2944.

# EQUILIBRIUM

*Paul Grimshaw*

When a body is in a state of rest (not moving) or is in a state of constant velocity (moving but not accelerating or decelerating) it is said to be in a condition of equilibrium. When the body is at rest it is in a state of static equilibrium and when it is moving with constant velocity it is in a state of dynamic equilibrium. The first condition of equilibrium is concerned with the sum of forces acting on a body ($\Sigma F = 0$). When the sum of all the external forces acting on a body is equal to zero and the body does not translate (linear motion), the object is in static equilibrium.

Considering *Figure A3.1*, when an object is placed on a table the external forces that are acting consist of the weight of the object acting downward (caused by gravity) and the reaction force from the table acting upwards on the object. The sum of the external forces acting on the object is equal to zero and the object does not move (static equilibrium).

The object shown in *Figure A3.2* is subjected to several forces simultaneously and the sum of these forces acting on the object is zero. In this case the object will not move and it will stay in its state of equilibrium. The forces can be expressed as a **polygon** (a plane figure formed by three or more segments (lines) – or they can also be expressed by any shape (and in any order)

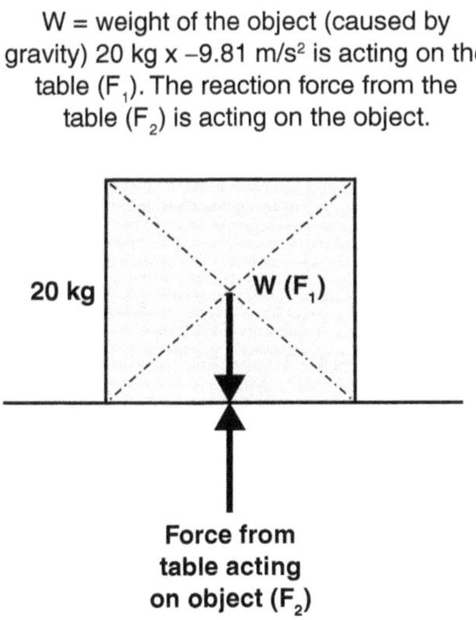


**First condition of Equilibrium ($\Sigma F = 0$)**

$W(F_1) = 20 \times -9.81 = -196.2$ N

Negative sign indicates force is directional (i.e. acting downwards because of acceleration due to gravity)

Calculate the reaction force from the table acting on the object

$\Sigma F = 0$

$-196.2 + F_2 = 0$

**$F_2 = +196.2$ N (acting upward)**

**Figure A3.1** Static equilibrium

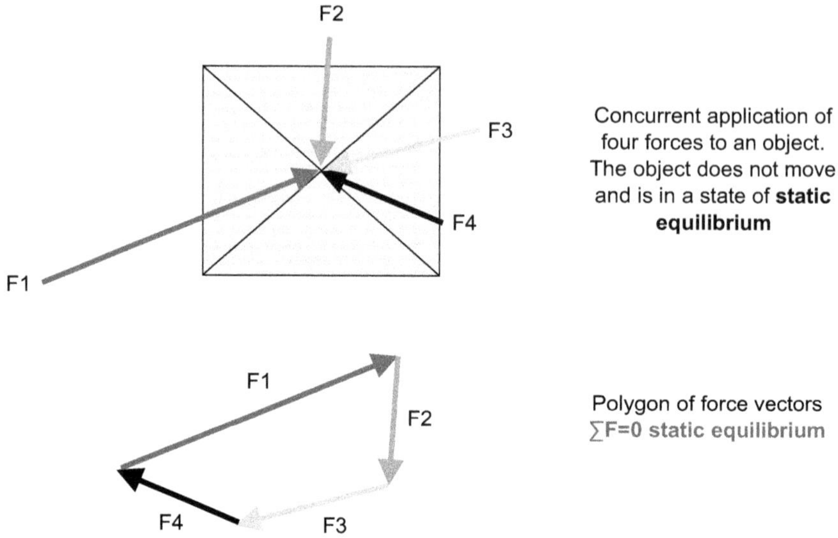


**Figure A3.2** Static equilibrium (polygon of forces)

where they are presented in the "tip to tail" closed method of representation of force vectors). If two or more forces act on an object and the object does not move, the first condition of equilibrium can be used to solve the situation for the resultant of these two or more forces.

In *Figure A3.3*, two coplanar (occupying the same plane) forces act on an object and the object does not move. Hence, there must be a third force that acts (unless the forces applied are

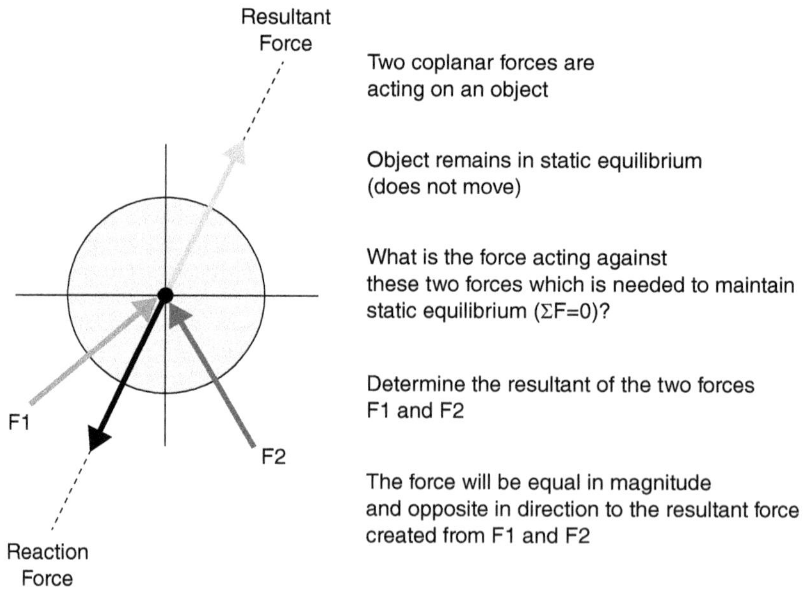


**Figure A3.3** Coplanar forces acting on an object (F1 and F2) with resultant and reaction forces shown

equal and opposite) in order to maintain equilibrium. To calculate the third force, the **resultant** (one force that would have the equivalent effect of the two forces, i.e. the sum of two or more **vectors** (having both magnitude and direction)) of the two forces must be determined. This can either be solved graphically (by actually drawing a scaled diagram and using the "tip to tail" method by joining the two forces with a third (the resultant)) or mathematically by resolving the forces and then using the first condition of equilibrium ($\Sigma F = 0$) to determine the third force (the force needed to maintain equilibrium).

In order to determine the resultant force (R) of these two forces (F1 and F2) in *Figure A3.3*, the concurrent forces are expressed at a point using a **free body diagram** (which is a diagram where all external forces on the body/object are represented by vectors). The resultant of these two forces can now be determined graphically or mathematically (*Figure A3.4, Figure A3.5, Figure A3.6* and *Figure A3.7*).

## FIRST AND SECOND CONDITION OF EQUILIBRIUM

In the previous examples the forces that were examined were represented within the linear force system with concurrent, coplanar forces. However, within human motion, forces do not always act in the same plane and they often act on bodies as force couples (pairs of equal and opposite parallel forces). In this latter case, we use the parallel force system to determine **static equilibrium.** The forces that act at parallel positions will often cause rotations of objects about specific axes. Hence, in this example we use **moments** (force × perpendicular distance (from the axis of rotation)) to solve the equilibrium condition. The **second condition of equilibrium** states that the sum of the moments (torques) acting on an object is equal to zero (Sum of the clockwise moments ($\Sigma$CWM) + Sum of the anti-clockwise moments ($\Sigma$ACWM) = 0).

*Figure A3.8* illustrates this in more detail using an example of the flexion and extension movement of the arm. This figure shows a position (approximately 100 degrees of elbow flexion with the lower arm horizontal) of the arm during the flexion/extension movement. The biceps muscle is exerting a force ($F_1$), which holds the arm in this position (it creates an anti-clockwise moment which is considered positive (+ve)). The mass of the arm and hand create a force ($F_2$), acting at a perpendicular distance from an axis of rotation (the elbow),

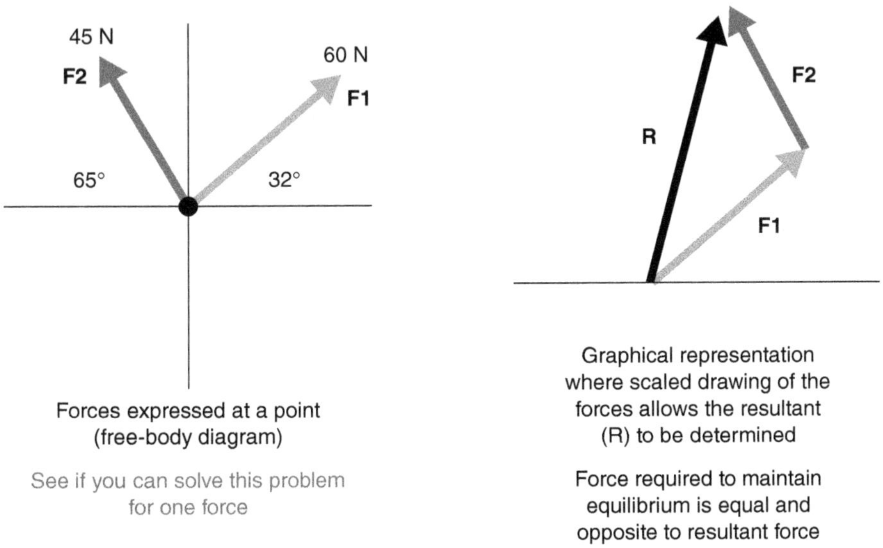


**Figure A3.4** Free body diagram of concurrent forces

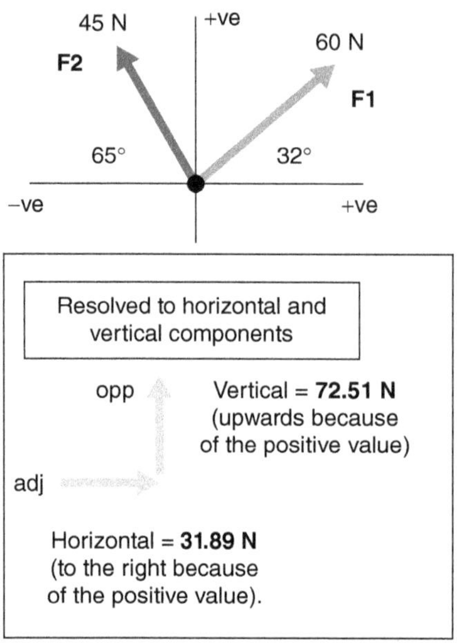


Sum of vertical components

Using F Sin θ
(Sin θ = **opp**/hyp) (hyp x Sine θ = **opp**)

= 45 Sin 65° + 60 Sin 32°
(plus sign because both forces are going in the same direction vertically i.e. upwards)

= 40.77 + 31.74

**= 72.51 N**

Sum of horizontal components

Using F Cos θ
(Cos θ = **adj**/hyp) (hyp x Cos θ = **adj**)

= 60 Cos 32° – 45 Cos 65°
(minus sign because they are going in different directions horizontally i.e. right and left)

= 50.88 – 18.99

**= 31.89 N**

**Figure A3.5** Mathematical resolution of forces

Resultant force

R = √ V² + H²
determined from Pythagoras

R = √ **72.51² + 31.89²**

R = √ 6274.6

**R = 79.21 N**

Angle of application

Tan θ = opp/adj
Tan θ = V/H
Tan θ = 72.51/31.89
Tan θ = 2.273

**Inv Tan θ = 66.25°**

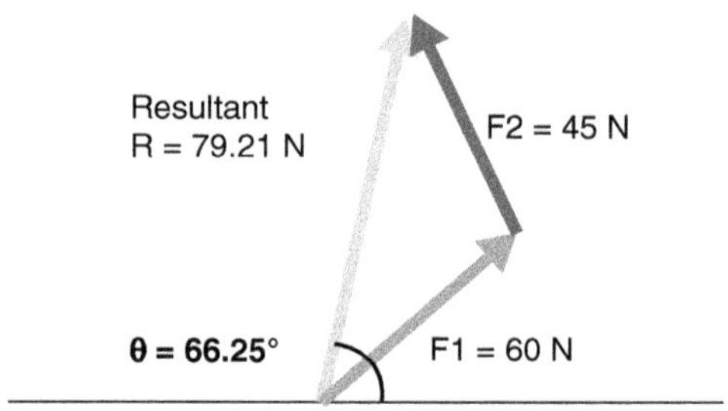


The resultant force (R) is equal to 79.21 N acting at an angle of 66.25° to the horizontal. This is the one force that is the equivalent of the other two forces (F1 and F2)

Force determined by ΣF=0
+79.21N – F = 0
F = **–79.21 N** (acts opposite to resultant)

**Figure A3.6** Mathematical resolution of forces

which opposes the moment created by the muscle force (it creates a clockwise moment which is considered negative (–ve)). The weight of the arm and hand creates a clockwise moment (i.e. it would have a tendency to rotate the arm in a clockwise direction) whereas the muscle force creates an anti-clockwise moment or torque. *Figure A3.9* shows the second condition of equilibrium (ΣMoments = 0), which is used to solve the condition of equilibrium (no movement) and calculate the muscle force needed to hold the arm stationary in this position.

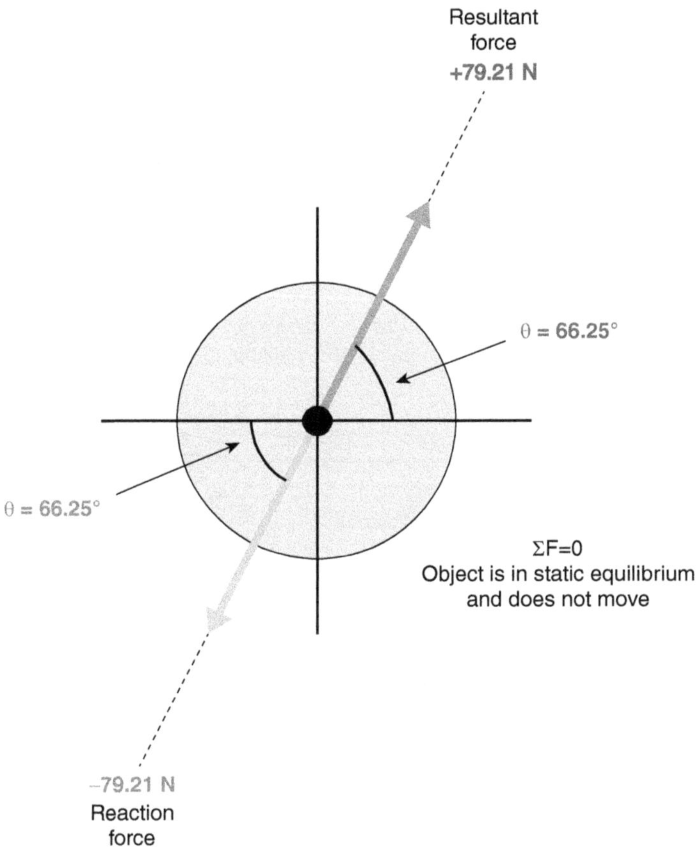


**Figure A3.7** Resultant and reaction forces resolved

The muscle force needed to hold the arm in this static equilibrium is 382.59 N. This force creates an anti-clockwise moment that balances the clockwise moment created by the weight of the arm and hand. The result is that the arm remains stationary and the limb is held in a position of **static equilibrium.**

When a body is in motion and it is not accelerating (constant velocity) it is said to be in **dynamic equilibrium.** The same conditions as for static equilibrium apply and can be re-written to include this motion component. This approach can also be used to determine the **force necessary to accelerate an object** (hence with the inclusion of the acceleration component). These equations will be developed later in the text.

## TASK

*Figure A3.10* shows an example of an arm holding an 8kg gym weight (technically a mass) in the hand. Determine the muscle force that is needed in the Biceps Brachii muscle in order to hold the arm and the weight (8 kg) still (i.e. in static equilibrium). Note: that the perpendicular distance of the muscle force is not provided, and you will need to calculate this to solve this problem (the perpendicular distance of the muscle force = sin 58° × 0.03 m). Also, express your answer in terms of absolute newton values and also in multiples of body weight for this athlete given that the person has a mass of 75 kg. *Figure A3.11* shows the free body diagram that you should use in solving this problem.

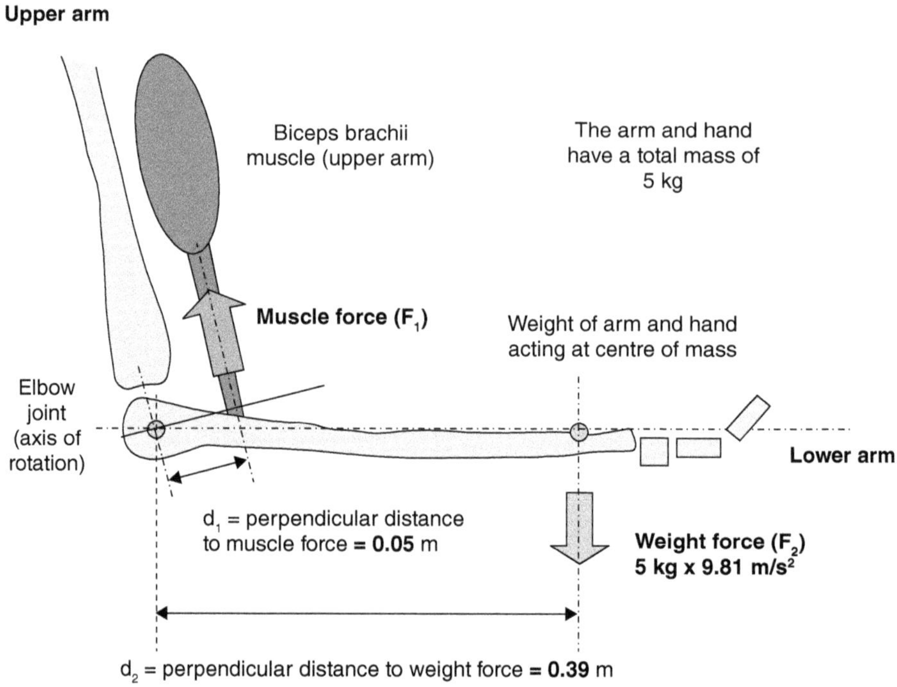


**Figure A3.8** Elbow flexion/extension position (static equilibrium)

Determine the muscle force that is needed to hold this static position of the arm (equilibrium)

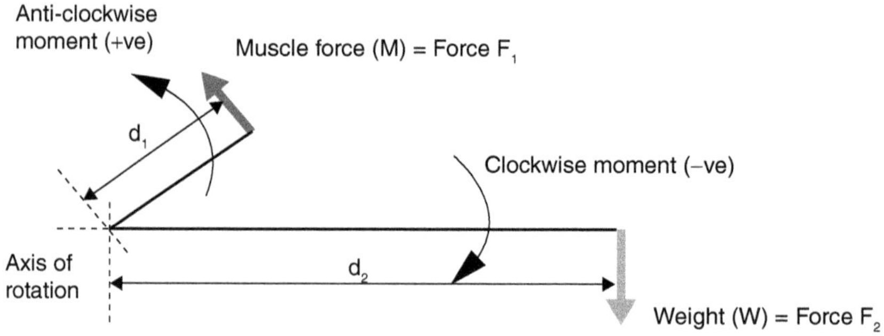


$\underline{\Sigma CWM + \Sigma ACWM = 0}$

$-((5 \times 9.81) \times 0.39) + (M \times 0.05) = 0$

$W \times d_2$ = clockwise moment (negative)

$M \times d_1$ = anti-clockwise moment (positive)

$$M = \frac{((5 \times 9.81) \times 0.39)}{0.05}$$

**Muscle force = 382.59 N**

**Figure A3.9** The second condition of equilibrium ($\Sigma CWM + \Sigma ACWM = 0$)

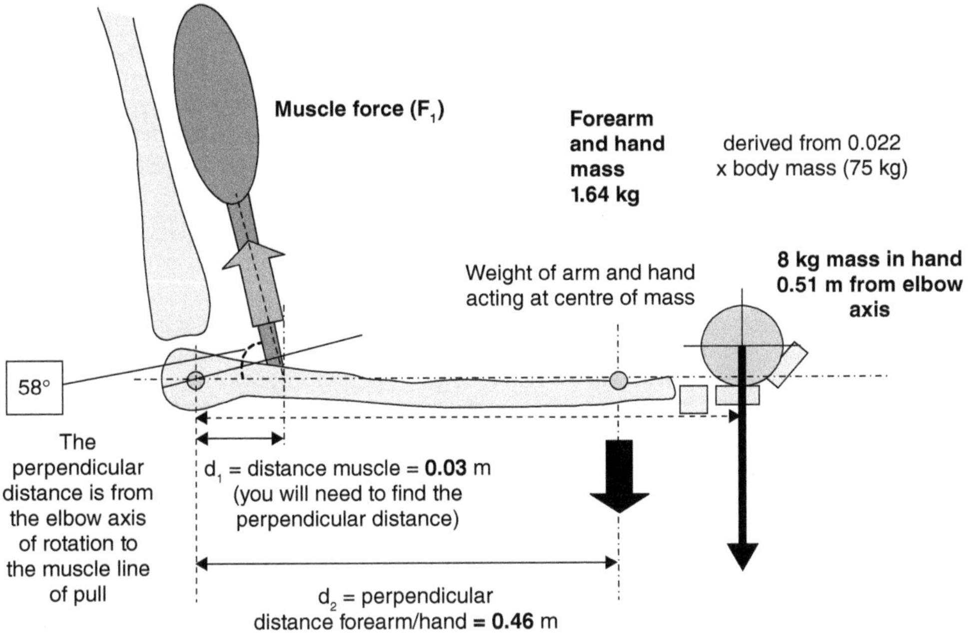


**Figure A3.10** Elbow flexion/extension position with additional mass (gym weight) in the hand (static equilibrium)

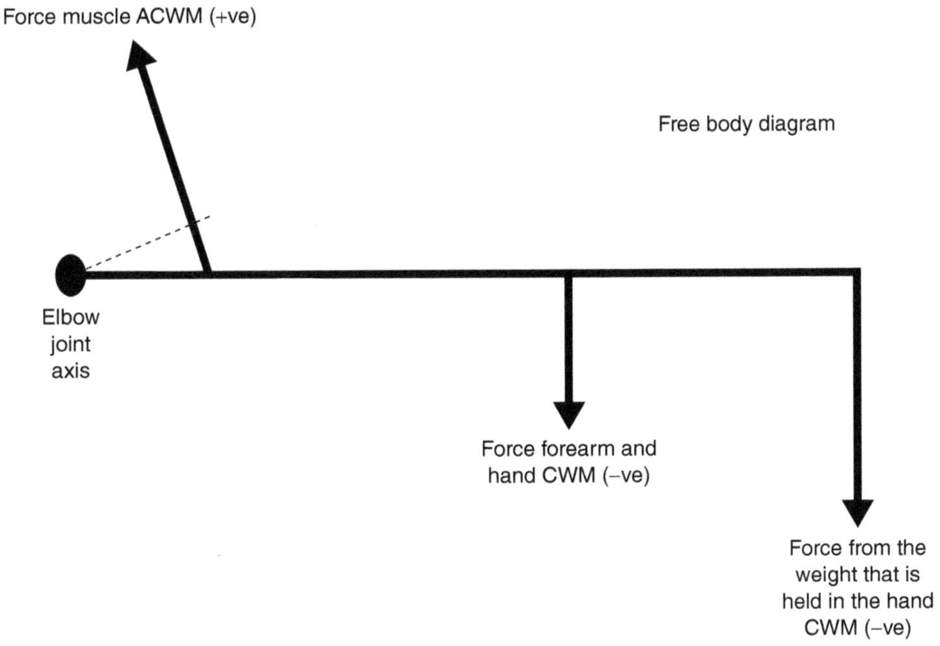


**Figure A3.11** Free body diagram showing the second condition of equilibrium ($\Sigma CWM + \Sigma ACWM = 0$) applied to the elbow example

| *Key notes* | |
|---|---|
| **Equilibrium** | When a body is in a state of rest or is in a state of uniform motion (i.e. constant velocity) it is in a condition of equilibrium. Static equilibrium is concerned with when the object is not moving and dynamic equilibrium is a condition where the object is in a state of movement with constant velocity (i.e. not accelerating or decelerating). |
| **The first and second condition of equilibrium** | These conditions can be used to solve for forces that are needed to maintain equilibrium. The first condition of equilibrium is concerned with when the sum of the external forces acting on an object is zero ($\Sigma F = 0$) whereas the second condition of equilibrium is concerned with when the sum of the moments (force × perpendicular distance from the axis of rotation to the applied force) acting on a system is zero ($\Sigma CWM + \Sigma ACWM = 0$). |

## FURTHER READING

The following two resources provide additional reading on the concept of static equilibrium in sport.


1 Quaine, F., & Martin, L. (1999). A biomechanical study of equilibrium in sport rock climbing. *Gait & Posture*, *10*(3), 233–239.
2 Kibele, A., Granacher, U., Muehlbauer, T., & Behm, D. G. (2015). Stable, unstable and metastable states of equilibrium: Definitions and applications to human movement. *Journal of Sports Science & Medicine*, *14*(4), 885–887.

# KINEMATIC DATA

## Differentiation

*Adrian Burden*

**Numerical differentiation** is a mathematical process that quantifies the change in one variable with respect to another. In the case of sport and exercise biomechanics, these variables are often displacement, velocity and acceleration with respect to time. The biomechanical study of human motion requires an understanding of the precise relationship between the changes in these variables. As outlined elsewhere in this text, it is shown that the **average velocity** of any moving object is given by the **change in position (displacement) divided by the time** over which the change takes place. If position is represented by the letter $p$ and time by the letter $t$, the average velocity ($v$) between *time 1* and *time 2* may be determined using:

$$v = \frac{p_2 - p_1}{t_2 - t_1} \text{ or } \frac{\Delta p}{\Delta t}$$

The capital Greek letter delta (Δ) is used to denote the change in a variable. The average velocity is also called the rate of change of displacement. Remember, velocity is a vector quantity and therefore this represents the average velocity in a specific direction; if the direction is not specified or is unimportant to the situation then the above equation is preferably termed the **average speed.** Similarly, the average acceleration ($a$), or rate of change of velocity, between *time 1* and *time 2* may be determined from the equation:

$$a = \frac{v_2 - v_1}{t_2 - t_1} \text{ or } \frac{\Delta v}{\Delta t}$$

Returning to the calculation of velocity, *Figure A4.1a* graphically represents the change in position (displacement) of a moving object plotted against time. From this it can be seen that the equation for the average velocity between $p_1$ and $p_2$ is, in fact, the same equation that gives the **slope or gradient** of the line between the points marked A and B, which correspond to the times $t_1$ and $t_2$, respectively. Similarly, the **gradient** of the line between points C and D must be the **average velocity** of the object over the smaller time interval ($\delta t$). The Greek lower-case letter delta (δ) is used here to denote a small change in some quantity, in this case a small change in time. The same relationships would also exist when obtaining average acceleration from a velocity *vs* time graph.

Knowing the average velocity or acceleration of a moving object is only of limited value. The value at a particular point in time is usually of greater interest; this is known as the **instantaneous velocity** or **acceleration.** As the change in time becomes smaller and smaller, the average value becomes the instantaneous value. This is demonstrated for velocity in *Figure A4.1a* as the gradient of line C–D becomes the **tangent** to the curve at the instant in time in *Figure A4.1b*

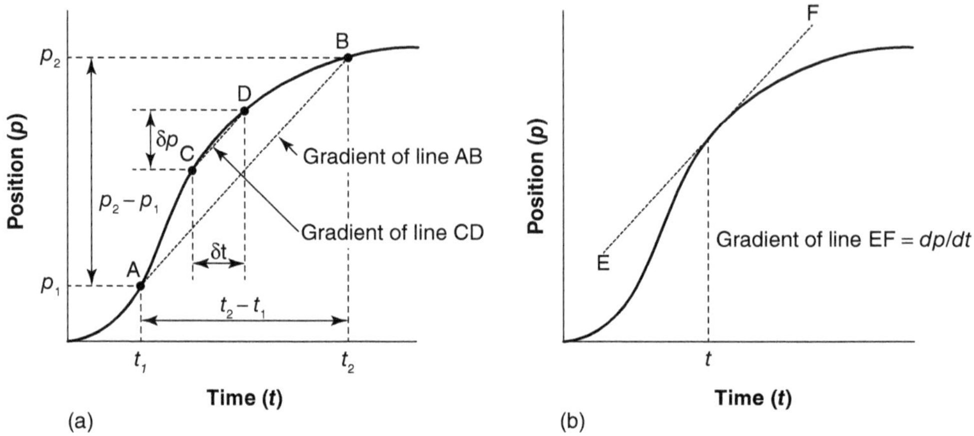


**Figure A4.1** An illustration of how the gradient between two points becomes the tangent to the line as the time interval between the two points reduces to zero

(line E–F). Mathematically this is represented as the instantaneous velocity ($v$), and is the **differential of displacement** (***dp***) with respect to the **differential of time** ***(dt)***:

$$v = \frac{dp}{dt}$$

Following similar reasoning, the **instantaneous acceleration** is given by the gradient of the tangent to the velocity curve at that instant in time and therefore the instantaneous acceleration, $a$, is said to be the **differential of velocity** (***dv***) with respect to the **differential of time** (***dt***). This is written mathematically as:

$$a = \frac{dv}{dt}$$

As this term contains velocity, which is itself a rate of change of displacement with respect to time, acceleration is said to be the **second differential of displacement** with respect to time.

## INTERPRETING KINEMATIC GRAPHS

In *Figure A4.1a*, the displacement and, therefore, the average velocity between time $t_1$ and $t_2$ will be positive because $p_2$ is greater than $p_1$ and therefore subtracting $p_1$ from $p_2$ will produce a positive outcome. In this example, the **sign of the gradient** of the line between A and B is **positive,** and the gradient of the tangent to the curve in *Figure A4.1b* is also positive. An example of this would be a 100-metre sprint.

Consider now *Figure A4.2*. Here $p_2$ is less than $p_1$, therefore subtracting $p_1$ from $p_2$ will produce a negative displacement and, thus, velocity. Similarly, the line between points A and B and the tangent to the curve (C–D) have a **negative gradient.** Because displacement and velocity are vector quantities, their sign tells us about the direction of travel. In *Figure A4.1*, the object is moving away from the reference point (i.e. its displacement and velocity are positive).

In *Figure A4.2*, however, the object is getting closer to the origin, so its displacement and velocity are negative. Importantly, the negative sign tells us that the object is now moving in the opposite direction.

In a 100 m sprint where the person is always moving away from the starting point, their displacement and velocity are positive. When the gradient of the velocity *vs* time graph is positive, the acceleration is positive (i.e. they are speeding up). Conversely, when the gradient of the velocity *vs* time graph is negative, the acceleration is negative (i.e. they are slowing down).

However, when an object is moving in two directions (e.g. forwards/backwards or upwards/downwards), acceleration must be interpreted carefully. In a standing vertical jump depicted in *Figure A4.3*, during phase A the person is moving down (negative displacement and velocity) and their acceleration is also negative. Thus, they are speeding up in a downwards direction during the start of the counter-movement. During phase B, they are still moving down (negative displacement and velocity), but their acceleration is now positive. This means that they are decelerating or slowing down in the negative or downwards direction. In phase C they have previously reached the bottom of their counter-movement and are now moving upwards (positive displacement and velocity). This is caused by their continued positive acceleration, which is also causing their positive velocity to increase. Finally, in phase D, the short phase that occurs up until their feet leave the ground, the acceleration now reverts to being negative, or downwards. While the displacement and velocity remain positive, the effect of this negative acceleration is to slow the body's motion while still travelling upwards.

In the example of the standing vertical jump above, when the gradient of the position *vs* time graph is positive, then the velocity is positive, and vice versa. Similarly, when the gradient of the velocity *vs* time graph is negative, then the acceleration is negative, and vice versa.

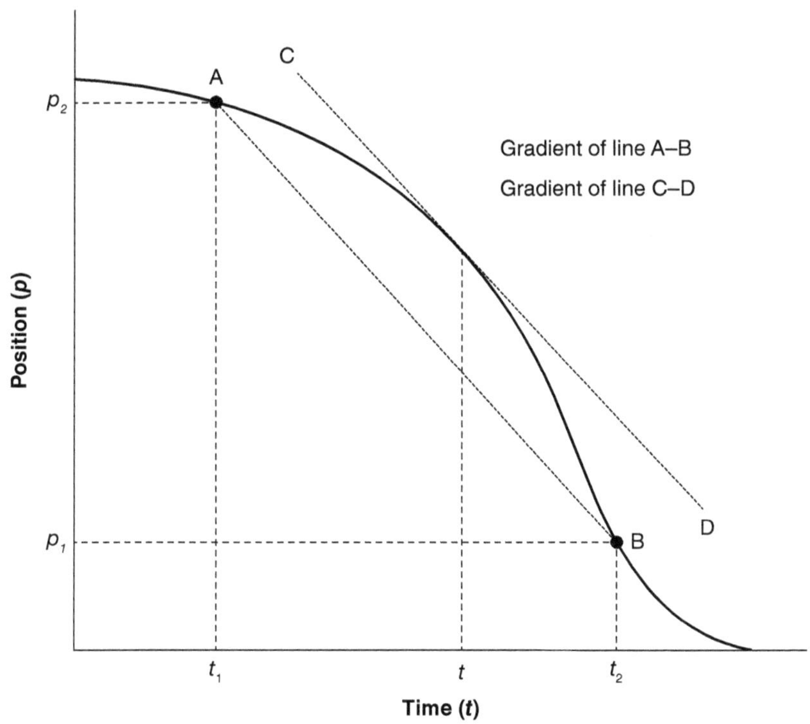


**Figure A4.2** An example of a negative gradient and negative tangent

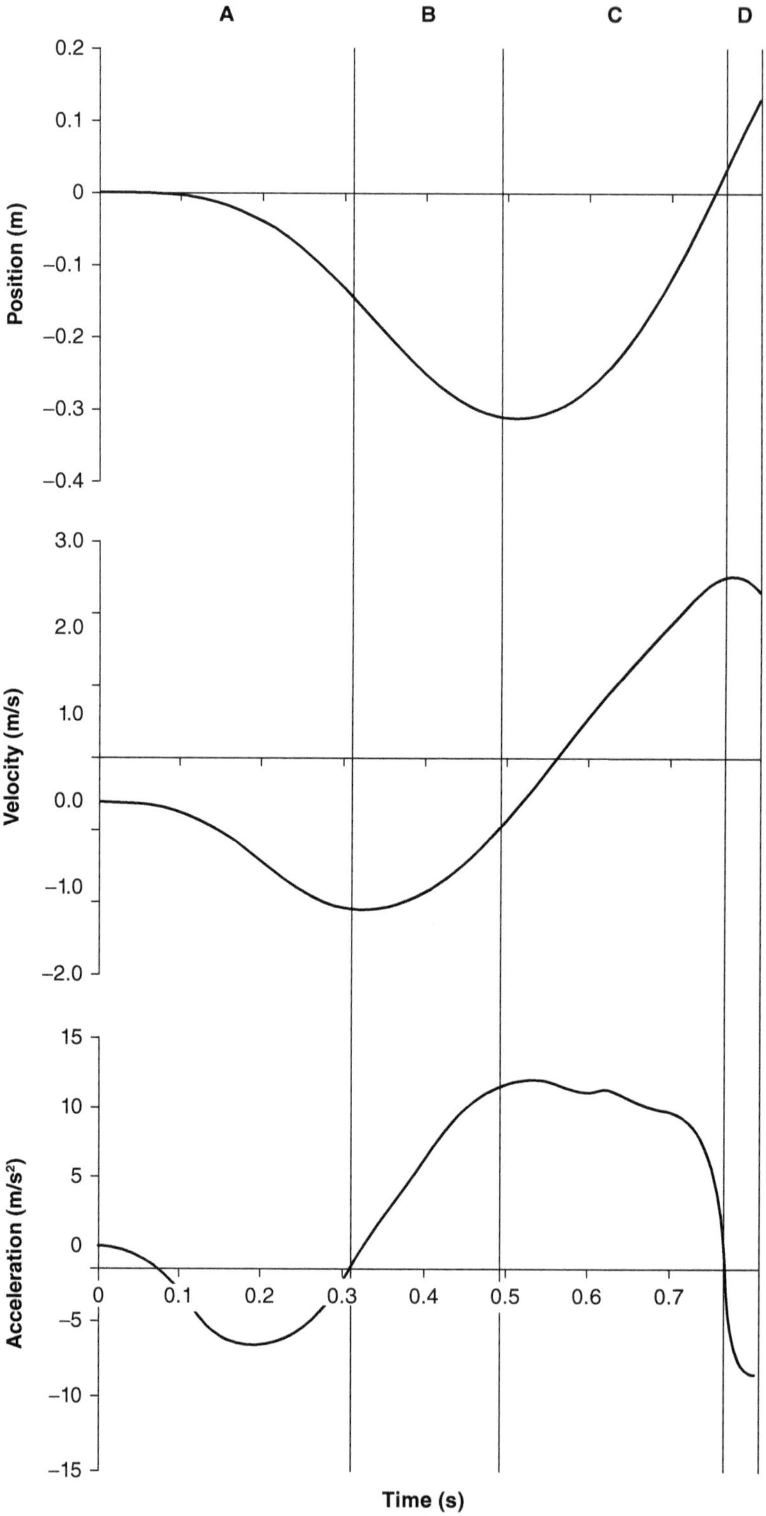


**Figure A4.3** Centre of gravity position, velocity and acceleration profiles for a standing vertical jump

Often when plotting the motion of an object on a position *vs* time graph, we see localised **points of maximum** (point A, *Figure A4.4a*) or **minimum** (point B, *Figure A4.4a*) displacement. At these points the gradient of the curve is neither positive nor negative because the tangent is horizontal. Here the velocity must be zero. Similarly, on a velocity *vs* time graph, localised maxima or minima represent times when the acceleration is zero.

**Points of inflection** often also occur when the curve moves from concave to convex (point C, *Figure A4.4b*) or from convex to concave (point D, *Figure A4.4b*). These represent localised maximum and minimum gradients respectively and hence points of maximum or minimum velocity on a position *vs* time graph. Following the same reasoning, points of inflection on a velocity *vs* time graph indicate local maximum or minimum acceleration.

*Figure A4.5* represents the flexion angle, angular velocity and angular acceleration of the knee joint during a normal walking stride from heel strike to heel strike. Note that the angular position, velocity and acceleration curves follow the same rules as linear position, velocity and acceleration curves. The points of inflection on the angular position and velocity–time curves also indicate local maximum or minimum velocity and therefore points of zero acceleration.

In sport and exercise biomechanics, experimental data often consist of a series of discrete data values, so to obtain the instantaneous velocity or acceleration it is necessary to use a method of **numerical approximation.** There are a number of **numerical differentiation methods** we can use but the simplest (and most frequently used) technique is the **finite difference method.**

*Table A4.1* shows the position *vs* time data for an athlete during a 100 m sprint race. The true instantaneous velocity of the athlete can never be known but if the time interval was sufficiently small it may be possible to estimate the velocity of the athlete during the race by calculating the average velocity between the known data points using the equation:

$$v = \frac{p_{i+1} - p_i}{t_{i+1} - t_i}$$

where $v$ is the approximated instantaneous velocity, and $p_i$ and $t_i$ are displacement, and times at instant $i$, and $p_{i+1}$ and $t_{i+1}$ are the displacement and times at the next data value. For example, between time 0.0 s and 1.0 s the velocity is given by:

$$\text{v} = \frac{4.20 - 0.00}{1 - 0} = 4.20\,\text{m/s}$$

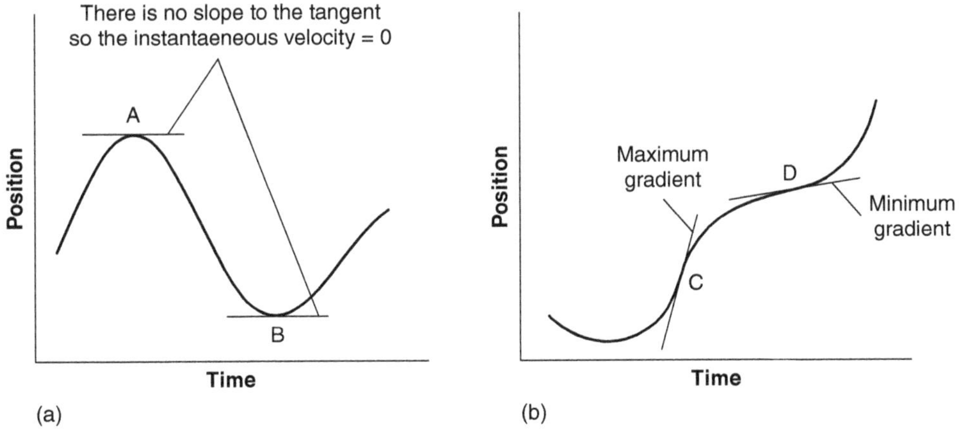


**Figure A4.4** Points of maxima, minima and inflection

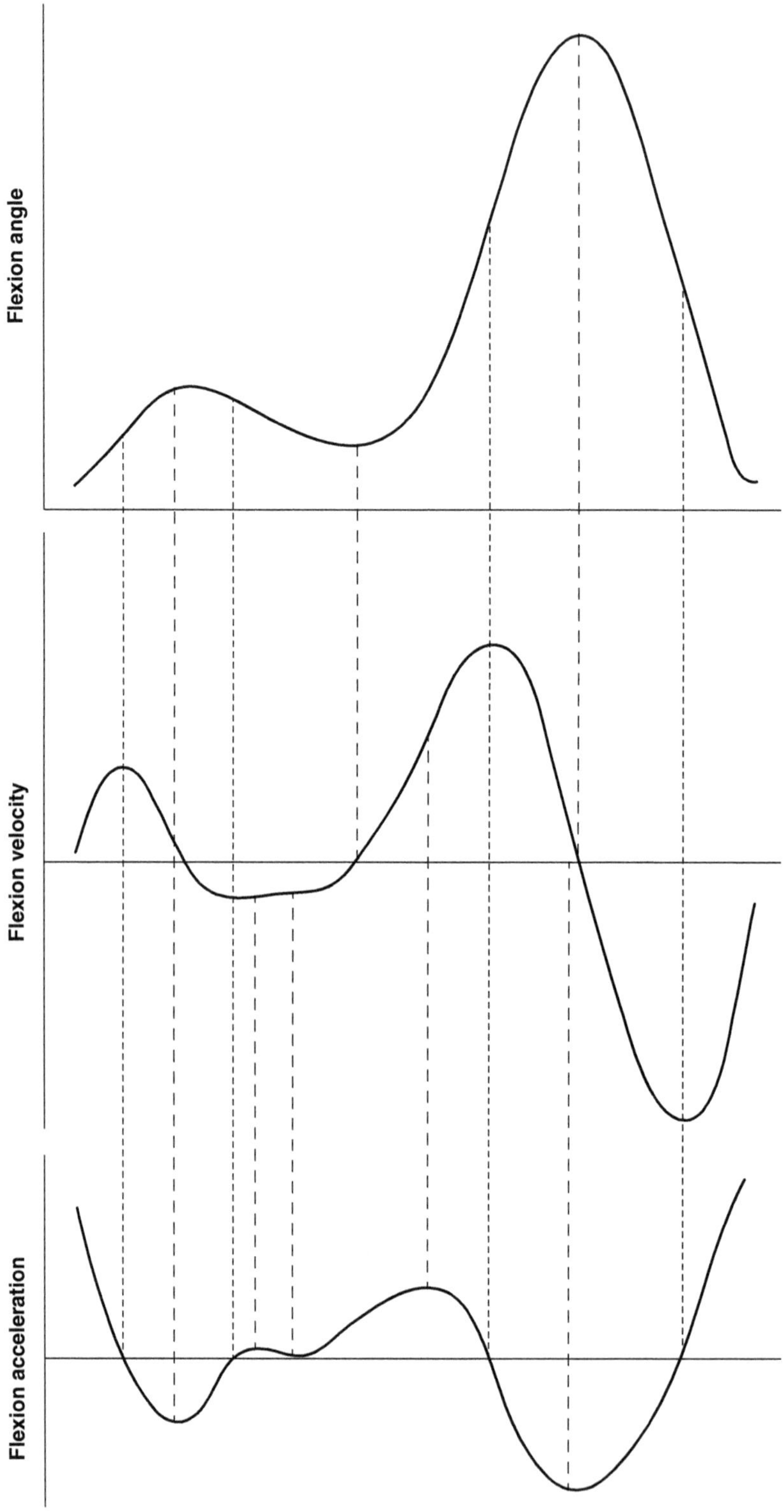


**Figure A4.5** An illustration of knee joint kinematics (angular position, velocity and acceleration) during one stride of walking (i.e. heel contact to heel contact of the same foot)

It is important to note that the velocity value is ascribed to the mid-point between the two displacement values (i.e. for position at times 0 s and 1 s the velocity is attributed to 0.5 s). This is because the velocity is assumed to be constant between the two position points so it makes sense to pick a point midway between them. Acceleration is calculated in a similar way, and the acceleration value is ascribed to the mid-point between the two velocity values. Thus, end-point data are lost when velocity is calculated, with even more data lost when acceleration is determined.

The estimate of the instantaneous velocity becomes better and better the smaller the value of *t* becomes. It is common practice in sport and exercise biomechanics to use video or other motion capture systems. This gives the possibility of reducing the time interval to, for example, 0.02 s (for a video image rate of 50 Hz). With high-speed cameras and other online systems a higher image rate (and, hence, a smaller time interval) can be achieved. Under these circumstances the "approximated" instantaneous velocity will be close to the actual instantaneous velocity.

**Table A4.1** A worked example of numerical differentiation for a 100 m race

| t *(s)* | p *(m)* | v *(m/s)* | a *(m/s²)* |
|---|---|---|---|
| 0.0 | 0.00 | | |
| *0.5* | | 4.20 | |
| 1.0 | 4.20 | | 3.03 |
| *1.5* | | 7.23 | |
| 2.0 | 11.43 | | 1.68 |
| *2.5* | | 8.91 | |
| 3.0 | 20.34 | | 1.51 |
| *3.5* | | 10.42 | |
| 4.0 | 30.76 | | 1.10 |
| *4.5* | | 11.52 | |
| 5.0 | 42.28 | | 0.61 |
| *5.5* | | 12.13 | |
| 6.0 | 54.41 | | 0.16 |
| *6.5* | | 12.29 | |
| 7.0 | 66.70 | | −0.29 |
| *7.5* | | 12.00 | |
| 8.0 | 78.70 | | −0.75 |
| *8.5* | | 11.25 | |
| 9.0 | 89.95 | | −1.20 |
| *9.5* | | 10.05 | |
| 10.0 | 100.00 | | |

| *Key notes* | |
|---|---|
| **Numerical differentiation** | Numerical differentiation is the name given to a method for calculating the rate of change of one variable with respect to another, usually time. In sport and exercise biomechanics the variables most widely used are displacement, velocity and acceleration. The rate of change of position with respect to time is called velocity while the rate of change of velocity with respect to time is called acceleration. |
| **Gradient of a curve** | The gradient of a curve representing data gives the rate of change and is calculated from the slope between two data points. The process is best illustrated graphically. |

*(continued)*

*(continued)*

| | |
|---|---|
| **Average and instantaneous velocity** | The gradient of a displacement–time curve gives the average velocity. If the time interval between the two data points reduces to a very small value the average velocity becomes the instantaneous velocity. |
| **Positive and negative gradients** | A positive gradient indicates a positive rate of change. For a position–time curve this represents a positive velocity, meaning that the object is travelling in a positive direction with respect to the measuring axes (usually forwards or upwards). A negative gradient indicates a negative velocity and means the object is travelling in the negative direction (usually backwards or downwards). |
| **Points of minima, maxima and inflection** | The position–time curve will have points of inflection and localised minima and maxima. Points of minima and maxima indicate the object has zero velocity. Points of inflection indicate a minimum or maximum velocity has been reached. |
| **Finite difference method for numerical differentiation** | The finite difference method is an algorithm for performing numerical differentiation. In practice, this is a simple method and is based on the equation for average velocity (when calculating velocity from displacement data) or average acceleration (when calculating acceleration from velocity data). |

## FURTHER READING

The following resource provides additional reading on the use of kinematic analysis to improve 100-metre sprint performance:

1 Maćkala, K. (2007). Optimisation of performance through kinematic analysis of the different phases of the 100 metres. *New Studies in Athletics*, 22(2), 7–16.

# KINEMATIC DATA

## Integration

*Adrian Burden*

Previously in this text it was shown that the process of differentiation allows the velocity of a body to be determined from displacement, and its acceleration to be determined from velocity. **Integration** is the reverse of differentiation and allows the determination of velocity from acceleration and displacement from velocity. In other words:

| | | | | | |
|---|---|---|---|---|---|
| Differentiation allows | Displacement | ⇨ | Velocity | ⇨ | Acceleration |
| Integration allows | Acceleration | ⇨ | Velocity | ⇨ | Displacement |

Whilst differentiation measures the gradient of the appropriate curve at a given instant, **integration measures the area under a curve.** This is demonstrated in the velocity–time graphs presented in *Figure A5.1.*

*Figure A5.1a* represents an object moving with constant velocity. Between two points in time, $t_1$ and $t_2$, we know that:

$$v = \frac{p_2 - p_1}{t_2 - t_1} \text{ or } \frac{\Delta v}{\Delta t}$$

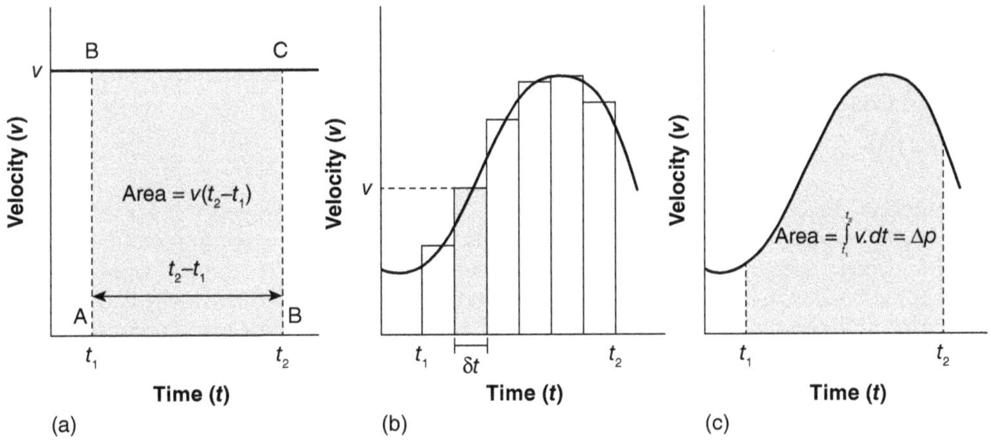


**Figure A5.1** An illustration of the area under the velocity–time curve for different conditions

where $v$ is the object's velocity, $p_2 - p_1$ is the change in position ($\Delta p$) between time points 1 and 2, and $t_2 - t_1$ is the change in time. Therefore, multiplying both sides of this equation by $(t_2 - t_1)$ gives:

$$p_2 - p_1 = v \times (t_2 - t_1) = \text{area of the rectangle ABCD}$$

The velocity data in *Figure A5.1b* is more complex, although this may be approximated by a number of smaller rectangles that have a width that is equal to the time interval ($\delta t$). The Greek lower-case letter delta ($\delta$) is used here to denote a small change in some quantity, in this case a small change in time. The area of a single rectangle (given by $v \times \delta t$) is approximately equal to the displacement over the time $\delta t$. The area under the curve, and hence the displacement over the period from $t_1$ to $t_2$ can then be approximated by adding together the areas of all such rectangles between $t_1$ and $t_2$. This is only an approximation as the velocity is assumed to be constant during each small time interval $\delta t$, but the approximation gets closer to reality as $\delta t$ becomes smaller and is exact if $\delta t = 0$.

The ideal case, when the time intervals become so small that their sum reflects the exact area under the curve, is illustrated in *Figure A5.1c*, and this equals the change in displacement that occurs between $t_1$ and $t_2$. This is written as:

$$\Delta p = \int_{t_1}^{t_2} v \times dt$$

where $\Delta p$ is the displacement, $v$ is velocity and $t$ is time. The symbol $\int$ is the symbol for integration. The letter *d* indicates the variable over which the change is being measured (in this case time, $t$) and $t_1$ and $t_2$ are termed the limits of integration. Using similar reasoning, it can be shown that the area under an acceleration time curve between two points in time must be equal to the change in velocity that occurs and hence:

$$\Delta v = \int_{t_1}^{t_2} a \times dt$$

where:

v = velocity

a = acceleration

t = time

In sport and exercise biomechanics the data that are most commonly available for integration are data from force platforms and accelerometers. Data from force platforms can be used to compute acceleration following Newton's second law ($F = m \times a$). The process of integration enables the velocity to be obtained, and the process can be repeated on this velocity data in order to obtain displacement. Integration of this type of data is best done using numerical integration. The term **numerical integration** describes a process of finding the **area under the data curve,** represented by the thin columns, such as those in *Figure A5.1b*, which are summed together to provide an approximation for the true area under the curve. There are a number of numerical integration techniques that can be used but the most commonly used is the **Trapezium Rule.**

*Figure A5.2* illustrates a velocity–time graph in which the velocity data have been sampled at equal intervals ($\Delta t$). The curve may be represented by a series of trapeziums. A trapezium is a rectangular shape with one side longer than the other. The area under the curve may then be considered to be equal to the sum of the areas of the trapeziums. The area of a trapezium is equal to half the sum of its two sides multiplied by its base. If, for a set of discrete data, the base of each trapezium is equal to the time between samples ($\Delta t$), and the sides are defined by the magnitude of adjacent samples ($v_i$ and $v_{i+1}$), the area under a single trapezium = $\Delta t \times (v_i + v_{i+1})/2$, and for n samples, the total area under the velocity curve is given by:

$$\text{Area} = \Sigma(\Delta t \times (v_i+v_{i+1})/2) \quad \text{for } i = 1 \text{ to } n-1$$
$$= \Delta p$$

This is the **displacement** ($\Delta p$) from $t_1$ to $t_n$. The term $\Sigma$ means sum all terms between the stated limits, here from 1 to n–1. Similarly, if we were working with acceleration–time data:

$$\text{Area} = \Sigma(\Delta t \times (a_i+a_{i+1})/2) \quad \textbf{for } \mathbf{i = 1 \text{ to } n-1}$$
$$= \Delta v$$

This is equal to the **change in velocity** ($\Delta v$) from $t_1$ to $t_n$. An example is given below which shows how the increase in displacement can be calculated based on the above equations.

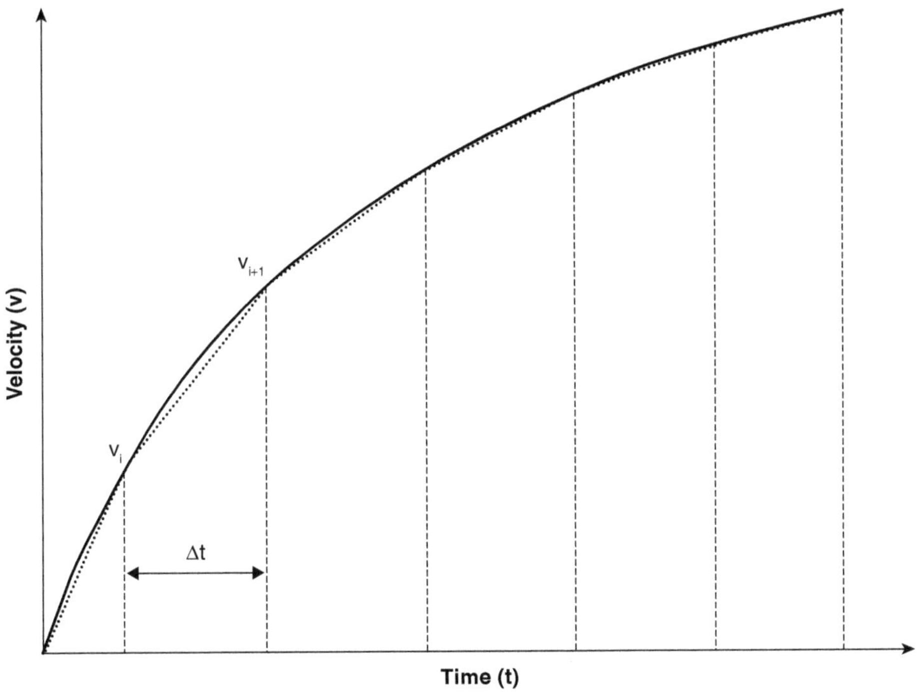


**Figure A5.2** Instantaneous (solid line) and trapezoidal approximation (dotted line) showing that as the interval reduces the two lines become similar

## TASK

Determine the distance travelled (i.e. the change in position Δp) of the sprinter over the first three seconds given the velocity data in *Table A5.1*.

## Solution

From the Trapezium Rule: $\Delta p = \Sigma(\Delta t \times (v_i + v_{i+1})/2)$ for i = 1 to 6

$$\begin{aligned}\Delta p &= ((0.0 + 2.2)/2) \times 0.5\\ &+ ((2.2 + 3.3)/2) \times 0.5\\ &+ ((3.3 + 4.0)/2) \times 0.5\\ &+ ((4.0 + 4.6)/2) \times 0.5\\ &+ ((4.6 + 5.1)/2) \times 0.5\\ &+ ((5.1 + 5.5)/2) \times 0.5\\ &= 0.55 + 1.375 + 1.825 + 2.15 + 2.425 + 2.65\end{aligned}$$

Therefore:

$$\Delta p = 10.825 \text{ m}$$

The Trapezium Rule is only an approximation of the true area under the curve because it assumes that the curve between two adjacent samples is a straight line. If the curve is not a straight line this will result in an error. These errors will be reduced if the time interval between samples is reduced, i.e. the narrower the trapezium, the more closely the curve between the sides of the trapezium will approximate a straight line. Although some error will always persist, for most practical purposes in sport and exercise biomechanics this method of integration is considered suitable.

**Table A5.1** Velocities of a sprinter over the first three seconds of the race

| | | | | | | | |
|---|---|---|---|---|---|---|---|
| Time (s) | 0.0 | 0.5 | 1.0 | 1.5 | 2.0 | 2.5 | 3.0 |
| Velocity (m/s) | 0.0 | 2.2 | 3.3 | 4.0 | 4.6 | 5.1 | 5.5 |

*Key notes*

| | |
|---|---|
| **Numerical integration** | Numerical integration is the name given to a method for calculating the total change of one variable with respect to another, usually time. In sport and exercise biomechanics the variables most widely used are acceleration and velocity. The integration of acceleration with respect to time gives the total velocity change. The integration of velocity with respect to time gives the total displacement. There are several algorithms for performing this calculation but a common one is the Trapezium Rule. This process is best illustrated graphically. |

| | |
|---|---|
| **Area under a curve** | The area under a data curve between two points in time gives the total change in that variable from the first point in time to the second point in time. For a velocity–time graph, the area under the curve between two points in time gives the displacement during this time period. |
| **Accurate and estimated displacements – the Trapezium Rule** | The area under a velocity–time curve can be broken down into small slices representing the displacement during each small time interval. Each of these slices can be represented by a trapezium (a rectangular shape with one side longer than the other). The area of a trapezium is easily calculated and so the total area under the curve is given by the sum of the areas of all of the trapezia. This area is an estimate of the total area so the displacement computed represents an estimate of the change in position. As the time interval for each small slice reduces, the estimate becomes more accurate. |

## FURTHER READING

The following resource provides additional reading on the use of integration to examine the velocity and displacement of the centre of mass during a vertical jump.

1 Linthorne, N. P. (2001). Analysis of standing vertical jumps using a force platform. *American Journal of Physics, 69*(11), 1198–1204.

A6

# UNIFORM ACCELERATION AND PROJECTILE MOTION

*Adrian Burden*

For examples of movements that occur with the body in contact with the ground (e.g. take-off phase of a standing vertical jump (SVJ)) the body's acceleration is rarely constant, or uniform, because of the changing forces that act on it. Such forces and their effect on the motion of the body are covered elsewhere in this text. However, in situations where the forces acting on the body are constant, it experiences a constant acceleration. An obvious example of this is when a body is in flight (e.g. the time when a shot-put or a long jumper is airborne) and the only force acting on the body is the attractive or gravitational force that exists between the body and the Earth. This is assuming that the effect of air resistance is negligible, which it can be for bodies of large mass travelling at low speeds. The acceleration that a body experiences as a consequence of the gravitational force varies slightly depending on its position on the Earth (it is slightly greater at the poles than the equator) but is generally agreed to be equal to 9.81 $m/s^2$. It should also be referred to as negative (i.e. –9.81 $m/s^2$) because the acceleration acts in a downwards direction, towards the centre of the Earth.

Other constant acceleration situations can occur when a body is not airborne. For example, a cyclist who stopped pedalling on a flat road would experience a fairly constant horizontal deceleration. Similarly, providing it was travelling up or down a smooth incline, a bobsleigh would also experience an approximately constant deceleration or acceleration.

When a body is moving in one direction in a straight line under constant acceleration (e.g. a car experiencing approximately constant acceleration at the start of a race) its velocity increases in a linear fashion with respect to time and thus its position changes in a curvilinear (exponential) manner (*Figure A6.1*). The situation is more complicated when a body moves in two directions, again in a straight line. An example of this is when someone jumps directly up and then lands back in the same place (e.g. an SVJ) and experiences the constant acceleration due to gravity during both the ascent and descent. In this situation the velocity of the body decreases linearly to zero at the apex of the jump and then increases in the same manner until landing. Their position changes in a curvilinear fashion (*Figure A6.2*).

The changes in position and velocity of a constantly accelerating body were first noted by an Italian mathematician named Galileo in the early seventeenth century. Galileo also derived the following **equations of uniformly accelerated motion** that can be used to generate the curves shown in *Figure A6.1* and *Figure A6.2*, and therefore to describe the motion of bodies experiencing constant acceleration.

$$v_2 = v_1 + at \qquad \text{(Equation A6.1)}$$

$$d = v_1 t + \tfrac{1}{2} at^2 \qquad \text{(Equation A6.2)}$$

$$v_2^2 = v_1^2 + 2ad \qquad \text{(Equation A6.3)}$$

$$d = \tfrac{1}{2} (v_1 + v_2)t \qquad \text{(Equation A6.4)}$$

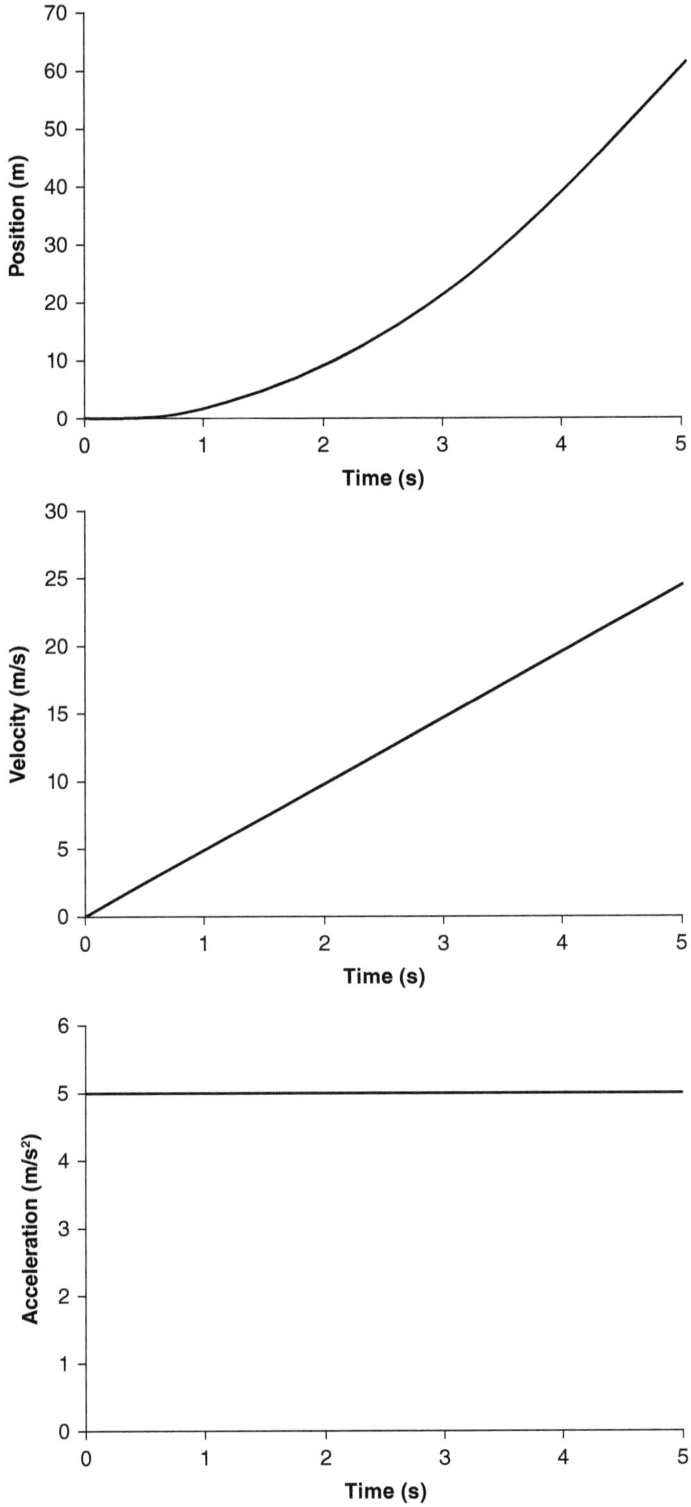


**Figure A6.1** Horizontal position and velocity of a body when experiencing a constant acceleration

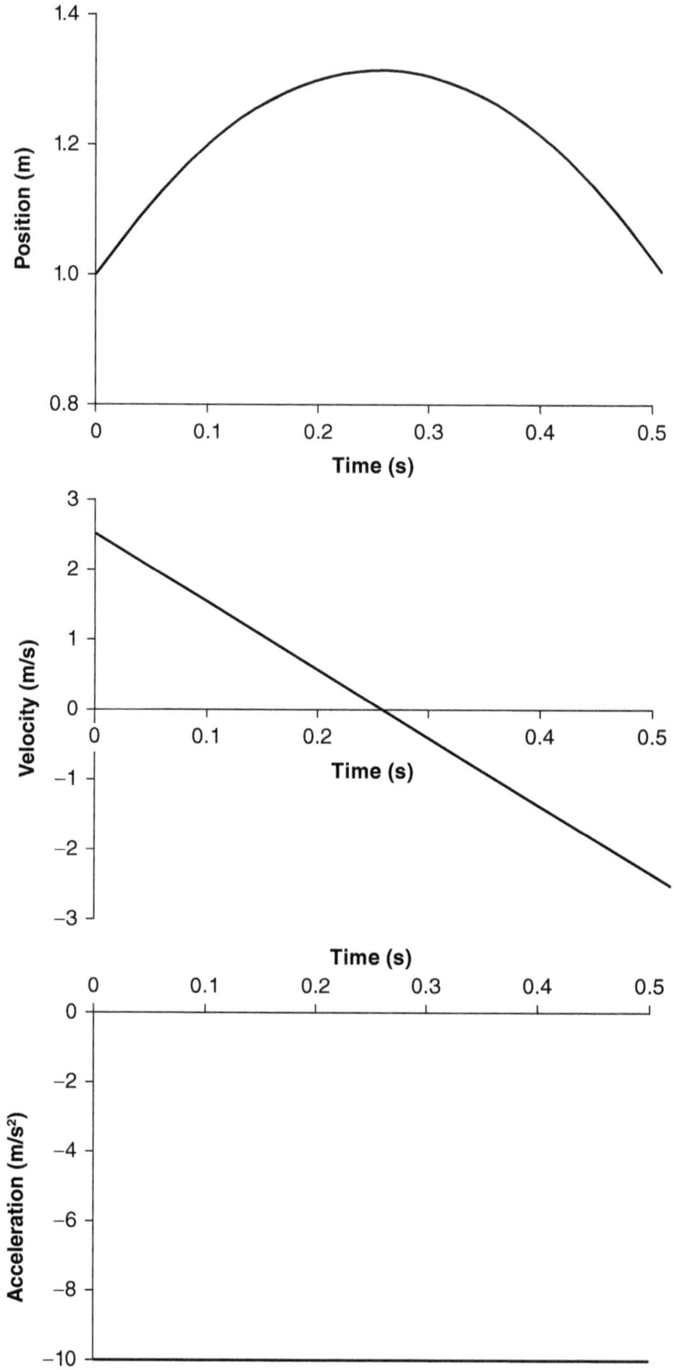


**Figure A6.2** Vertical position, velocity and acceleration of the centre of mass (COM) during the flight phase of a standing vertical jump

where:

$v_1$ = initial velocity

$v_2$ = final velocity

d = change in position or displacement

t = change in time

Sport and exercise biomechanists often wish to analyse the motion of a body while it experiences constant acceleration. It may be important to know, for example, how high somebody jumped, what velocity they would experience after a certain time, or how long it would take them to reach that velocity. Any of the equations above (A6.1–A6.4) can be used to answer such questions. For example, consider someone performing a SVJ with a take-off velocity of their centre of mass (COM) of 2.4 m/s. What would be the displacement of their COM between the instant of take-off and the highest point of their COM (i.e. how high do they jump)? The best way to answer this question is to break it down into a series of steps:

### *First, it is necessary to decide which of the equations (A6.1–A6.4) is appropriate:*

In this case we have the take-off velocity ($v_1$) and we also know that the COM will have a velocity of zero when it reaches the apex of its flight before the person falls back to the ground; so $v_2 = 0$. We also know that the acceleration of the CoM during flight is –9.81 m/s². Thus, the only equation that we can use to determine the displacement (*d*) of the CoM is Equation A6.3. All of the other equations either do not include what we wish to find (i.e. Equation A6.1 does not include *d*) or include variables that are not available to us (i.e. Equations A6.2 and A6.4 include *t*).

### *Next, rearrange the equation, if necessary:*

As the variable that we want to find is not always isolated on the left-hand side of the equation, it may be necessary to rearrange the equation. In our situation, we need to rearrange the equation to isolate *d*; the stages of which are as follows:

$$v_2^2 = v_1^2 + 2ad$$

Subtract $v_1^2$ from both sides of the equation to give:

$$v_2^2 - v_1^2 = 2ad$$

Divide both sides of the equation by 2a to give:

$$\frac{v_2^2 - v_1^2}{2a} = d$$

Once the equation has been rearranged, insert the known variables and calculate the unknown variable:

$$d = \frac{v_2^2 - v_1^2}{2a}$$

$$d = \frac{0^2 - (2.4)^2}{-19.62}$$

$$d = \frac{-5.76}{-19.62}$$

$$d = 0.29 m$$

Thus, the COM rose by 0.29 m between the instant of take-off and the highest point of the jump.

Suppose that the equipment required to obtain take-off velocity (e.g. a force platform or video camera) was not available but the time that the person spent in the air was able to be recorded (e.g. using a jump mat) and was 0.488 s. Assuming that the time the COM spent rising the instant after take-off was the same as the time that it spent lowering the instant before touchdown, Equation A6.2 can be used to calculate the displacement of the COM instead. Thus, $v_1$ becomes the velocity of the COM at the apex of the flight phase (i.e. zero), $t$ is half of the flight time (i.e. 0.488/2 = 0.244 s), and therefore $d$ is the displacement of the COM between this point and landing. The equation does not need rearranging so $d$ can be calculated by inserting the known variables into Equation A6.2:

$$d = v_1 t + \tfrac{1}{2} at^2$$

$$\mathrm{d} = (0 \times 0.244) + (\tfrac{1}{2} \times -9.81 \times (0.244)^2)$$

$$\mathrm{d} = -0.29 \text{ m}$$

Unlike the (upward) displacement calculated before using the take-off velocity, this value is negative as it is the downward displacement of the COM between the apex and landing.

In the examples above, the body in question was moving in the same direction (e.g. the horizontal motion of a car during the start of a race) or along the same line (e.g. the vertical motion of the COM during a SVJ). However, in many situations (e.g. the flight of a soccer ball or javelin) the body has both a horizontal and a vertical component of velocity at the point of release or take-off, and thus moves horizontally and vertically in flight. The movement of a body or object that is unsupported (i.e. in flight) and only affected by the forces associated with gravity and air resistance is known as **projectile motion**. If air resistance is ignored, as it often is for bodies of relatively large mass travelling at low speeds, the flight path or trajectory of a projectile follows that of a parabola, which is symmetrical about its highest point. The greater the vertical in relation to the horizontal component of velocity that the projectile has at release or take-off, then the more peaked its trajectory will be (*Figure A6.3*).

If the effect of air resistance is neglected and there are no other horizontal forces acting on a projectile, then its horizontal acceleration is zero. Thus, a shot-put or someone doing a standing broad jump will have the same horizontal velocity when landing as they do at release or take-off. Conversely, the vertical motion of a projectile is affected by the gravitational force which, as stated above, on Earth provides an acceleration of –9.81 m/s$^2$. Thus, if a projectile has a positive velocity at take-off or release, the effect of this downward acceleration will be to decrease this velocity to zero at the apex of the trajectory. The projectile then gains negative (i.e. downward) velocity until landing. If the projectile lands at the same height that it was released from, then it will have the same magnitude of velocity at the start and end of its trajectory. *Figure A6.4* shows how the pattern of both horizontal and vertical velocity of a projectile alter throughout its trajectory.

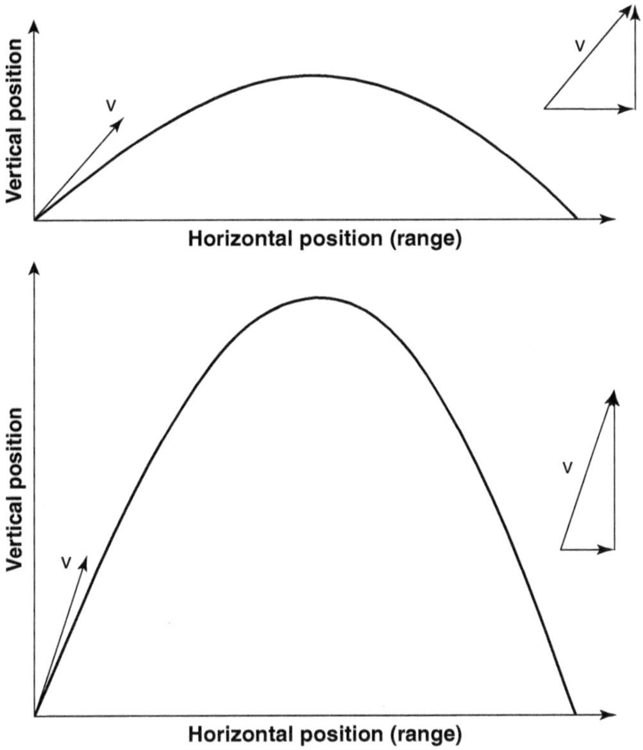


**Figure A6.3** Effect of different horizontal and vertical components of take-off velocity on the trajectory of the projectile

Often in sport, the aim is to maximise the range, or horizontal displacement, of a projectile (e.g. a goal kick in soccer). The range of a projectile that lands at the same height as it is released from is given by the following equation, which can be derived from the equations of uniform acceleration:

$$R = \frac{v^2 \sin 2\theta}{g} \qquad \text{(Equation A6.5)}$$

where:

R = range or horizontal displacement

v = resultant take-off or release velocity

θ = take-off or release angle, defined as the angle between the horizontal component of velocity and the resultant velocity vectors at take-off or release

g = acceleration due to gravity (i.e., $-9.81\ m/s^2$)

From Equation A6.5 it is evident that, for a given take-off angle, the range of a projectile will increase in proportion to the square of the take-off velocity (*Figure A6.5*).

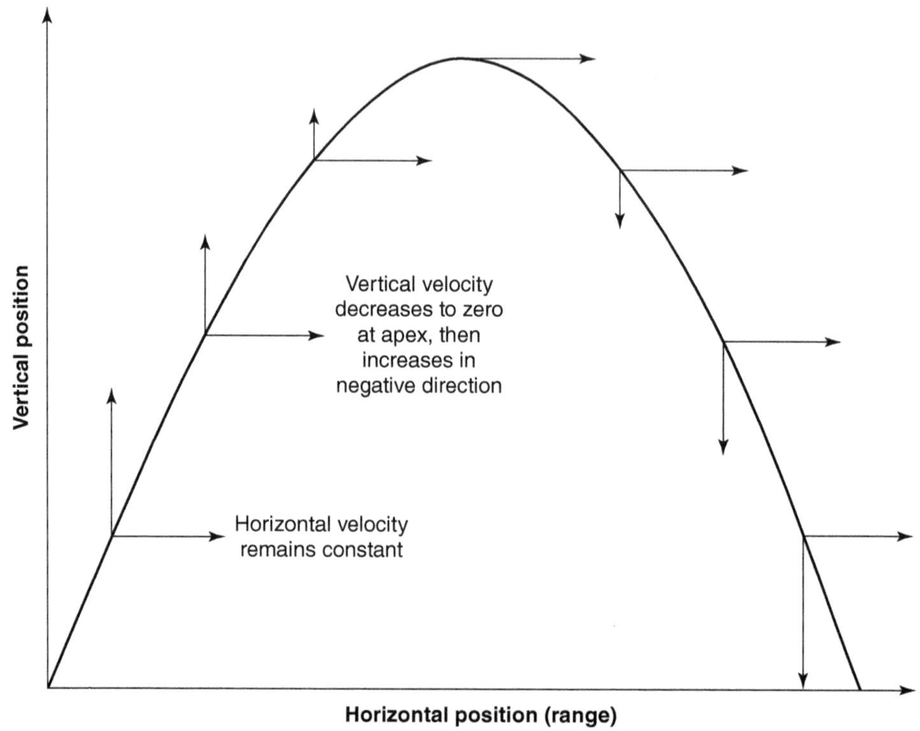


**Figure A6.4** Change in horizontal and vertical velocity of a projectile during flight

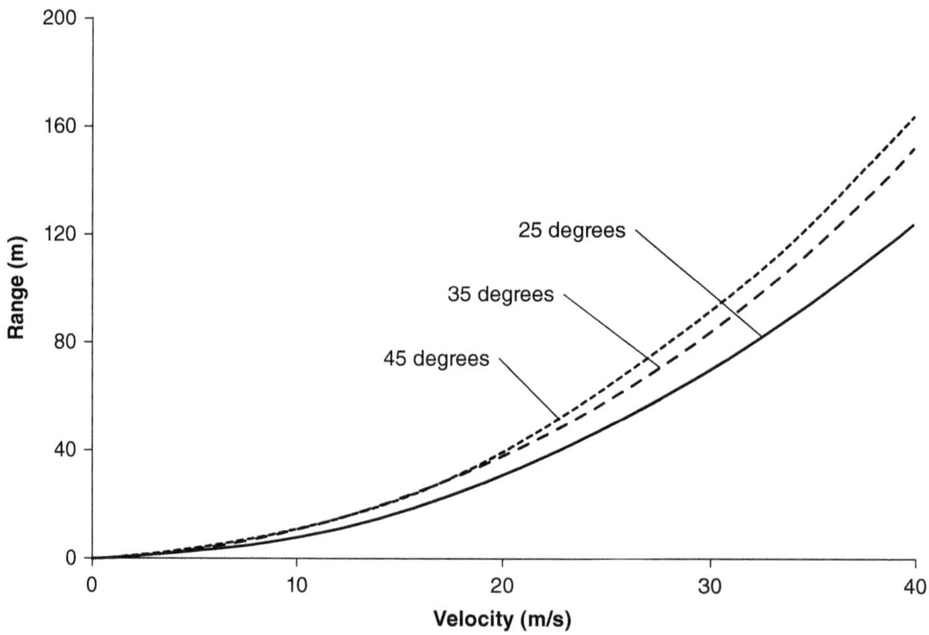


**Figure A6.5** Effect of take-off velocity on the range of a projectile, at three different take-off angles

The same equation also reveals the optimum take-off angle for a projectile (i.e. the angle that will result in the greatest range). The sine of 90° is equal to 1, and any angle either smaller or larger than this will result in the sine of the angle being less than 1. Thus, as Equation A6.5 contains the expression 'sin 2θ', a take-off angle of 45° will result in a value for this expression of 1 and, therefore, produce the optimum range (*Figure A6.6*). For any given take-off velocity, a take-off angle that is a particular number of degrees less than 45° will result in a range that is identical to that produced by an angle that is the same number of degrees greater than 45°. For example, if the take-off velocity of a projectile is 20 m/s its range will be 38.3 m if the take-off angle is either 35° or 55° (*Figure A6.6*).

Often, in sports biomechanics, it is the height (*h*) that a projectile reaches (e.g. volleyball blocker) or the time (*t*) that it is in the air (e.g. a soccer pass) that is of greater importance. These variables can be calculated using Equations A6.6 and A6.7, which also can be derived from the equations of uniform acceleration:

$$h = \frac{(v\sin\theta)^2}{2g}$$

(Equation A6.6)

$$t = \frac{2v\sin\theta}{g}$$

(Equation A6.7)

Thus, for a volleyball blocker who has a take-off angle of 80° and velocity of 3.2 m/s, the height raised by the COM during flight can be calculated by inserting the known variables into Equation A6.6:

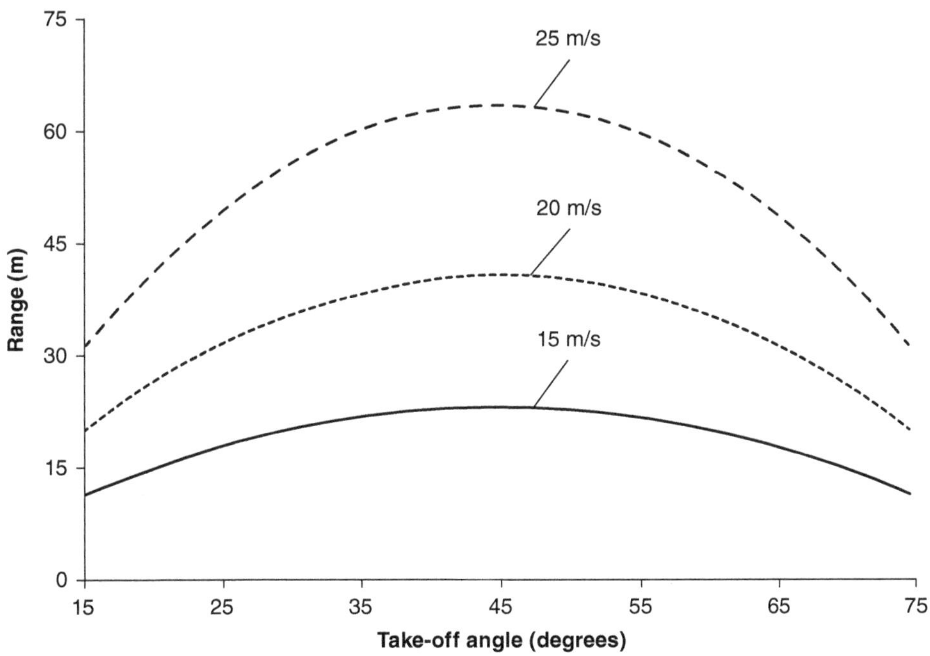


**Figure A6.6** Effect of take-off angle on the range of a projectile, at three different take-off velocities

$$h = \frac{3.2^2 \times \sin^2 80}{2 \times 9.81}$$

$$h = \frac{10.24 \times 0.97}{19.62}$$

$$h = 0.51 \text{ m}$$

Using Equation A6.7, the flight time of a soccer ball kicked with a velocity of 16 m/s at an angle of 56° can be calculated:

$$t = \frac{2 \times 16 \times \sin 56}{9.81}$$

$$t = \frac{32 \times 0.829}{9.81}$$

$$t = 2.70 \text{ s}$$

A number of smartphone applications (apps) are now available that allow the user to alter the input parameters of a projectile to see the effect that this has on its range, height and flight time. A few of these apps are covered in the Biomechanics Apps (Applications) section of this text.

Most of the projectiles used as examples previously in this section landed at the same height as they were released from (e.g. a soccer goal kick or pass that is not intercepted by another player). However, in many sports the projectile is more commonly released from a greater height than it lands at (e.g. shot-put, long jump) or less frequently lands at a greater height than it was projected from (e.g. basketball free throw). Equations A6.5–A6.7 can only be used in situations where the release height is the same as the landing height, and two new equations (A6.8 and A6.9) are needed to calculate the range and flight time of a projectile that has different release and landing heights:

$$R = \frac{v^2 \sin\theta \cos\theta + v\cos\theta\sqrt{(v\sin\theta)^2 + 2gh}}{g} \quad \text{(Equation A6.8)}$$

$$t = \frac{v\sin\theta + \sqrt{(v\sin\theta)^2 + 2gh}}{g} \quad \text{(Equation A6.9)}$$

In such situations the optimal angle of take-off or release is no longer 45°, as it is for projectiles that have the same take-off and landing heights. For projectiles that have a higher take-off than landing height (e.g. shot-put) the angle that will result in the greatest distance is always less than 45°. Conversely, bodies that land higher than they are released from (e.g. basketball free throw) have an optimal release angle of more than 45°. The actual optimal angle of projection in either situation is dependent on both the difference in height between take-off and landing, and the take-off velocity (Equation A6.10):

$$\cos 2\theta = \frac{gh}{v^2 + gh} \quad \text{(Equation A6.10)}$$

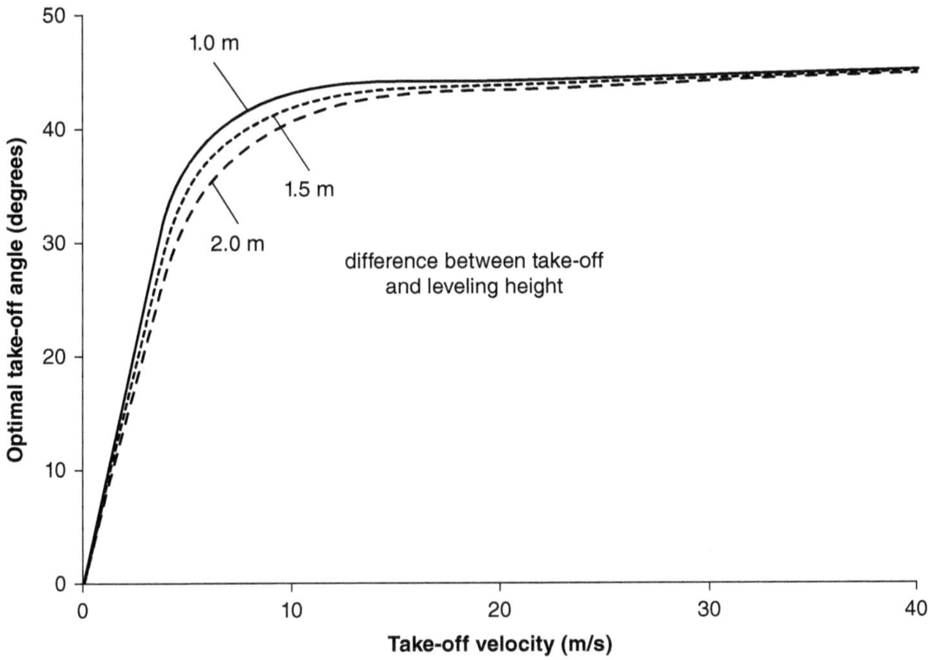


**Figure A6.7** Effect of take-off velocity on optimal take-off angle for projectiles that have greater release than landing height

In the more common situation experienced by long jumpers, shot-putters and so on, the optimal angle of take-off or release decreases with the difference between take-off and release height. For example, for a given release velocity, shot-putters with a high release height will have a lower optimal release angle than those athletes who release the shot from a lower height. The optimal angle also depends on the take-off or release velocity of the projectile. The higher the velocity, the closer the angle gets to (but never reaches) 45°. *Figure A6.7* shows the effect that both the height and velocity of release have on the optimal release angle. This figure shows that at low velocities (e.g. less than 5 m/s), small changes in velocity have a large effect on the optimal angle of projection. At higher velocities, that are more realistic for shot-putters (e.g. 10–15 m/s), the same changes in velocity or release height have a much smaller influence on the optimal angle of release (*Figure A6.7*).

*Key notes*

| | |
|---|---|
| **Introduction** | Newton's Second Law of Motion dictates that bodies that experience a constant force also accelerate at a constant rate. The most common example of this occurring on Earth is when a body is airbourne, where the attractive force between the body and the Earth provides an acceleration equal to $-9.81$ m/s$^2$. |
| **Effects of constant acceleration** | A body subjected to a constant acceleration will experience a linear change in velocity and a curvilinear change in position when viewed over time. At any point in time the motion of a body (e.g. its position or velocity) that is accelerating constantly can be calculated using one of Galileo's equations of uniformly accelerated motion. These equations can be used to find, for example, the height raised by an athlete's centre of mass (COM) during a jump. |

*(continued)*

*(continued)*

| | |
|---|---|
| **Projectile motion** | A projectile is a body that is unsupported (e.g. a ball in flight) and is only affected by the forces associated with gravity and air resistance. Projectiles generally have both horizontal and vertical velocity components during flight. If air resistance can be ignored, the horizontal velocity remains constant and the vertical velocity is affected by the constant acceleration due to gravity, which results in the projectile having a parabolic flight path. |
| **Maximising the range of a projectile** | For a body that lands at the same height that it was projected from, its range is dependent upon both its velocity and angle of projection at take-off. More specifically, range is proportional to the square of the take-off velocity, so higher velocities will result in proportionally greater gains in range. Whatever the velocity, a take-off angle of 45° will result in the greatest range. |
| **Projectiles with different take-off and landing heights** | For projectiles that are released from and land at different heights the optimal angle of projection is dependent upon both the take-off height and velocity. As is often common in sport, where the height of take-off is greater than landing (e.g. shot-put), the optimal angle is always less than 45°. The smaller the distance between take-off and landing heights, the closer the optimal angle gets to 45°. Similarly, for greater velocities, the optimal angle approaches 45°. |

## FURTHER READING

The following resource describes the influence of release angle on performance measures in the shot-put.

1 Linthorne, N. P. (2001). Optimum release angle in the shot put. *Journal of Sports Sciences, 19*(5), 359–372.

## SECTION B

# Kinetics of linear motion

# FORCES



*Paul Grimshaw*

The consideration of forces within human movement can be broadly divided into two categories: attractive and contact forces. **Attractive forces**, such as in the case of gravity, are the result of any two masses acting upon each other. In the case of the Earth's gravitational force, the Earth is acting on the human body. Similarly, although with much less affect, the human body is acting on the Earth. **Contact forces** can be used to describe most other forms of force encountered within human movement, such as the frictional force between the foot and the ground when walking. Contact forces can involve both "pushes" and "pulls" and they cause some form of change in direction or movement (speeding up or slowing down). However, it is important to point out that although **all forces will produce or alter motion**, this actual motion is not always apparent (i.e. it is the **net** result of all forces acting that will result in movement or not). Examples of contact forces within human movement include the reaction force between the feet and the ground during landing from a jump, the impact force between two players colliding in soccer, the force applied to a tennis ball by the racket of the tennis player or the force exerted across the knee joint when the quadriceps muscle contracts during movement.

Within the human body these forces can be further identified as being either external or internal forces. **External forces** include all the forces that exist outside of the body, such as in the case of kicking a soccer ball, and **internal forces** include all those that act inside the body, such as the forces through the anterior cruciate ligament in the knee when a player is tackled in rugby. Often many of the external forces are responsible for the internal forces that are experienced by the muscles, bones, joints, ligaments and tendons. Both external and internal forces act at various places all over the body during movement. Without the existence of these forces we would not be able to move with any velocity or acceleration. Similarly, to achieve a specific performance, such as in the case of sprinting 100 m in less than 10 seconds, requires the careful execution and control of force. However, such forces can also cause injury and a more thorough definition of them will help with both performance improvement and future injury prevention. Consider *Figure B1.1* and see if you can identify some of the different external and internal forces that are occurring in these activities.

Any change in a body's motion (considered without rotation at this stage) that is brought about by the application of a force will incorporate a change in speed and/or a change in direction (i.e. changing its velocity). In order to start an object moving we need to apply a force to the object. **Inertia** is defined as the reluctance of an object to undergo any change in velocity (that is, to either change its current state of velocity or begin any state of movement). The amount of inertia possessed by a body is directly proportional to the amount of mass possessed by the body and it is measured in kilograms (kg). **Mass** is defined as the quantity of matter (atoms and molecules) present in a body (the term body is used to describe both the human body and any objects associated with it). The mass of a body remains relatively constant over time (relative of course to how much you eat and drink in the case of the human body) and it is the same for an object that is on the Moon as it is for the same object when it is on the Earth.

**Figure B1.1** Forces acting within sport

For example, if you have a mass of 55 kg on the Earth you will also have a mass of 55 kg on any other planet or indeed a mass of 55 kg when you are not on any planet at all.

In the context of understanding the term inertia, try to push a book across a table surface. Initially the book will be stationary but as you continue to apply a force (or more force) the book will begin to move (accelerate). The resistance you feel to your "pushing" of the book is a measure of the frictional force that exists between the book and the table and the resistance offered by inertia (ignoring air resistance). The frictional force is derived from the weight (remembering weight is a force) of the book and the two types of materials in contact (book cover and table surface). The more mass the book has, the more weight the book will have. Since the inertia possessed by the book is related to how much mass the book has, you will need to apply more force to accelerate the book at a greater rate. At the beginning of this action you applied a force and yet the book did not move. This would be because you did not apply enough force to overcome the inertia and the frictional force between the book and the table. Now, place another book on the same table but this time use a book that is much heavier. Now try the experiment again. This time you will see that you need to exert a much greater force to accelerate this new book across the table at the same rate as the first, much lighter, book. The new book has more mass and, depending on the friction associated with the book cover and table, a greater reluctance to change its current state of motion (i.e. inertia or the reluctance of the book to accelerate across the table).

This same understanding of inertia and force applies to the movement of the human body. If you try to push (or accelerate) another individual, you will feel a resistance to this "pushing" action. This resistance will be dependent upon the mass of the person you are trying to push and the frictional force between them and the ground. Trying to push someone who is 110 kg is much more difficult than trying to push someone who is 52 kg (although the relative heights and positions of their respective centre of gravity will also have an effect on this exercise).

As we have already observed, **mass** is the term used to describe the quantity of matter in an object (a measure of the number of atoms and molecules in the object) and it is relatively constant. **Weight** (on the planet Earth) is the effect of the Earth's gravitational force acting on a body (again the term body can be used for the human body or any object). Mass and weight are different quantities and the units of measurement for each quantity are also different. Mass is measured in kilograms and weight (by virtue of the fact that it is a measure of the force acting on a body) is measured in newtons (N). The unit of 1 newton (named after the English mathematician Isaac Newton, 1643–1727) is derived from the force required to accelerate a mass of 1 kg at a rate of 1 $m/s^2$.

$$1 \text{ newton}\left(\text{N}\right) = 1 \text{ kilogram}\left(\text{kg}\right) \times 1 \text{ m/s}$$

The terms mass and weight are different and should not be used to describe the same quantity. How often have you heard people say, "How much do you weigh?" The correct answer to this would be to work out the force acting on your body (your weight) by virtue of the fact that you are being pulled to the centre of the Earth by the gravitational force of the planet (because it and you have mass). Since the acceleration due to gravity at sea level is given as 9.81 $m/s^2$ we should, if we know our mass, be able to accurately work out our weight.

## TASK

Using the following example as a guide, determine the individual weights of a 75 kg, 95 kg and 51 kg person while standing at sea level on the planet Earth. Once you have determined their individual weights, calculate their combined weight while standing at sea level on the planet Earth.

### Solution for Earth

Weight is the force acting on a person by virtue of the gravitational pull of the planet Earth and, at sea level, this is expressed as an acceleration due to gravity which is 9.81 $m/s^2$.

$$W = mg$$

where:

W = weight

mg = mass × acceleration due to gravity

### For the 75 kg person

$$\text{Weight} = 75 \text{ kg} \times 9.81 \text{ m/s}^2$$

$$\text{Weight} = 735.75 \text{ newtons (N)}$$

Note: The Moon has a much smaller mass than that of the Earth and therefore it will have a much smaller gravitational effect on your body. Although you will still have the same mass on the Moon as you did on the Earth you will actually weigh much less (this is discussed further in the Gravity, Weight and Vertical Projection section of this text).

The product of mass multiplied by the acceleration (acceleration due to gravity in this calculation in order to determine weight, or the vertical effect of gravity) of the object is a measure of the force.

$$F = ma$$

where:

F = force (measured in Newtons (N))

m = mass (measured in kilograms (kg))

a = acceleration (measured in metres per second squared ($m/s^2$))

And, as we have seen from our calculations of weight, this equation is often re-written to express the calculation of the weight of a body that is being acted upon by gravitational acceleration (i.e. the vertical effects).

$$W = mg$$

where:

W = weight (measured in newtons (N))

m = mass (measured in kilograms (kg))

g = acceleration vertically downward due to gravity (measured in metres per second squared ($m/s^2$))

Note: ***a*** and ***g*** are often used interchangeably in certain equations but in this context they both mean the acceleration due to gravity. However, when we are referring to acceleration in general, such as horizontal motion, for example (as in ***F = ma***), then the letter ***a*** is used to represent acceleration.

Next, think back to the experiment of pushing the book across the table; once the book is moving and you are still pushing it, it will be accelerating (i.e. it was stationary before you pushed it and now it is moving across the table (***F = ma***) – hence its velocity is changing and therefore it is accelerating). The product of the mass of the book multiplied by the acceleration possessed by the book as it changes its velocity will be a measure of the amount of force that you are exerting on the book to move it.

Note: to keep the book accelerating you will need to continue to apply force (***F = ma***). This same understanding is applicable to many aspects of human movement. If you use a force to accelerate your leg/foot down onto the ground quickly, you will need a larger force than if you tried to move it towards the ground slowly. Similarly, if you use a heavier racket or bat to hit a ball, you will impart more force on the ball and potentially increase its velocity and acceleration. However, in both cases you would have been required to generate more force to accelerate the heavier mass at the same or a faster rate. This recognition of force and its relationship with mass and acceleration is a very important concept in human movement and will be expanded on in more detail in the sections concerned with Impulse and Momentum, and Newton's Laws of Motion: Linear Motion.

Since a force has both magnitude (the amount of force you exert or is exerted upon you) and direction (the specific direction in which the force occurs or is applied) it is a vector quantity and can be expressed either mathematically or graphically on paper using straight lines. A vector quantity, such as force, can be identified as an arrow that has both magnitude (length) and direction (angle – position).

When two forces are acting vertically upwards, the composition (or resultant) of these two forces can simply be the summation of their magnitudes. Similarly, when two forces are acting horizontally (or in any direction that is the same) the resultant can be expressed as the sum of their individual parts. The same principle applies for any number of forces that are **parallel** to each other and acting in the same direction (see *Figure B1.2*). Similarly, if two forces are acting in direct opposition to each other the sum of their respective parts (i.e. upward minus downward or right minus left) will indicate the resultant of the two and the direction in which it occurs (i.e. positive or negative value).

Next, imagine trying to move an object with a force (ignoring any frictional or gravitational forces at this stage). If you push the object with **ONE force** it will (provided you apply enough force) move off in the direction of this single force. But if you push the object with **TWO forces** simultaneously it will now move off in the resultant direction of the two forces. Similarly, it will move off with an acceleration (changing from rest to movement) that is proportional to the magnitude of the resultant of the two forces that you are applying to the object (see *Figure B1.3*).

In order to determine the resultant of these two forces (or any number of forces) it is necessary to solve this problem either mathematically or graphically. Graphically the solution is achieved by drawing each force to a specific scale on a piece of paper and using the "tip to tail" method of resolution. This method (which works for any number of coplanar forces) is achieved by carefully drawing each force with the next force drawn on the tip of the previous force. In addition, it does not matter in what order you draw the force vectors, as the solution (resultant – joining the beginning and final end of the vectors (i.e. the first and the last) and

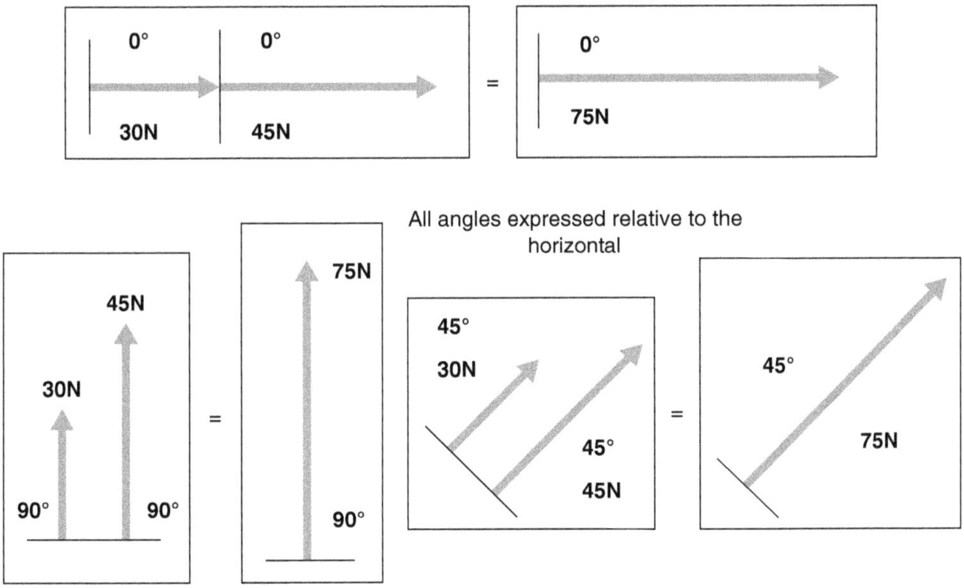


**Figure B1.2** Composition of force expressed as vector quantities (not drawn to scale)

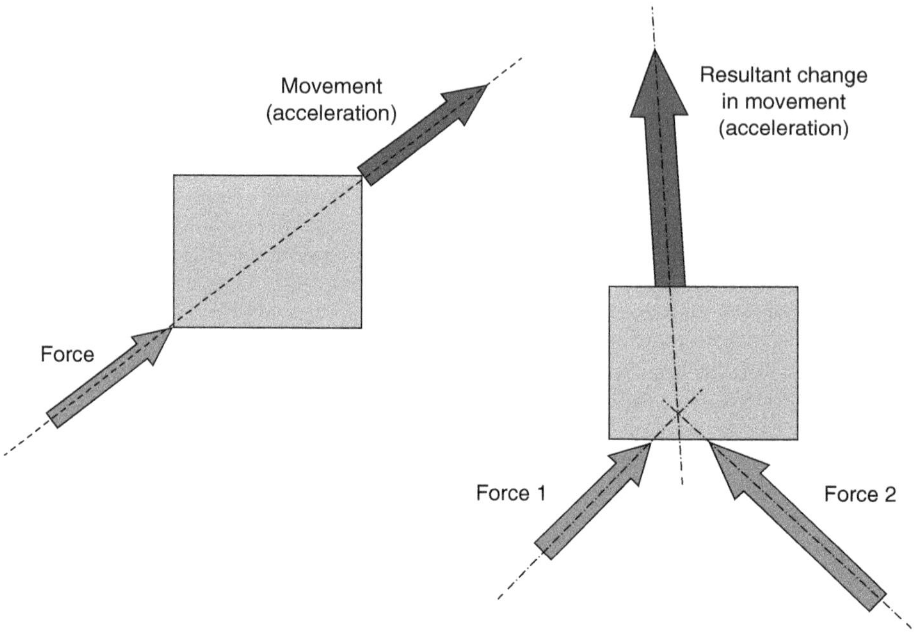


**Figure B1.3** Resultant of force application (not drawn to scale)

measuring the length and direction of the line) will always be the same. In this context, it is important to point out that the resolution of forces will also work for forces that are acting in several planes simultaneously (i.e. three-dimensional). *Figure B1.4* illustrates the graphical "tip to tail" method in more detail.

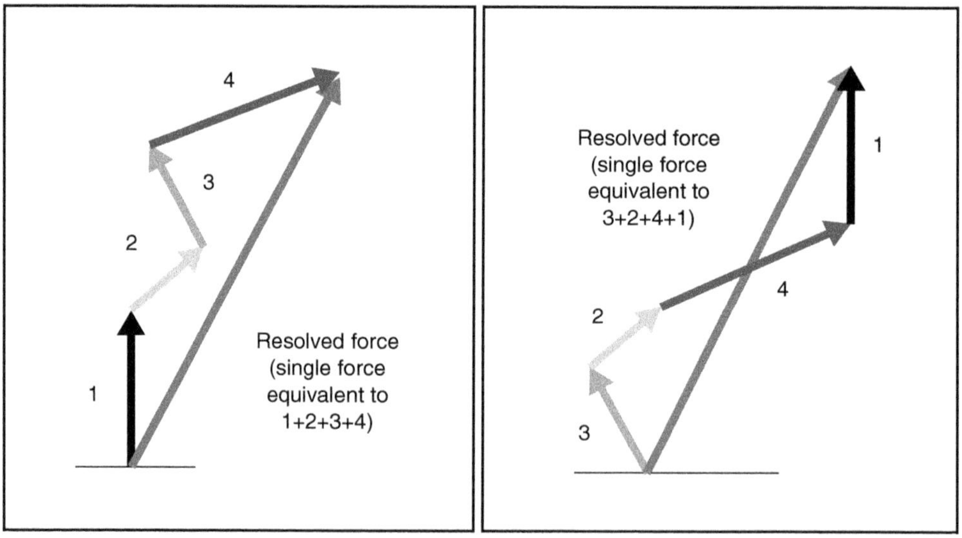


**Figure B1.4** Resolution of force vectors (graphical solution)

These force vectors can also be solved mathematically using trigonometry and this involves first resolving the forces into a single vertical and horizontal component. Next, the vertical and horizontal components are resolved into one resultant force. *Figures B1.5a to B1.5c* illustrate this method in more detail.

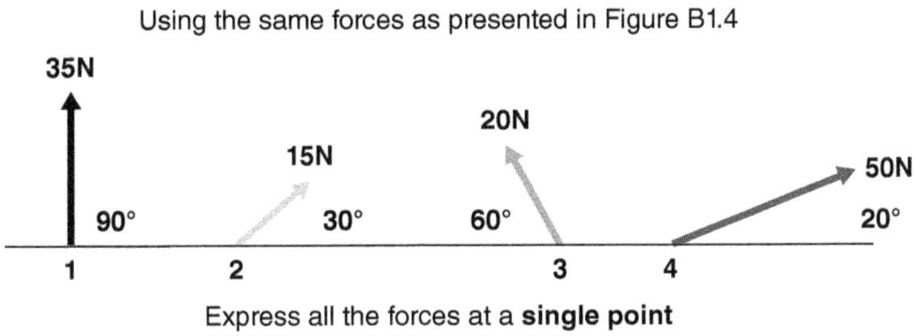


**Figure B1.5a** Mathematical resolution of force application (composition of force vectors solution)

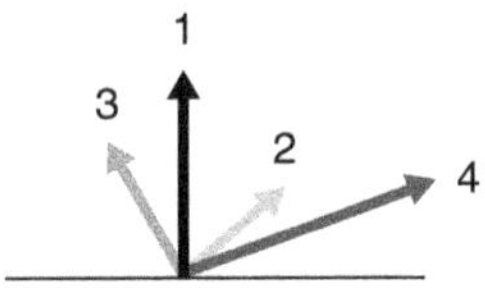


Forces opposing each other should be subtracted and forces acting in the same direction summated. In this case ALL forces have an upward component (with force 1 being perfectly vertical). However, force 3 has a left component while forces 2 and 4 have a right component. Hence the horizontal components of these forces (2+4 and 3) should be subtracted. Note force 1 has no left or right component.

**Figure B1.5b** Mathematical resolution of force application (composition of force vectors solution)

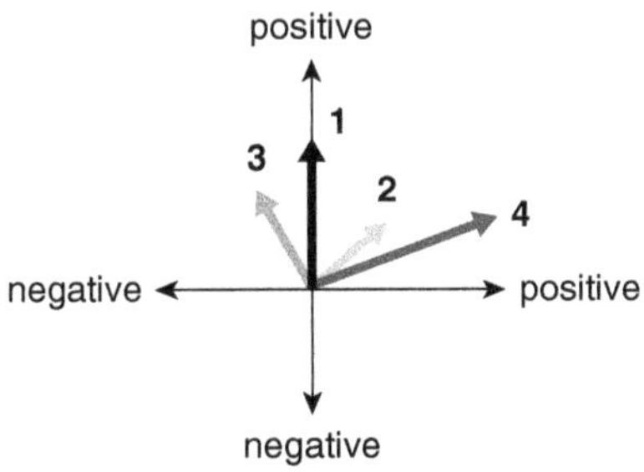


In this configuration it is possible to divide the diagram into positive and negative components as identified above. This process will help determine the direction and magnitude of the resultant force at the end of the calculation

**Figure B1.5c** Mathematical resolution of force application (composition of force vectors solution)

Next, we resolve these forces mathematically into vertical and horizontal components and then finally into a single force resolution. *Figures B1.5d* to *B1.5g* shows this process in detail.

All forces are acting upwards hence all positive

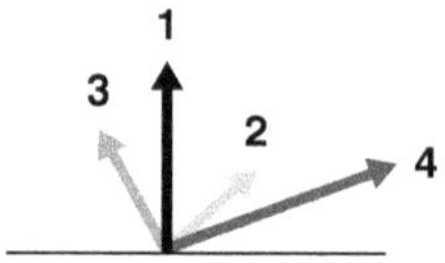


**1. 35N at 90°**
**2. 15N at 30°**
**3. 20N at 60°**
**4. 50N at 20°**

<u>Sum of all forces (vertically) = F Sin θ</u>

F Sin θ

= 35 Sin 90° + 15 Sin 30° + 20 Sin 60° + 50 Sin 20°
= (35 x 1.0) + (15 x 0.50) + (20 x 0.866) + (50 x 0.342)
= 35 + 7.5 + 17.32 + 17.1
= +**76.92 N** (vertical component of all the above forces)

**Figure B1.5d** Mathematical resolution of force application (composition of force vectors solution)

Force 3 is acting left and forces 2 and 4 are acting right; hence, consider force 3 as negative

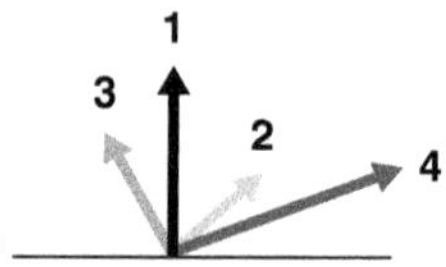


**1. 35N at 90°**
**2. 15N at 30°**
**3. 20N at 60°**
**4. 50N at 20°**

<u>Sum of all forces (horizontally) = F Cos θ</u>

F Cos θ

= 35 Cos 90° + 15 Cos 30° + ( - 20 Cos 60° ) + 50 Cos 20°
= (35 x 0) + (15 x 0.866) + (- 20 x 0.50) + (50 x 0.939)
= 0 + 12.99 + (-10) + 46.95
= +**49.94 N** (horizontal component of all the above forces)

**Figure B1.5e** Mathematical resolution of force application (composition of force vectors solution)

The following two equations for vertical (a) and horizontal (c) resolution of forces were derived from the following trigonometric functions in a right-angled triangle (b would equal the force vectory).

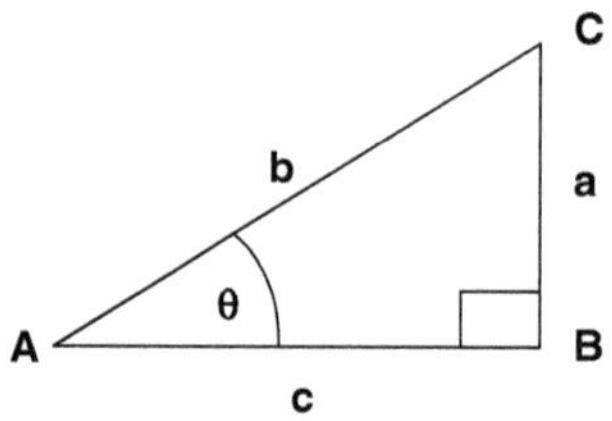


Sine 90° = 1 and Cosine 90° = 0

Sine = $\frac{\text{opposite}}{\text{hypotenuse}}$

Cosine = $\frac{\text{adjacent}}{\text{hypotenuse}}$

Sin θ = $\frac{a}{b}$

Cos θ = $\frac{c}{b}$

**Figure B1.5f** Mathematical resolution of force application (composition of force vectors solution)

Determination of the resultant force (using Pythagoras' theorem)

Resultant force = $\sqrt{FV^2 + FH^2}$

where:
FV = vertical force
FH = horizontal force

$$R = \sqrt{FV^2 + FH^2} = \sqrt{76.92^2 + 49.94^2} = \sqrt{5916 + 2494} = \underline{\mathbf{91.71\ N}}$$

Angle of application of resultant force =

$$\tan\theta = \frac{FV}{FH}$$

$$\tan\theta = \frac{FV}{FH}$$

$$\tan\theta = \frac{76.92}{49.94}$$

$$\tan\theta = 1.54$$

$$\theta = \text{INV TAN}(1.54)$$

$$\theta = \underline{57.0}°$$

**INV = Inverse and often appears as $Tan^{-1}$ on your calculator**

**Figure B1.5g** Mathematical resolution of force application (composition of force vectors solution)

It is now possible to see that the graphical representation of the resultant force (*Figure B1.4*) is the same as the mathematical representation of the resultant force. The resultant of **91.71 N** is acting at an angle of **57.0°** (to the right horizontal) upwards to the right (as both the vertical and horizontal components of it were positive – see *Figure B1.5d* and *Figure B1.5e*). This force vector (magnitude and angle of application) would be the one force that is the equivalent of the forces 1, 2, 3 and 4, as shown in *Figure B1.5a*.

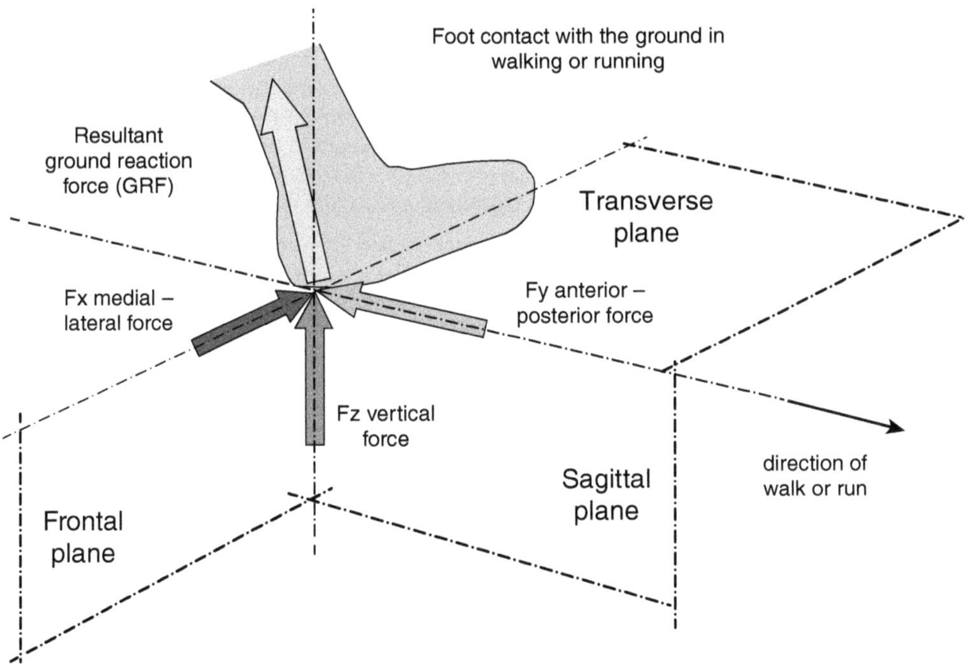


**Figure B1.6** Application of force vectors during heel strike with the foot and ground during walking or running

## APPLICATION

Consider the forces acting underneath the foot as a person walks or runs along the ground. *Figure B1.6* identifies the forces that are acting in the sagittal, transverse and frontal planes (three-dimensional) during walking when a person's foot contacts the ground. From *Figure B1.6* it is possible to see that there are two forces acting in the sagittal plane (anterior–posterior and vertical), two forces acting in the frontal plane (medial–lateral and vertical) and two in the transverse plane (medial–lateral and anterior–posterior). The forces in the sagittal plane are classified as the **vertical force** (acting straight upwards) and the **anterior–posterior force** (acting posteriorly as a braking force when the foot hits the ground at heel strike and the runner is moving forwards). In the frontal plane it is possible to also see another force, which is classified as the **medial–lateral force** (depending on whether it is going medially or laterally with respect to the foot). In the same diagram the resolution (resultant) of all these three forces is also shown. This force is known as the **ground reaction force** and it is the force that is acting at a specific direction and with a specific magnitude. This is the force that can be important for injury considerations. The ground reaction force is the resultant force, which is derived from the composition of the three (now including the transverse plane which would be medial–lateral and anterior–posterior) planar forces described previously. Speed of running, running shoes, type of running style, type of surface of contact, type of foot contact and previous injury can all affect these forces. From both an improvement in performance and an injury prevention perspective it is important that the development and attenuation of these forces is fully recognised by the student of biomechanics.

*Key notes*

| | |
|---|---|
| **Attractive and contact forces** | Forces can be divided into either attractive or contact forces. An example of an attractive force is gravitational force, which exists between any two objects that have mass. Examples of contact forces would be the contact of the foot with a soccer ball or with the ground while running. All forces produce or alter motion, although this motion (i.e. acceleration/deceleration) will not always be apparent. A book cannot be pushed across a table unless you apply enough force to overcome the inertia of the book and frictional force between the book and the table. |
| **External and internal forces** | Forces can also be both external and internal. An external force would be the force on the tennis ball that is exerted by the racket, whereas an internal force would be the forces occurring in the elbow joint when you hit the ball with the racket. |
| **Force and inertia** | An application of force is required to cause a change in movement of an object (i.e. an acceleration or deceleration). Inertia is the reluctance of an object to change its state of motion. Inertia is directly related to mass. The more mass an object has the more inertia it has. If you push a "massive" book across a table it will take more force to accelerate the book at a rapid rate than it would to accelerate a "less massive" book at the same rate. |
| **Mass and weight** | Mass is the measure of the number of atoms or molecules in an object. Mass is relatively constant over time. If you have a mass of 75 kg on the planet Earth, you will have a mass of 75 kg wherever you are. Weight is the measure of force acting on an object. It is dependent on the position of the object on the planet and also on which planet the object is located. An object on Earth will weigh more than the same object when it is located on the Moon. |

| | |
|---|---|
| **Vector quantities** | Vector quantities have both magnitude and direction. A force can be expressed as a vector quantity. This can be either mathematically expressed or graphically represented on paper by a line drawn to scale. The length of the line would be the magnitude of the force and the position/angle/orientation of the line would be its direction. |
| **Resolution of forces** | Several forces acting simultaneously on an object can be resolved into one force. This resolution of force can be carried out by a mathematical method or by using a graphical technique called "tip to tail". Forces can be resolved in both two and three dimensions. The resolution of forces is critical for a recognition of both performance and injury prevention. |

## FURTHER READING

The following two resources provide additional reading on the concept of forces in sport.

1 Besier, T. F., Lloyd, D. G., Cochrane, J. L., & Ackland, T. R. (2001). External loading of the knee joint during running and cutting maneuvers. *Medicine and Science in Sports and Exercise*, *33*(7), 1168–1175.
2 Dixon, S. J., Collop, A. C., & Batt, M. E. (2000). Surface effects on ground reaction forces and lower extremity kinematics in running. *Medicine and Science in Sports and Exercise*, *32*(11), 1919–1926.

# B2 GRAVITY, WEIGHT AND VERTICAL PROJECTION

*Paul Grimshaw*

In addition to developing the three laws of motion (described elsewhere in this text), Sir Isaac Newton (English mathematician, 1643–1727) also formulated the **universal law of gravitation.** This law states:

> Any two objects exert a gravitational force of attraction on each other. The magnitude of this force is proportional to the masses of the two objects and is inversely proportional to the square of the distance between them.

Numerically, this attractive force that each mass exerts on the other can be expressed by the following:

$$F = \frac{GMm}{r^2}$$

where:

G = the Newtonian gravitational constant ($6.67 \times 10^{-11}$ $Nm^2/kg^2$)

M = mass 1 (measured in kg)

m = mass 2 (measured in kg)

r = the distance between the centres of the two masses (measured in metres (m))

It is important to point out that we have seen that the **inertia** of an object (by virtue of its mass) determines the force needed to produce a given acceleration of the object. **Gravitational mass** determines the force of attraction between two bodies. In mechanics the value referred to as the Newtonian gravitational constant is the gravitational force that exists between two 1 kg objects separated by a distance of 1 m. Hence $6.67 \times 10^{-11}$ newton of force is an exceptionally small attractive force.

This equation for the force of gravity can be further developed to express the value for the acceleration (*a*) of mass 2 (***m***) as it is pulled towards mass 1 (***M***):

$$F = ma \text{ or } mg$$

where:

F = force

m = mass 1 (measured in kg)

a = acceleration

g = acceleration due to gravity

$$F = \frac{GM}{r^2} = mg$$

As we have seen mg = ma, and m is common to both, so therefore: g = a

$$F = \frac{GM}{r^2} = g$$

where:

F = force

G = the Newtonian gravitational constant ($6.67 \times 10^{-11}$ $Nm^2/kg^2$)

M = mass 1 (measured in kg)

g = acceleration due to gravity

r = the distance between the centres of the two masses (measured in metres (m))

next:

$$g = a$$

where:

g = acceleration due to gravity

a = acceleration

$$a = \frac{GM}{r^2}$$

where:

a = acceleration

G = the Newtonian gravitational constant ($6.67 \times 10^{-11}$ $Nm^2/kg^2$)

M = mass 1 (measured in kg)

r = the distance between the centres of the two masses (measured in metres (m))

This shows that the acceleration of a free-falling object acting in a gravitational field of mass M falls at a rate independent of its own mass (m). This law is stating that **any two objects that have mass** will exert an attractive gravitational force on each other. Although it is hard to imagine this being the case, you will be exerting an attractive force on the Earth as well as the one you can feel from the Earth acting on you (i.e. your weight). Similarly, you will (because you have mass) be exerting an attractive force on this textbook or your computer as you read this text. The textbook or computer will also exert an attractive gravitational force on you. All objects that have mass will obey this universal law. *Figure B2.1* helps to illustrate this attractive gravitational force with examples from human motion.

As we stand and move about on the Earth we can experience the attractive force of gravity quite regularly and very obviously, because we are subjected to its gravitational field strength. Getting up from a chair requires an effort because the force of gravity from the Earth is pulling us downwards (although, as we are clearly aware, gravity will be acting on us all the time, even when we are just sitting in the chair and not moving). As we walk, gravity holds us to the Earth's surface so that we are able to generate forces to overcome external forces (such as friction) and move forwards. As we throw a ball in the air and then try to catch it again we experience gravity: first, in holding the ball stationary in our hand and trying to get it into the air, then by trying to catch it and finally by holding it in our hand again (i.e. gravity is acting all the time). The gravitational attractive force of the Earth on our body will affect all activities we perform on this planet or on any objects we choose to use while we are on it (or within its gravitational field strength). The reason we can feel this force so obviously is that the Earth's mass (current estimate is 5.9725 billion trillion tonnes (or $5.9725 \times 10^{24}$) – where 1 metric tonne is 1000 kg) is very large in relation to the mass of our bodies (or indeed all other masses on its surface or close to its surface). Similarly, the reason that we do not feel the pages of this text pulling us towards them is because the two masses (the pages of the text and our body) and the attractive forces involved are relatively small (compared with the mass of the Earth). Remember from earlier how small the attractive force was between two 1 kg masses 1 m apart ($6.67 \times 10^{-11}$ or 0.000000000667 newton).

All these bodies will exert an attractive force on each other that is proportional to their net mass and inversely proportional to the distance between their mass centres. Although this **gravitational attraction** between two bodies is defined as **a force**, the effects of this force, or technically the effects of the net force acting on the bodies, are usually expressed in the form

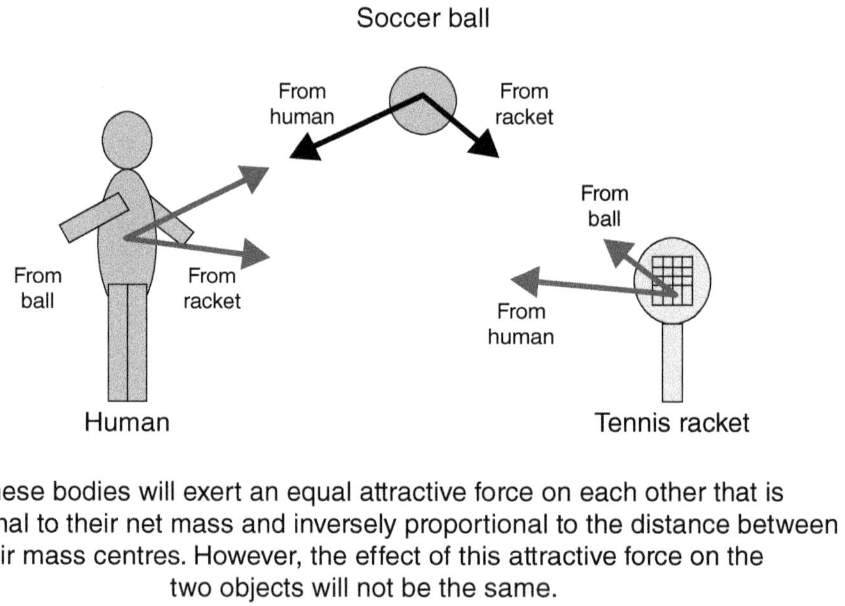


**Figure B2.1** Gravitational attractive forces between masses

of an **acceleration:** for example, how much is one object accelerating (or pulling) the other one towards it?

At or close to the surface of the Earth (remembering the equation from earlier, $a = GM/r^2 = F$) the gravitational acceleration (the pull towards its centre) on our bodies and all other objects that have mass (pens, books, soccer balls, tennis balls, javelins and so on) is given as being close to 10 m/s². In England, the acceleration due to the pull from the mass of the Earth is said to be –9.81 m/s² (downwards towards its centre). This means that any object in a state of free fall that is dropped towards the centre of the Earth (or towards the ground in our case) will increase its velocity by 9.81 m/s for every second of motion (obviously neglecting air resistance). On the surface of the Earth this acceleration is considered to be **constant** and it is the same for an object with 100 kg mass as it is for an object that is 0.5 kg in mass.

## TASK

To illustrate this, try dropping two or more objects of obviously different masses (such as a hammer and a pen) from the same height towards the ground and see which one hits the floor first (*Figure B2.2*). In this example the force exerted from the Earth on the object that you are dropping is the same as the force the object exerts on the Earth. However, the **effect of the force of attraction** between the two masses (Earth and falling object) **is not the same.** The Earth pulls the object towards it at an acceleration of 9.81 m/s² and although the Earth is being pulled towards the object with the same force, because the Earth is so much more massive than the object, there is no (or very little) effect on the Earth. As the different masses of the hammer and the pen are introduced into the example and both are in free fall under the Earth's gravitational

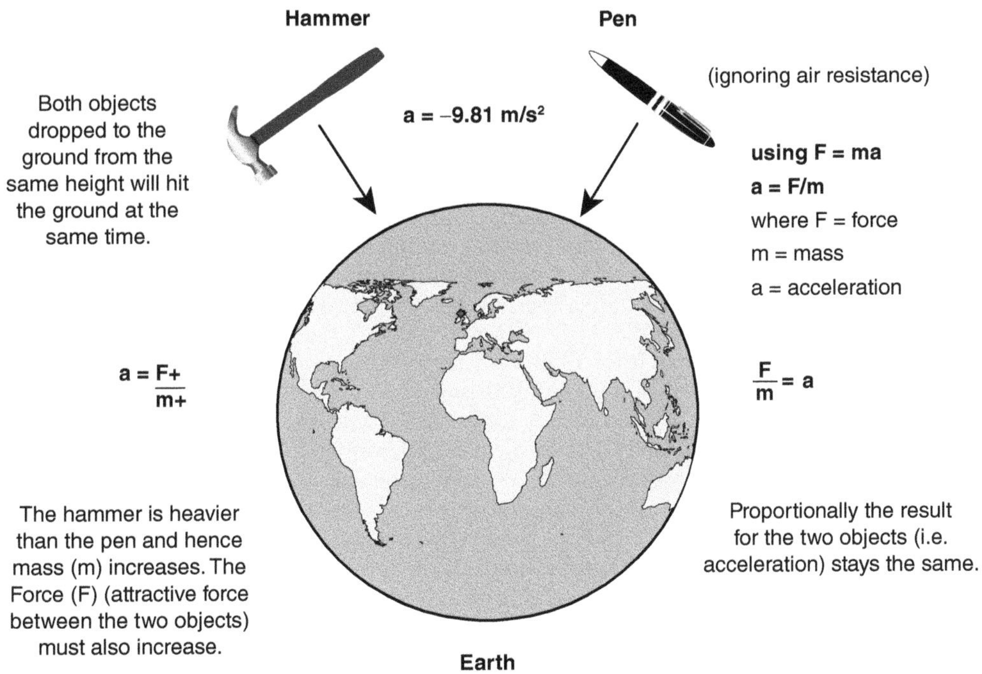


**Figure B2.2** Acceleration due to gravity acting on two objects of different masses

strength, the forces between the Earth and the hammer and the Earth and the pen change. As we have seen previously, the net attractive force that is produced between two objects that have mass is proportional to the masses of the two objects (the larger combined mass of the Earth and the hammer will have a larger attractive force than the combined mass of the Earth and the pen). However, as described elsewhere in this text, objects with larger mass will also have more inertia, or a greater resistance to change in the state of motion (i.e. acceleration or deceleration). The combination of an increased attractive force (due to increased net or combined mass) and an increased resistance to acceleration means that the hammer will continue to fall at the same acceleration rate as the pen (as the mass increases, the attractive force increases and acceleration, which is the ratio of the force divided by the mass, stays the same). Provided both objects (hammer and pen) are dropped at the same distance from the centre of the Earth, the two objects will hit the ground at the same time. This results in a constant acceleration of 9.81 $m/s^2$ for all objects that have mass close to the Earth's surface. The same experiment has also been performed on the Moon and it shows that two objects of different masses dropped into free fall under the Moon's gravitational field strength also hit the surface of the Moon at the same time (although, as we know, they will be accelerating towards the surface of the Moon at a lesser rate than they would be on the surface of the Earth because the Moon has a smaller mass than the Earth).

The gravitational force from the Earth acting on all objects is directed vertically downwards (or to be more correct it is directed towards the Earth's mass centre) and as we sit or stand on the Earth we are (in the balance of net external forces) being accelerated vertically downwards at 9.81 $m/s^2$. However, the reason we do not continue to move (or be accelerated) downwards is because we have the ground to stand on and the ground will be exerting an equal and opposite reaction force on us (in an upwards direction).

The gravitational acceleration from the Earth (its gravitational field strength) will vary slightly depending on where you are on the planet's surface. As we have seen already, in England it is specifically 9.81 $m/s^2$ and because the Earth varies from its width at the equator to its width at the poles (the Earth is approximately 43 kilometres wider at the equator than it is at the poles) the gravitational acceleration will also vary (because we will be nearer or further away from its centre, which is where the Earth's mass is primarily concentrated). For example, the acceleration varies by 1 part in 200 from the equator to the poles (i.e. 0.5%). Similarly, it will also vary if you are either at sea level or standing on the top of a mountain (because on the top of the mountain you should technically be further away from its centre and the gravitational force or acceleration should be very slightly less). However, such variation (due to being at sea level or on a mountain at altitude) is even smaller than that described previously because of the shape of the Earth. This variation is said to be not more than a maximum value of 0.001 $m/s^2$ (or 1/1000th of a $m/s^2$) for all locations on the surface of the Earth. For example, in human movement it is often argued that it is easier to jump higher at altitude than it is to jump for height at sea level. As such, many Olympic records in athletics have been set at high-altitude meetings such as in Mexico City (Olympics 1968) which is approximately 2250 m above sea level. While it is true that the gravitational field strength will be very slightly less at altitude, it is unlikely that this small difference (variation) will have any effect on your ability to jump higher into the air. However, if you were located on the surface of the Moon, you would be able to jump much higher into the air because the Moon has only 1.23% of the Earth's mass and hence causes a gravitational acceleration of only 1.62 $m/s^2$ (about one-sixth of the gravitational acceleration at the surface of the Earth). So as we can see, on the surface of the Moon you would have the same mass (because this is a measure of the quantity of matter, i.e. atoms and molecules, in your body) and you would also have the same strength (related to muscle mass, size and girth) but you would weigh much less (because of the reduced downwards gravitational pull from the Moon on your body).

As we described in the Forces section of this text, the **weight of a body** is defined as the gravitational force acting on your body. Since the effect of this force is expressed as an acceleration value we can use this to calculate our weight. Using the equation proposed by Isaac Newton to demonstrate this force of gravity we can see the following:

$$F = \frac{GMm}{r^2}$$

If this is considered in the context of human movement where we are concerned with the effects of the Earth's gravity on our bodies or on the movement of our bodies, the components of this equation would be:

F = the gravitational force acting on us because of the Earth's mass and the fact that we move on or near to its surface (i.e. our weight)

G = the gravitational constant provided by Newtonian mechanics

M = the mass of the Earth (a constant value)

m = the mass of our body or object

r = the distance between the centre of mass of our body and the centre of mass of the Earth. This value will also remain relatively constant even if we jump into the air or are on the top of a mountain – as we have seen the variation on gravity by virtue of position on the Earth is no more than 0.5% depending on where you are on its surface or 0.001 $m/s^2$ depending on how far away you are from its centre of mass (i.e. at sea level or on top of a mountain)

Since we have several constants in this equation we can now use this knowledge to develop the equation to calculate our weight (or the force acting on our body due to the gravity of the Earth), which is more relevant to our study of human motion.

$$F = \frac{GMm}{r^2}$$

Since ***G*** is a constant value, ***M*** is also a constant value that creates an acceleration at the surface of the Earth of 9.81 $m/s^2$. Furthermore, we can consider ***r*** as a constant value; hence, we can rearrange this equation to represent our weight on the surface of this planet.

## Weight (or force) at the surface of the Earth

$W = m \times g$

or

$F = m \times g$

where:

W = weight (gravitational force)

F = force

m = mass of our body (kg)

g = acceleration due to gravity

This is measured in newtons (N). One newton (1 N) is defined as the force required to accelerate a 1 kg mass at 1 $m/s^2$.

$$N = kg \times m/s^2$$

For a 75 kg person standing on the surface of the Earth in England their weight would be calculated as follows:

$$W = m \times g$$

$$W = 75 \times 9.81\ m/s^2$$

$$W = 735.75 \text{ newtons (to two decimal places)}$$

As an example it is also possible to calculate (in order to illustrate how weight changes because of different gravitational forces) the weight of the same person standing on the surface of the Moon. In this case their weight would be calculated as:

$$W = m \times g$$

$$W = 75 \times 1.62\ m/s^2 \text{ (the Moon's gravitational acceleration)}$$

$$W = 121.5 \text{ newtons}$$

In both cases the subject's mass would be exactly the same (75 kg) and the number of atoms and molecules that make up the person (the measure of their mass) would also be exactly the same. However, this is a clear illustration of why it is easier for astronauts to jump higher while they are on the surface of the Moon (i.e. the reason why you see them able to take large leaps and bounds while moonwalking). However, for the purpose of study in biomechanics, the Earth's gravitational acceleration should be considered as –9.81 $m/s^2$.

**Gravity**, as we have seen previously, is an **external force** that affects only the **vertical component** of projectile motion. In other sections of this text, we have discussed the fact that gravity does **not** affect the horizontal component of projectile motion. The effect of the force of gravity in the balance of the net forces acting is often expressed as an acceleration value (9.81 $m/s^2$) and in the understanding of vertical projection it is important to represent velocities and accelerations with directional components (as they are vector quantities that have both magnitude and direction).

If, while standing on the Earth's surface, we were able to throw a ball perfectly vertically upwards into the air (although in practice this is not so easy to achieve) gravity would be acting on the ball (actually gravity is acting on both us and the ball all the time). The acceleration due to gravity in this case would be expressed as –9.81 $m/s^2$. The minus sign would denote that gravity is acting vertically downwards (i.e. trying to pull the ball downwards towards the Earth's mass centre or trying to slow down its vertical ascent when we throw it into the air). *Figure B2.3a* and *Figure B2.3b* helps to illustrate this exercise in more detail.

In *Figure B2.3a* the ball leaves our hand with a specific amount of upwards vertical velocity (+Vv). This is created from how much net force was eventually applied to the ball and for how long it was applied (i.e. net vertical impulse = force × time = change in momentum (vertical momentum)). The amount of this vertical velocity will determine how high the ball will travel (since the acceleration caused by gravity is considered **constant** at the surface of the Earth regardless of the mass). Hence, the ball with the largest vertical velocity at the point of release from the hand will travel to a higher point in its flight path vertically upwards.

As the ball leaves the hand, the force that was applied to the ball to make it leave the hand becomes zero and gravity will be the only force still acting on the ball (ignoring air resistance).

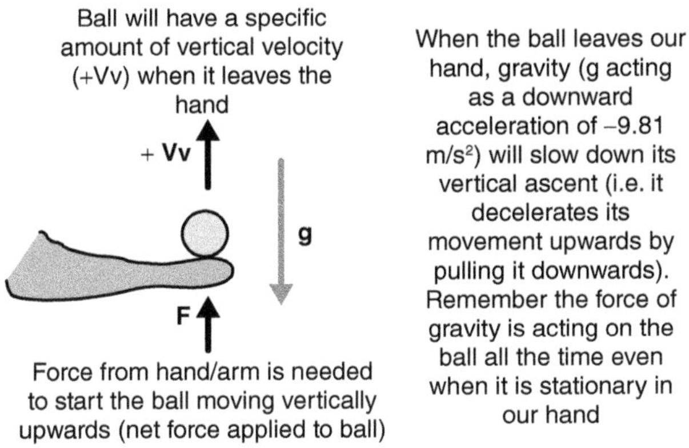


**Figure B2.3a** Gravitational acceleration from the Earth

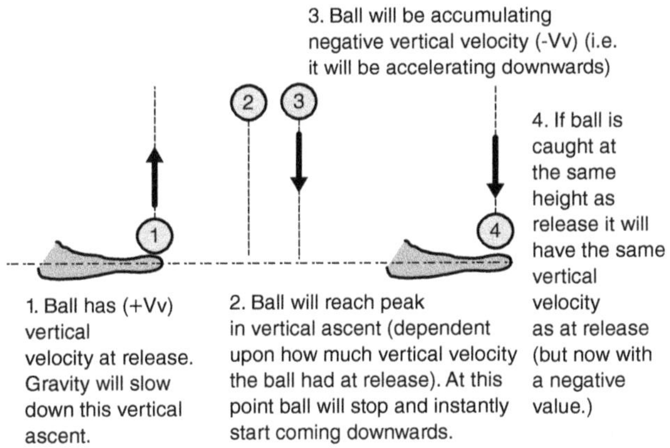


**Figure B2.3b** The motion of a ball thrown vertically upwards

Although the ball will still travel upwards, gravity will be acting by slowing down its vertical ascent (i.e. pulling it back downwards). Eventually, gravity will bring this ball's upwards movement to a stop (it will have slowed it down such that there will be no positive vertical velocity upwards). Its vertical velocity at this point will become zero and it will instantly change its direction of motion (zero positive vertical velocity at the peak height of the flight path upwards). The ball will now start to move downwards, and it will do so at an acceleration rate of –9.81 m/s$^2$, although it is important to remember that throughout this action (flight) it has always been accelerated downwards (i.e. from when (and before) it left our hand, gravity was affecting its vertical (upwards) movement). Similarly, it will also have an accumulating negative vertical velocity coming downwards (negative indicating it is moving downwards). If you manage to catch the ball at the exact same height as that at which you released it, you will find that at this point the ball will have the **same** vertical velocity (ignoring air resistance) as it had when it left your hand (although it will now have a negative sign indicating a downwards movement). *Figure B2.3b* identifies this in more detail.

If the ball is not caught and it is allowed to continue to fall until it hits the ground, it will continue to accelerate at a rate of –9.81 m/s$^2$ in this direction. The ball will accelerate at –9.81 m/s$^2$

until it is acted upon by an external force such as contact with the ground (the force from the ground acting on the ball) or any other object, or until it reaches "terminal velocity" (where the net forces from gravity and air resistance, when not negligible, are balanced).

Now, if we take the same ball and this time throw it with both a vertical velocity (the same as in our previous experiment) and a horizontal velocity (i.e. it would now project into the air at an angle) we could demonstrate exactly the same effect from the force of gravity. *Figure B2.4* illustrates this in more detail.

The ball will travel to the same height as it did in our experiment where we just threw it vertically (because it has had the same vertical force applied to it), but now because it also has a component of horizontal velocity it will travel in a **parabolic flight path** (upwards and forwards). If the ball is caught at the same height as it was released we know that it will have the same vertical velocity as when it was released (even though it is now also travelling in a parabolic flight path with added horizontal displacement). In addition, we can see that from Newton's first law of motion (explained further in the section of this text on Newton's Laws) that this ball will travel forwards with constant (no acceleration) horizontal velocity (in the absence of an external force; remember gravity is not considered to act as an external force on the horizontal component of motion) until it hits the ground or any other object in its flight path.

Similarly, the space shuttle or satellites (i.e. an object in orbit) continue to orbit the Earth for the same reason (i.e. the space shuttle travels around the Earth at a speed of about 17,500 miles per hour (mph)). That is, although it is constantly being pulled downwards towards the centre of the Earth, because the gravitational effect from the Earth is balanced with the forward momentum (mass multiplied by velocity) of the object it is always falling at an angle to the Earth rather than directly downwards towards its centre (the pull of gravity downwards and the

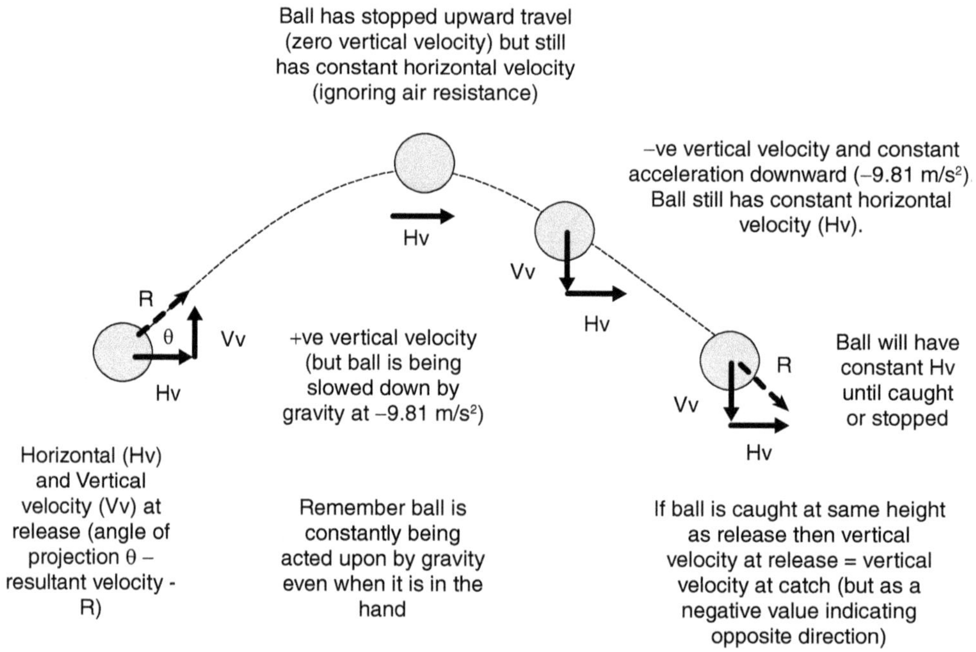


**Figure B2.4** The motion of a ball thrown with both vertical and horizontal velocity

forward momentum are perfectly balanced). Although the horizontal component of motion (the forward momentum) is completely independent of the vertical component of motion (gravity), the object will continue this way until something changes (i.e. another or different external force acts upon it). Graphically this independence of vertical and horizontal motion for the ball experiment is shown in *Figure B2.5*.

Considering *Figure B2.5* in terms of the vertical component of the ball's motion, we can see that it travels upwards and downwards (displacement/time graph: *Figure B2.5*, graph 1) with a decreasing vertical velocity (positive value) as it travels upwards (*Figure B2.5*, graph 2). The ball then reaches the peak vertical height of the flight path and the velocity changes direction (i.e. it stops going upwards and instantly starts coming downwards). Throughout this action it is being accelerated downwards at a constant rate (–9.81 m/s$^2$: *Figure B2.5*, graph 3), which caused the decreasing positive vertical velocity at the beginning of the flight path and the increasing negative vertical velocity towards the end (*Figure B2.5*, graph 2). This is exactly the same as with the ball that was thrown perfectly vertically upwards (providing the vertical release velocity was the same in both experiments).

Horizontally, the ball will be displaced as shown *Figure B2.5*, graph 4 (i.e. increasing its horizontal displacement). It will travel forwards with constant horizontal velocity (*Figure B2.5*, graph 5) in accordance with Newton's first law and it will do so with zero horizontal acceleration (constant velocity horizontally and zero horizontal acceleration, as shown in *Figure B2.5*, graphs 5 and 6 (i.e. no acceleration in this direction)). Hence, vertical and horizontal motions during projectile flight are independent of each other and gravity affects only the vertical component.

In the understanding of vertical projection it is worth making a comment about the effects of **air resistance**. Normally, in human motion we consider the effects of air resistance to

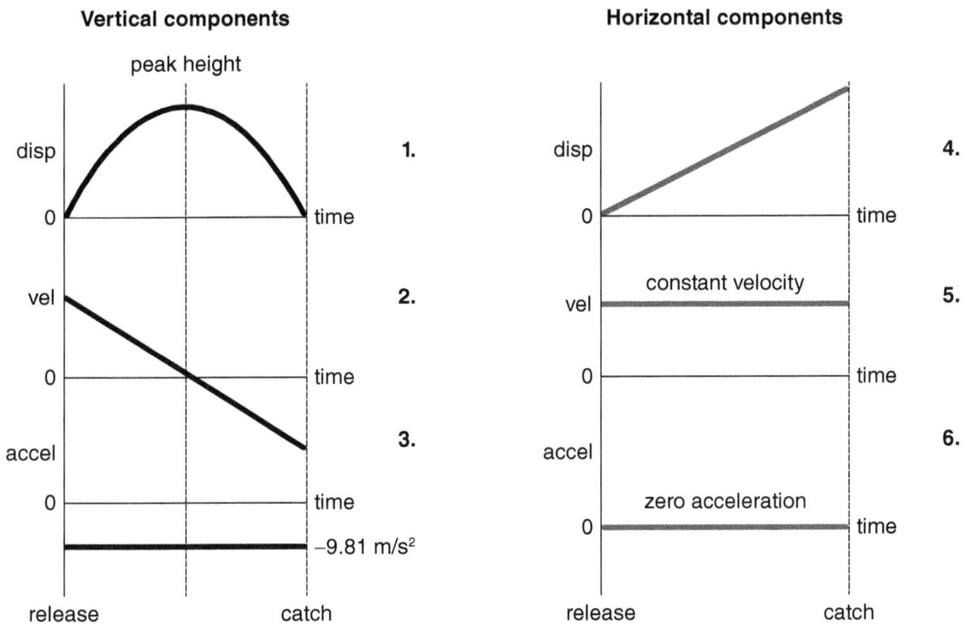


**Figure B2.5** Graphical representation of the motion of a ball thrown with vertical and horizontal velocity

be negligible (particularly on the human body as it travels as a projectile through the air). However, in certain applications the effects of air resistance will not be negligible and will be considered as an **external force** that affects motion (such as in extended free fall). For example, in the case of dropping objects vertically on Earth, we know from Newton's law of gravitation that any object near to or on its surface, regardless of its mass, will accelerate towards the ground (or centre of the Earth) at a constant rate (as we saw previously, two objects of different masses when dropped at the same height will both hit the ground at the same time). However, if you take the case of dropping a piece of paper and a golf ball you will see that the golf ball may hit the ground first. In this case air resistance (an external force) could affect the piece of paper by a significant amount such that its descent towards the Earth will be slowed down as air resistance becomes an external force acting in the opposite direction (upwards) against gravity. Similarly, in sports such as javelin, hammer throwing and discus, and even to an extent in long jumping when there are "head and tail" winds, air resistance will and can have a significant effect. Often long jumps that are wind assisted (usually with a +2 m/s or more wind) are not legitimate jumps that are used for record purposes (in this case the tail wind would be an external force of assistance). Hence, in certain sports and movements it may be the case that the air resistance effects should be considered to be more than negligible.

## TASK

Experiment with dropping different objects from the same height to see if you can now demonstrate the effects of air resistance on the vertical downwards acceleration of objects.

*Key notes*

| | |
|---|---|
| **Newton's law of gravitation** | This law states that any two objects that have mass exert an attractive force on each other. This force is directly proportional to the mass of the objects and inversely proportional to the distance between the objects. |
| **The force of gravity** | The pages of this text and the person who is reading it will be exerting an attractive force on each other. However, because of the relatively small masses involved (i.e. the pages of the text and the human body) the force will be very small and it will not be possible to observe its effects. The Earth (due to its very large mass) exerts a significant force on the human body. This force produces the weight of an individual or object. The attractive force of the Moon on your body would be less than the attractive force you experience on the Earth (because the Moon has much less mass than the Earth). Hence you will weigh less on the Moon although your mass will be exactly the same. This is the reason that astronauts are able to jump large distances when they are on the surface of the Moon.<br>The force of gravity is constantly acting on all bodies. The effect of the force of gravity (in the balance of all external forces that are acting, i.e. the net force) results in an acceleration of the body or object. The acceleration on the Earth is given as $-9.81$ m/s$^2$ (presented as a negative value because the force of gravity from the Earth acting on the body will be trying to pull the body downwards towards the centre of mass of the Earth). |

| | |
|---|---|
| **Acceleration caused by the force of gravity** | On the planet Earth, the effects of the force of gravity due to the mass of the Earth act on all objects that have mass. However, this effect will only act on the vertical component of any movement. Any horizontal component of movement will be independent of the external force of gravity. For objects that are at or close to the surface of the Earth the acceleration of $-9.81\ m/s^2$ is considered to be constant. This acceleration (because of the position of the object in relation to the centre of the Earth and because of the Earth's relatively large mass compared to the object's small mass) will act on all objects with the same rate regardless of their mass. Hence, dropping a hammer and a pen from the same height while on or at the surface of the Earth will result in both objects hitting the floor at the same time (neglecting air resistance). |
| **Air resistance** | In some situations, within human movement and sport, the effects of air resistance (as an external force) are **not** negligible. Air resistance will affect the trajectory of a golf ball and the trajectory of a javelin during flight. Often long jumpers who have a strong "tail wind" during their jump are not allowed to claim the distance that they have achieved for record purposes because of the contribution of this external force. However, because the "wind" assistance is technically the same for all athletes the jump is recorded for the purpose of the competition. |

## FURTHER READING

The following resources provide additional reading on the concept of gravity in sport.

1 Linthorne, N. P., Guzman, M. S., & Bridgett, L. A. (2005). Optimum take-off angle in the long jump. *Journal of Sports Sciences*, *23*(7), 703–712.
2 Hunter, I. (2005). The effect of venue and wind on the distance of a hammer throw. *Research Quarterly for Exercise and Sport*, *76*(3), 347–351.

B3

# FRICTION

*Adrian Burden*

Biomechanics is concerned with the study of forces and the effects of these forces on living things. Most of the forces we are concerned with in biomechanics tend to be external forces that are acting on the body or object of interest (i.e. the forces that cause the body to move). External forces are outside the body and can be both contact and non-contact forces (e.g. gravity is a non-contact external force). Internal forces are those that are within the body and these usually result from the net effect of the external forces. The net force on the player's foot as he/she kicks a soccer ball would be an external force, whereas the force on the anterior cruciate ligament in the knee caused by the kicking action would be an internal force.

Forces can be resolved into individual component parts, such as vertical and horizontal forces. *Figure B3.1* shows the contact forces between the foot and the ground at heel strike during walking (sagittal plane only). The ground reaction force (GRF) that results from the foot contacting the ground at heel strike is the result of all reaction forces acting between the foot and the ground at foot contact (i.e. in three dimensions). When viewed in the sagittal plane (i.e. in two dimensions), the GRF can be resolved into two components, which are shown as a vertical and a horizontal component (*Figure B3.1*). It is worth remembering that in reality the forces acting will be in three dimensions and there will also be a medial–lateral (side to side) GRF component.

The force that is perpendicular to the surface (vertical) is called the **normal force** and this always acts at 90 degrees to the contact surface. The force acting parallel to the surface of contact (horizontal) is termed the **friction force**. Friction forces act between any two surfaces that are in contact and oppose motion or sliding between two objects.

The friction force is essential to human movement and locomotion, and without frictional forces between two objects, it would be very difficult to initiate and maintain movement. For

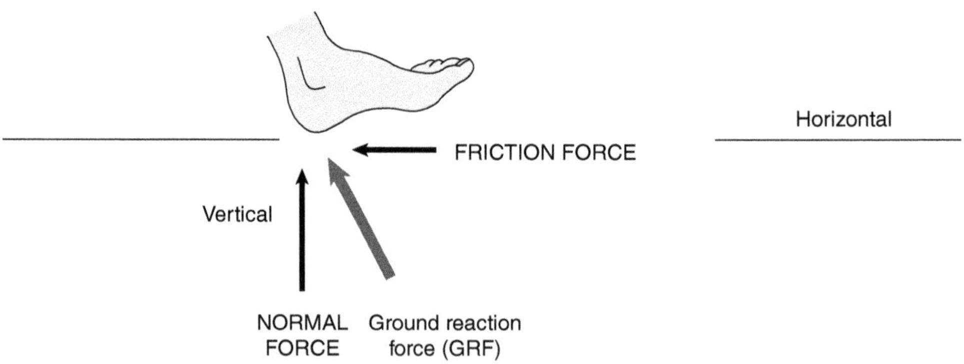


**Figure B3.1** Normal and frictional forces at heel strike during walking (sagittal plane components only are shown (two-dimensional))

example, imagine trying to run across an ice rink in normal shoes. The frictional force between the ice and the shoe is very small and the result is a slipping of the foot during locomotion. The relationship between the two surfaces in contact that gives rise to friction can be described by what is termed the **coefficient of friction.** This is represented by the Greek letter mu (μ).

To illustrate the coefficient of friction, *Figure B3.2* describes the friction force that develops as a stationary bobsleigh on a horizontal surface is brought to motion. In *Figure B3.2a* the stationary bobsleigh has only two forces acting on it, its weight (W) and the ground reaction force (GRF). Because the surface is flat, the GRF is perpendicular, or normal (N) to the surface. A small horizontal force (Fa) is applied to the bobsleigh in *Figure B3.2b* and, in opposition to this force, a small friction force exists between the bobsleigh and the ice. Because these two forces are equal in magnitude, but opposite in direction, the bobsleigh remains stationary. In *Figure B3.2c* the bobsleigh driver increases the applied horizontal force and the frictional force increases accordingly. If the driver applies any more force the bobsleigh will slide forward and the frictional force will decrease slightly. Thus, the friction force in *Figure B3.2c* is termed the **maximal frictional force (Fmax)**. The point immediately before motion occurs is the **coefficient of *static* friction** ($\mu_s$), which describes the interaction between the two surfaces. For a given normal force, the greater the coefficient of static friction between two surfaces, the higher the maximal friction force will be, as shown in the equation below:

$$Fmax = \mu_s \times N$$

where:

Fmax = the maximal frictional force

$\mu_s$ = the coefficient of static friction

N = the normal reaction force

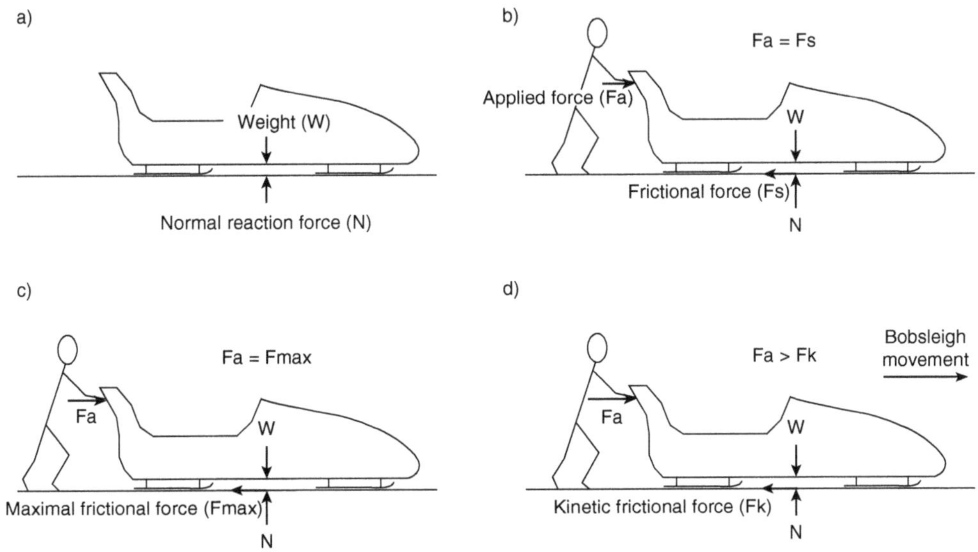


**Figure B3.2** The forces acting on a bobsleigh, when it is stationary (a and b); when motion is impending (c); and when it is in motion (d)

In *Figure B3.2d*, the horizontal force applied by the driver is now greater than the maximal friction force, and the previously stationary bobsleigh experiences a forward acceleration. While in motion, the interaction between the bobsleigh and the ice is represented by the **coefficient of *kinetic* friction** ($\mu_k$), which is slightly lower than the static coefficient ($\mu_s$). Consequently, the frictional force that exists while the bobsleigh is in motion, known as the **dynamic or kinetic friction force,** is slightly lower than the maximal frictional force, and is equal to:

$$Fk = \mu_k \times N$$

where:

Fk = the kinetic friction force

$\mu_k$ = the coefficient of kinetic friction

N = the normal reaction force

A summary of how the friction force changes as the (horizontal) force applied to it increases is shown in *Figure B3.3*. While the bobsleigh is stationary, the friction force increases in proportion to the horizontal force applied to it, until motion is impending when the maximal friction force is reached. As the applied force further increases, the bobsleigh begins to accelerate and friction force reduces slightly.

In examples such as the one above, the **coefficient of static friction** could be determined by placing the object on a force platform and measuring the horizontal (friction) and vertical (normal) components of the ground reaction force as a horizontal force is applied to it. As shown in *Figure 3.2c*, the horizontal GRF component measured immediately before the object begins to move is the maximal friction force. Using Fmax = $\mu_s \times N$, the coefficient of static friction between the object and surface can be derived by dividing the maximal friction force by the normal reaction force at this time.

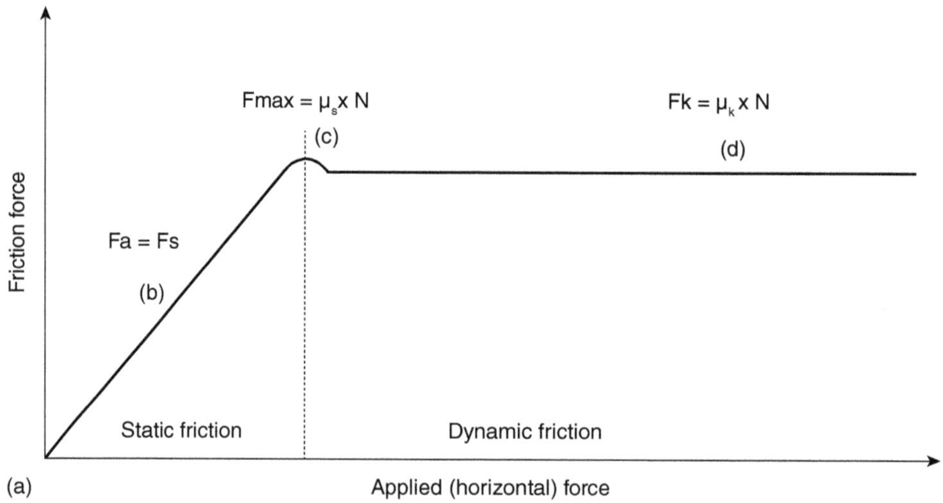


**Figure B3.3** The relationship between the horizontal force applied to the bobsleigh in Figure B3.2 and the friction force between its runners and the ice (points (a) to (d) correspond with the separate images in Figure B3.2)

A simpler way to determine the coefficient of static friction is to tilt the surface on an angle (as shown in the bottom right diagram of *Figure B3.4*). As the surface is tilted upward, the component of the weight of the block acting parallel to the surface increases. This force is trying to slide the block along the surface. As described previously, this (applied) force is matched by an increase in the friction force acting in the opposite direction. When the surface is tilted to the point when any further inclination will result in the block sliding (i.e. motion is impending) the magnitude of the applied force will be equal to the maximal friction force. Similarly, the component of the weight of the block acting perpendicular to the surface will be equal in magnitude and opposite in direction to the normal reaction force. The components of the weight of the block (i.e. the applied forces – top right diagram) and the reaction forces (i.e. maximal friction force and normal force – bottom right diagram) when motion is impending are shown in *Figure B3.4*. At this point, the angle between the vector describing the component of the weight of the block acting perpendicular to the surface and the weight vector is the same as the angle of tilt of the surface (see *Figure B3.4*, top right diagram). Thus, using Fmax = $\mu_s \times$ N, and the trigonometry shown on the left of *Figure B3.4*, the tangent of the angle (i.e. tan $\theta$) at which the surface is inclined immediately before the block begins to slide gives the **coefficient of static friction** ($\mu_s$). Further detail of how this can be used to calculate the maximal friction force is shown in the following example.

## TASK

The angle of inclination immediately before a 20 kg block starts sliding down a plastic covered surface is 35 degrees. Calculate the coefficient of static friction ($\mu_s$) and the maximum frictional force (Fmax) that exists between the two surfaces in contact.

As Fmax = $\mu_s$ x N

∴ $\mu_s$ = Fmax/N

As Fmax = W x Sin $\theta$ *and* N = W x Cos $\theta$

∴ $\mu_s$ = (W x Sin $\theta$)/(W x Cos $\theta$)

As Sin $\theta$/Cos $\theta$ = Tan $\theta$

∴ **$\mu_s$ = Tan $\theta$**

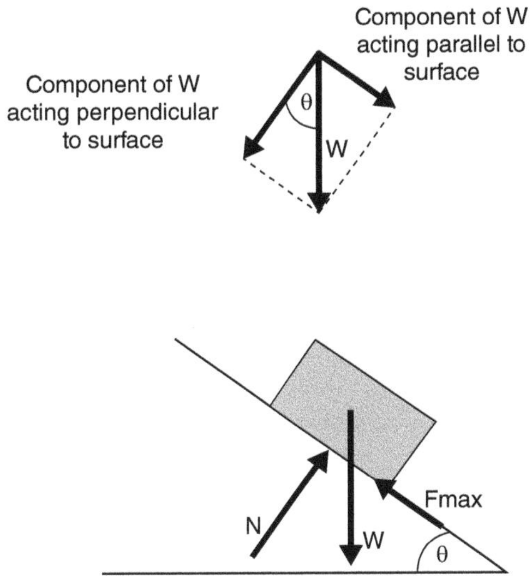


**Figure B3.4** Use of an inclined surface to determine the coefficient of static friction (W = the weight of the block, N = the normal reaction force, Fmax = the maximal friction force and $\mu_s$ = coefficient of static friction)

## Solution

First, we need to establish the coefficient of static friction ($\mu_s$) by tilting the surface until motion is impending, as shown in *Figure B3.5*. Calculation of the weight of the block (W) is also shown in *Figure B3.5*. In order to calculate the maximum frictional force (Fmax) we need to use the equation Fmax = $\mu_s \times$ N, but first we need to establish the normal reaction force (N) acting between the two surfaces.

When the normal reaction force (N), weight (W) and maximal friction force (Fmax) are transposed into a right-angled triangle (*Figure B3.6*), the adjacent side forms the normal reaction force. Using trigonometry (see *Figure B3.6*) N = W × Cos $\theta$, where $\theta$ is the angle of surface tilt.

Finally, now that we know the coefficient of static friction and the normal reaction force, we can calculate the maximal friction force as follows using Fmax = $\mu_s \times$ N:

$$\begin{aligned} \text{Fmax} &= 0.700 \times (\text{W} \times \text{Cos } \theta) \\ &= 0.700 \times ((20 \times 9.81) \times \text{Cos } \theta) \\ &= 0.700 \times (196.2 \times 0.819) \end{aligned}$$

$\mu_s$ = Tan θ

$\mu_s$ = Tan 35°

**$\mu_s$ = 0.70**

W = m x g

W = 20 kg x 9.81 m/s$^2$

**W = 196.2 newtons**

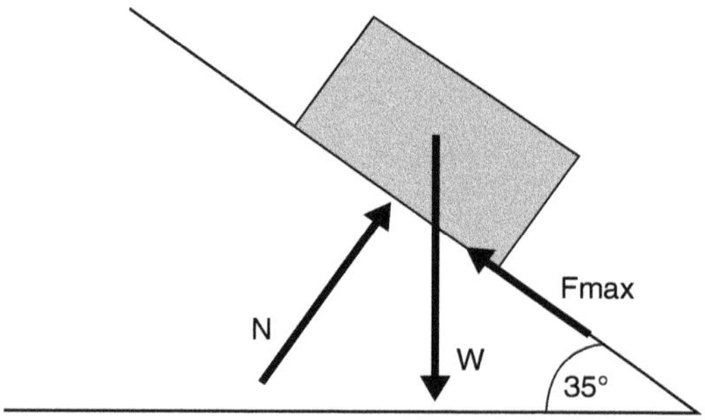


**Figure B3.5** Determination of the coefficient of static friction ($\mu_s$) and weight of the block (W) (see *Figure B3.4* for other nomenclature)

Cos θ = adjacent/hypotenuse

Cos θ = N/W

∴N = W x Cos θ

N = 196.2 newtons x Cos 35°

N = 160.7 newtons

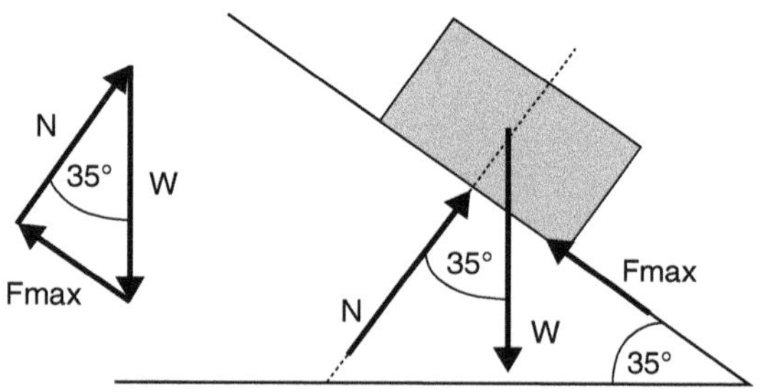


**Figure B3.6** Determination of the normal reaction force (N) (see *Figure B3.4* for other nomenclature)

$= 0.700 \times 160.72$

$=$ **112.53 newtons (N)**

This is the maximum frictional force that exists between the two surfaces in contact. This force would need to be overcome before one object could be pushed across the surface of the other object. Hence, you would need to exert a force of more than **112.53 N** to start the block sliding (in the flat condition).

The friction force depends on the type and nature of each surface in contact. For example, different surfaces in contact will have different **coefficients of friction**. Similarly, different roughness of surfaces in contact will also have different frictional properties: steel and plastic (as used in artificial hip joint replacements) have very low coefficients of friction and move easily over each other. A rough surface acting on another rough surface will have frictional properties different from two smooth surfaces acting together and it should be easier to slide or move the smooth surfaces across each other. While it may be easy to visualise how two rough objects have a high coefficient of friction and hence maximal frictional force, due to the coarseness of the surfaces in contact, this is not so evident for smoother surfaces. However, when viewed microscopically, smooth surfaces also have jagged protrusions that interact with each other and cause a high coefficient of friction (see *Figure B3.7* for a representation). Many of these examples can be seen throughout sport and human movement, for example the type of grip on the javelin; the chalk used by weightlifters or gymnasts for better grip; the table tennis bats with rough and smooth surfaces; and even soccer boots with modified uppers for better contact with and control of the ball.

The frictional force that is created between the contact of two objects is independent of (not connected with) the **surface area of contact**. For example, place a book on a table and try to

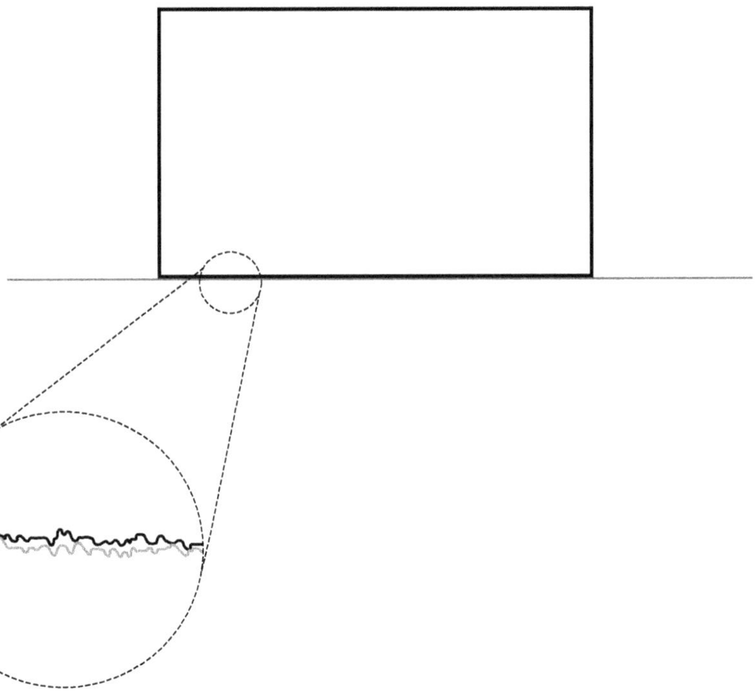

**Figure B3.7** Representation of the interaction between two smoother surfaces

push it horizontally. Now open the outside covers of the book, place it flat on the table and try to push it again, as shown in the diagrams on the left of *Figure B3.8*. The book with its covers closed will create the same frictional force as the book open with both its outside covers in contact (effectively doubling its contact area). The reason for this is that although there is increased surface area of contact (i.e. when you opened the book), the same mass is distributed over a larger area of contact. In other words, you have maintained the mass of the book but spread it over a larger area, thus reducing each of the small contact forces.

From consideration of **Fmax** = $\mu_s \times$ **N** the maximal frictional force is proportional to the **normal reaction force.** Hence, if you increase the normal reaction force you will increase the maximal frictional force between the two objects. In the case of the open and closed book, the normal reaction force did not increase; the same force was just spread over a larger surface area. The frictional force between the two objects remains the same in both the open and closed book situations because the mass and the normal reaction force remain the same.

As a further example of this, adding another book on top of the initial closed book will cause the maximum frictional force to increase, thereby requiring more force to slide the books across the table. Again, as the mass has increased, the normal reaction force, and hence the frictional force between the books and the table, has also increased (see diagram on the right of *Figure B3.8)*.

Another example of this can be seen by placing your hand flat on a table and then seeing how easy it is to initiate movement (i.e. slide it across the table). Next, repeat the same experiment but this time press hard down onto the table. Now it will be more difficult to slide your hand across the table because you have increased the normal reaction force and thus the frictional force existing between the two surfaces in contact (i.e. the hand and the table).

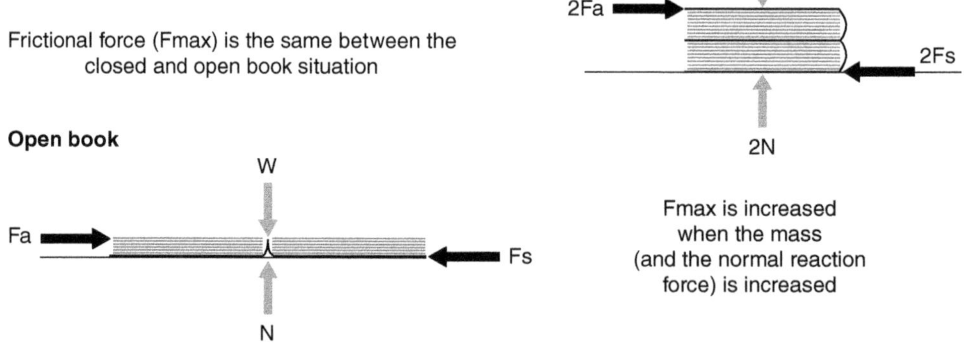


**Figure B3.8** Area of contact between two surfaces and the normal reaction force

**Table B3.1** Coefficient of static friction values for different surfaces in contact

| *Surfaces* | | *Coefficient of static friction* ($\mu_s$) |
|---|---|---|
| Rubber on concrete | (dry) | 0.60–0.85 |
| Rubber on concrete | (wet) | 0.45–0.75 |
| Polystyrene (plastic) on steel | (dry) | 0.35–0.5 |
| Wood on wood | (wet) | 0.20 |

## APPLICATIONS IN SPORT AND EXERCISE

Running shoes with rubber soles are designed to grip the floor so that the athlete can push into the ground with a large enough force to cause a reaction force from the ground, resulting in a net force applied to move the body forward with speed. Artificial joints, such as knees and hips, within the human body require low coefficients of friction so that they will last for a number of years before they begin to wear down. *Table B3.1* presents some of the coefficient of friction values for a number of different surfaces in contact.

For a car tyre, the coefficient of dynamic friction is much less than the coefficient of static friction. The tyre provides the best traction with the road when it is not sliding. However, in the case of the car skidding, the tyres become less effective because of the reduced sliding coefficient of dynamic friction. The coefficient of kinetic friction for metal on metal (same type) can be as low as 0.15, which is lower than any of the values presented for limiting or static coefficient of friction shown in *Table B3.1*.

Heat is often generated at sites of friction between two objects in contact. Within the human body this heat can cause damage to the soft tissue structures. Blisters would be an example of excessive amounts of friction between two surfaces in contact in the human body. The body would respond by producing a layer of fluid between the superficial and deep layers of skin, thus trying to protect the deeper layers. In the long term, the superficial layer of skin often becomes thickened, as in the case of the skin on the ball of the foot. Lubrication of the surfaces in contact helps reduce the amount of friction between objects in the dry condition. Articulating joints within the body that are lubricated with synovial fluid can produce a sliding system that is five times as slippery as ice on ice. As a result human joints can last for well in excess of 70 years before significant wear and tear issues occur (such as arthritis and joint degeneration).

| *Key notes* | |
|---|---|
| **Friction force** | Friction forces act between any two surfaces in contact. This friction force opposes the motion or sliding between the two objects. The frictional force that exists between objects is essential for human movement. The maximal friction force that exists between two surfaces occurs when motion is impending and is equal to the product of the normal reaction force and the coefficient of static friction. |
| **Coefficient of static friction** | The relationship that exists between the two surfaces in contact that gives rise to the frictional force can be described by what is termed the coefficient of friction. The symbol $\mu$ (mu) is used to denote the coefficient of friction between two surfaces in contact. Increase the coefficient of static friction ($\mu_s$) between the two objects in contact and there will be an increase in the maximum frictional force. Similarly, decrease the coefficient of friction and the maximum frictional force is reduced. |

*(continued)*

*(continued)*

| | |
|---|---|
| **Factors affecting the maximal friction force** | Altering the normal reaction force or the coefficient of static friction between two surfaces can affect the maximal friction force. Providing these two factors remain unchanged the friction force is, however, unaffected by the surface areas in contact. |
| **Applications in sport and exercise** | Countless examples exist in sport and exercise when the friction force needs to be increased or decreased. For example, the maximal friction force between a running or basketball shoe and the ground needs to be as high as possible to prevent slipping. Conversely, the friction force between the runners of a bobsleigh or a pair of skis and the ice and snow needs to be as low as possible to encourage sliding over the surface of contact. |

## FURTHER READING

The following resource provides additional reading on the effects of friction.

1 Dowling, A. V., Corazza, S., Chaudhari, A. M. W., & Andriacchi, T. P. (2010). Shoe-surface friction influences movement strategies during a sidestep cutting task: Implications for anterior cruciate ligament injury risk. *The American Journal of Sports Medicine, 38*(3), 478–485.

# IMPULSE AND MOMENTUM

*Paul Grimshaw*

The impulse–momentum relationship develops from Newton's second law of motion ($\Sigma F = ma$) and it allows us to apply this law to situations where forces are continually changing over time. For example, in many cases involving human motion, forces will continuously change (i.e. they are applied over a period of time). Two rugby players who contact each other in a tackle situation will exert changing forces over time. Similarly, when you run and jump on the ground you will apply forces that vary over time depending on a number of related variables: the speed of running, the surface of contact, the shoe type, the body position and many other aspects.

In human movement, it is usually the effect of these changing forces which we are concerned with and it is such forces and their effects that will be used to determine performance characteristics or injury potential (outcome measures). As we apply a force over the ground with our foot during the contact phase in running, what will be the outcome of this application for our running speed? Does our speed increase or do the forces acting on our legs increase to a level that could develop a potential for injury? Similarly, in the athletic event of the shot-put, how does the athlete apply enough force to project the 16 lb (7.27 kg) object through the air? *Figure B4.1* illustrates some other examples of this varying force application over time within human movement.

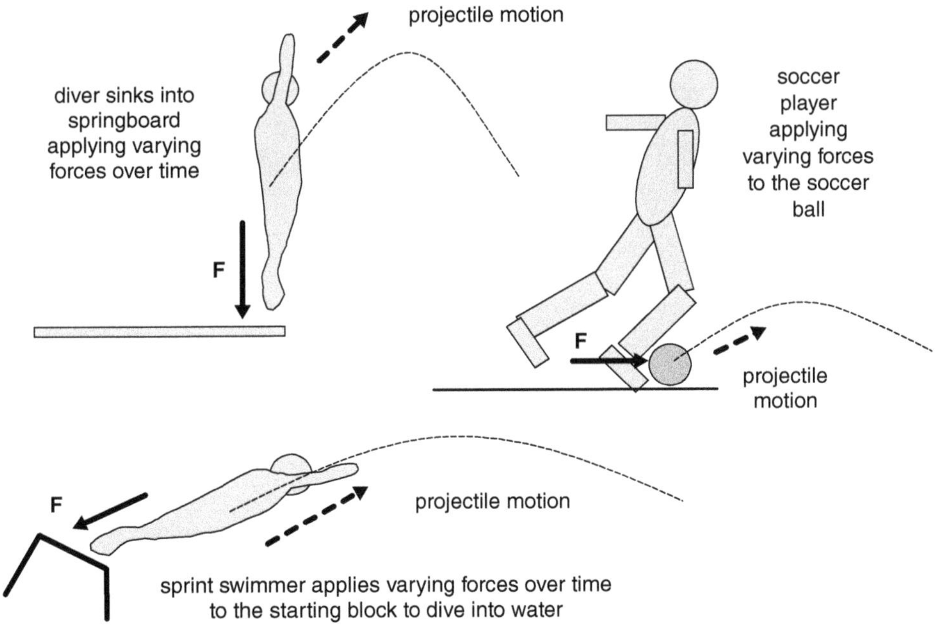


**Figure B4.1** Varying force application over time in human movement

Newton's second law of motion allows us to be able to understand this application of varying force over time (the impulse–momentum relationship) in more detail. From Newton's second law:

$$\Sigma F = ma$$

where:

$\Sigma F$ = sum of forces (net force)

m = mass

a = acceleration

But linear acceleration (a) is also expressed as:

$$a = \frac{v_2 - v_1}{t_2 - t_1}$$

where:

a = linear acceleration ($m/s^2$)

$v_2$ = final velocity (m/s) measured at position $t_2$

$v_1$ = initial velocity (m/s) measured at position $t_1$

$t_2$ = time at position $t_2$ (seconds) for final velocity ($v_2$)

$t_1$ = time at position $t_1$ (seconds) for initial velocity ($v_1$)

Now substitute for "a" into the F = ma equation from Newton's second law:

$$F = m\left(\frac{v_2 - v_1}{t_2 - t_1}\right)$$

In order to cancel out the division component (on the right-hand side) we multiply both sides of the equation by ($t_2 - t_1$) or $t$ (since ($t_2 - t_1$) will produce a single value for $t$ (time).

$$Ft = m\,(v_2 - v_1)$$

Multiply out the brackets, and we now have the equation for impulse:

$$Ft = mv_2 - mv_1$$

where:

Ft represents impulse measured in newton second (Ns)

$mv_2 - mv_1$ represents change in momentum in kilograms.metres per second (kg.m/s)

**Impulse** is defined as the force multiplied by the time (duration) for which the force acts. Impulse can also be determined by using the average force acting over the time period. **Linear**

**(translational) momentum** is defined as the object's mass (kg) multiplied by the object's linear velocity (m/s). Hence, the faster an object is moving, or the more velocity it has, the greater will be the object's linear momentum. Similarly, if you could increase the object's mass you would also produce the same effect and increase the object's linear momentum. Linear momentum can be represented by the following equation:

$$p = m \times v$$

where:

p = linear momentum

m = mass (kg)

v = velocity (m/s)

In the current impulse example ($Ft = mv_2 - mv_1$) we can see that the right-hand side of the equation ($mv_2 - mv_1$) is referring to a **change in momentum**, which in the case of human movement is primarily brought about by a change in velocity (because the mass will remain relatively constant). Similarly, by considering the left-hand side of the equation ($Ft$) we see that this change in momentum can be affected by either increasing or decreasing either the force or the time for which the force acts (**either increase or decrease *F* or *t***). If we increase the amount of force applied (say in the example from *Figure B4.1* for the diver) the change of momentum would also increase (and we would have a greater change in velocity). Similarly, if we increased the amount of time over which the force was applied we could also increase the change in momentum and hence also increase the velocity (since the mass remains relatively constant). At this point it is also important to identify that by using this principle we can also decrease either of these components and thus cause a decrease in the change in momentum and hence a reduced velocity of movement. *Figure B4.2* helps to illustrate this in more detail.

Considering *Figure B4.2*, where the soccer player applies a force to a soccer ball (with the foot) for a specific period of time (contact time), we see that there are two components to this

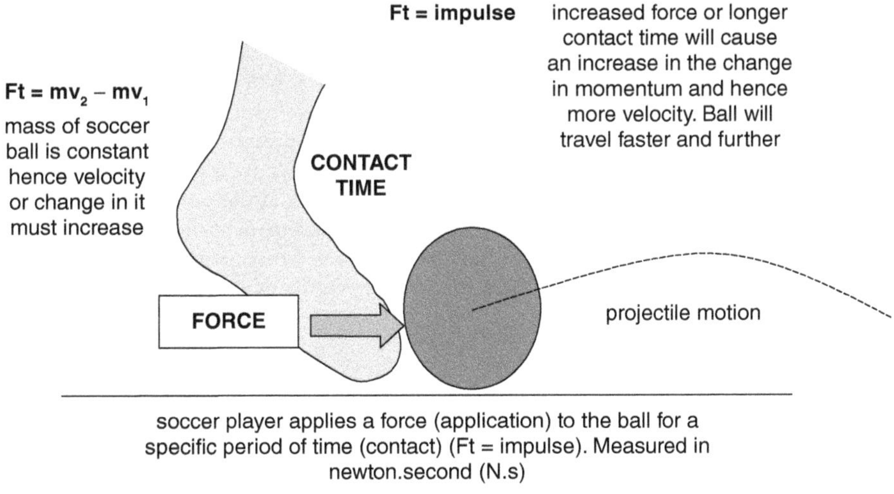


**Figure B4.2** Soccer player applying contact force to a ball to change its state of motion

application. For example, how would the soccer player either increase his/her force applied to the ball or how would they increase the contact time? Both aspects are important in the understanding of the impulse–momentum relationship. Increasing the force applied is generally achieved by either increasing the muscle strength of the player (and hence generally the mass of the striking object or segment), although this can also be achieved by moving the leg much faster ($\Sigma F = ma$). The player can try to move the leg faster towards the ball and thus impart a greater net kicking force to the ball at impact. However, how is it possible to increase the contact time or time of application of this force? The player would use technique and skill in attempting to increase contact time. For example, they may try to hit the ball in such a position that it allows them to follow through the kick, remaining in contact for longer – or indeed they may even intentionally impart a degree of spin to the ball through lateral contact with the soccer boot. In all these methods it is important to point out that it is not simply just a matter of being able to apply a large force for a long period of time. Anyone who has ever kicked a ball will understand that the contact time for the kicking of a ball is very short and it happens in only a fraction of a second. Similarly, it is not possible to apply the large net kicking force all the time throughout this contact phase. The longer you try to stay in contact with the ball the smaller will be the average force applied and hence the smaller will be the impulse generated (impulse = force × time). As the foot makes contact with the ball an external net force (essentially from the foot) is applied to the ball. As the contact time increases, this force applied is averaged over a longer period of time. Therefore, it becomes a careful, skilled compromise of force application and contact time to execute an efficient kick. As an example, try kicking a soccer ball with a pillow tied to your foot (not very practical) and you will see that you are unable to kick the ball with any great speed (velocity) or very far at all. In this example you will have reduced the impact force (applied force) because the pillow is acting as a shock absorber for this applied force, and although you have probably increased the contact time you have much less average force and therefore much less impulse acting on the ball. The result is that the ball has a much smaller change in momentum (less velocity because the mass is constant) and it will not travel as far or as quickly. Although this example is probably not very practical (i.e. it is not easy to fix a pillow to your foot) it does, however, clearly demonstrate the point.

## TASK

Experiment while wearing different shoes and kicking a soccer ball. The same effect can be achieved (although it is not as obvious) by changing your shoes: kick a ball with soccer boots, wellingtons, walking shoes, walking boots, barefoot or kick a ball with large novelty slippers (a comfortable slip-on shoe that is worn indoors) and see what happens.

Consider the example when you are performing a vertical jump from the ground (for maximum height jumped as in, for example, volleyball). As you prepare for the jump (from a stationary standing position) you sink down into the ground while at the same time swinging your arms backwards and forwards (and downwards towards the bottom of the sink). At the bottom of the sinking downwards period you would then drive your arms forwards and upwards and push off with your legs, propelling yourself into the air vertically. *Figure B4.3* illustrates this action in more detail.

In order to have achieved this action and jump into the air you will have applied a force to the ground for a period of time (contact with the ground). The ground reaction force (i.e. from the ground and acting on the person) would be the force that is used to determine the amount of impulse that is acting on the body (impulse = force × time). This impulse would provide a change in momentum (because the two are related by $\boldsymbol{Ft} = \boldsymbol{mv}_2 - \boldsymbol{mv}_1$). Now, since your mass is constant throughout this activity, this change in momentum will result in a change in velocity. The greater the impulse (the more positive the net result) the greater will be the change in velocity. Since at the beginning of the jump you are not moving (zero velocity – stationary) the more net vertical impulse you can generate the greater will be the take-off velocity in a vertical

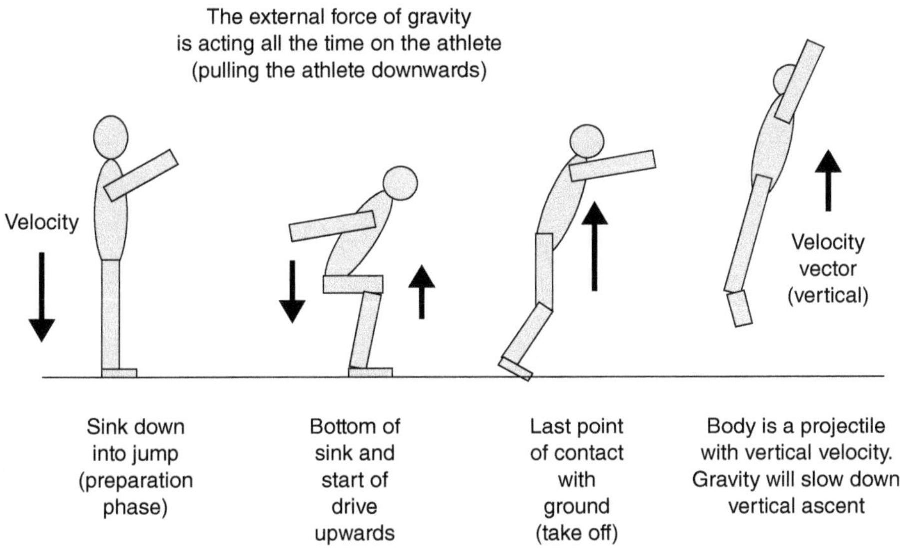


**Figure B4.3** The vertical jump action

direction (important since we are considering the vertical impulse generated). The more vertical take-off velocity you have, the higher you will jump, although, as we know, gravity, which is acting throughout this whole activity, will begin to slow you down at a constant rate as soon as you take off. However, if you have more vertical velocity to begin with, it will take longer for gravity to slow you down at this constant rate and hence you will jump higher.

Now, let us use the equation ($Ft = mv_2 - mv_1$) to look at this example in more detail. *Figure B4.4* identifies this vertical jump example in a subject jumping from a force platform (in order that we can actually measure the amount of impulse that is created). In this figure it is important to identify that we are only considering **vertical impulse.** As we have seen, gravity will affect the vertical components of movement and throughout this exercise (vertical jump) gravity will continue to act on the person. At the beginning of the jump (as the person sinks down towards the ground) the weight of the body is not being supported and the body will accelerate downwards (this appears as a negative force effect (i.e. a drop in the force from the 1 × body weight line) as seen on the graph; in this period of time, the body is assisting gravity to help pull the body downwards and the net effect is a downwards acceleration of greater than −9.81 m/s$^2$ which causes an unweighting situation). At the point where the force trace returns to the 1 × body weight line, the body will have maximum downwards velocity (remember our acceleration–time graphs from other sections in this text; this force–time trace could also be viewed as an acceleration–time graph since the mass (the body) is constant ($F=ma$ so hence $F$ could equal $a$)). Next, the deceleration (the stopping) of this downwards velocity will need to occur in order to stop the body at the lowest point in the downwards movement prior to the body beginning the drive upwards phase (this is marked on the graph). In the case of the force trace shown in *Figure B4.4*, the force from gravity (an acceleration downwards) is inclusive in this net impulse calculation because the trace is presented about the subject's body weight ($W = m \times g$) where gravity is acting on the subject throughout the jump. The fact that the trace (vertical force trace) is plotted about the body weight line accounts for this effect from gravity (i.e. the negative part, under the body weight line, of the vertical force trace). Note: at this point it is important to clarify that in the consideration of **horizontal impulse** (that would also be created in some capacity in a vertical jump, as we never jump perfectly vertically) the effect of gravity is not considered as an external force (i.e. it only affects the vertical component).

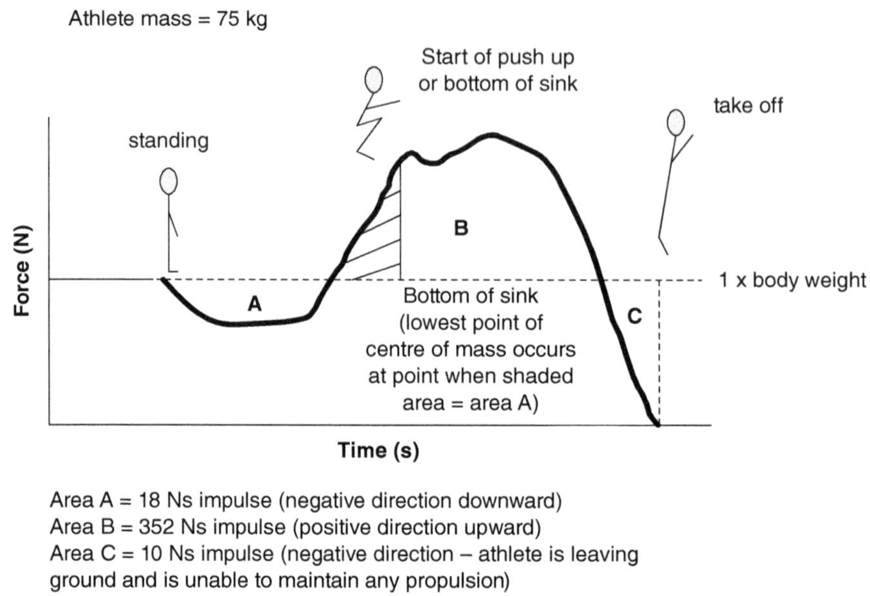


**Figure B4.4** Vertical force-time trace of a standing vertical jump measured using a force platform

In *Figure B4.4* we can see that it is important to identify that in the impulse–momentum equation, force is a vector quantity (i.e. it has magnitude and direction). An increase in impulse will cause a change in momentum in a specific direction (i.e. the direction of the force). For example, if you create a force downwards (which is necessary to initiate a vertical jump) the change in velocity (change in momentum, but indicated as velocity because mass is constant) will also be in the downwards direction. In the example of the vertical jump it is therefore possible to see that impulse will be created in both positive and negative parts (or directions). In the case of you sinking down into the jump you are creating a negative impulse that is not contributing to the vertical component of your jump. However, it is a necessary part for you to be able to initiate the push-off propulsive phase to move you vertically upwards. Hence, the downwards (negative) impulse that is created is subtracted from the positive impulse that is created and the result will be either a positive or negative net impulse. Considering *Figure B4.4*, it is possible to calculate the vertical velocity of take-off from the net vertical impulse that is produced.

Athlete mass = 75 kg

Net impulse = 352 – (18 + 10) N.s

= [B – [A + C]] positive and negative components (= net impulse)

= 324 N.s (positive impulse)

Area A = 18 Ns impulse (negative direction downwards)

Area B = 352 Ns impulse (positive direction upwards)

Area C = 10 Ns impulse (negative direction – athlete is leaving ground and is unable to maintain any propulsion)

Now substitute this impulse value and the athlete's mass into the equation for impulse–momentum:

$$Ft = mv_2 - mv_1$$

$$324 = 75\ (v_2 - v_1)$$

Considering that $v_1$ (initial velocity) is zero (because you started from a standing stationary position on the plate (or ground)) we can now see the following:

$$324 = 75\ (v_2)$$

Divide both sides by 75 to get $v_2$ (final velocity) on its own:

$$\frac{324}{75} = v_2$$

$$4.32 = v_2$$

$$4.32\ \text{m/s} = v_2 \qquad \text{(vertical velocity at take-off)}$$

In this example (*Figure B4.4*), the impulse derived was from the application of **vertical forces** (i.e. from the body, from the ground and from gravity – although there will also be other forces acting in different directions: anterior–posterior forces, medial–lateral forces, and obviously friction) acting over a period of time. In order to demonstrate the importance of this generation of impulse in human movement it is possible to adjust the values from *Figure B4.4* in order to see what would happen if it were possible for us to create more positive impulse. This could be achieved by either increasing the force applied or changing our technique such that the application of the force was for a longer period of time (providing the average force was not significantly less). Alternatively, we could also have changed our technique such that we had less negative impulse (perhaps by modifying the descent phase or our arm movement).

Below are the revised values for the net impulse calculations (note the areas of A, B and C have all increased):

Area A = 22 Ns

Area B = 400 Ns

Area C = 15 Ns

What is the effect on the vertical velocity if with this new technique we manage to create more positive vertical impulse?

$$\text{Net impulse} = 400 - (22 + 15)$$

$$= [B - \{A + C\}]$$

$$= 363\ \text{Ns}$$

Now substitute this value and the athlete's mass into the impulse–momentum equation.

$$\mathbf{Ft} = mv_2 - mv_1$$

$$363 = 75\ (v_2 - v_1)$$

$363 = 75\ (v_2)$ since $v_1 = 0$ (stationary starting position)

$$\frac{363}{75} = v_2$$

$$4.84\ \text{m/s} = v_2$$

This value of 4.84 m/s is an increase on the previous value of 4.32 m/s; hence we will now jump higher.

## TASK

Use the equations of uniform motion from the Uniform Acceleration and Projectile Motion section of this text to see how much higher we actually jumped with this new vertical velocity.

In human movement there are many other examples of where increasing the impulse will result in a greater change in momentum and hence greater velocity. In the case of the shot-put, the athlete applies a larger force in a certain direction (by virtue of their strength (muscle size and muscle mass)) for a longer period of time (by virtue of their technique). The athlete would start by leaning over the back of the throwing circle and then by jumping backwards and rotating in the middle of circle to finally leaning out over the front of the throwing circle. This would allow the athlete to apply a force to the shot for a much longer period of time and potentially (providing the average force was not substantially reduced) create more impulse (that is, acting on the shot in a certain direction (i.e. ideally in the direction in which they are throwing it)). This would result in a greater change in momentum of the shot and hence more directional shot velocity at release. The same applies in javelin, where the athlete would also try to apply a force to the javelin for a long time period by leaning back in the run-up before rotating and leaning forward during the throwing phase (thus creating more impulse).

However, within human movement it is not always desirable to create large amounts of impulse and it is sometimes the case that the net force needs to be reduced (or averaged over longer time) in order to minimise the potential for injury. Imagine trying to catch a cricket ball that is thrown at you. If you stand still and hold your arms outstretched (and rigid) you will feel the large force of impact acting on your hands and arms as you catch the ball. Why? As it contacts your hand, the ball will require an equal impulse in the opposite direction to that which has been applied to the ball at release in order for it to be stopped (i.e. to change its momentum). The amount of impulse that will be required to change the momentum possessed by the ball will be related to force and contact time applied to stop the ball (i.e. it will go from travelling quickly to almost a sudden stop). Hence, the reason you feel a large force is that you have allowed the contact period (between your hands and the ball) to be a very small period in time (by holding out your hands rigidly the ball will just hit your hands and stop suddenly). The force that is applied to the ball in order to stop it (and consequently to your hand – Newton's third law) is high because it is acting over a very short period of time (contact time). Therefore, if you now try to catch the cricket ball by moving your hands in the direction the ball is travelling, this time as you begin to catch it you will increase the contact time and thus average out the force (i.e. it will feel much easier to catch the ball this way). *Figure B4.5* illustrates some other examples where the control of impulse (force and time) is important in human movement.

As an example, an object with 50 kg.m/s of momentum (e.g. a cricket ball) must experience 50 Ns of impulse (from the hands) in order for it to come to a stop, remembering that impulse and momentum are related ($Ft = m(v_2 - v_1)$). Any combination of force and time could be used to provide the 50 Ns of impulse needed to stop the ball. In this case if the contact time were 2 seconds, the force would need to be 25 newtons. Similarly, if the contact time is increased to 4 seconds, the force would be reduced to 12.5 newtons (both producing the net result of 50 Ns). The same principle applies for many other situations in human movement where it is important to increase contact time to reduce potentially damaging impact forces.

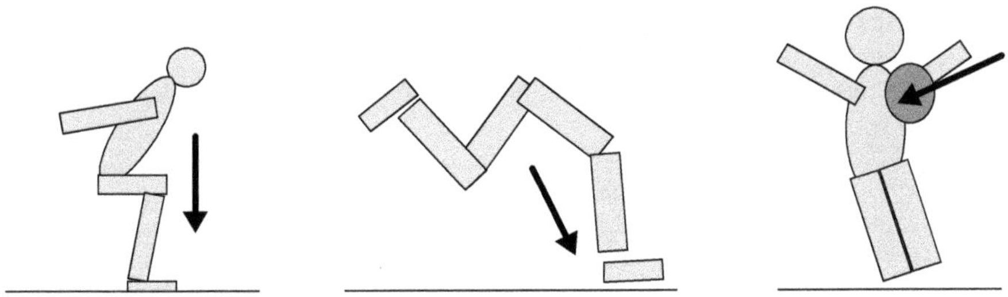

Landing from a drop jump with knees bending increases contact time and reduces impact force (shock absorbing)

Wearing shock-absorbing running shoes or flexing your knee upon contact will help reduce the impact forces

Moving with the direction of the ball in soccer when receiving a pass on the chest

**Figure B4.5** Impulse–momentum within sport (increasing contact time to reduce impact force)

The impulse–momentum equation is one of the most important principles in biomechanics and it provides a method for the recognition of improvement in performance and injury prevention within human movement.

*Key notes*

| | |
|---|---|
| **The impulse–momentum relationship ($Ft = m\,(v_2 - v_1)$)** | This develops from Newton's second law of motion ($\Sigma F = ma$) and is an important relationship within biomechanics. |
| **Momentum** | The linear momentum possessed by an object is a measure of the object's mass multiplied by the object's linear velocity. Since the mass of an object remains relatively constant, the change in momentum experienced by an object represents a change in its velocity (increase or decrease). |
| **Impulse** | Impulse is defined as the force applied multiplied by the time of force application ($Ft$). It is equal to the change in momentum possessed by an object ($Ft = m\,(v_2 - v_1)$). Impulse can be increased either by increasing the applied force or increasing the time of force application. In certain situations within human movement, it is necessary to have a large force and small time of application (contact) and in other examples it is valuable to have the opposite situation. |
| **Application** | The shot-putter in athletics applies a force to the shot for a long period of time in order to give the shot more impulse and hence a greater change in momentum (i.e., more velocity at release because the shot's mass is constant). The vertical (high) jumper applies a force to the ground in order to jump off the ground. The ground applies a reaction force to the jumper in order for them to be able to leave the ground. The net vertical impulse created during the preparation of the vertical jump will affect how high the athlete is able to jump. When catching a ball, it is often necessary to increase the time of contact with the ball in order to reduce the force of impact (between ball and hand). This is achieved by following the ball's direction with your hands as you make the catch (as in the case of catching a cricket ball). |

## FURTHER READING

The following two resources provide additional reading on the concepts of impulse and momentum in sport.

1 Hendricks, S., Karpul, D., & Lambert, M. (2014). Momentum and kinetic energy before the tackle in rugby union. *Journal of Sports Science & Medicine*, *13*(3), 557–563.
2 Hunter, J. P., Marshall, R. N., & McNair, P. J. (2005). Relationships between ground reaction force impulse and kinematics of sprint-running acceleration. *Journal of Applied Biomechanics*, *21*(1), 31–43.

# CONSERVATION OF LINEAR MOMENTUM



*Adrian Burden*

**Linear momentum** is defined as the product of the mass of the body multiplied by its linear velocity:

Linear momentum = mass × linear velocity

(kg.m/s) = (kg) × (m/s)

In order to increase the linear momentum possessed by a body it would be necessary to increase either its mass or its linear velocity. Generally, within human motion it is difficult to increase the mass of the body (as this remains relatively constant), so in order to increase linear momentum we would increase the object's linear velocity.

In human movement there are many situations where collisions between objects or bodies occur. For example, two rugby players (or American football or Australian Rules football players) collide with each other in a tackle situation, and a soccer player will collide with the ball on numerous occasions throughout a game, such as in the case of receiving a chest pass or heading the ball. *Figure B5.1* illustrates some examples of collisions within human movement.

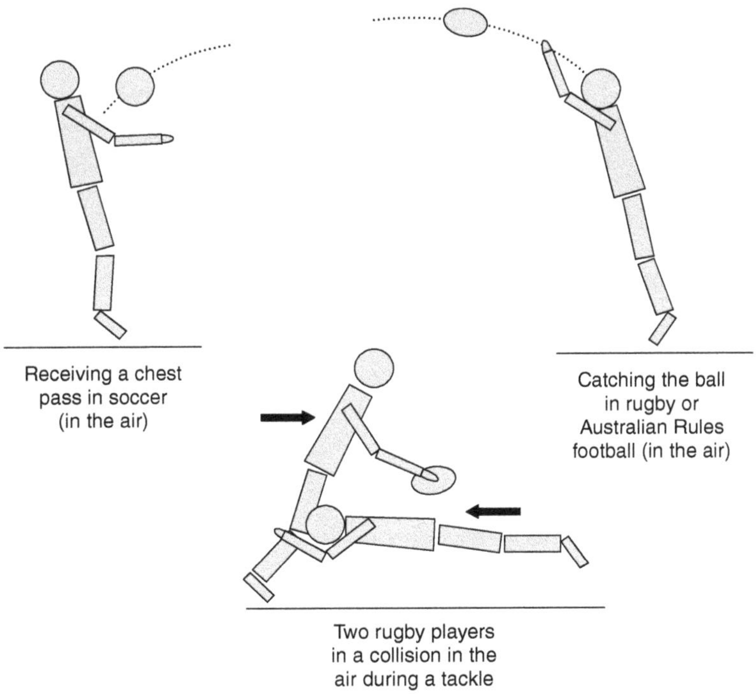


**Figure B5.1** Collisions between bodies within human movement

Consider Newton's first law of motion, the law of inertia, where a body will remain in a state of rest or constant velocity unless it is acted upon by an external force. We can now extend this law for examples involving collisions to explain the **principle of conservation of linear momentum.**

The principle of conservation of linear momentum is an extension of Newton's first law of motion. It states that in any system where bodies collide (and there can be more than two bodies) or exert a force upon each other, the total momentum in any direction remains constant unless some external force acts on the system in that direction. In this context, the term "system" can be used to describe two or more bodies in motion that exert forces on each other. *Figure B5.2* illustrates this in the catching of a ball during goalkeeping in soccer (the goalkeeper is shown catching the ball in the air because of the external forces that would need to be considered if they were on the ground).

As we can see from *Figure B5.2*, the momentum of the ball and the goalkeeper (i.e. the system) before the collision (i.e. the catching of the ball) is equal to the momentum of the system (i.e. the ball and the goalkeeper together) after the collision. For simplicity the ball and the goalkeeper before contact could be termed System 1 and the ball and goalkeeper together after contact (when the goalkeeper holds the ball) System 2.

To calculate the momentum in System 1 before the collision, we use:

Momentum of the ball + momentum of the goalkeeper

$$[\text{mass}_{\text{ball}} \times \text{horizontal velocity}_{\text{ball}}] + [\text{mass}_{\text{goalkeeper}} \times \text{horizontal velocity}_{\text{goalkeeper}}]$$

It must be noted that we are considering **linear momentum in a horizontal direction** and hence we can neglect the effects of gravity (an external force). After the collision, the momentum in System 2 can be determined as:

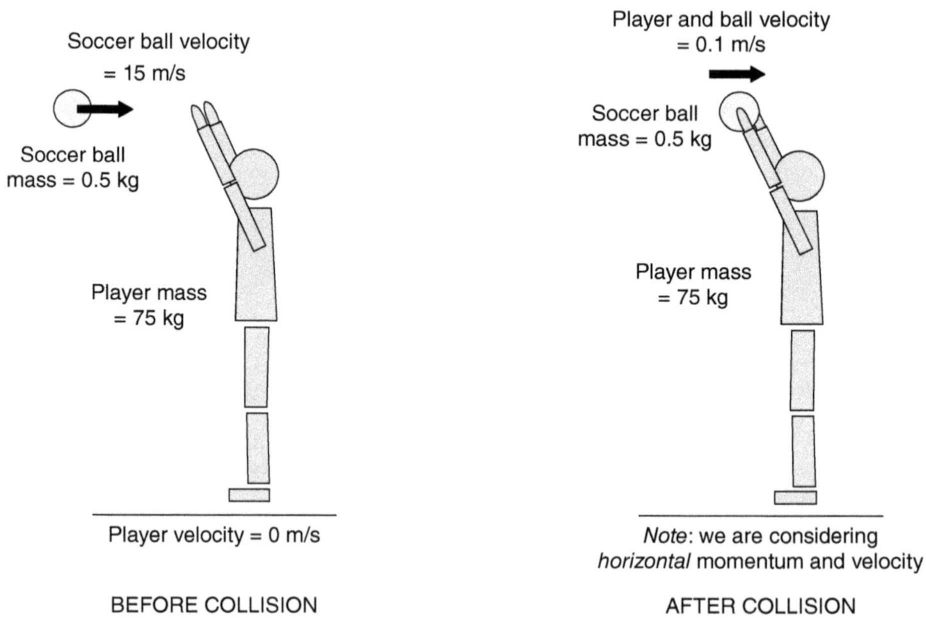


**Figure B5.2** A system of forces before and after collision: goalkeeper catching a soccer ball in the air

Momentum of the ball + goalkeeper combined

[mass of ball and goalkeeper × combined horizontal velocity of ball and goalkeeper]

Now according to the principle of conservation of linear momentum, the momentum possessed by the system before the collision equals the momentum possessed by the system after the collision (the amount of momentum is constant – it is conserved). In order to prove this we can use the values for mass and velocity as shown in *Figure B5.2*.

Momentum before collision = momentum after collision

$$(0.5 \times 15) + (75 \times 0) = (75.5 \times 0.1)$$

$$7.5 + 0 \text{ kg.m/s} = 7.5 \text{ kg.m/s}$$

Again, it should be noted that this is **horizontal** linear momentum that we have determined. In addition, it is also possible to see that if we did not already know the velocity for the ball and the goalkeeper after collision we could use this equation to calculate the combined velocity.

## APPLICATION

It is important to remember that the linear momentum possessed by a system will remain constant in both magnitude and direction, and that **the principle of conservation of linear momentum is valid only if the following conditions are met:**

1 There is no external impulse (as impulse = force × time and it is related to a change in momentum); in other words no external force.
2 The total mass of the system (bodies that are colliding) remains constant.

To illustrate this principle in a more simplified form (i.e. not involving human bodies or projectiles), *Figure B5.3* identifies a more practical example. Considering *Figure B5.3*, we can see that Ball A has a mass of 2 kg and is moving towards Ball B with a horizontal velocity of 8 m/s. Ball B is also moving in the same direction away from Ball A but with a horizontal velocity of 2 m/s. Ball B has a mass of 1 kg.

As the balls collide, there will be an impulse exerted by each ball on the other (i.e. a force applied for a period of time). In this case the contact time is expected to be small and therefore it is likely that the force will be relatively high. There will be a change in momentum brought about by the impulse but the total amount of momentum (before and after collision) will remain constant. In accordance with Newton's third law, Ball A will experience a force, and therefore an impulse, from Ball B in the direction from right to left. Ball B will experience an impulse in the opposite direction, from left to right, from Ball A. Each ball will experience a change in momentum that is equal to the amount of impulse that is created. We know that momentum is the product of mass and velocity, so in the case of the ball with more mass (2 kg) there will be less change in velocity for a given momentum. Similarly, for the ball with a smaller mass (1 kg) there would be a greater change in velocity in a given direction. It is important to point out that this change in momentum (or velocity in each case since the mass is constant) will take place in the direction of the impulse or applied force. Thus, for Ball A there will be a change in momentum in the direction from right to left, whereas for Ball B the momentum change will be from left to right. Again, it is important to reiterate that we are referring to **horizontal linear momentum** in this application.

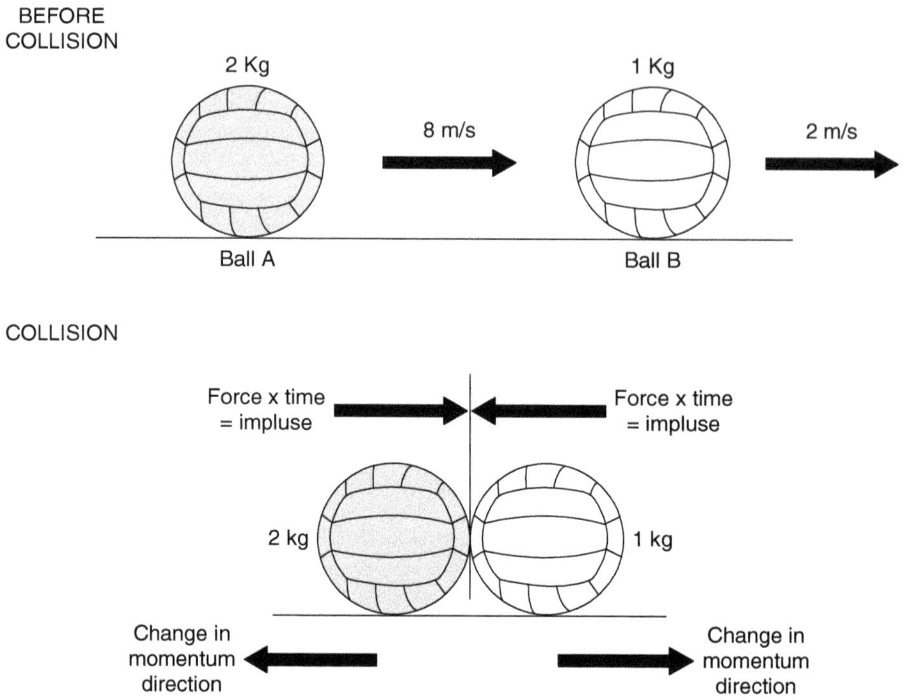


**Figure B5.3** Two balls colliding to demonstrate the principle of conservation of linear momentum

The impulse acting on Ball A (i.e. from right to left) is given a **negative** sign to identify its direction, and is equal to the change in momentum of Ball A (i.e. the difference between momentum before and after collision), which can be expressed as:

$$-Ft = (m_A \times v_A) - (m_A \times u_A)$$

where:

$-Ft$ = impulse in the right–left direction (negative value indicates direction)

$m_A$ = mass of Ball A

$v_A$ = final velocity of Ball A (after collision)

$u_A$ = initial velocity of Ball A (before collision)

The impulse acting on Ball B (i.e. from left to right) is given a **positive** sign since, again, it is important to identify direction, and is equal to the change in momentum of Ball B, which can be expressed as:

$$+Ft = (m_B \times v_B) - (m_B \times u_B)$$

where:

+Ft = impulse in the left–right direction (positive value indicates direction; it is not necessary to express the + sign)

$m_B$ = mass of Ball B

$v_B$ = final velocity of Ball B (after collision)

$u_B$ = initial velocity of Ball B (before collision)

Now, considering that the impulses acting on the two balls are of **equal** magnitude (i.e. the forces acting on each ball are the same [action–reaction] and the contact time is the same) we can now express the equation to demonstrate the principle of conservation of linear momentum:

$$Ft = -(m_A \times v_A) - (m_A \times u_A) = (m_B \times v_B) - (m_B \times u_B)$$

That is, impulse equals the change in momentum of Ball A, which equals the change in momentum of Ball B. **Note:** the minus sign preceding the first part of this equation, which is related to Ball A's momentum, denotes the direction of momentum change. Rearrange this equation and we have:

$$(m_A \times u_A) + (m_B \times u_B) = (m_A \times v_A) + (m_B \times v_B)$$

Momentum before collision = momentum after collision

This confirms the conservation of linear momentum principle, as the momentum in the system before the collision equals the system's momentum after the impact. *Figure B5.4* shows that the two balls (A and B) have continued to move forwards but with different velocities.

Each ball will undergo an equal change in the magnitude of momentum but in opposite directions, even though they will experience a different change in velocity because their respective masses are different. The conservation of linear momentum equation in this example (where both balls continue, but with different velocities) therefore leaves us with one equation to find two unknown quantities (i.e. the final velocities of Ball A and Ball B). Therefore, we need one other equation to solve this problem, and hence we must know either of the following:

1 The final velocity of Ball A or Ball B, or;
2 That there is no rebound and the two balls will continue forwards coupled together with a common velocity ($v_A = v_B = v$).

We can now rewrite the conservation of linear momentum equation for this specific case of two bodies that continue to travel, coupled together (i.e. no rebound), with a common velocity (which is often the case in human movement applications).

$$(m_A \times u_A) + (m_B \times u_B) = (m_A + m_B) \times v$$

*Figure B5.5* illustrates the many different collision situations that can occur within human movement: objects travel on with different velocities, objects rebound and objects or bodies travel on with a velocity that is combined.

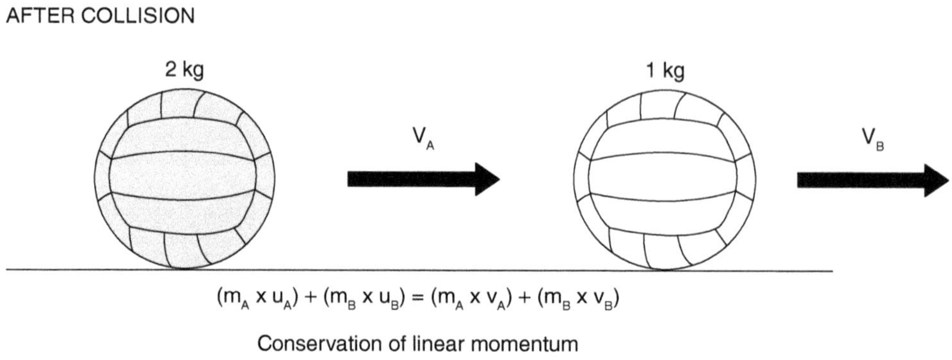


**Figure B5.4** The two balls after collision indicating the conservation of linear momentum principle

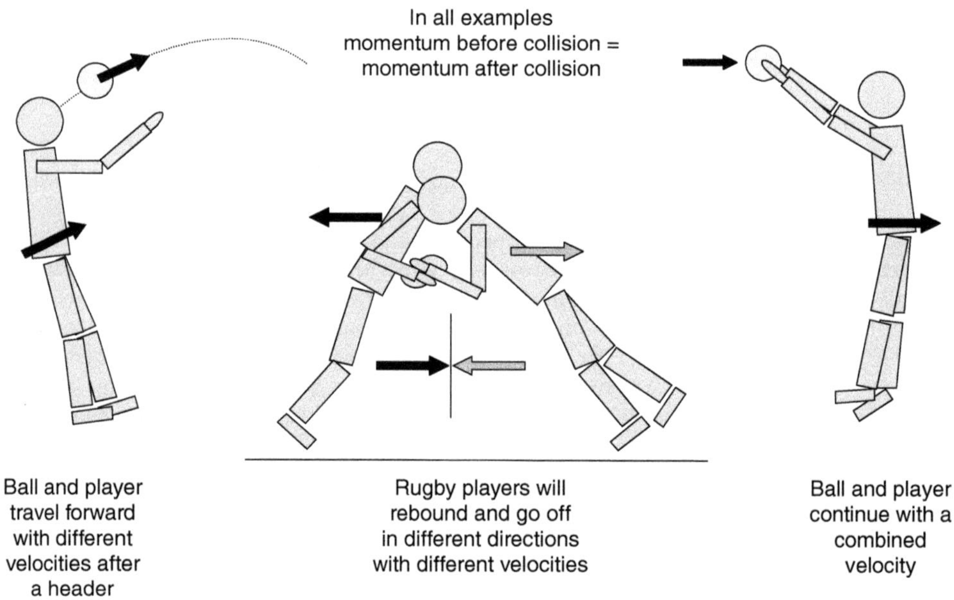


**Figure B5.5** Collision of bodies in human movement

We can now see that it is possible to use these equations, which have been developed from Newton's first law and the conservation of linear momentum principle, in many applications in human movement. For example, it is possible to work out the momentum or change in momentum experienced by bodies that collide or indeed the forces that are exerted due to the impulses that are created in such collisions.

*Key notes*

| | |
|---|---|
| **Linear momentum** | Linear momentum is the product of the mass of the body multiplied by its linear velocity. To increase the linear momentum possessed by an object we could either increase its mass or increase its velocity. Generally, within human movement the component of mass remains constant and therefore it is the change in velocity that is used to change the momentum possessed by a body. |
| **Principle of conservation of linear momentum** | This principle states that in any system where bodies (or objects) collide or exert a force upon each other, the total momentum in any direction remains constant unless an external force acts on the system. The term **system** is used to describe two or more bodies that are in motion and that exert a force on each other. In determining linear momentum it is therefore important to specify the direction in which the momentum is considered (i.e. consider all the forces that are acting in that direction: vertically, anteriorly and posteriorly, or medially and laterally). The principle of conservation of linear momentum is valid only when: 1) there is no external impulse acting on the system (i.e. no external force); and 2) the total mass of the system remains constant (before and after collision). |
| **Application** | When a goalkeeper in soccer catches a ball in the air the momentum in the system before the collision (where collision equals contact between goalkeeper and ball) is equal to the momentum in the system after the contact or collision (i.e. when goalkeeper and ball are together). The example is given as "in the air" because when the goalkeeper is in contact with the ground there would be other external forces acting on the system. |

## FURTHER READING

The following resource provides additional reading on the topics of linear momentum and kinetic energy, as it applies to tackling in rugby union.

1 Hendricks, S., Karpul, D., & Lambert, M. (2014). Momentum and kinetic energy before the tackle in rugby union. *Journal of Sports Science and Medicine, 13*(3), 557–563.

B6

# NEWTON'S LAWS OF MOTION

## Linear motion

*Paul Grimshaw*

Sir Isaac Newton (1643–1727) formulated three laws of motion that created the basis of Newtonian mechanics and which can be directly applied to human movement and the study of biomechanics. These are summarised as follows:

> **Law 1 – The law of inertia:** Every object will remain at rest or continue with uniform motion unless it is acted upon by an unbalanced force.
>
> **Law 2 – The law of acceleration:** When a force acts on an object, the change of motion (momentum) experienced by the object takes place in the direction of the force, is proportional to the size of the force and inversely proportional to the mass of the object.

This law indicates that if a net external force acts on an object it will accelerate (i.e. speed up, slow down, start moving, or stop moving) in the direction of the net external force. This acceleration is proportional to the net external force and is inversely proportional to the mass of the object.

> **Law 3 – The law of reaction:** Whenever an object exerts a force on another there will be an equal and opposite force exerted by the second object on the first.

In order to define these laws in more detail, and in particular their relevance to human movement, it is necessary to consider an application of each of the three laws separately.

Application of the **law of inertia** (Newton's first law) to human movement states that a body at rest will remain at rest and a body in motion will continue in motion unless it is acted upon by an unbalanced (net) external force. In situations where **an object is not moving (i.e. it is stationary)**, Newton's first law states that an object that is *not* moving will remain in a non-moving state (at rest), unless the object is acted upon by an unbalanced (net) external force. In human movement, it is difficult to see how this law directly applies. For example, all objects that are on this planet will be subjected to the vertical external force of gravity. If we did not have the ground on which to stand, we would accelerate downwards towards the centre of the Earth at a rate of $-9.81$ m/s$^2$. When we try to jump off the ground, gravity will immediately pull us back down. Although, as we know, this effect is actually happening all the time, it is more obvious as soon as we are in the air (unless of course you were able to exert enough force to overcome the force of gravity and get away from the Earth's gravitational pull – such as in the case of a rocket and space shuttle travelling into space). Hence, in this static example of Newton's first law, it is difficult to see how it can apply. Consider the following examples: a soccer ball placed on the penalty spot, a book placed on a table, a hurdle on the track in a 400 m hurdle race or the human body sitting in a chair; and it is obvious that all these are continuously subjected to the force of gravity (i.e. they all have weight) but none of them are moving. This is because the external forces are balanced and the result (net) is zero. The weight of the soccer ball acting downwards onto the penalty spot is balanced by the ground reaction force acting upwards on the ball from the ground contact.

In contrast, **when a body is in motion (i.e. moving)**, it is useful to use the example of the long jumper to understand the application of Newton's first law of motion to human movement. During a long jump, an athlete will leave the ground with both vertical and horizontal velocity. This combination of velocities determines the angle of take-off, the resultant velocity and primarily the horizontal distance jumped by the athlete. This horizontal and vertical velocity together will produce the **projectile motion** (subjected to only the external force of gravity, ignoring air resistance) of the athlete during the flight phase, which is illustrated in more detail in *Figure B6.1*.

As the athlete leaves the ground he/she will have both vertical and horizontal velocity. Once in the air the athlete will be a projectile and the flight path of the athlete is pre-determined. The parabolic flight path (see *Figure B6.1*) is the result of the combination of vertical and horizontal velocities present at the take-off point. Since the vertical motion of a body is affected by gravity, it is necessary to consider the velocities (motions) separately in order to see an application of Newton's first law of motion during this movement. *Figure B6.2* shows this in more detail.

With respect to the vertical motion of the athlete (or vertical velocity), the athlete will travel both upwards and downwards while at the same time travelling forwards (horizontal velocity). Gravity will affect the vertical component thus allowing the athlete to reach a peak height in the parabolic flight path while constantly being pulled back towards the ground (at a rate of $-9.81$ $m/s^2$). However, because the athlete also has horizontal motion (or velocity) we can see that the athlete travels forwards and has the characteristic parabolic curved flight path (see *Figure B6.2*).

The body in flight during the long jump is considered to be a projectile (like a ball in soccer or a javelin in athletics). In this example it could be argued that other forces (as well as gravity) act on these objects during flight. For example, air resistance will also affect the long jumper's parabolic flight path; however, in this example, the air resistance can be considered negligible. In the context of the horizontal motion of the long jumper it is possible to see that the athlete will travel in a straight line (this can be seen more clearly when viewed from above (plan view) in *Figure B6.3*). In addition, during this straight-line motion the athlete will also travel with a **constant horizontal velocity**. Hence, according to Newton's first law of motion, in the absence of an external force (as gravity will only affect the vertical component) the body will continue in a straight line with uniform (constant) velocity.

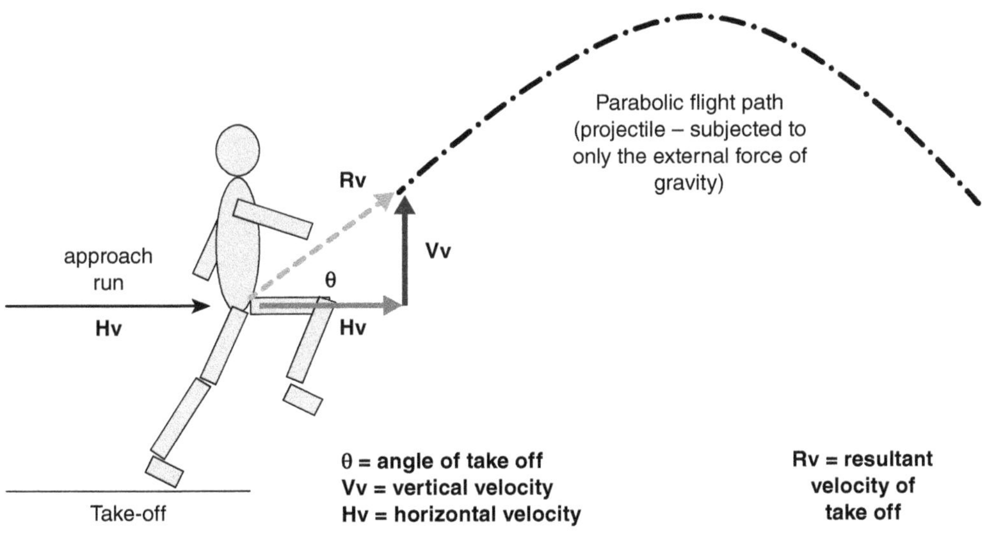


**Figure B6.1** Long jump take-off – horizontal and vertical velocities and parabolic flight path

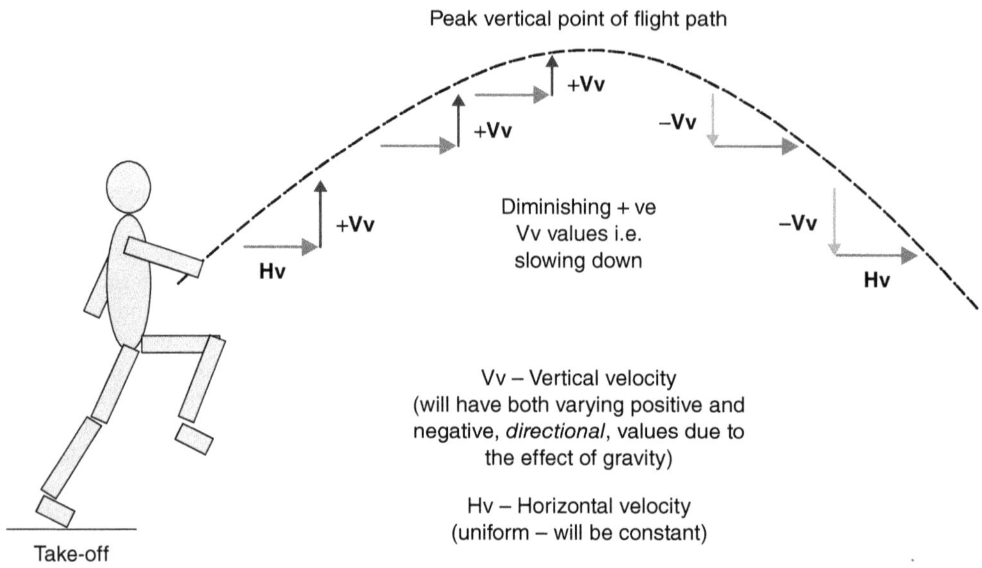


**Figure B6.2** Long jump take-off – constant horizontal velocity

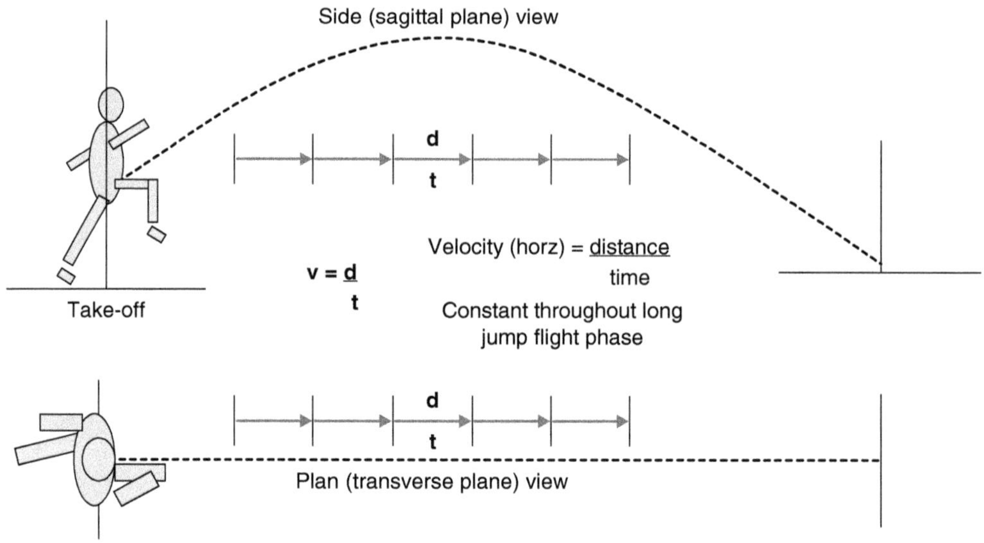


**Figure B6.3** Long jump take-off – constant horizontal velocity (sagittal and transverse plane (plan) views)

Although it does not seem logical that Newton's first law can apply to the long jumper in flight because the athlete will actually come to rest (a stop) in the sand, it is in fact correct. The horizontal velocity of the long jumper is constant, and the path of the athlete will be in a straight line (when viewed from above). The jumper will not speed up or slow down (horizontally) and, no matter what the athlete does during the flight phase, the flight path will be pre-determined from the point of take-off. The reason the athlete comes to rest (in the sand) is because the vertical component of the jump is affected by gravity and this will pull the athlete back towards the ground as soon as they leave it. Eventually, the athlete will hit the sand pit and stop, and it is the force from the sand pit on the athlete that would stop that motion. Therefore, in the absence of gravity the long jumper would continue to travel both upwards and forwards in a straight line with a constant velocity (note: upwards and forwards because the athlete would follow the resultant take-off velocity vector).

All projectiles that are thrown with horizontal and vertical velocity and that are only subjected to the external force of gravity will have a parabolic flight path that is pre-determined and they will all obey this law. A soccer ball when kicked, a basketball when thrown at the hoop, a tennis ball hit across court (without spin) and even something as simple as a pen that is thrown a short distance will all obey and demonstrate Newton's first law of motion. For a practical example, consider when you are travelling on a bus and the bus suddenly comes to a stop. In this case you will continue forwards with the same velocity towards the front of the bus even though the bus has stopped. Hence, in order to stop yourself from travelling forwards you will need to hold on to something inside the bus like a handrail (thus applying an external force). According to Newton's first law, you continued forwards with uniform velocity unless acted upon by a net external force (i.e. the gripping of the handrail to stop yourself moving forwards or you contacting the front of the inside of the bus).

**The law of acceleration** (Newton's second law) states that when a force is applied to an object and the result is a net force of greater than zero (therefore unbalanced), the change in motion of the object (i.e. change in velocity (acceleration)) is proportional (as one quantity increases in value so does the value of the other quantity) to the force applied to the object. This movement takes place in a straight line and in the direction in which the net force was applied. In addition, the law also states that this change in motion (acceleration or rate of change in velocity) is inversely proportional (as one value increases the other will decrease) to the mass of the object. *Figure B6.4* helps to illustrate this law.

This law is the most powerful of Newton's three laws of motion because it allows us to use this in the calculation of dynamics (movement). **For example, how do the velocities of objects change when forces are applied to them?** Using Newton's second law gives us the opportunity to calculate these changes. In the recognition of this law, it is, however, important to identify that the force applied to an object only causes a **change** in velocity (an acceleration

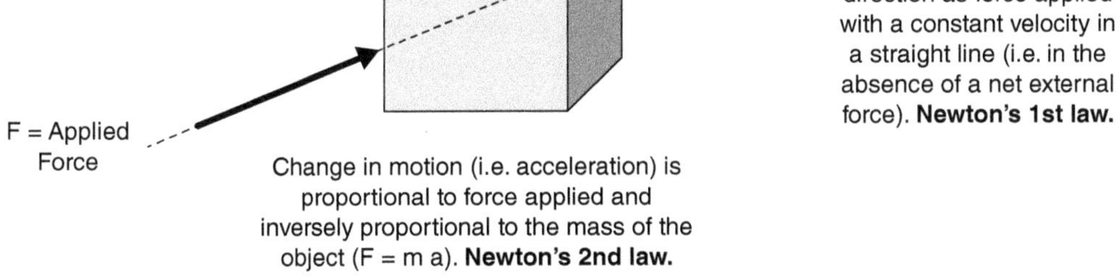


**Figure B6.4** Newton's second law of motion (acceleration)

or deceleration) and it **does not maintain** this velocity. In order to put Newton's second law of motion into a formula that we can use in understanding human movement it is necessary to identify the following equation:

$$\Sigma F = ma$$

where:

$\Sigma F$ = sum of forces or net external force (N)

m = mass of the object (kg)

a = acceleration of the object ($m/s^2$)

Using the previous example of the long jumper (*Figures B6.1 to B6.3*), which was used to illustrate Newton's first law, we observed that the horizontal velocity of the athlete during flight was constant. In addition, we also learned that once the athlete had left the ground (at take-off) the flight path was pre-determined (fixed). **How can this example be used to illustrate Newton's second law?**

At the point of take-off in the long jump we have seen that the athlete will have both vertical and horizontal velocity (*Figure B6.5*). In order to create the changes in velocity, such as the change from horizontal during the run-up to horizontal and vertical at take-off, the athlete will have applied a force to the ground in order to drive from the ground into the air. This force application is developed from the change in the athlete's stride patterns at or just before the take-off point (athletes usually lower their centre of gravity, lengthen the second-to-last and shorten the last stride into the take-off board). This adjustment of the body allows the athlete to be able to push his/her foot into the ground at the take-off board. This will create a reactive force from the ground that acts on the athlete. This ground reaction force will propel the athlete upwards and forwards. The resulting velocity of the athlete (upwards and forwards) is demonstrated in horizontal and vertical velocity at the point of take-off (*Figure B6.6*). Remembering that forces are vectors, it is possible to see that this propulsive force from the ground will have both vertical and horizontal components, and it is these two components that create the horizontal and vertical velocities used to determine the angle and resultant velocity of take-off.

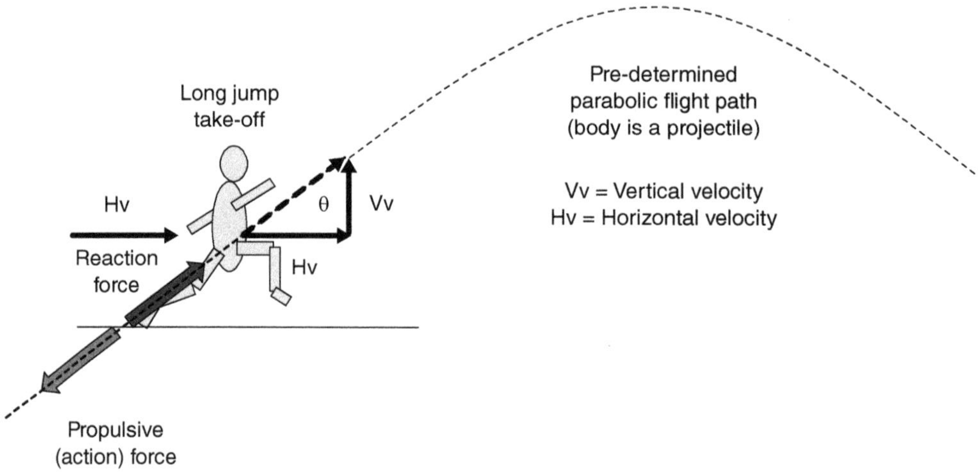


**Figure B6.5** Newton's second law of motion (long jump take-off and flight)

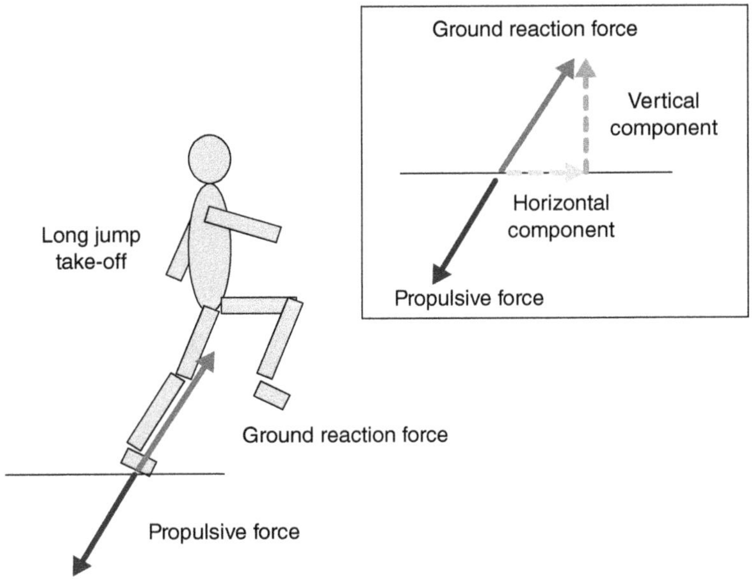


**Figure B6.6** Ground reaction force during take-off in the long jump

Now let us consider the point of take-off and in particular look at only the vertical motion (change in velocity) of the long jumper (since we know from Newton's first law of motion that the horizontal velocity is constant during flight). As soon as the athlete leaves the ground, the force of gravity will pull the athlete back down to the ground (remember, however, that it has always been doing this). As soon as the athlete is airborne (at take-off) the only external force acting on him/her (neglecting air resistance) is the force of gravity (i.e. what causes the athlete to have weight). As the athlete travels upwards (remember we are only considering the vertical component of the parabolic flight path at this stage) the downward pull (acceleration) of gravity is immediately slowing the vertical ascent of the athlete. Even though the athlete is travelling upwards he/she is actually being slowed down (decelerated). As the athlete is slowed down (vertically) they will eventually come to a stop at the highest point in the flight path (*Figure B6.2*). The acceleration of the athlete throughout his or her flight is **downwards,** even though they are going upwards at the beginning (i.e. they are being slowed down or always being pulled downwards at a constant rate – so their rate of ascent is lessened until it is stopped at the peak of the flight path).

Since we know that the weight (the force acting on the athlete because he/she has mass) of the athlete does not change and that the mass of the athlete remains the same, we can therefore say that the acceleration of the athlete is constant (derived from the previous equation $\boldsymbol{F} = \boldsymbol{ma}$). Next, we also know that because of the mass of the Earth and its gravitational pull on objects towards its centre, this acceleration will be –9.81 m/s$^2$ (remembering that it does vary slightly between different positions on the Earth's surface – depending on how close or how far away you are from its centre).

This means that as an athlete travels upwards at the beginning of the flight phase, they will be slowed down at a rate of 9.81 m/s for every second of movement (–9.81 m/s$^2$). Similarly, as the athlete begins to come back downward again, in the second half of the flight phase they will be accelerated downwards (and speed up) at a rate of –9.81 m/s for every second of motion (i.e. a *constant* acceleration).

This constant vertical acceleration acts on all objects and will accelerate a heavy object and a lighter object at exactly the same rate, which is why a heavy object and lighter object dropped from the same height will hit the ground at the same time (again obviously neglecting the effects of air resistance). Further, this downward acceleration is independent of any horizontal motion (as in the case of the long jumper); that is, it is not affected by horizontal motion, nor does it have an effect upon horizontal motion (the other reason why the horizontal velocity of a long jumper in flight is constant – Newton's first law of motion).

## TASK

This can be demonstrated by placing a pen on a table and also at the same time holding another pen at the same height as the table. Next, get someone to push the first pen off the table with a large force (i.e. accelerate the pen rapidly off the table horizontally). At the same point as they push the pen off the table (at the same moment in time that it leaves the table) drop the pen that you are holding. The pen on the table (that has now been pushed off) will have horizontal (the push) and vertical (gravity) velocity and it will have projectile motion towards the floor. The pen you have dropped should only have vertical motion and should drop to the floor in virtually a straight line. However, both pens will hit the ground at the same time. Hence, horizontal motion (velocity) does not affect vertical motion (velocity).

In the example of the long jumper, Newton's second law of motion is used to identify and explain constant vertical acceleration and how and why a long jumper is pulled back towards the ground immediately after they have left it (jumped into the air). However, there are many other applications of Newton's second law that are applicable to the understanding of human motion and these are discussed in more detail in other sections of this text.

**The law of reaction** (Newton's third law) states that for every action (a force) there will be an equal and opposite reaction (another force). In other words, push on an object and you will feel the object push back on you with an equal and opposite force. For example, if you push on a wall you will feel an equal and opposite force that is coming from the wall and acting along your hands and arms. Similarly, if you stamp your foot against the ground you will feel a force through your leg that is exerted from the ground on your foot. *Figure B6.6* shows the ground pushing back on the athlete from the athlete pushing into the ground at the point of take-off in the long jump.

However, when attempting to understand Newton's third law of motion it is important to be aware of two important facts. First, the effects of the forces are not cancelled out (i.e. there is not a net effect of zero) because one acts on one object while the reaction (or other force) force acts on the other. Second, although the forces are both equal in magnitude and opposite in direction the effects of the forces on the two objects are not the same. In some cases, one or both of the objects may accelerate (change their state of motion) and in other examples one or both of the objects will not accelerate.

One of the most common applications of Newton's third law is seen in the consideration of the ground reaction force during walking. As you walk across the ground your foot exerts a force on the ground and consequently the ground exerts an equal and opposite force back on your foot. As your foot hits the ground at heel strike during walking it will do so with a force which is derived from the mass of the foot and how it is accelerated into the ground at impact ($\boldsymbol{F} = \boldsymbol{ma}$).

As you hit the ground at heel strike there are a number of forces that are acting in this example. For example, there will be the force from the mass of your foot and its acceleration into the ground, there will also be the force of gravity (pulling your foot directly vertically downward), then there will be the ground reaction force from the ground acting back on your foot (which will consist of the friction force between the foot and the ground (the anterior–posterior force), the normal reaction force (vertically) and the medial–lateral force). All these forces act together, and it is not simply a case of one force (**the action**) being opposed by an equal and opposite force (**the reaction**). These are all external forces and it is the net sum of all these external forces that will cause the whole body (i.e. the person) to accelerate or decelerate. In this context it is also important to express that it is the external forces that will generally cause the internal forces within the joints. Hence, it is the external forces that cause the resulting change in motion (acceleration or deceleration).

Although the terms **action** and **reaction** are widely used within biomechanics in the context of Newton's third law, there is a slight confusion when these forces are applied in sporting and/or other human movement situations. For example, it is sometimes difficult to determine which of the forces constitute the action and which constitute the reaction. In addition, there is a degree of confusion in that these forces (or the terms action and reaction) when classified in this way could be misinterpreted to be movement rather than force. For example, when the racket hits the tennis ball during the "ground stroke" in a tennis match, there will be a force exerted on the ball by the racket. There will also be a force exerted from the ball onto the racket. These forces are equal and opposite, but it is the net effect of all the external forces that produces the change in movement (i.e. the acceleration or deceleration of one or both of the objects). The mass of the tennis ball is relatively small compared to the mass of the racket and once all the net forces are determined the net effect should be a force that causes the tennis ball to accelerate in the direction that the player intended to hit the ball.

This equal and opposite force principle from Newton's third law (**action–reaction**) appears to falsely apply to movement as well as force. For example, if you are in the air during the flight phase of the long jump and you rapidly move your arms down towards your legs (essentially caused by a muscle force within your body), the reaction is that your legs will move upwards towards your arms (what appears to be an equal and opposite action–reaction (i.e. it appears as a movement rather than a force)). However, this process is actually achieved by the equal and opposite torques (moments of force) that are applied to the body in order to cause this movement to happen. The torque that caused the trunk and arms of the athlete to move downwards created an equal and opposite torque that caused the legs to move upward. *Figure B6.7* illustrates some examples of Newton's third law of motion in human movement.

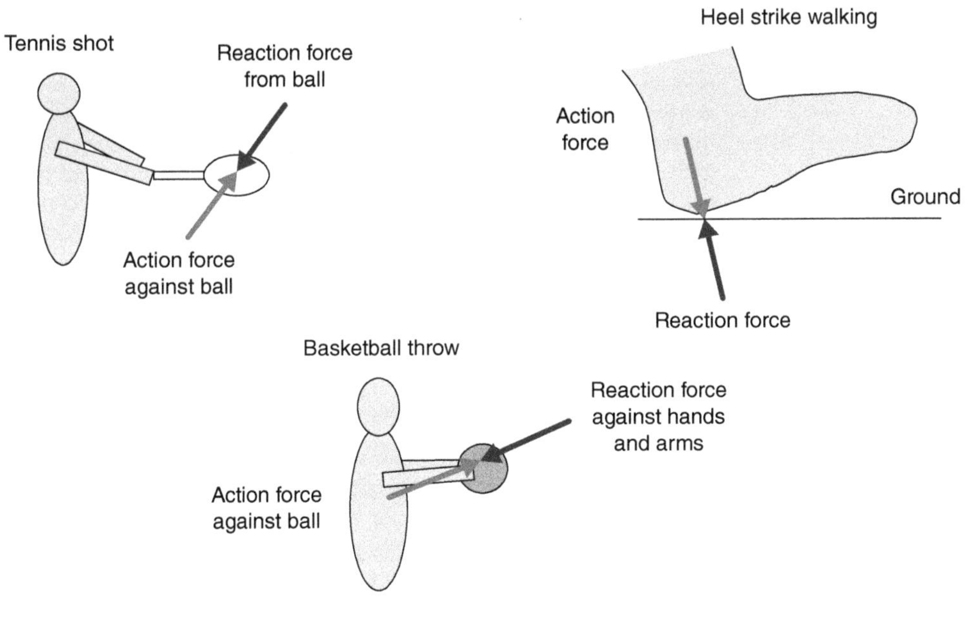


**Figure B6.7** Newton's third law (action–reaction)

This is the same principle used by a rocket or space shuttle to propel it into space. Although the rocket does not have anything to push against the external vertical force that is being exerted downward (exerted by the jet engines) onto or into the air causes an opposite (reaction) force to be exerted on the rocket (forces always come in pairs). The corresponding result (which is the net force) considering all the external forces (i.e. the vertical force acting downward from the rocket, the opposite force acting on the rocket upwards, the force of gravity pulling the rocket downward and possibly air resistance, friction and drag) accelerates the rocket vertically into the air and eventually into space. Similarly, the jet engines of a plane that are used to propel it horizontally through the air (although there are other forces such as aerodynamic lift and drag forces that also have a significant effect on the principle of flight) are acting backwards to its direction of motion (which is forwards).

The application of Newton's laws of motion is seen in many examples of human movement and a recognition of these laws can be important with regard to both injury prevention and improvement in performance. For example, how can a person reduce the potentially damaging impact force that is created during the heel strike in running (i.e. the impact force that is experienced throughout the leg during running that can be between four and five times your body weight) or how can an athlete increase the vertical component of the ground reaction force acting on the athlete at the take-off in a high jump, so that they can potentially jump higher?

*Key notes*

| | |
|---|---|
| **Newton's laws** | Sir Isaac Newton formulated three laws of motion that can be applied to the study of human movement. |
| **1 Law of inertia** | Every object will remain at rest or continue in its state of uniform motion unless acted upon by an external force. For example, a ball on the penalty spot in soccer is exerting a force on the ground, which is balanced by the force acting from the ground on the ball (net affect = zero). In the absence of gravity an object that is thrown or projected into the air will continue along the line of projection (resultant velocity) with constant uniform velocity (unless it is acted on by an external force). |
| **2 Law of acceleration** | When a net force acts on an object the change in motion that is produced (the acceleration) takes place in the direction of the net force. This acceleration is proportional to the net force and inversely proportional to the mass of the object. For example, push an object across a table with a force. The acceleration of the object will be dependent on the size of the net force that was applied to the object. The more massive an object the more net force is required to accelerate it at the same rate as an object with less mass. |
| **3 Law of reaction** | When a force is exerted on an object there will be an equal and opposite force exerted by the second object on the first. For example, when the head of a soccer player contacts the soccer ball there will be a force exerted on the soccer ball from the player and an equal and opposite force exerted on the head of the soccer player from the ball. |
| $\Sigma F = ma$ | The formula derived from Newton's second law. The summation of all the forces acting on an object (i.e. the net force) is equal to the mass of the object multiplied by the acceleration of the object. |
| **External forces** | The net force represents the sum of all the external forces acting on an object and the net force will be the reason why the object does or does not accelerate (i.e. speed up, slow down, start moving or stop moving). |

## FURTHER READING



The following resources provide additional reading around the application of Newton's three laws of motion to sport.

1 Dawes, J., & Lentz, D. (2012). Methods of developing power to improve acceleration for the non-track athlete. *Strength & Conditioning Journal*, *34*(6), 44–51.
2 Frost, D. M., Cronin, J., & Newton, R. U. (2010). A biomechanical evaluation of resistance. *Sports Medicine*, *40*(4), 303–326.
3 O'Shea, M. J. (2004). Snowboard jumping, Newton's second law and the force on landing. *Physics Education*, *39*(4), 335–341.

# SECTION C

# Kinematics and kinetics of angular motion

# ANGULAR MOTION

*Paul Grimshaw*

**Angular motion** is rotatory movement about an **imaginary** or **real axis** of rotation and where all parts on a body (note that the term body need not necessarily be a human body) or segment move through the same angle. **Angular kinematics** describes quantities of angular motion using such terms as angular displacement, angular velocity and angular acceleration. *Figure C1.1* identifies two examples of angular motion in more detail.

**Angular distance** or **displacement** (scalar or vector quantity) is usually expressed in the units of **degrees** (where a complete circle is 360 degrees). Similarly, **angular velocity** (depicted by the Greek letter omega ($\omega$)) and **angular acceleration** (depicted by the Greek letter alpha ($\alpha$)) are often expressed as degrees per second (°/s) and degrees per second squared (°/$s^2$ or degrees/second$^2$) respectively. However, it is more convenient within human motion to use the unit **radian** due to the large amount of angular displacement involved. The value for **1 radian** represents an angle of approximately **57.3°** and there are $2\pi$ (where pi is approximately 3.142) radians in 360 degrees (one circle).

As with the terms used to describe linear motion, within angular motion there exists both scalar and vector quantities. However, it is often possible and more easily understandable to describe angular movement using such definitions as clockwise or anti-clockwise rotation.

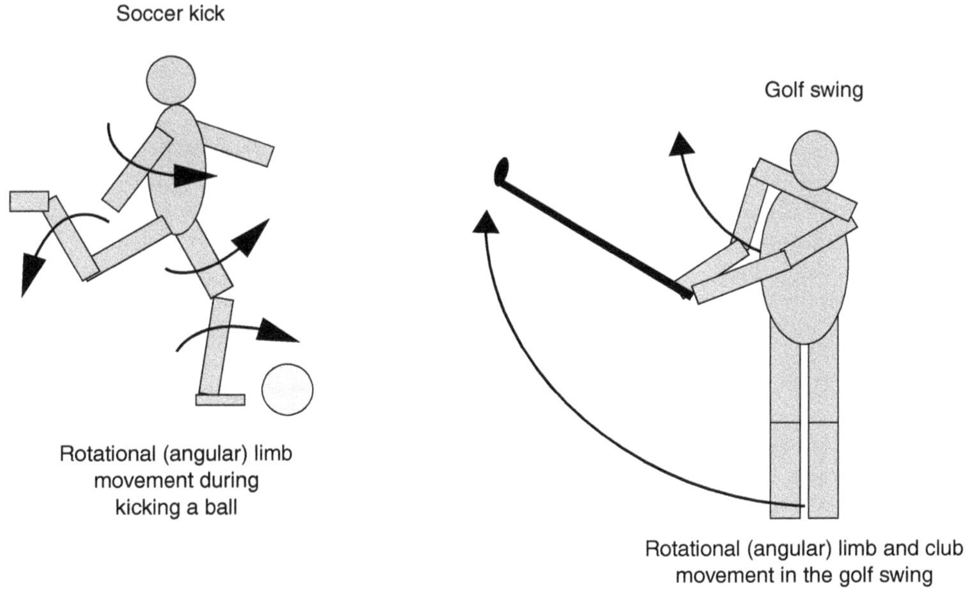


**Figure C1.1** Angular motion in sport

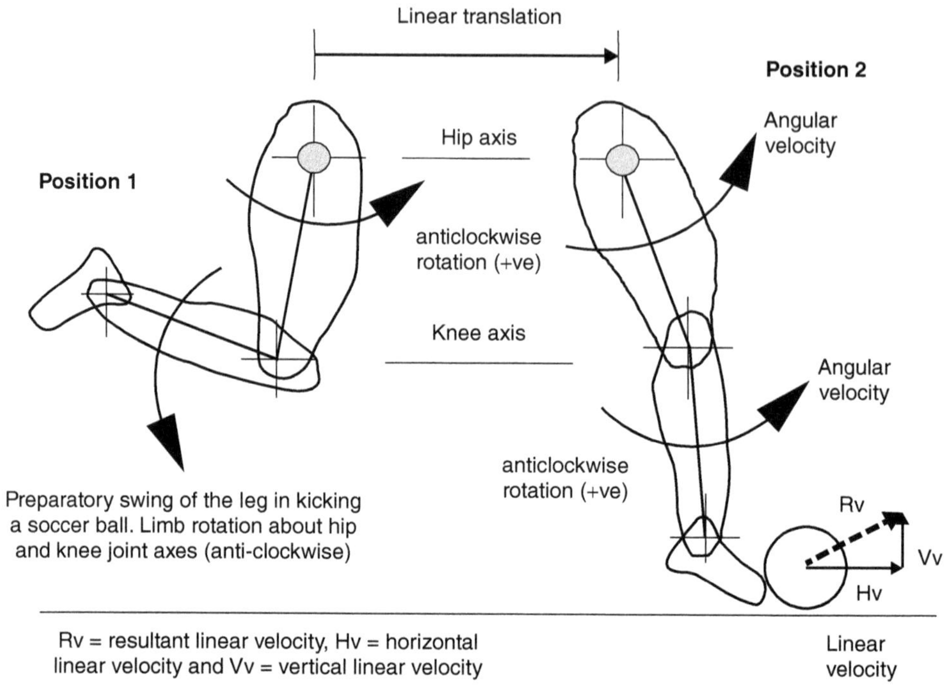


**Figure C1.2** Angular motion during kicking a ball

Again, positive and negative signs can be used to denote the different directions (e.g. **clockwise rotation** may be assigned a **negative sign** and **anti-clockwise rotation** a **positive sign** which is the common convention used within biomechanics).

Considering *Figure C1.2* it is possible to see the actions employed by the leg when kicking a soccer ball. The upper leg segment (thigh) moves with an anti-clockwise rotation between position 1 and position 2. The lower leg segment also moves in an anti-clockwise rotation between these positions. Note that these two actions happen simultaneously and they also move in association with the linear (forward) movement of the whole body (which is common when kicking a ball). From such a description and knowing the angles moved through and time taken for this movement it would be possible to calculate the angular velocity of each of these segments in kicking this soccer ball.

Considering the individual action of the upper leg segment in this example, we can see that the upper leg segment (represented by a single line in *Figure C1.3*) moved anti-clockwise through 30 degrees (10° before the vertical line and 20° after the vertical line). If we know that the upper leg moved through this angle in 0.5 seconds, it is possible to calculate the **average** angular velocity of this limb segment during this movement.

It is important to point out that for every part (or point) of the limb that is along the limb segment shown in *Figure C1.3*, the average angular velocity will be the same (providing it is moving with the limb). All the parts along this limb travel through the same angle of 30 degrees in 0.5 seconds (10° before the vertical line and 20° after the vertical line) and as such their average angular velocities (ω) will be determined using the formula:

$$\omega = \frac{\text{angular displacement (degrees or radians)}}{\text{time taken (seconds)}}$$

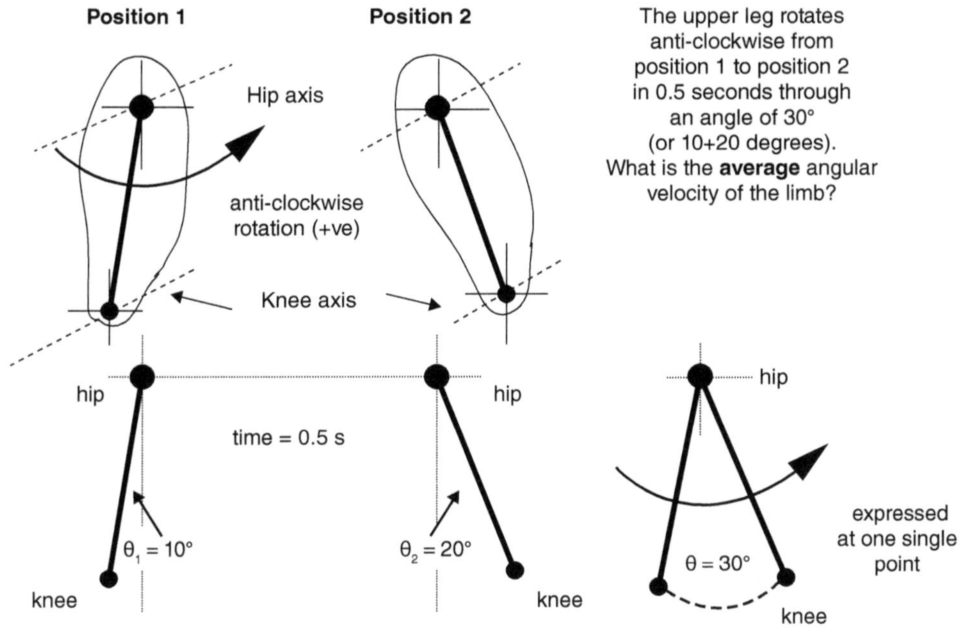


**Figure C1.3** Angular motion of the upper leg

where angular displacement = angular movement between the initial and final angular position (which is 30° anti-clockwise (+ve) in this case):

$$\omega = \frac{30^\circ}{0.5\,\text{s}}$$
$$= 60^\circ/\text{s}$$

**Average angular velocity = 1.05 radians/s** (which is derived from 60/57.3=1.05)

However, it is important to point out again that this is the **average angular velocity** that occurs between position 1 and position 2 (i.e. **it is not an instantaneous value**) for every point along the upper limb or segment. In this case the more the time interval or angular displacement tends towards a zero (or small) value the more the instantaneous measure of angular velocity. In this example, **angular displacement** is defined as the **difference** between the **initial and final angular position** of the object or segment (in either a clockwise or an anti-clockwise rotation).

**Angular acceleration** (the Greek letter alpha (α)) is calculated by dividing the **angular velocity** (ω) by the **time taken.** It is defined as the **rate of change of angular velocity** and is expressed between two points of interest (i.e. an interval of time ($t_1$ and $t_2$) or position 1 and 2 as in our example). In the case of the example shown in *Figure C1.3*, it is, however, not possible to accurately determine the angular acceleration value for the limb because we do not have two accurate angular velocity values for the two points in time (i.e. we have an average value expressed between two points). In order to accurately determine angular acceleration of the limb, it would be necessary to measure angular velocity at regular intervals.

Considering *Figure C1.4*, when kicking a ball it is possible to see that the upper segment of the leg (the thigh or hip-to-knee segment) rotates about the hip joint (or hip axis of rotation) in an anti-clockwise direction (hence making it a displacement (vector quantity) because we now

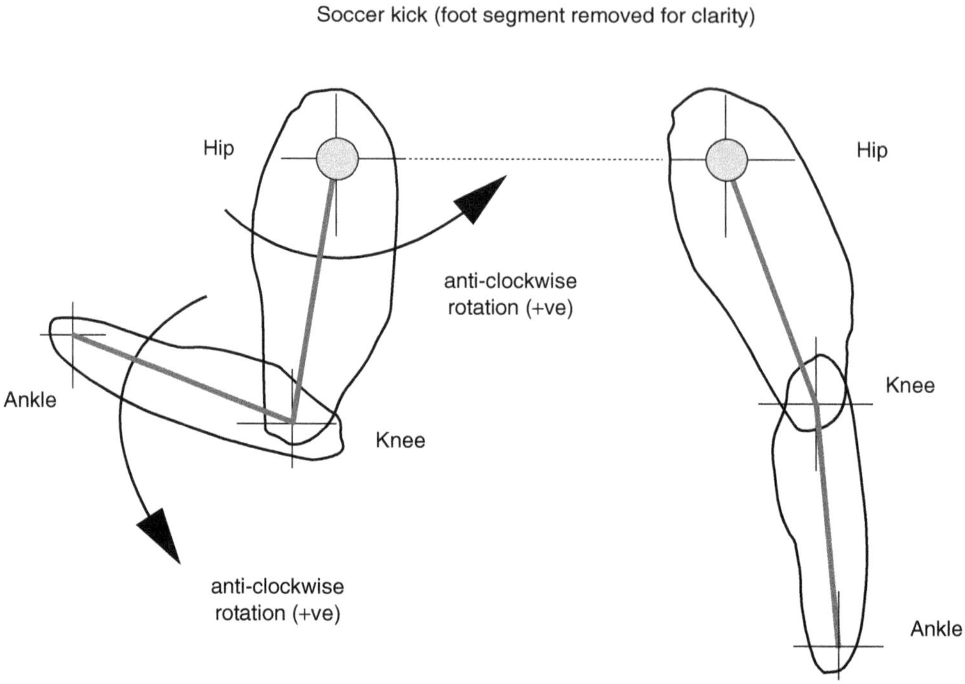


**Figure C1.4** Angular motion of the leg during a soccer kick

have a directional component). Similarly, and at the same time, the lower segment of the leg (the shank or knee-to-ankle segment) rotates about the knee joint (knee axis of rotation) also in an anti-clockwise direction. Both these actions occur simultaneously, and both contribute to generating the important **linear velocity** and **linear acceleration** generated by the limb at contact with the ball (summation of speed), which is required in order to kick in soccer.

*Key notes*

| | |
|---|---|
| **Angular motion** | All the parts of a body (i.e. all parts on a rigid object or all parts on a segment of the human body) move through the same angular displacement. |
| **Angular kinematics** | Describes quantities such as angular displacement, angular velocity and angular acceleration. |
| **Angular displacement and distance** | Angular displacement is the difference between the initial and the final angular position of a rotating body (it is expressed with both magnitude and direction); for example, 36 degrees (or 0.63 radians) anti-clockwise. Angular distance is expressed with magnitude only. |
| **Degrees and radians** | Units that are used to measure angular displacement (where a circle = 360 degrees or $2\pi$ radians). 1 radian is approximately 57.3 degrees. |
| **Angular velocity and angular acceleration** | Angular velocity is the angular displacement divided by the time taken. Angular acceleration is the rate of change of angular velocity and is calculated by change in angular velocity (final – initial) divided by the time taken. |

| | |
|---|---|
| **Clockwise and anti-clockwise rotation** | Clockwise rotation is movement in the same direction as the hands of a clock when you look at it from the front. Clockwise rotation is given a negative symbol (–ve) for representation. Anti-clockwise rotation is the opposite movement to clockwise rotation and it is given a positive symbol (+ve) for representation. |

## FURTHER READING

The following resources provide additional reading around the assessment of angular motion in sport and human movement.


1 Gholipour, M., Tabrizi, A., & Farahmand, F. (2008). Kinematics analysis of lunge fencing using stereophotogrammetry. *World Journal of Sport Sciences, 1*(1), 32–37.

2 Hiroyuki, N., Wataru, D., Shinji, S., Yasuo, I., & Kyonosuke, Y. (2002). A kinematic study of the upper-limb motion of wheelchair basketball shooting in tetraplegic adults. *Journal of Rehabilitation Research and Development, 39*(1), 63–71.

# C2 LINEAR-ANGULAR MOTION

*Paul Grimshaw*

The **linear** and **angular components** of movement are linked by a **mathematical relationship.** Specific formulae exist that show how the linear translation of points on a rotating object (or segment) can be determined. Often within biomechanics it is necessary to understand and apply this relationship. For example, in the case of the soccer kick, it is the angular movement of the leg that creates the resultant linear velocity (derived from the horizontal and vertical components) that is applied to the ball in order to give it trajectory and motion. Similarly, in golf, it is the angular movement of the arms and the club that imparts resultant linear velocity to the golf ball to give it an angle of take-off and a corresponding parabolic flight path. *Figure C2.1* shows an example of this angular–linear relationship.

In *Figure C2.1* the arm (AB) moved from position 1 to position 2 in 0.45 seconds. The **angular displacement** in this example was 35 degrees (i.e. **change in angular position**). The **average angular velocity** of the arm AB can be determined by the **change in angular position** (angular displacement) divided by the **time taken.**

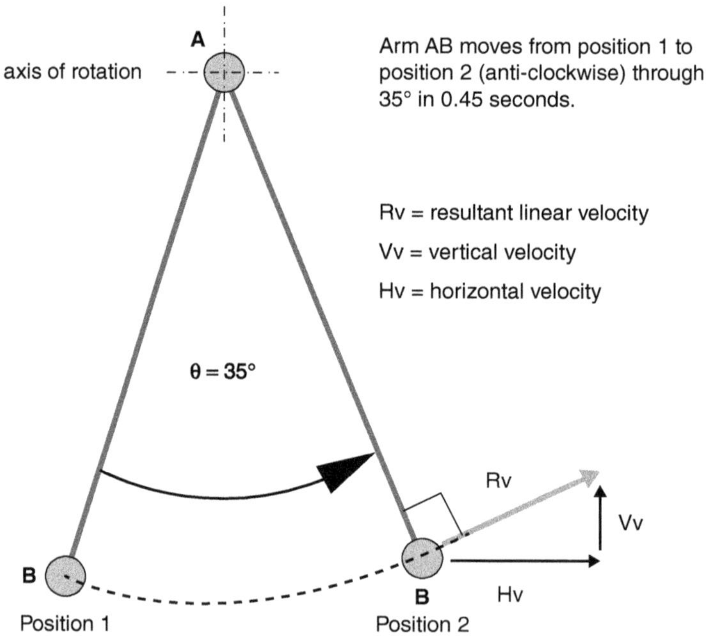


**Figure C2.1** Angular and linear motion

$$\omega = \frac{\text{angular displacement}}{\text{time taken}} \quad \text{(between position 1 and position 2)}$$

$$= \frac{35^\circ}{0.45\,\text{s}}$$

Average angular velocity = **77.8°/s (or 1.36 rad/s)**

This value represents the **average angular velocity** of the arm AB between position 1 and position 2. The angular displacement is the angle that is formed between the final and the initial position of the line/segment/arm (i.e. 35° in this example). Any point on this line AB (or segment) will move through the same angle in the same time. Within biomechanics, however, it is often the case that we refer to points or objects on a rotating line or segment. In order to express the angular displacement of an object that is on a rotating line or segment it would be necessary to choose any two points that are on the object and join them with a line. The new line on the object will rotate (if the original line/segment/arm is a rigid body) in exactly the same manner as the original segment or line that the object is located on. *Figure C2.2* and *Figure C2.3* help to illustrate this in more detail.

In *Figure C2.3* it is possible to see that all the points (parts) on a rotating body (A, B and C collectively in this case (i.e. the additional line)) will travel through the same angle (60° anticlockwise). Since the angular displacement of the original line/segment is also 60 degrees, and it moves through this angle in the same time, the angular velocities of all the parts on the arm will be the same. In this context it is important to indicate that in relation to angular displacement we are technically referring to a collection of points or parts (i.e. the arm/line A, B and C) as it is not possible for a single point to have an angular orientation (see *Figure C2.3*). However, if we were to consider the linear distance/displacement and linear velocities (technically linear speed without the directional component) of each point A, B and C

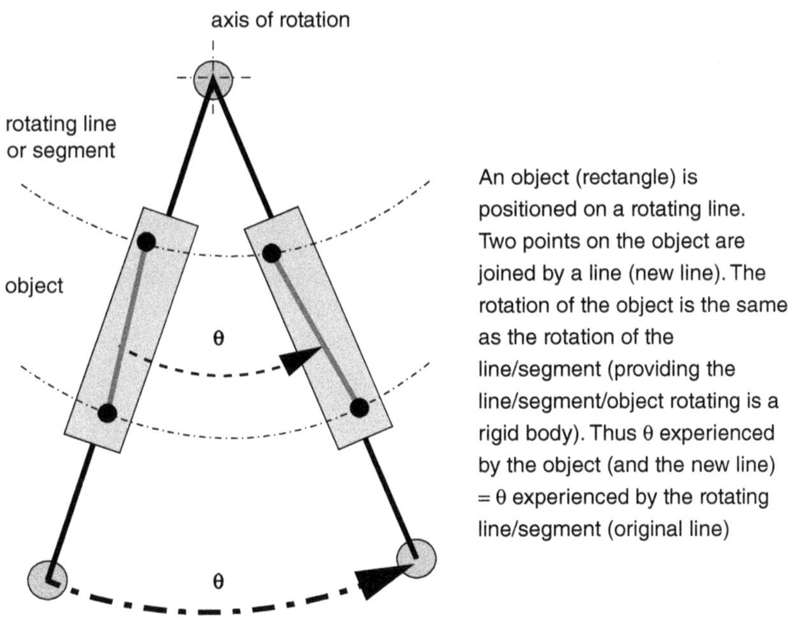


**Figure C2.2** Points and objects on a rotating arm

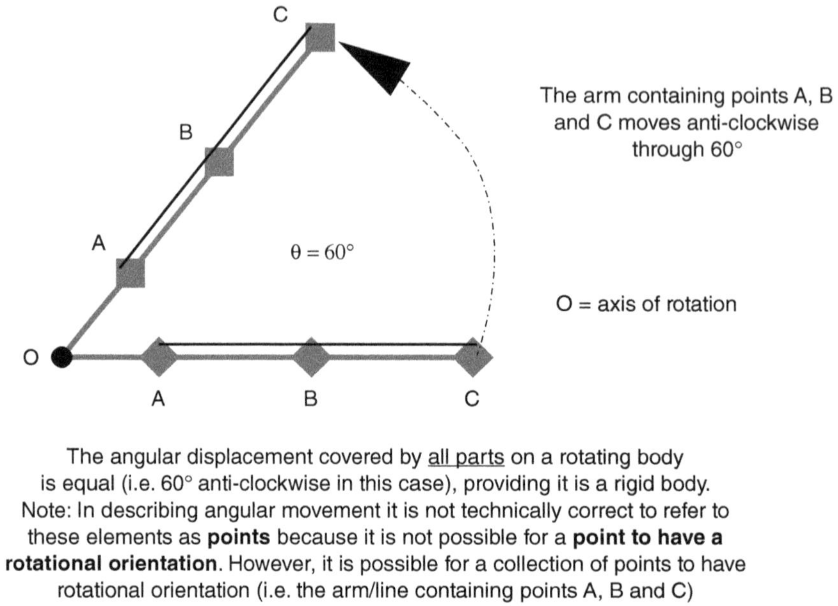


**Figure C2.3** Angular displacement of points on a rotating arm

(which we can because a point can have a linear orientation/translation) we can determine that the linear distances covered by each of these points and hence their linear velocities will be different (because each will have a different radial (radius) distance from the point of rotation). The further away the point is from the centre of rotation (as in the case of point C in *Figure C2.3*), the greater will be the linear distance/displacement covered by that point.

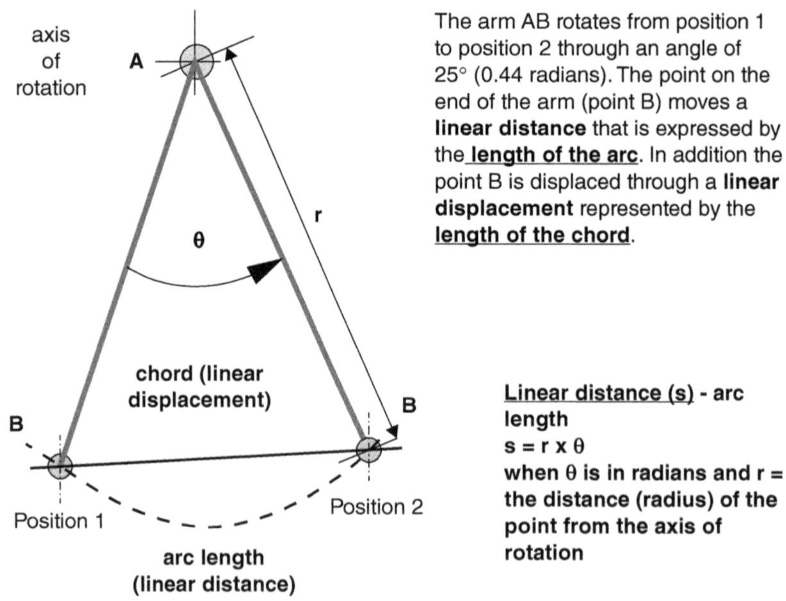


**Figure C2.4** Linear displacement and linear distance of a point on a rotating arm

The point B on the rotating arm (AB) in *Figure C2.4* will move through a **linear distance** (scalar quantity) that can be expressed by determining the **length of the arc.** However, the **linear displacement** (vector quantity) of the point B can be determined by calculating the **length of the chord** that is contained from the position of point B at the beginning of the movement to the position of point B at the end of the movement (a straight line). A **chord** is defined as a straight line connecting two points on a curve or surface that lies between the two points. To calculate the **linear distance** (scalar) covered by the rotating point (i.e. the arc length) we can use the relationship that is presented in *Figure C2.4* which links the angle (only when it is **expressed in radians**) with the distance of the point from the axis of rotation (the radius). In *Figure C2.4*, if point B was located 0.34 m from the axis of rotation (i.e. the radius) what would be the linear distance covered by the point if the arm rotated through an angle of 25 degrees (or 0.44 radians)? The following formula (also shown in *Figure C2.4*) is used to calculate the linear distance (i.e. the length of the arc) moved by the point in this example.

The arm AB rotates from position 1 to position 2 through an angle of 25 degrees (0.44 radians). The point on the end of the arm (point B) moves a **linear distance** that is expressed by the ***length of the arc.*** In addition, the point B is displaced through a **linear displacement** represented by the ***length of the chord.***

> *Linear distance (s)* – **arc length $s = r \times \theta$ when $\theta$ is in radians and r = the distance (radius) of the point from the axis of rotation**
>
> $s = r\,\theta$

where:

> $s$ = linear distance (arc length)
>
> $\theta$ = angle or angular displacement (expressed in radians only)
>
> $r$ = distance of point from axis of rotation (radius)

**Please note:** for convenience the symbol s is used for linear distance in this example – SI units for linear distance are normally l and d.

For point B located 0.34 m from the axis of rotation and moving through 25 degrees (0.44 radians)

> $s = r\,\theta$
>
> $s = 0.34 \times 0.44$
>
> $s$ = **0.15 m (to two decimal places)** – arc length or linear distance moved by point B

In order to calculate the length of the **chord** (or the **linear displacement**) it is necessary to also use the distance of the point from the axis of rotation (the radius) and the angular displacement (i.e. the angle $\theta$ that the arm has been displaced through). However, this relationship is not as straightforward as the calculation for linear distance (the arc length) because although the **chord length** is directly proportional to the radius it is not directly proportional to the angle or angular displacement. *Figure C2.5* shows the calculation for chord lengths for angles up to 90 degrees. For angles greater than 90 degrees there are more complex tables of chords that are used within the area of mathematics (i.e. Ptolemy's table of chords).

From *Figure C2.6* it is possible to see that **a point** (point A) **that is closer** to the **axis of rotation** will **travel with less linear distance** (arc length) **than a point** (point B: when both points are located on the same line/segment/rotating arm) **that is further away from the axis of rotation.** This is the same application as with muscles and their points of attachments to bones. The

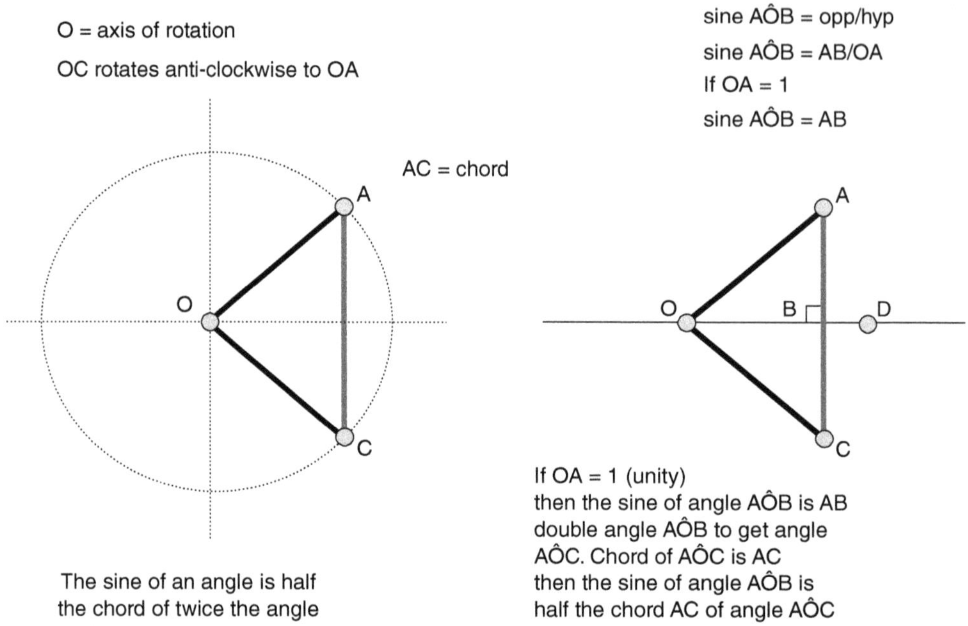


**Figure C2.5** Determining the chord length of a rotating arm

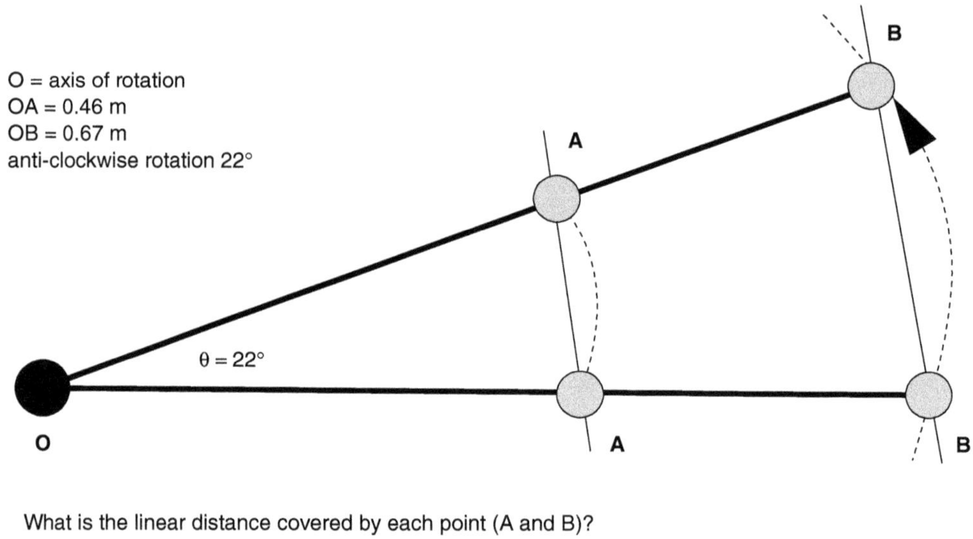


**Figure C2.6** Linear distance moved by two points on a rotating arm

muscle (muscle tendon and muscle contraction which causes linear movement) will move with a small amount of linear distance (i.e. a small contractile element) but it will cause a large amount of movement at the end of the segment or arm (i.e. flexing your arm in a bicep curl causes a large linear movement at the end of the arm (i.e. at the hand) and a small linear movement at the muscle point of attachment (which is ideal because the muscle is only able to move a small linear distance in contraction)). In *Figure C2.6*, point A is located 0.46 m from the axis of rotation (O) and point B is located 0.67 m from the axis of rotation. If the angular displacement of the limb (rotating arm) in this case is 22 degrees (0.38 radians) anti-clockwise what is the linear distance (arc length) covered by point A and point B?

### *Linear distance covered by point A*

$s = r\,\theta$ (where $\theta$ is expressed in radians)

$s = 0.46 \times 0.38$

$s = \mathbf{0.175\ m}$

### *Linear distance covered by point B*

$s = r\,\theta$

$s = 0.67 \times 0.38$

$s = \mathbf{0.255\ m}$

All points (parts) on this rotating arm in *Figure C2.6* will move through the same angle (angular displacement) in the same time. Hence all these points (or this collection of points) will have the same average angular velocity. However, point A has moved through a linear distance of 0.175 m and point B has moved through a linear distance of 0.255 m. Both these movements occurred in the same time and therefore the two points will have a different average linear speed (because B has moved through 0.255 m in the same time that A has moved through 0.175 m).

By algebraic manipulation we can take the formula used to determine linear distance and develop the equations to be used to calculate average linear and average angular velocity components of rotational motion.

Angular velocity ($\omega$) (average) $= \frac{\theta}{t}$

Linear velocity (v) (average) $= \frac{d}{t}$ or $\frac{s}{t}$

Linear distance (s) = $r\,\theta$ → So $v = \frac{\theta\, r}{t}$ → Hence $v = \omega\, r$

where $\theta$ = change in angular position

Average linear velocity (v) = average angular velocity ($\omega$) x radius (r)

please note certain symbols (i.e. "s") are used for convenience and are not SI units

**Figure C2.7** The relationship between linear and angular velocity

Considering *Figure* C2.7 it is possible to see that we can take the formula used to determine the linear distance (s = r θ) moved by a point on a rotating body (the arc length) and by algebraic manipulation we can develop the formula that is used to link average linear velocity (v) with average angular velocity (ω). Technically, this should be referred to as linear and angular speed as we are using distance rather than displacement (i.e. scalar rather than vector quantity). *Figure* C2.7 shows that by algebraic manipulation we can derive that the relationship between average linear velocity and average angular velocity is as follows:

$$\mathbf{v} = \omega\, \mathbf{r}$$

where:

v = average linear velocity

ω = average angular velocity

r = radius or distance from point of rotation to point of interest

As we have seen, the **average linear velocity** (or speed without a directional component) of a point on a rotating arm is derived from the equation of **average angular velocity** multiplied by **the radius** or distance of the point of interest from the axis of rotation (v = ω r). This is expressed as an average value because it is determined from the movement between two points (i.e. an angular distance/displacement). However, the linear velocity of point B at any **instant in time** throughout the movement will act at a **tangent** to the curve (arc of movement). This will be expressed as velocity at an instant and is known as an **instantaneous velocity.** In this context it is expressed as a **tangential linear velocity** and is given the expression $\mathbf{v}_T$. A **tangent** is defined

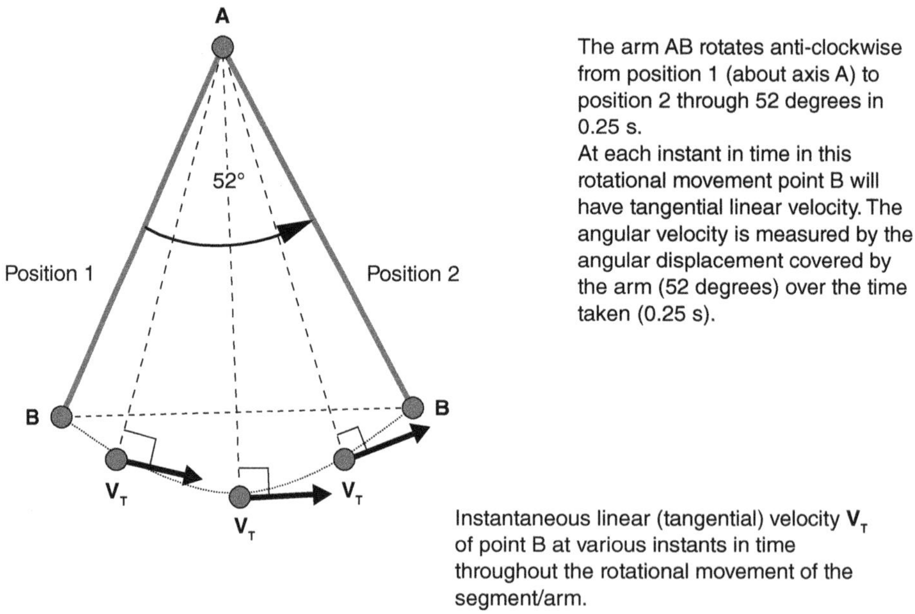


**Figure C2.8** Tangential linear velocity

as a line that touches a curve at a point. The line will have the same gradient (slope of a line measured as its ratio of vertical to horizontal change) as that of the curve at that point. The tangent will act at 90 degrees (perpendicularly) to the rotating arm that the point is contained on (*Figure C2.8*). The **direction** of the velocity of this point will be **perpendicular** to the rotating arm (radius of the point) and at a **tangent** to the curve (circular path of the object).

Considering *Figure C2.8* it is possible to see that at any instant in the rotational movement of the arm AB the point B will have a tangential linear velocity (a velocity that acts at a tangent to the curve (arc) of rotation). This linear tangential velocity will be derived from how much linear distance (the arc length) the point has moved through in a given time or from the angular velocity in a given time. The smaller the time intervals the more the value will tend towards an instantaneous value. **Instantaneous** is defined as **occurring at a given instant** or limit as the time interval approaches zero. In biomechanics it is important to understand, and determine, this linear velocity so we can assess how effective our rotational movements (such as in the golf swing) are in relation to producing linear motion (such as the resultant linear velocity of the golf ball).

As with velocity, a mathematical relationship exists that will link **average linear acceleration** with average angular acceleration and this is portrayed in *Figure C2.9*. From *Figure C2.9* we can see that **average angular acceleration** is expressed as **change in angular velocity over time** (or the rate of change of angular velocity). This is portrayed as $\alpha = \omega/t$. In order to link average linear acceleration with average angular acceleration we use the same method of algebraic manipulation that we used for determining the relationship between angular and linear velocity. Thus, the formula that links average linear acceleration with average angular acceleration is as follows:

$$a = \alpha\, r$$

where:

$a$ = average linear acceleration

$\alpha$ = average angular acceleration

$r$ = radius or distance of the point of interest from the axis of rotation

Average angular acceleration ($\alpha$) = $\frac{\omega \text{ (change in angular velocity)}}{t \text{ (time)}}$

Average linear acceleration (a) = $\frac{v \text{ (change in linear velocity)}}{t \text{ (time)}}$

From figure C2.7

Average linear velocity (v) = average angular velocity ($\omega$) x radius (r)

Hence

Average linear acceleration (a) = $\frac{\omega r}{t}$ ➘ $a = \alpha r$

Average linear acceleration (a) = average angular acceleration ($\alpha$) x radius (r)

**Figure C2.9** The relationship between linear and angular acceleration

In this context it is important to also clarify that if this **linear acceleration** was to be determined at an **instant in time** then it would also be classified as an **instantaneous value** that is acting at a tangent to the curve (or arc of motion). This would be represented by the expression $a_T$ or **tangential linear acceleration.** Similarly, as with linear tangential velocity this acceleration would act in the direction of the tangent to the curve at that instant in time.

As we have seen, linear and angular movement have an important relationship with each other. In human movement we use rotational motion of an arm to generate linear translation of a point. For example, hitting a tennis ball requires rotational movement of the arm and racket; throwing a basketball requires rotational movement of the upper body, arms and hands; and obviously in golf we use rotational displacements of the club and arms to generate the high linear velocity of the club head and hence the golf ball (in excess of 45 m/s linear velocity in professional golfers). Therefore, a good recognition of this relationship is valuable in working towards providing an effective knowledge of human movement.

Finally, it may be important to note that for smaller angles of segmental movement, which are often found in human movement (i.e. generally less than 180 degrees) the arc and chord are very similar in magnitude, which would account for the confusion of using the distance and/or displacement terms. For example, the chord length for an angle of 120 degrees with a radius of 0.37 m would be equal to 0.640 m whereas the arc length for the same conditions would be equal to 0.773 m. Hence the distance value (the arc length) could be used interchangeably for either speed or velocity calculations.

## *Chord length*

Sine 60° = opposite / 0.37

0.37 × Sine 60° = **0.320 m or 0.640 m for 120 degrees**

## *Arc length*

$s = r\,\theta$

$s = 0.37 \times 2.09$ radians (or 120°/57.3)

$s =$ **0.773 m**

| *Key notes* | |
|---|---|
| **Linear and angular movement** | Linear and angular movement are linked by a mathematical relationship. Specific formulae exist that are used to show the linear translation of a point that is caused by an object that is rotating. The angular movements of arms, segments and implements cause the linear motion of an end point that is applied to objects such as soccer, golf and tennis balls. |
| **Objects and points on a rotating arm (segment)** | To examine the angular displacement of an object that is located on a rotating arm (rigid body), it is necessary to join two points on the object with a line. The new line on the object will have the same angular displacement as the original line segment (rigid body). A point does not have a rotational orientation and hence it cannot be expressed as having angular displacement. However, a collection of joined points (i.e. a line) or an arm/segment can be used to determine this angular displacement. |

| | |
|---|---|
| **Linear distance and displacement of a point on a rotating arm** | The linear distance covered by a point on a rotating arm is the arc length (curve of motion). The linear displacement of a point on a rotating arm is expressed by determining the chord length. The linear distance (arc length) is calculated by the formula $s = r\,\theta$ where $\theta$ is expressed in radians. A point that is further away from the axis of rotation will cover a greater linear distance (arc length) than a point that is nearer to the axis of rotation. Hence this point (which is further away) will also have greater linear velocity and acceleration. |
| **Relationship between linear and angular velocity and between linear and angular acceleration of objects/points on a rotating arm** | The formula $v = \omega\,r$ is used to relate the average linear velocity of a point or object on a rotating arm with the average angular velocity of the arm. The formula $a = \alpha\,r$ is used to relate average linear acceleration of a point or object on a rotating arm with the average angular acceleration of the arm. |
| **Instantaneous linear velocity and acceleration** | Instantaneous velocity or acceleration is the velocity or acceleration expressed at an instant in time. Average velocity or acceleration is determined between successive time points. Instantaneous is defined as occurring at a given instant or limit, as the time interval approaches zero. As the time interval between points gets smaller (i.e. approaches zero) the value determined for velocity and acceleration tends towards an instantaneous value. |
| **Tangential** | The linear velocity and acceleration of a point on a rotating body/arm/segment acts at a tangent to the curve (arc of motion). A tangent is defined as a line that touches a curve at a point. This line has the same gradient as that of the curve at that point. It acts perpendicular (at 90°) to the rotating arm or segment. |

## FURTHER READING

The following resources provide additional reading on the concept of linear and angular motion in sport and exercise.

1 Burden, A. M., Grimshaw, P. N., & Wallace, E. S. (1998). Hip and shoulder rotations during the golf swing of sub-10 handicap players. *Journal of Sports Sciences*, *16*(12), 165–176.
2 Hume, P. A., Keogh, J., & Reid, D. (2005). The role of biomechanics in maximising distance and accuracy of golf shots. *Sports Medicine*, *35*(5), 429–449.

# C3 TORQUE AND MOMENTS

*Paul Grimshaw*

A **torque** is defined as a twisting or turning moment. The term **moment** is the force acting at a perpendicular distance from an axis of rotation. Torque can therefore be calculated by multiplying the **force** applied by the **perpendicular distance** at which the force acts from the axis of rotation. Often the term torque is referred to as the **moment of force.** The moment of force is the tendency of a force to cause rotation about an axis. Torque is a **vector quantity** and as such it is expressed with both **magnitude and direction.** Within human movement or exercise science **torques cause angular acceleration that result in the rotational movements of the limbs and segments.** These rotational movements take place about axes of rotation. For example, the rotational movements created in the leg while kicking a soccer ball would occur about the ankle (the foot segment), the knee (lower leg segment) and the hip (upper leg segment) joints or axes of rotation. If an object is pushed with a force through its centre of mass it will move in a straight line (linear motion) in the same direction as the applied force. However, if an object is pushed with a force at a perpendicular distance away from its centre of mass it will both rotate (about an axis of rotation) and its centre of mass will translate (move in a straight line). *Figure C3.1* and *Figure C3.2* illustrate this in more detail.

In *Figure C3.2* it is possible to see that when a **force** is applied at a **perpendicular distance** from the centre of mass (which in this example is considered to be the axis of rotation), the box

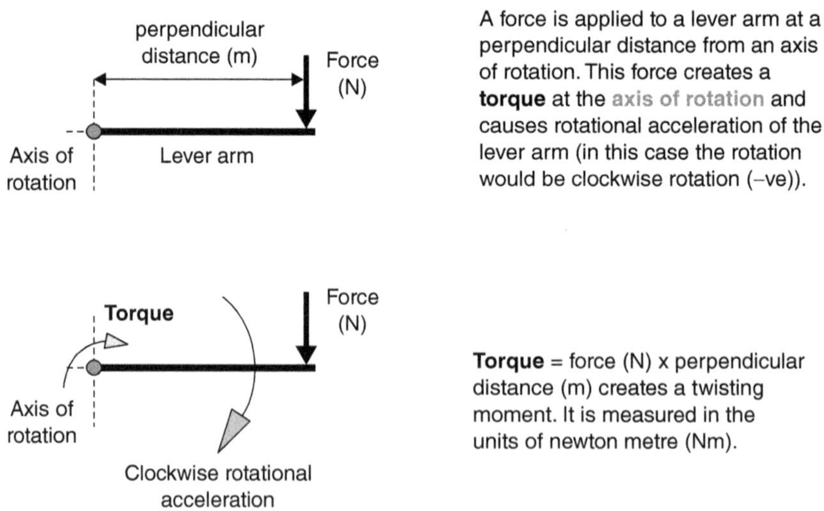


**Figure C3.1** Torque and moments

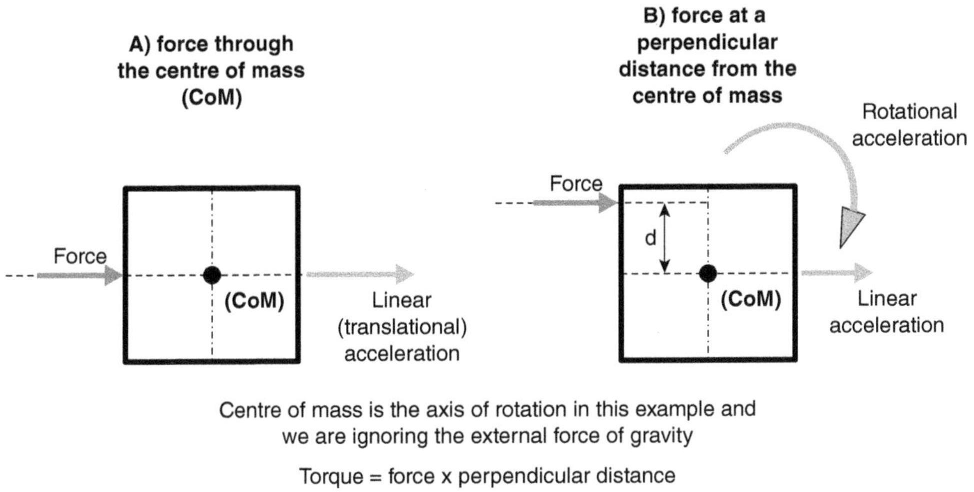


**Figure C3.2** Force and rotation

(object) will both rotate and move forwards. The torque that is created as a result of applying this force at a perpendicular distance will cause the box to rotate about its **axis of rotation.** However, the box will also move forwards (translate) as the force is applied in a horizontal direction. Although the force is acting at a perpendicular distance from an axis of rotation and it will create a torque or twisting moment, it will also have a horizontal component of force acting on the box (because the force is being applied horizontally). When we apply a **force** at a **perpendicular distance** to an **axis of rotation** we have seen that we create a torque or twisting moment (a tendency to rotate). The perpendicular distance from the centre of rotation is called the **moment arm.** The torque that is created causes a **potential** for the **rotational acceleration** and thus the resulting rotation of the limb, lever or segment to which it is being applied. This rotation can be described as being either **clockwise rotation** or **anti-clockwise rotation** (described by reference to the direction of the movement of the hands on a clock or watch). Within biomechanics, **clockwise rotation** is usually given the **negative symbol** (–ve) whereas **anti-clockwise** rotation is given the **positive symbol** (+ve). In many situations within biomechanics it is often the case that pairs of forces act about a segment and about an axis of rotation. Two equal and opposite forces that are acting on a system create what is termed a **force couple.** The term **couple** is therefore defined as a pair of **equal** and **opposite parallel forces.** *Figure C3.3* and *Figure C3.4* clarify these terms in more detail.

In *Figure C3.4* it is possible to see the effect of a couple on two objects (a box and a lever arm). In each example, a couple (a pair of equal and opposite parallel forces) is seen applied to the objects. In both cases the objects (a box and a lever arm) rotate (and in these cases the objects rotate solely because the only forces acting on them are equal and opposite – obviously in this example we have ignored the external force of gravity) in both a **clockwise** (the box) and **anti-clockwise** (the lever arm) direction. The couples create rotation about the axes of rotation. In these examples there is no translation (linear motion) because the total net **linear** force on the systems is zero (i.e. the forces are equal and opposite). According to Newton's first law of motion an object will remain at rest or continue with uniform **linear** motion unless it is acted upon by an external force. In both these cases the net linear force on the objects from the couples would be zero and the objects would remain in the same positions (i.e. they would not move linearly). However, they would rotate about their respective axes of rotation because the

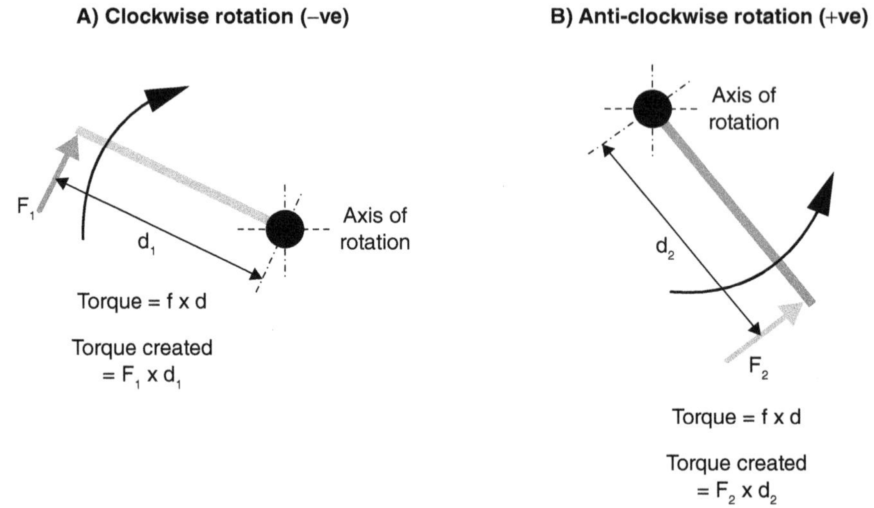


**Figure C3.3** Clockwise and anti-clockwise rotation

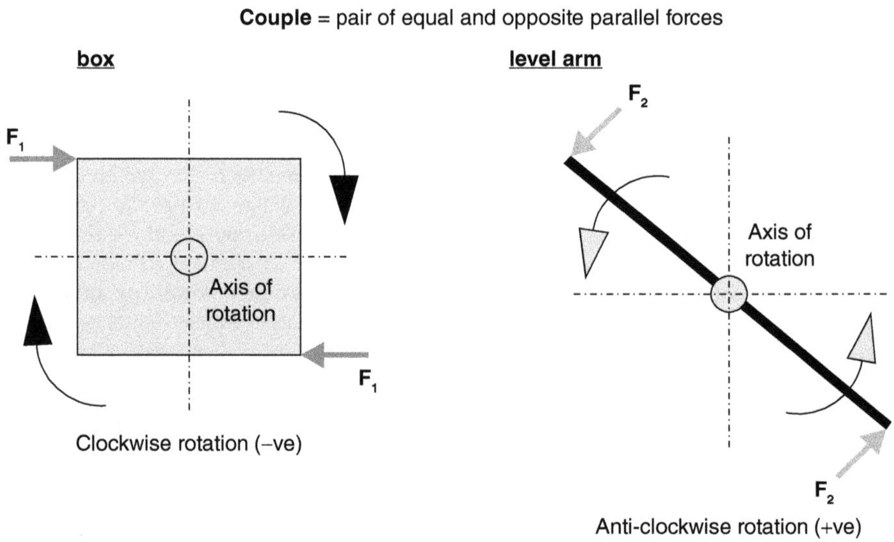


**Figure C3.4** A couple (a pair of equal and opposite parallel forces)

couples cause torques (that are in the same direction) and, hence, a combined rotational effect. Since torque is expressed as force multiplied by perpendicular distance from an axis of rotation it can be expressed mathematically as follows:

Torque = force × perpendicular distance

$T = \mathbf{F} \times \mathbf{d}$

where:

F = force (measured in newtons (N))

d = perpendicular distance from axis of rotation (measured in metres (m))

T = torque (measured in newton metres (Nm))

Considering this equation, it is possible to see that in order to **increase** the **torque**, and hence increase the turning moment (or rotational effect) applied to an object, we can either **increase** the **force** applied or **increase** the **distance** from which the force is applied (i.e. from the axis of rotation).

**Force applied = 20 N and the distance is 0.3 m**

T = F × d

F = 20 N

d = 0.3 m

T = 20 × 0.3

**T = 6.0 Nm**

**Force applied = 35 N and the distance is 0.3 m**

T = F × d

F = 35 N

d = 0.3 m

T = 35 × 0.3

**T = 10.5 Nm**

**Force applied = 20 N and the distance is 0.5 m**

T = F × d

F = 20 N

d = 0.5 m

T = 20 × 0.5

**T = 10.0 Nm**

Similarly, in order to **decrease** the amount of **torque** applied to a system we can either **reduce** the **force** applied or **shorten** the **distance** from the axis of rotation from which the force is applied. Both of these applications have particularly important implications within human movement and exercise science.

## APPLICATION

There are many applications within human movement where is it beneficial to increase and decrease the amount of torque or twisting moment acting on an object (or within a system where a system could be the interconnected segments of the lower limb). For example, in the case of swimming freestyle you may notice that many Olympic-level swimmers now use a pronounced bent elbow action during the pull phase of the freestyle stroke. Indeed, after the swimmer's hand has entered the water, the swimmer will immediately bend the elbow and pull through the stroke almost entirely in this bent elbow position. This technique is designed to allow the athlete to be able to create as much propulsion as possible and yet at the same time protect the shoulder joint from excessive torque and loading (which was the case when previously using an extended arm to pull in freestyle (i.e. a long lever arm)). In recent Olympics you may have seen many 1500 m freestyle swimmers adopt this technique due to the possible effects of injury on the shoulder caused by an extended arm pull over this 1500 m distance of swimming.

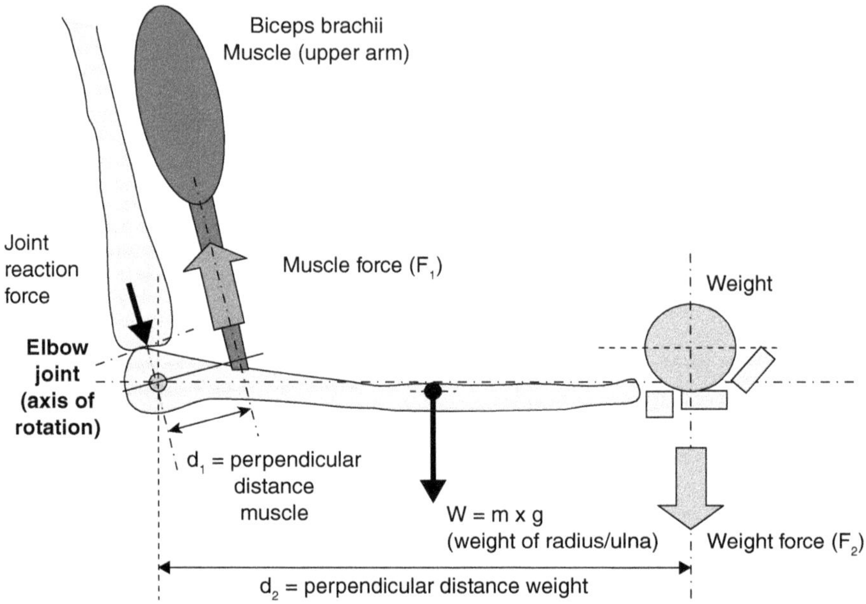


**Figure C3.5** Torque and moments during an arm curl with a weight

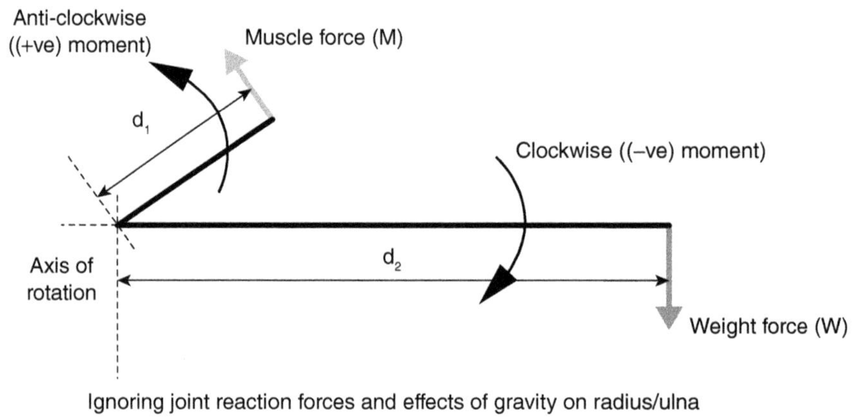


**Figure C3.6** Forces and moments during the arm curl example

Similarly, within gym exercises (see *Figure C3.5*), using the action of "arm curling" with weights requires the consideration of torque. The arm curl is where the elbow is flexed and extended while the athlete holds a weight in the hands. This action requires the biceps brachii muscle in the upper arm to create a torque (or turning moment) in order to resist the turning effect (or torque) created by holding the weight at a distance from the elbow (i.e. in the hand). In order to perform the action, the athlete must exert a force in this muscle that creates a turning moment (or torque) that overcomes the turning effect (the torque) created by the weight at the hands (due to its position in the hand from the axis of rotation (i.e. the elbow joint)). *Figure C3.5* and *Figure C3.6* help to illustrate this in more detail.

As vector quantities, torques are described based on their magnitude and direction of angular motion that they produce (i.e. clockwise or anti-clockwise) and they can be summated and resolved. In the example within *Figure C3.6*, it is possible to see that the system (the arm curling example with weights) has two moments acting (in the sagittal plane and about the transverse axis): a clockwise moment (–ve) which is created by the weight and the distance this weight acts from the elbow axis of rotation and an anti-clockwise moment (+ve) which is created by the muscle force (biceps brachii) and the distance that this force acts from the elbow joint axis of rotation. In this case, it is important to note that we are ignoring the moment created by the weight of the radius and ulna, which is seen in *Figure C3.5*. Note that it is important to point out that both moments are expressed with reference to the same axis of rotation (i.e. the elbow joint).

If the turning moment created by the muscle force equals the turning moment created by the weight then the system will be in what is called **equilibrium** (balanced forces and moments – resulting in no rotational acceleration – a constant or no velocity situation). However, if the muscle force creates a larger torque (turning moment) then the result is that the weight will be lifted and the arm will move in flexion. Alternatively, if the weight creates the greater turning moment then the arm will drop and the elbow will extend. This latter case happens when athletes lower weights down in a controlled manner (i.e. there is a mechanical advantage in favour of the weight). Throughout the action of arm curling with weights, the muscle force exerted by the biceps brachii will need to continually change. The reason for this change is due to the position of the arm at various points in the flexion–extension movement. In this case both the moment arm of the weight and the moment arm of the muscle force will continually change. Since the "weight" has a constant mass and therefore a constant weight (force = mass × acceleration due to gravity), the muscle force will need to change to accommodate the different torque or twisting moment created by the differing moment arms. *Figure C3.7* shows two positions of the arm in this example which illustrate the changing lengths of the respective moment arms.

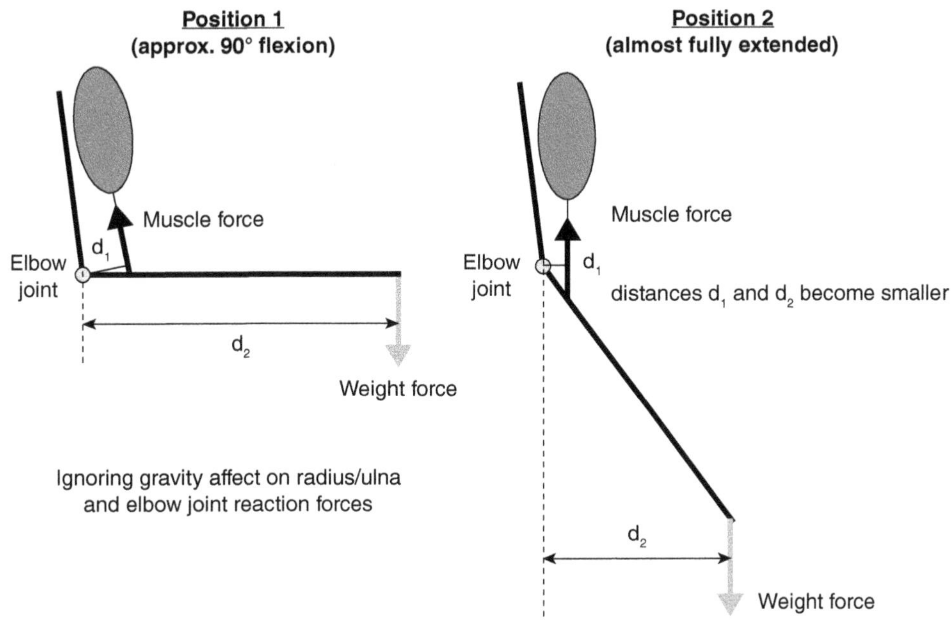


**Figure C3.7** Torque and moments during exercise

The first condition of equilibrium that is derived from Newton's first law of motion states that an object in a state of equilibrium does not accelerate (i.e. change its state of motion). In this case the sum of all the forces acting on the object is zero and the object is in a state of balance (i.e. it does not accelerate linearly in any direction). Similarly, an object that is in a state of equilibrium does not change rotational (or angular) velocity and **the second condition of equilibrium** states that the sum of all the torques acting on an object is also zero. Since the second condition of equilibrium is related to torque or turning moments we can write the second condition of equilibrium as:

> **The sum of the anti-clockwise moments and clockwise moments about a point (rotational point) is equal to zero.**

Since we know that torques cause rotation and are vector quantities (with magnitude and direction) we can use this knowledge to expand the second condition of equilibrium as follows:

$$\Sigma \mathbf{ACWM} + \Sigma \mathbf{CWM} = 0$$

where:

$\Sigma$ = the sum of

ACWM = anti-clockwise moments (+ve)

CWM = clockwise moments (–ve)

In this context, it is important to point out that this is referring to a system of torques about a common axis of rotation. Hence, it is important to use the same origin (point of rotation) for anti-clockwise and clockwise moments in the calculation of the second condition of equilibrium. In the example in *Figure C3.5*, we had a system with two moments (remember that we are ignoring the moment caused by the weight of the radius and ulna) that were acting about a single axis of rotation (a common axis or origin), which was the elbow joint.

Consider the example in *Figure C3.5*, with the arm held in the 90-degree flexion position. The moments for this example (ignoring radius/ulna weight) can be seen in *Figure C3.6*. If we now add some values to this example, we can work out the muscle force (M) that would be needed in order to balance the system (i.e. hold the weight in the hand in a stationary position).

$d_1$ = the perpendicular distance to the muscle force (**0.05 m**)

$d_2$ = the perpendicular distance to the weight force (**0.45 m**)

M = the muscle force (**unknown**)

W = the weight force.

In this case the weight is a 5 kg dumb-bell. Hence the force acting would be determined by the mass multiplied by the acceleration due to gravity ($F = m \times g$). For the weight this would be equal to 5 kg × 9.81 m/s$^2$ = **49.05 N.**

Next, we use the formula for torque ($T = f \times d$) and the second condition of equilibrium to solve the problem for the muscle force needed to hold this weight stationary.

Clockwise moment or torque (–ve) = $F \times d$

$= W \times d_2$

$= 49.05 \times 0.45$

$= \mathbf{-22.07\ Nm}$

Anti-clockwise moment (+ve) = $F \times d$

$= M \times d_1$

$= \mathbf{+M \times 0.05}$

Now substitute these into the second condition of equilibrium equation (note: the clockwise moment is given a negative sign):

$$\Sigma\ \mathbf{ACWM} + \Sigma\ \mathbf{CWM} = 0$$

$$\mathbf{M} \times 0.05 + (-22.07) = 0$$

Now rearrange the equation to get **M**:

$$\mathbf{M} = \frac{22.07}{0.05}$$

Note: 22.07 has now become positive because we have moved it to the other side of the equation and hence changed its sign.

$\mathbf{M = 441.4\ N}$ (muscle force)

Therefore, it is clear that we have to exert 441.4 N of force in the biceps brachii muscle to hold the weight in this static 90-degree flexion position. In order to overcome this weight and flex the joint further we would have to exert more force than this and thus create a larger turning moment (anti-clockwise). In this example it is useful to note that because the moment arm of the muscle force is small (0.05 m) we have to exert a large force in the muscle to balance the effect from the weight because it is acting at a much longer moment arm (0.45 m).

Probably the simplest way to understand clockwise and anti-clockwise moments in action and their relationship with equilibrium is to imagine the see-saw you probably sat on as a child. In order to balance the see-saw it was necessary to move either person further in or further out from the centre of the device (the fulcrum or pivot point). The central point of the see-saw in this case would be the axis of rotation. The weight (technically their mass) of each person sitting on the see-saw created the **forces** and the distances from the centre of the see-saw at which each person sat created the **moment arm**. One person would create a clockwise rotation (moment) of the see-saw and the other would create an anti-clockwise rotation (moment). In order to move the see-saw or balance the see-saw you either had to move in, move out or push off the ground with a force (hence changing the moment or torque created and moving the see-saw up or down).

Throughout human movement and exercise science there are numerous situations of the use and application of torques, moments and levers. For example, the torques created at the joints during the pull phase in swimming; the torques on the lower back during the golf swing; the levers and torques created by the canoeist and paddle in white-water slalom; and the torques needed in the limbs of the soccer players to kick the ball effectively with both speed and accuracy. In many of these examples it is often desirable to both increase and decrease the torques that are created. The use and application of levers is one example where the consideration of torque and moments is clearly applied and within human movement there are many applications of different types of levers.

| Key notes | |
|---|---|
| **Torque** | A torque is a twisting or turning moment that is calculated by multiplying the force applied by the perpendicular distance (from the axis of rotation) at which the force acts (the moment arm). Torques cause angular accelerations that result in rotational movement of limbs/segments. |
| **Clockwise and anti-clockwise rotation** | Clockwise rotation is the rotary movement of a limb/lever/segment in a clockwise direction (–ve). Clockwise is referring, in this case, to the direction of movement of the hands of a clock or watch. Anti-clockwise rotation is rotary movement in the opposite direction (+ve). |
| **Force couple** | A force couple is a pair of equal and opposite parallel forces acting on a system. |
| **Equilibrium** | This is a situation in which all the forces and moments acting are balanced, and which results in no rotational acceleration (i.e. a constant angular velocity situation). |
| **Second condition of equilibrium** | This states that the sum of all the torques acting on an object is zero and the object does not change its rotational velocity. Re-written, this condition can be expressed as that the sum of the anti-clockwise and clockwise moments acting on a system is equal to zero ($\Sigma ACWM + \Sigma CWM = 0$). |
| **Application** | Swimmers are now utilising a pronounced bent elbow underwater pull pattern during the freestyle arm action. This technique change allows the swimmer to acquire propulsive force and yet prevent excessive torques being applied to the shoulder joint (which were previously caused by a long arm pull underwater pattern). Large torques are needed at the hip joint (hip extensor and flexor muscles) to create the acceleration of the limbs needed to kick a soccer ball. |

## FURTHER READING

The following resources provide additional reading on torques and moments in sport and exercise.

1 Flanagan, S. P., & Salem, G. J. (2007). Bilateral differences in the net joint torques during the squat exercise. *Journal of Strength and Conditioning Research, 21*(4), 1220–1226.
2 Hewett, T. E., Stroupe, A. L., Nance, T. A., & Noyes, F. R. (1996). Plyometric training in female athletes. *The American Journal of Sports Medicine, 24*(6), 765–773.

# LEVERS

*Paul Grimshaw*

The description of **levers** is really only an applied practical example of moments. The problems and considerations concerned with levers can be solved by using **clockwise** and **anti-clockwise moments** and the second condition of equilibrium (as discussed elsewhere in this text). The student of biomechanics should hence recognize the application of moments rather than the classification of simple lever systems. Nevertheless, as biomechanists and exercise scientists we are often required to explain principles (i.e. to coaches, to clinical practitioners, to athletes, and even to students) in a language that is not technical. This application of levers would be a non-technical example of expressing the principle of moments.

**Levers** can be defined as **rigid bars** that rotate about **axes** or **fulcrums**. In the human body the **bone** or the limb/segment would act as the **rigid bar**, and the **joint** would act as the **axis of rotation** or **fulcrum**. A **fulcrum** can be defined as the pivot about which a lever turns. Generally, a force is applied to a lever in order to move or overcome a resistance (i.e. another force). Within the human body the muscles often provide the force while the resistance (i.e. the other force) is provided by other external forces acting on the system (e.g. gravity, collision, contact and load). *Figure C4.1* illustrates this in diagrammatic form.

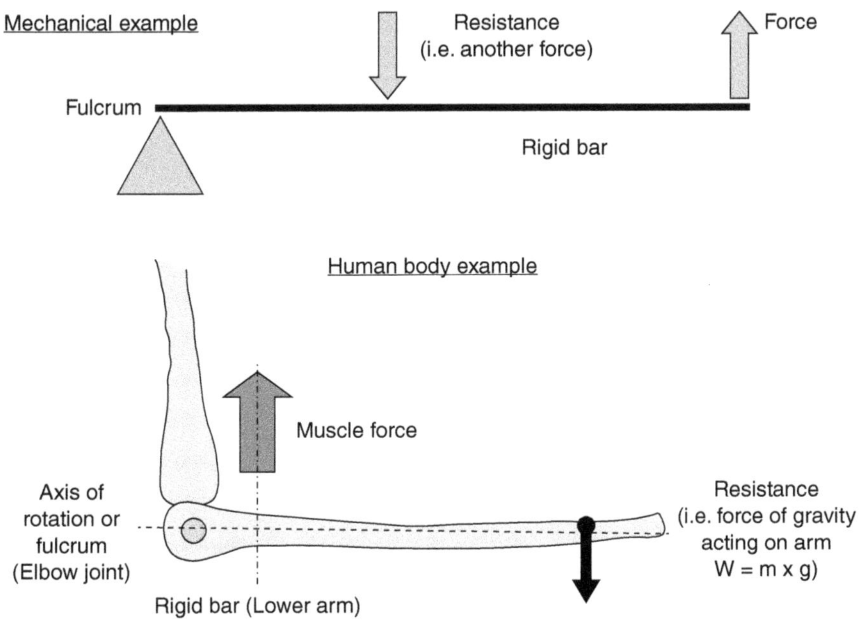


**Figure C4.1** Examples of levers, as they apply to both mechanical and human body applications

Levers are classified into one of three types. These are termed **first-**, **second-** or **third**-class levers. A **first-class lever** is when the force and the resistance are located at separate sides of the fulcrum. A **second-class lever** is when the force and the resistance are located on the same side of the fulcrum. In this case the force is further away (at a greater distance) from the fulcrum than the resistance. A **third-class lever** is similar to the second-class lever (with the force and resistance on the same side from the fulcrum position) but this time the force is nearer to the fulcrum than the resistance. In the first-class lever system the distances of the force and the resistance that act either side of the fulcrum do not need to be equal. *Figure C4.2* shows the three lever systems in more detail.

Within the human body, levers play an important role in the application of force and overcoming resistance (another force) in order to initiate or continue movement. Each lever classification (first, second and third) can have different applications. *Figure C4.3* identifies some of the applications both in general and within human movement.

The **first-class lever** classification is similar to the see-saw that children would play on in the park or playground. In this situation you would observe that a child is sitting at each side from the point of rotation of the see-saw (i.e. the fulcrum). The children (depending on their respective mass) would move further in or further away from the fulcrum in order to balance the see-saw. If they both sat at equal distances from the fulcrum the see-saw would move downward at the end with the heavier child. In more general terms a pair of scissors would represent a situation in which a first-class lever is used. The fulcrum would be the axis of rotation of the scissors (i.e. approximately in the middle), and the force would be provided at one end by the hands in order to overcome a resistance at the other end (i.e. the cutting of the paper or object). Within the human body there are many applications of the first-class lever.

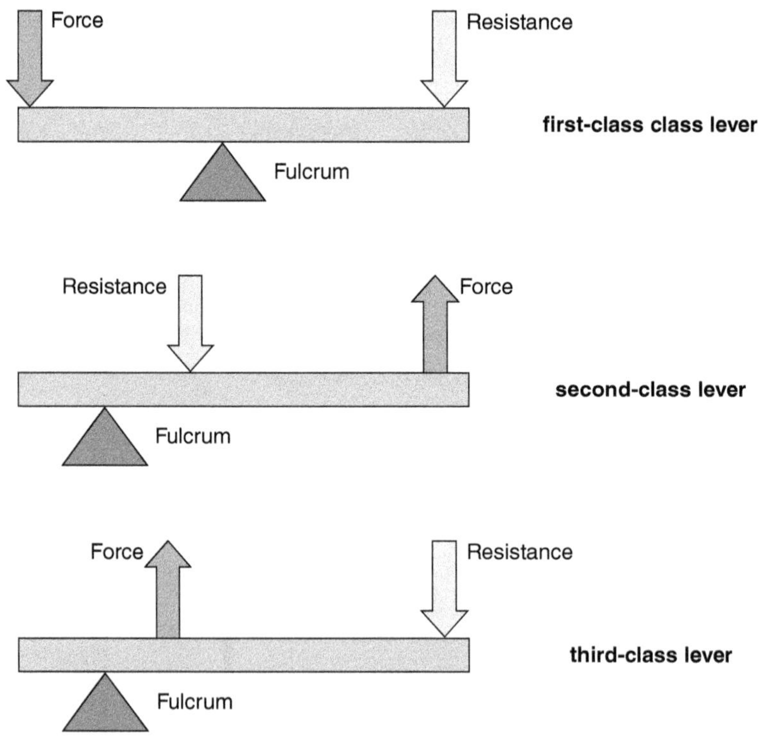


**Figure C4.2** Levers (classification)

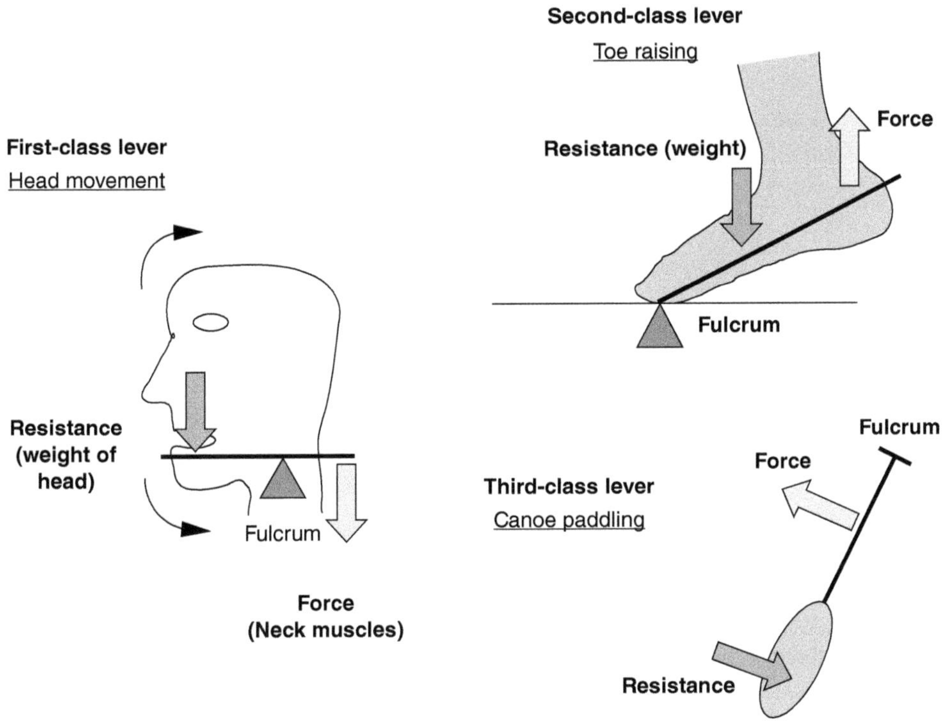


**Figure C4.3** Examples of first-, second- and third-class levers in human movement

For example, rotating the head (or "nodding") forwards and backwards would be an example of a first-class lever within the body.

In the **second-class lever** the force and the resistance are located on the same side (from the fulcrum position) but the force moment arm (distance) is greater than the resistance moment arm. For an equal and opposite resistance this would mean that the **force required** to **move the resistance** would be **less than the resistance.** In this case there would be a **mechanical advantage** in the favour of the **force** being applied. Within human movement examples of second-class levers include rising up onto your toes. However, the more common general example of the second-class lever shown in many biomechanics texts would be the use of a wheelbarrow to move a load.

In the **third-class lever** system again we can see that the force and the resistance are located on the same side from the fulcrum but, this time the force is nearer to the fulcrum than the resistance. This would mean that more force would be required to move a given resistance. One of the most common examples of this within human movement is the action of **flexion of the elbow joint.** The bicep muscle acts at a position that is close to the fulcrum (i.e. the elbow joint) whereas the resistance acts at a point further away from the fulcrum (usually a load held in the hand such as an object or weights when doing arm curl exercises and the weight of the forearm and the hand). In more general terms the use of a shovel would be a good example of a third-class lever and the use of a paddle while canoeing would be another example within human movement. *Figure C4.4* shows this human flexion example in more detail.

**Mechanical advantage** is defined as the ratio of the force moment arm (distance from the fulcrum) divided by the resistance moment arm (distance from the fulcrum). When the moment arm (perpendicular distance) of the **applied force** is **greater than** the moment arm of

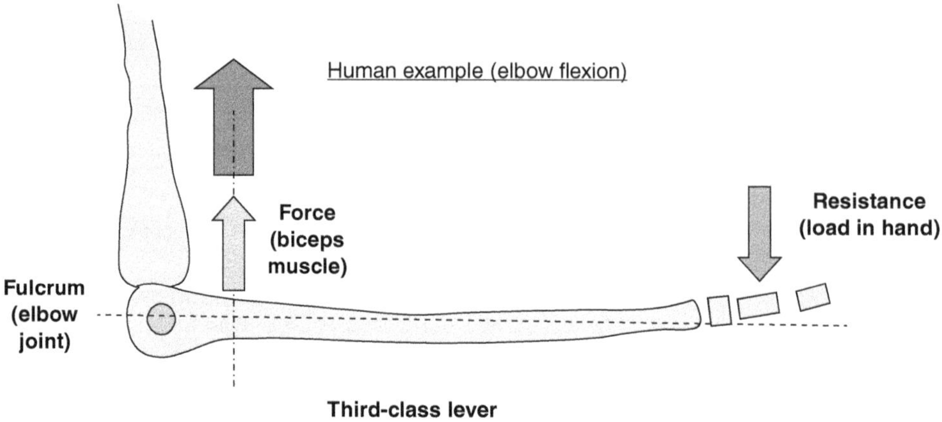


**Figure C4.4** Third-class lever (flexion of the elbow joint)

the **resistance** (i.e. for a given resistance/force) then the force needed to move the resistance is less than the force offered by the resistance (i.e. there is a mechanical advantage). This is the case in the second-class lever example. Conversely, when the moment arm of the **applied force** is **less than** the moment arm of the **resistance** then more force is needed to move the given resistance (i.e. there is a mechanical disadvantage). This is the case in the third-class lever example. A further explanation of this can be observed using the previously identified example of the see-saw in the children's playground. *Figure C4.5* represents a first-class lever as in the case of the see-saw on the children's playground. *Figures C4.6* and *C4.7* respectively illustrate second- and third-class lever systems.

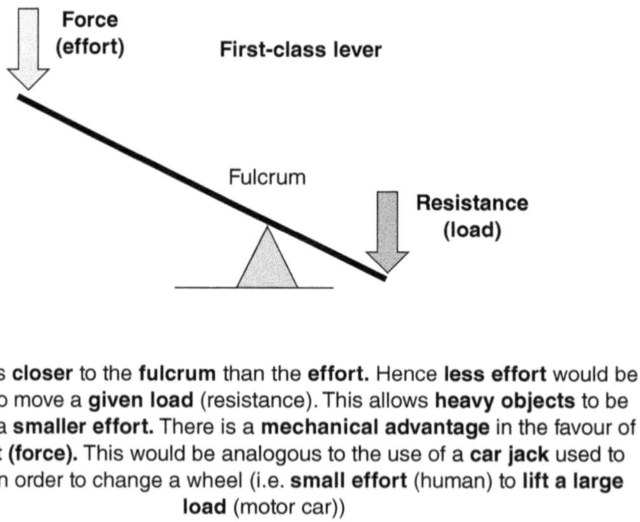


**Figure C4.5** First-class lever – mechanical advantage in favour of the force

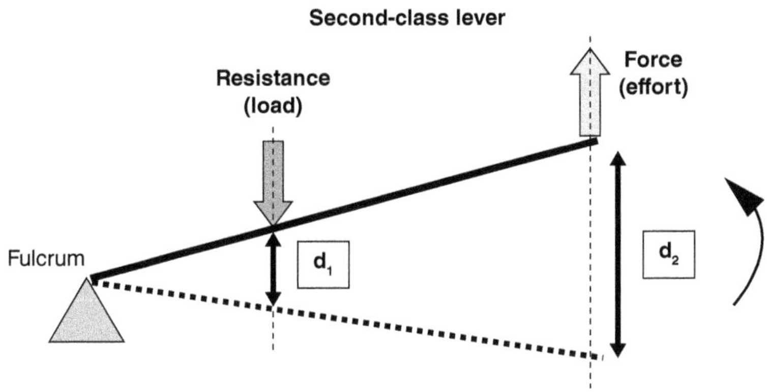


**Mechanical advantage** in favour of the **force** (**effort**). The moment arm of the **force** (distance from fulcrum) is **greater than** the **moment arm** of the **resistance.** The **force** required to move a given resistance would be **less than** the **resistance.** Note that the smaller **force** is required to move through a **larger <u>linear</u> range of motion** (ROM) than the **resistance**. Hence although the given force needed to move the resistance is less it would need to be applied through a larger ROM. Although the **angular motion** of all points on the rigid bar (Lever) is **the same,** the **linear motion** of the two points: load and effort ($d_1$ and $d_2$) is **different**

**Figure C4.6** Second-class lever – mechanical advantage in favour of the force

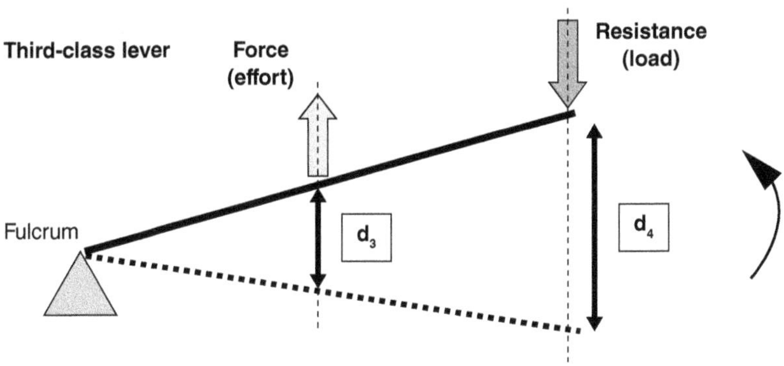


**Mechanical advantage** in favour of the **resistance** (**load**). The moment arm of the **resistance** is **greater than** the **moment arm** of the **force.** The **force** required to move a given resistance would be **more than** the **resistance.** Note that the **larger force** is only required to move through a **small linear ROM** (**$d_3$**) in order to move the **load a larger linear displacement** (**$d_4$**). Hence although the given force needed to move the resistance is more it has the effect of moving the resistance through a large ROM

**Figure C4.7** Third-class lever – mechanical advantage in favour of the resistance

## TASK

Two children sit at each end of a see-saw. One of the children has a mass of 28 kg and the other a mass of 35 kg. The child who has a mass of 35 kg is sitting at a point that is 1.2 m away from the fulcrum position. This child is sitting to the right-hand side of the fulcrum and would cause a clockwise moment (i.e. a tendency to cause a clockwise rotation of the see-saw). At what distance must the child who is 28 kg sit in order to balance the see-saw? This problem can be solved by using the second condition of equilibrium where the sum of clockwise and anti-clockwise moments equals zero ($\Sigma M = 0$).

## Solution

It is clear that the child who is sitting to the left must be sitting at a distance greater than 1.2 m in order to overcome or balance the moment created by the child who is sitting to the right (because the child on the left has much less mass). The child who is sitting at a distance of 1.2 m to the right has a mass of 35 kg, which would represent a weight of 343 N (weight = mass × acceleration due to gravity). This would create a clockwise moment (rotation) of 412 Nm (343 N × 1.2 m). In order to balance the see-saw we can use the second condition of equilibrium to solve this problem. The child who is sitting to the left and who is 28 kg will create a weight force of 274 N.

$$\sum \text{Clockwise moments} + \sum \text{Anti}-\text{Clockwise moments} = 0$$

$$F_1 \times d_1 + F_2 \times d_2 = 0$$

Re-arrange equation:

$$F_1 \times d_1 = F_2 \times d_2$$

$$d_1 = \frac{F_2 \times d_2}{F_1}$$

$$d_1 = \frac{412\,\text{Nm}}{274\,\text{N}}$$

$$\mathbf{d_1 = 1.5\,m}$$

We can now see that the child on the left must sit at a distance of 1.5 m from the fulcrum in order to balance the see-saw. In terms of mechanical advantage this can now be expressed in the context of the formula that is related to levers.

$$\textit{Mechanical advantage} = \frac{\textit{Force distance (moment arm)}}{\textit{Resistance distance (moment arm)}}$$

$$\textit{Mechanical advantage} = \frac{1.5}{1.2}$$

= 1.25 **(no units as metres (m) cancel out i.e. it is a ratio)**

When the ratio of the force moment to resistance moment arm is **greater than 1.0** then there is a **mechanical advantage** in favour of the **force.** In this case the ratio is 1.25 and the **force** required to move (or in this case balance) the **resistance** would be less than the resistance. As we can see by considering the actual figures this is correct as the force created by the child on the left (child 1) sitting at a distance of 1.5 m is only **274 N,** while the force created by the child on the right (child 2) who is sitting at a distance of 1.2 m is **343 N.** There is a mechanical advantage in favour of the child who is sitting on the left (child 1). This can be applied to any situation concerning levers and the movement of a resistance (load).

## APPLICATION

Within the body, muscles can often operate at different mechanical advantages (due to where they connect in the body) and there are applications when the muscle is required to operate at an advantage and also at a disadvantage. For example, the bicep muscle in the upper arm which causes a flexion rotation about the elbow joint will function throughout the movement creating both maximum and minimum amounts of torque, i.e. the moment that may cause rotation (flexion). This mechanical advantage and disadvantage can be seen in more detail by considering the action of raising and lowering a weight in the hand (arm curling). When the perpendicular distance of the muscle line of pull from the joint axis (elbow joint) of rotation is larger the more torque (at the elbow joint) will be created for a given effort. When the perpendicular distance to the muscle line of pull is less, the torque that is created (as in the case of an extended arm position) is much smaller. The moment arm of the biceps' force at different points in the movement of flexing the arm will change. This will create more or less torque at the elbow for a given effort. This principle of torque generation and mechanical advantage through levers is applied in the design of many modern exercise machines. For example, many devices will be able to accommodate the different torque-generating capacities of a muscle during exercise. The machines will often change the loading patterns at various points in the exercise in order to accommodate to the changing torques. Similarly, machines will often load muscles more at the points of their greatest torque-generating capacity.

*Key notes*

| | |
|---|---|
| **The study of levers** | A more detailed definition and application of levers can be explained through the study of moments and the second condition of equilibrium. This section provides an applied approach that is useful for explanation in non-technical terms (i.e. possibly for coaches, clinicians, athletes and even students). |
| **Levers** | Levers can be defined as rigid bars that rotate about axes or fulcrums. Generally, a force is applied to a lever in order to move or overcome a resistance. In the human body the bone or limb segment would be the rigid bar and the joint would be the fulcrum or point of rotation. The muscles would provide the force (effort) and the resistance (i.e. external force provided by gravity, contact or collision) would be the load. |
| **First-class lever** | A first-class lever is defined as when the force and resistance are located on separate sides of the fulcrum (axis of rotation). An example of this within human movement would be the nodding action of the head. The force provided by the muscles in the back of the neck is located at the opposite side of the fulcrum (i.e. the rotation point in the middle of the neck or the skeletal atlas point) from the resistance (which would be primarily provided by the weight of the head). |

*(continued)*

*(continued)*

| | |
|---|---|
| **Second-class lever** | A second-class lever is defined as when the force and resistance are located on the same side of the fulcrum but when the force is located further away from the fulcrum than the resistance. An application of this within human movement would include rising up onto the toes of your foot. The calf muscle (gastrocnemius) through the Achilles tendon provides the force and the resistance is provided by the weight of the foot/body acting through the ball of the foot area. The fulcrum in this case would be the point of rotation at the toes. |
| **Third-class lever** | A third-class lever is defined as when the force and resistance are located on the same side as the fulcrum but when the force is nearer to the fulcrum than the resistance. An application within human movement would be an arm flexion action during arm curling in weightlifting. The biceps brachii muscle would provide the force and the weight in the hand (and the lower arm) would provide the resistance. The fulcrum would be at the elbow joint. |

## FURTHER READING

The following two resources provide additional reading on the concept of levers in sport and human movement.

1 Jeran, J. J., & Chetlin, R. D. (2005). Training the shoulder complex in baseball pitchers: A sport-specific approach. *Strength & Conditioning Journal*, *27*(4), 14–31.
2 Russell, J. A., McEwan, I. M., Koutedakis, Y., & Wyon, M. A. (2008). Clinical anatomy and biomechanics of the ankle in dance. *Journal of Dance Medicine & Science*, *12*(3), 75–82.

# CENTRIPETAL FORCE AND ACCELERATION

*Adrian Burden*

**Newton's first law** states that an object will remain at rest or in uniform motion unless some external force acts on it to change its state of motion. This implies that if an object deviates from a straight line (i.e. moves in a curved path) then some force must act to cause this to happen. When an object moves in a circular path (e.g. when a hammer thrower rotates the hammer before release, *Figure C5.1*), the force causing circular motion is said to be the **centripetal** force ($F_{centripetal}$). This term describes that the force is a **centre-seeking** force.

The centripetal force will cause an acceleration in the direction of the force, according to **Newton's second law**; this is called **centripetal acceleration** ($a_{centripetal}$):

$$F_{centripetal} = \text{mass} \times a_{centripetal} \qquad \text{(Equation C5.1)}$$

This centripetal acceleration causes the object to move in a circle and acts in the same direction as the centripetal force, i.e. towards the centre of the circle or path of rotation. As described elsewhere in this text, any point along the radius of a rotating body will have a

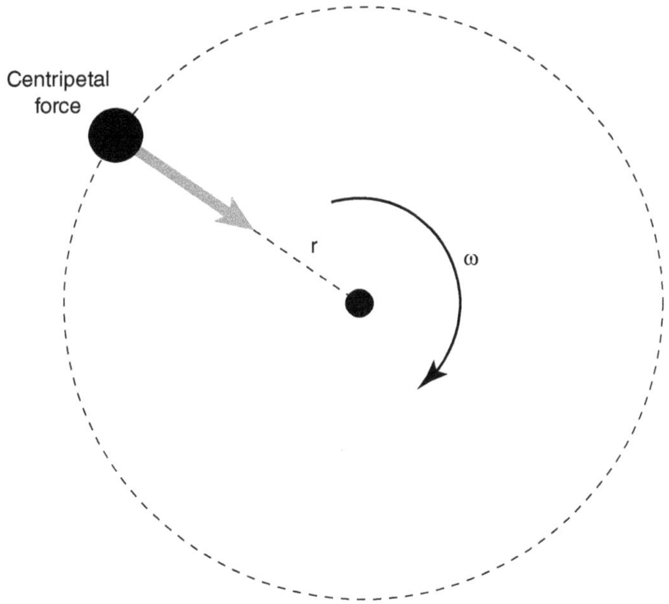


**Figure C5.1** Centripetal force during the rotation of a hammer (ω = angular velocity and r = radius of rotation)

linear, or tangential, velocity ($v_t$). Returning to the example of a hammer thrower, if the hammer rotates with a constant angular velocity ($\omega$) and radius (r), the magnitude of the tangential velocity will also be constant (as $v_t = r \times \omega$; see diagram on right in *Figure C5.2*). However, the direction of the two tangential velocities, shown in *Figure C5.2* taken at two points in time, are different as the hammer is travelling in a circular path. The change in direction of the two velocities is towards the centre of rotation (see the diagram on the left of *Figure C5.2*), hence the (centripetal) acceleration and force are in the same direction.

If an object rotates about a circle of **radius of rotation** (r) at a **constant angular velocity** ($\omega$), then the centripetal acceleration is given by the equation:

$$a_{centripetal} = r \times \omega^2 \qquad \text{(Equation C5.2)}$$

As $v = \omega \times r$, Equation C5.2 can be further developed to give an alternative expression for the centripetal acceleration:

$$a_{centripetal} = r(v^2 / r^2) = v^2 / r \qquad \text{(Equation C5.3)}$$

Equations C5.2 and C5.3 can be substituted into Equation C5.1 to provide two expressions for the centripetal force.

$$F_{centripetal} = m \times r \times \omega^2 \qquad \text{(Equation C5.4)}$$

$$F_{centripetal} = m \times v^2 / r \qquad \text{(Equation C5.5)}$$

How is the centripetal force applied? In sports and exercise this is normally provided by some linkage, which can physically apply a force to the object of interest. Consider the case of a hammer thrower (*Figure C5.3a*) at the start of the movement when the hammer is increasing

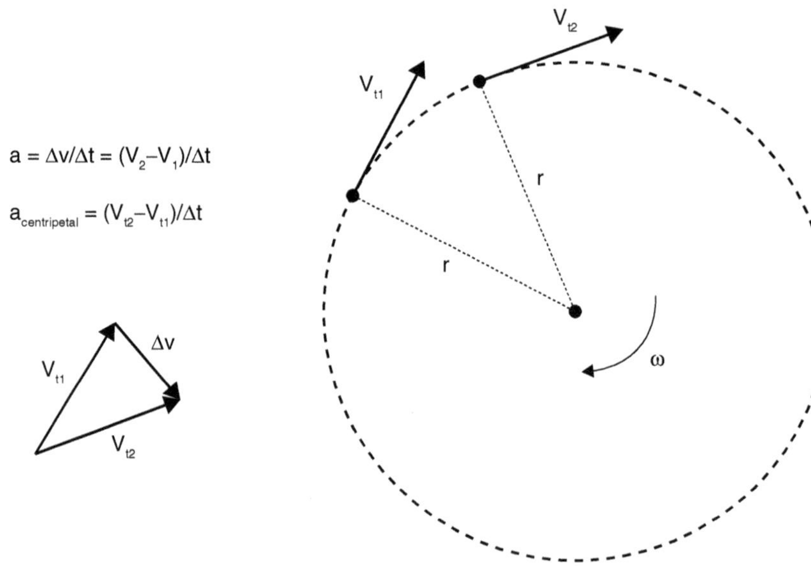


**Figure C5.2** Change in the direction of tangential velocity ($v_t$) during the rotation of a hammer ($\omega$ = angular velocity and r = radius of rotation)

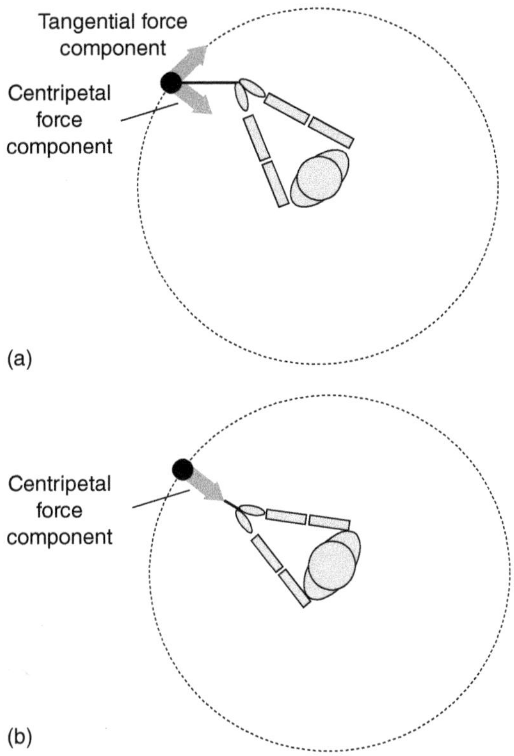


**Figure C5.3** The tension in the hammer wire and its components (a) during the increase in hammer head speed and (b) at maximum (constant) velocity before release

in speed. The force that causes the hammer to rotate around the athlete is provided through the wire. This force has two components, one acting in the direction towards the centre of rotation (the centripetal force) and the other tangential to the circular path. The centripetal force causes motion of the hammer head in a circle while the tangential force component causes its acceleration (i.e. increase in speed) around the circle. Once hammer head speed has been developed, the athlete must maintain that speed in preparation for release. Constant speed of rotation is achieved when the centripetal force alone acts (*Figure C5.3b*).

The term **centrifugal force** (centre-fleeing force) often crops up in the literature and much unnecessary confusion exists over the correct use of the term. From **Newton's third law** every force has to have an equal and opposite reaction force. The tension in the hammer wire supplies the centripetal force to the hammer head (*Figure C5.4*). The same tension also applies an equal and opposite reaction force to the thrower and, as this is directed away from the centre, it is called the centrifugal force. Thus, the hammer thrower experiences a force pulling away from the hands and it is this sensation which sometimes makes people think the force acting on the hammer is also in the same centre-fleeing direction.

## APPLICATION 1

In the performance of the **grand (or giant) circle in men's gymnastics**, the body rotates about the high bar (*Figure C5.5*). The force which keeps the body moving in a circle is supplied by the

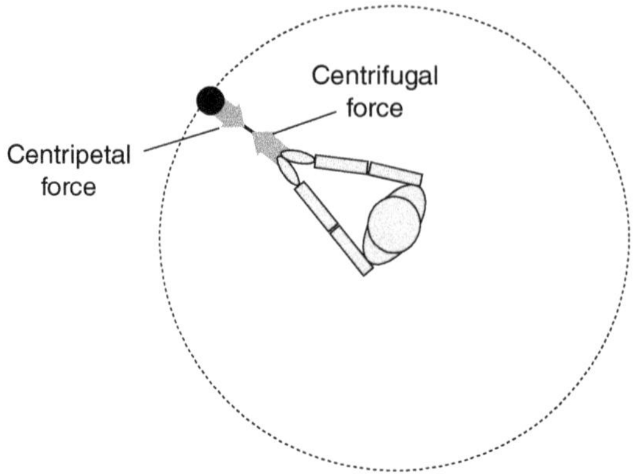


**Figure C5.4** The centrifugal force acting on the hammer thrower

tension in the arms and is the source of the centripetal force. This tension will vary during the performance, but will be a maximum when the angular velocity is maximum (Equation C5.4); that is, when the gymnast swings directly beneath the bar.

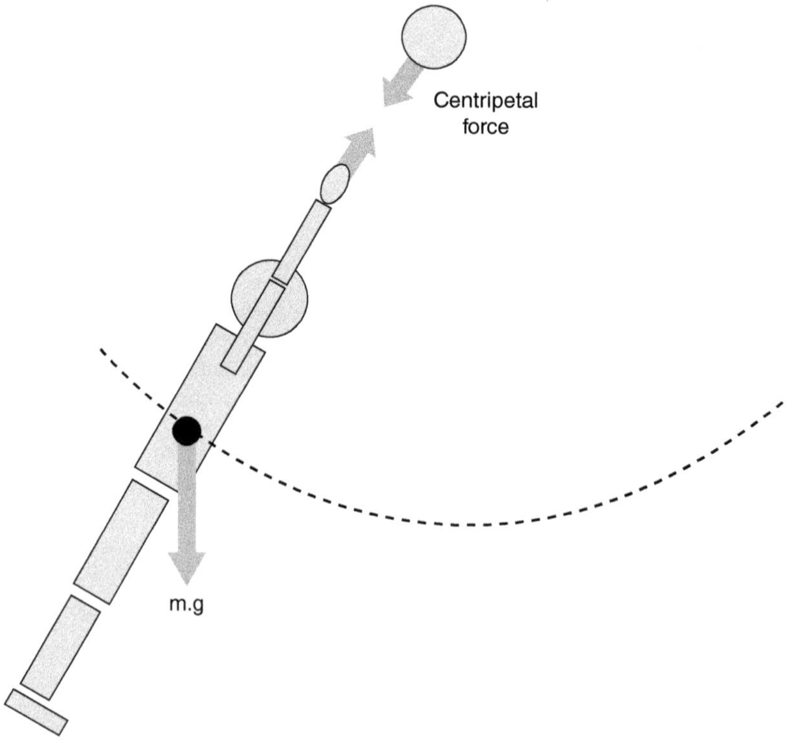


**Figure C5.5** Free body diagram of a gymnast

The centripetal force acting at the gymnast's hands pulls the body towards the bar and is given by Equation C5.4. The component of the gravitational force acting on the centre of mass and along the direction of the body inclined at an angle θ to the vertical (m × g × cos θ) pulls the gymnast away from the bar. This increases the tension (T) in the arms which is given by the sum of these two opposing forces as

$$T = m \times r \times \omega^2 + m \times g \times \cos\theta$$

where:

m = mass of gymnast

r = radius of rotation

ω = angular velocity

g = acceleration due to gravity (Equation C5.6)

Tension in the arms is maximal when the gymnast's body is parallel with the line of gravity (i.e. θ = 0), hence, the equation can be written in terms of body weight:

$$T = m \times g\,[r \times \omega^2/g + 1] \quad \text{(Equation C5.7)}$$

For typical values of r (1.3 m) and ω (5.5 rad/s) the maximum value of T is about five times body weight. Such a high force explains why the gymnast needs to be very strong, particularly in grip strength, and can sometimes fall off if the grip is not strong enough.

## APPLICATION 2

When **running around the bend on a flat track,** athletes are seen to lean into the bend. This leaning action causes a lateral frictional force, which is the source of the centripetal force. The lean will continue until the athlete reaches an angle θ where there is sufficient friction force produced to enable the athlete to round the bend comfortably. This situation can be analysed by considering the moments set up by the frictional force (F) and the normal reaction (N) force about the centre of mass.

The free body diagram of the runner is shown in *Figure C5.6*. As the athlete is balanced the moments about the centre of mass are equal:

Equating moments $N \times x = F \times y$

Therefore $F / N = x / y = \tan\theta$ (1)

Resolving vertically: $N - m \times g = 0$ (2)

The friction force is the source of the centripetal force, so

$$F = m \times v^2 / r \quad (3)$$

Combining (1), (2) and (3) gives

$$F / N = v^2 / r \times g = \tan\theta$$

Or $\theta = \tan^{-1} (v^2 / r \times g)$

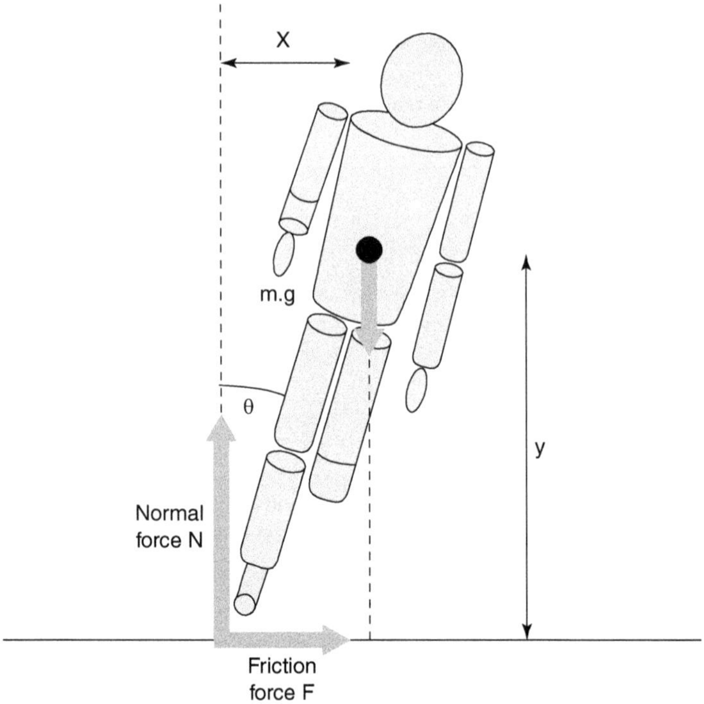


**Figure C5.6** Free body diagram of an athlete rounding a bend

This equation allows the prediction of the angle of lean of a runner for any velocity (v) and cornering radius (r). Typically, an athlete running at 10 m/s (a 200 m race) around an athletics track of radius of rotation 40 m, will have an angle of lean of 14 degrees. If the athlete performed on an indoor track of radius of rotation 20 m, then the angle of lean would increase to 27 degrees.

## APPLICATION 3

In indoor athletics, **the track is often banked on the bends** so that the athlete does not have to lean at large angles and run the risk of slipping and falling, possibly into other competitors. The ideal angle of banking is such that there is no lateral friction force at the athlete's foot. The analysis is performed in the same way as in the previous example using the following relationships from *Figure C5.7*.

Resolving vertically:

$$N \times \cos\theta - m \times g = 0 \quad (1)$$

The component reaction force is the source of the centripetal force, so:

$$N \times \sin\theta = m \times v^2 / r \quad (2)$$

Dividing (2) by (1) gives:

$$\tan\theta = v^2 / r \times g$$

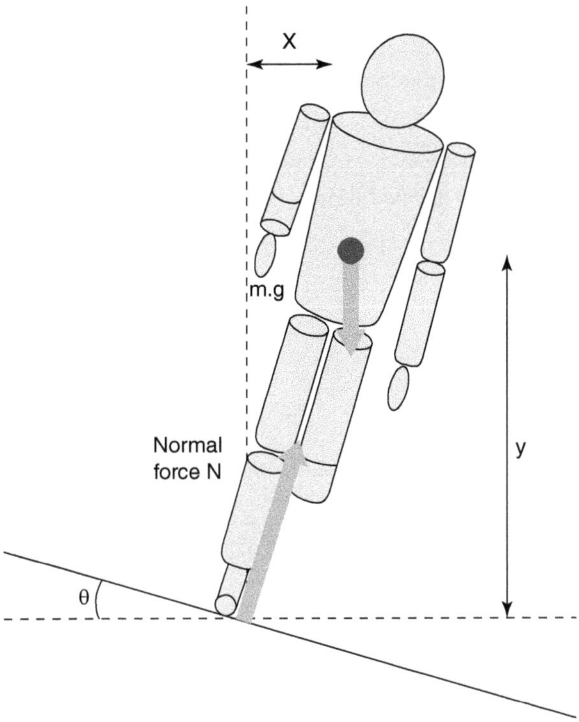


**Figure C5.7** Free body diagram of an athlete rounding a bend on a banked track

Which is the same as that which was found in the previous application. Note that the angle of banking indoors will be greater than the angle of lean outdoors, due to the reduced radius of the bend for indoor athletics. For indoor cycling, there is a critical speed for which the turn radius and the bank angle of the track has been designed. At this speed, the cyclist is not leaning relative to the track surface. However, at higher speeds the cyclist will lean towards the centre of the velodrome and at lower speeds they will lean away from the centre of the velodrome.

*Key notes*

| | |
|---|---|
| **Centripetal force** | The centripetal force ($\mathbf{F}_{\mathbf{centripetal}}$) is a force directed to the centre of rotation when objects move in a circular path. This force is always required to make objects move in a circular path. In other words, if an object moves in a circular path the centripetal force must be acting. |
| **Centripetal acceleration** | Newton's second law (F = ma) tells us that, whenever a force (F) acts, it produces an acceleration (*a*). So when the centripetal force acts it produces a centripetal acceleration. |
| **What is the centripetal force dependent on?** | There are two main formulae to describe the centripetal force. These are:<br>$\mathbf{F}_{\mathbf{centripetal}} = \mathbf{m} \times \mathbf{r} \times \omega^2$<br>$\mathbf{F}_{\mathbf{centripetal}} = \mathbf{m} \times \mathbf{v}^2/\mathbf{r}$<br>It can be seen that the centripetal force is dependent on the radius of rotation of the circle (r), and the angular (ω) or linear (*v*) velocity. The force is also dependent on mass (m). So the heavier, faster and more distant the object is to the centre of rotation, the greater the centripetal force needs to be in order to produce circular motion. |

## FURTHER READING

The following paper provides further reading around the topic of sprinting around curved running tracks.

1 Ishimura, K., & Sakurai, S. (2016). Asymmetry in determinants of running speed during curved sprinting. *Journal of Applied Biomechanics, 32*(4), 394–400.

# MOMENT OF INERTIA AND ANGULAR MOMENTUM

*Paul Grimshaw*

This section is concerned with the calculation of moment of inertia and angular momentum and it is related to angular kinetics (i.e. forces and the effect of these forces on angular movement). The linear momentum possessed by a body is defined as the product of its mass multiplied by its linear velocity and it is measured in the SI units of kg.m/s (kilogramme metre per second).

Linear momentum = mass × velocity (kg.m/s)

In angular terms, **angular momentum** is defined as the product of the moment of inertia of a body multiplied by its angular velocity, which is caused by the body's (or part of the body) mass and its distribution of mass in a circular motion about an axis of rotation. It is measured in the SI units of kg.$m^2$/s.

Angular momentum = moment of inertia × angular velocity (kg.$m^2$/s)

The angular momentum of an object about a particular axis will remain constant unless the object is acted on by an unbalanced eccentric force (such as another athlete, a ball or an implement) or a couple (a pair of equal and opposite parallel forces).

The importance of angular momentum within sport and exercise can be seen by considering how a soccer player learns to kick a ball effectively, how a golfer transfers angular movement of a club to the linear movement of the golf ball or indeed how a sprinter manages to move the limb quickly through the air in order to make the next contact with the ground that is needed to push off and move forward with speed.

The **inertia** of an object is referred to as the resistance offered by the stationary object to move linearly and it is directly proportional to its mass. The **moment of inertia**, however, is defined as the reluctance of an object to begin rotating or to change its state of rotation about an axis. Moment of inertia is related to the mass of the object (body or body part) and the location (distribution) of this mass from the axis of rotation. Without specific reference to a particular axis of rotation the moment of inertia value has little meaning.

*Figure C6.1* shows the moment of inertia values in some selected athletic situations during sport. It is important to reiterate that moment of inertia is specific to the axis of rotation about which the body is moving (e.g. either the centre of gravity (transverse) axis of the body as in diving or the high bar (transverse axis) in gymnastics). The greater the spread of mass from the rotation centre (axis) the greater will be the moment of inertia. In *Figure C6.1* the largest moment of inertia value is determined when the body is in the position when it is rotating about the wrist (hands) and the whole body is extended (i.e. the mass is distributed as far as possible away from the axis of rotation which in this case is about the hands (an axis of rotation through the hands)). Therefore, the moment of inertia of an object or body about a particular axis depends upon the mass of the object or body and the distribution of this mass about the axis of rotation. Specifically, an equation for moment of inertia about an arbitrary axis A can be given as:

Moment of inertia = mass × radius² (kg.m²)

(about an axis A)

$I_A = m \times r^2$

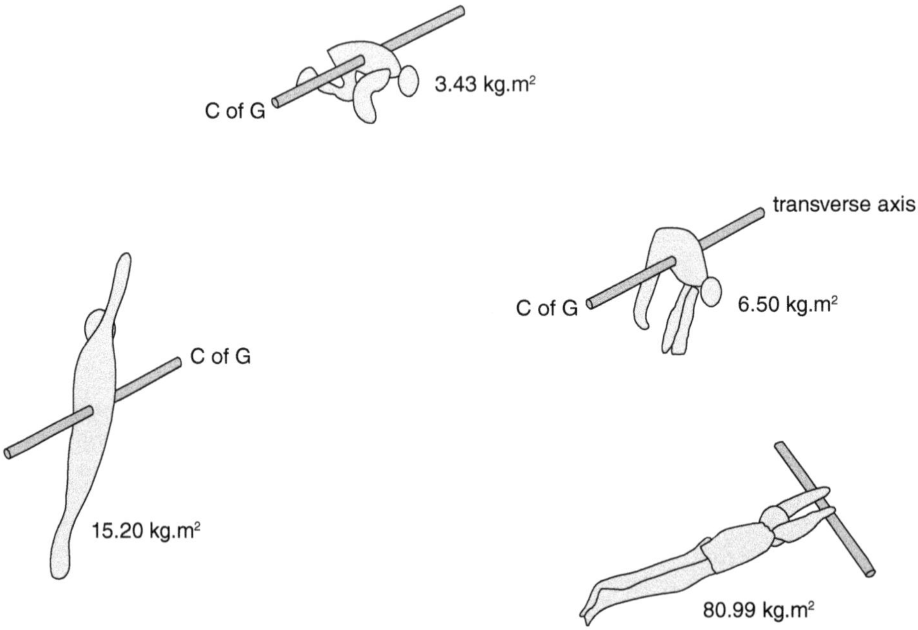


**Figure C6.1** Moment of inertia values in diving and gymnastic positions. Note: values were recalculated and figures redrawn from the originals presented in Hay (1978, p. 147), which is unfortunately now out of print (Hay, J. G. (1978). *The Biomechanics of Sports Techniques.* Englewood Cliffs, NJ: Prentice Hall, Inc.)

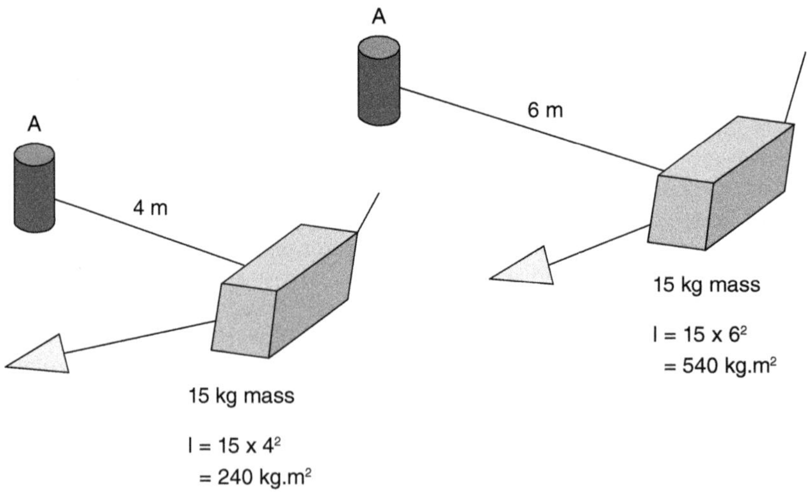


**Figure C6.2** Moment of inertia

*Figure C6.2* shows the moment of inertia calculation for a 15 kg mass (concentrated point mass) when it is rotating about two different axes of rotation (6 m and 4 m from the same axis of rotation). This clearly indicates that as the rotation axis changes the mass is located further away from the axis and as a result the moment of inertia changes. In this figure it is also possible to see that when the 15 kg mass is moved closer to the axis of rotation (4 m away instead of 6 m away) the moment of inertia value decreases. This has important implications in sport and it can be seen that a small moment of inertia value is achieved when the body forms a tight "tuck" about the centre of gravity axis of rotation. In this case the mass distribution is close to the axis of rotation (the centre of gravity) and the moment of inertia value is the least (3.43 $kg.m^2$ as opposed to 80.99 $kg.m^2$ in the extended position (*Figure C6.1*)). The calculation shown in *Figure C6.2* only works for a concentrated point mass such as the 15 kg mass used in this example. The distribution of mass in the human body is not considered to act as a concentrated point mass.

It is also important to note that because the equation for moment of inertia contains the $r^2$ (distance of the mass from the axis of rotation squared) component it is clear that changing the position of the mass has a much greater effect than changing the actual mass. For example, **doubling the distance** from the **axis of rotation** for a **given mass** will **quadruple the moment of inertia value.** This has important implications for human movement within sport and exercise.

The moment of inertia value of regularly shaped bodies about any arbitrary axis A is determined by taking a number of measurements of the mass distribution about the axis of rotation; then by summating the result the moment of inertia of the whole body is determined. *Figure C6.3* illustrates this is in a mathematically regular shaped body.

$$\textbf{Moment of inertia}_{(A)} = \mathbf{m_1 r_1^2 + m_2 r_2^2 + m_3 r_3^2 + \ldots + m_n r_n^2}$$

Where: n = the number of samples taken

$$I_A = \Sigma m_n r_n^2$$

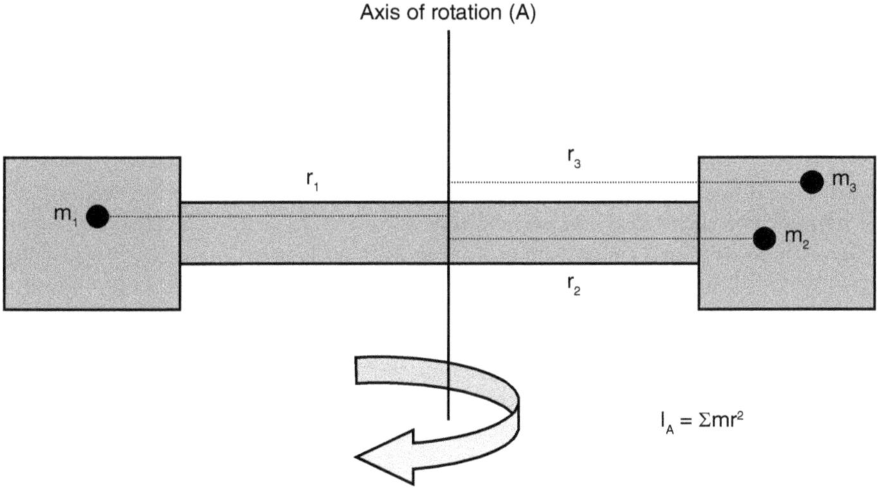


**Figure C6.3** Moment of inertia of regular shapes

**Table C6.1** Moment of inertia values of selected body segments about the transverse axis through the centre of gravity of the segment

| *Segment* | *Moment of inertia (kg.m²)* |
|---|---|
| Head | 0.024 |
| Trunk | 1.261 |
| Upper arm | 0.021 |
| Forearm | 0.007 |
| Hand | 0.0005 |
| Upper leg (thigh) | 0.105 |
| Lower leg | 0.050 |
| Foot | 0.003 |

Adapted and modified from Hay (1978, p. 145), which unfortunately is now out of print (Hay, J. G. (1978). *The Biomechanics of Sports Techniques*. Englewood Cliffs, NJ: Prentice Hall, Inc.)

This process of determining the moment of inertia is difficult to do mathematically and for complex shapes, like the limbs of the human body, these values have been computed by researchers so that they are available for use by others. The data presented in *Table C6.1* represents values of moments of inertia for human limbs about their own centre of gravity. These values can be used in further calculations.

Considering the moment of inertia of the whole body, it is clear that it is different depending on the axis about which the body is rotating. The moment of inertia of the whole body would be different if it was rotating about a longitudinal axis (such as in the case of a pirouette in ice skating) than it would be if it were rotating about the transverse axis (as in the case of a somersault). Similarly the moment of inertia of sports equipment can also vary for different axes and different planes of rotation. Imagine a young child trying to swing an adult golf club. In this case it is easy to see that the child has to move the hands down the club and shorten the lever in order to have any chance of swinging the club. In effect, the child is reducing the radius of rotation (i.e. the distance of the axis of rotation to the centre of gravity of the club) and thus decreasing the moment of inertia (the resistance to change). In essence this is why it is easier for an adult to swing a 9-iron club than it is to swing a longer club such as a driver. The moment of inertia of the 9-iron when it is rotating about the frontal (anterior–posterior) plane axis (although strictly speaking it is not a single-planar or single-axis movement) is less than that of a driver and hence it is easier to swing. The same principle applies to a soccer kick. The athlete flexes the leg before the kick, thus shortening the radius of rotation and decreasing the moment of inertia so it is easier to accelerate the limb quickly in order to impart greater velocity to the stationary ball. As the leg is brought to the ball it is slightly extended (more so after contact), which will slow down its rotation and allow more control for an effective and accurate contact.

The **parallel axis theorem** is one method that allows us to be able to calculate the moment of inertia of the whole body or the sum of several body parts (such as the leg of a soccer player before contact with the ball) about a particular axis and plane of rotation (e.g. rotation at the hip in the sagittal plane (transverse axis) as in the soccer kick). It is important to add, though, that these movements are never truly single-axis or single-planar activities, and even the soccer kick would involve rotation within two or more planes and about two or more axes of movement (i.e. it is a three-dimensional movement). However, the theorem identifies that the moment of inertia about an axis that is parallel to the axis for which the moment of inertia was derived equals the moment of inertia of the body segment in its original position (i.e. about its own centre of gravity (C of G)) plus the product of the mass and the square of the perpendicular distance between the parallel axes. Thus the equation for the moment of inertia of the body or body segments is rewritten as follows:

## Parallel axis theory of calculating moment of inertia

$$I_A = I_{C\ of\ G} + md^2$$

where:

$I_A$ = Moment of inertia of a body about an axis through a point, A

$I_{C\ of\ G}$ = Moment of inertia about a parallel axis through the body/segment centre of gravity

m = the mass of the body or segment

d = the distance between the parallel axes

*Figure C6.4* illustrates this theory in more detail and helps to show how the parallel axes are determined in the example of the leg in the position of knee flexion before contact with the ball during a soccer kick. Furthermore, *Table C6.1* identifies the moment of inertia values for the body segments, which represent the moment of inertia through an axis that is parallel to the axis of consideration and that is through the segment's centre of gravity ($\mathbf{I}_{CoG}$).

## Calculation of moment of inertia of a leg segment (using the parallel axis theory)

Considering *Figure C6.5*, it is possible to see that in the case of the leg in this position, which would be before the contact phase in a soccer kick, the centre of gravity of each segment is given as a distance from the centre of rotation (i.e. the hip joint). When the position of each segment's centre of gravity is given it is possible to use the parallel axis theorem to determine the moment of inertia about any axis if the moment of inertia of each segment about their respective parallel axis is known. This is an alternative method to using the radii of gyration measure

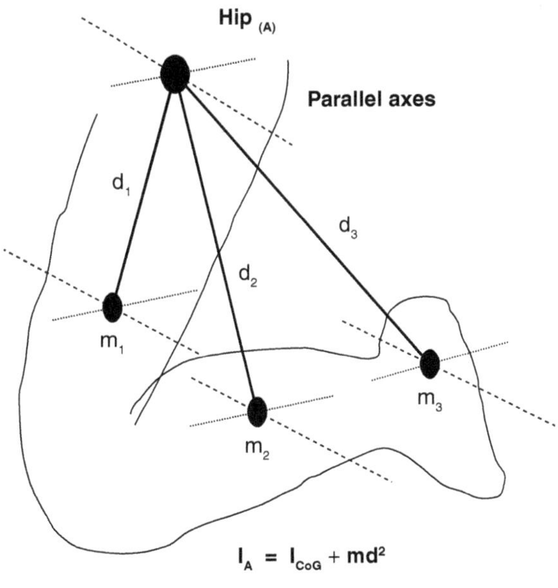


**Figure C6.4** The parallel axis theorem

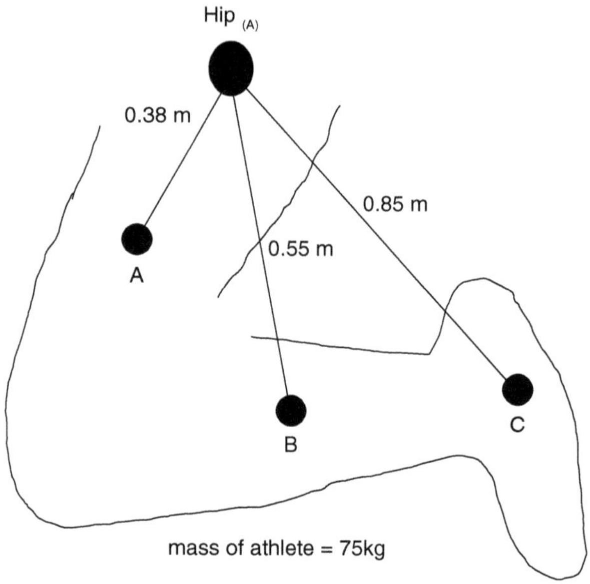


**Figure C6.5** Moment of inertia of the human leg

(which is essentially used for single segment moment of inertia calculations) and allows the calculation of the whole body or whole limb moment of inertia (such as in the case of the leg).

## Parallel axis theory of calculating moment of inertia

$$I_A = I_{C\,of\,G} + md^2$$

The above equation is applied individually to calculate the moment of inertia of the upper leg, lower leg and foot (*Figure C6.5*) and the three values are then summated to represent the moment of inertia of the whole leg in this position. However, in order to determine the mass of each segment in this 75 kg athlete it is necessary to use the anthropometric data presented in the work of Winter (1990). *Table C6.2* illustrates these values in more detail.

**Table C6.2** Anthropometric data where m is body mass in kg

| | |
|---|---|
| Hand | 0.006 × **body mass (m)** |
| Forearm | 0.016 |
| Upper arm | 0.028 |
| Forearm and hand | 0.022 |
| TOTAL ARM | 0.050 |
| Foot | 0.0145 |
| Lower leg | 0.0465 |
| Upper leg | 0.100 |
| Foot and lower leg | 0.061 |
| TOTAL LEG | 0.161 |

The values in the table are derived from data presented in Winter, D. A. (1990). *Biomechanics and Motor Control of Human Movement* (2nd ed.). New York: Wiley-Interscience Publishers (3rd edition published in 2004)

## Determine the moment of inertia of the leg in this position (*Figure C6.5*)

### *A. Upper leg*

$I_A = I_{C\ of\ G} + md^2$

$I_A = 0.105 + ((0.100 \times 75) \times 0.38^2)$

$I_A = 0.105 + 1.083$

$I_A = \mathbf{1.188\ kg.m^2}$ (upper leg moment of inertia)

### *B. Lower leg*

$I_A = I_{C\ of\ G} + md^2$

$I_A = 0.050 + ((0.0465 \times 75) \times 0.55^2)$

$I_A = 0.050 + 1.055$

$I_A = \mathbf{1.105\ kg.m^2}$ (lower leg moment of inertia)

### *C. Foot*

$I_A = I_{C\ of\ G} + md^2$

$I_A = 0.003 + ((0.0145 \times 75) \times 0.85^2)$

$I_A = 0.003 + 0.786$

$I_A = \mathbf{0.789\ kg.m^2}$ (foot moment of inertia)

### *D. Total moment of inertia of leg in this position*

$I_A = I_A \text{ (upper leg)} + I_{A\ (lower\ leg)} + I_{A\ (foot)}$

$I_A = 1.188 + 1.105 + 0.789$

$I_A = \mathbf{3.08\ kg.m^2}$

The moment of inertia of the leg in this position and about the transverse axis just before kicking a soccer ball would be **3.08 kg.m²**. It is important to understand that if the athlete could reduce the rotation axis (distribute the mass differently), by flexing the leg more, the moment of inertia would be reduced and the limb would be able to be moved (accelerated) much faster (rotationally) to generate potentially more velocity that could be imparted to the ball. The same principle will apply to a sprinter who wishes to bring the leg through quickly in order to make contact with the ground again, or indeed a golfer while swinging the golf club could bend the elbows to reduce the moment of inertia and hence increase the angular velocity of the swing (reducing the resistance to change). This will now be developed further with specific reference to angular momentum.

**Angular momentum** is represented by the **letter L** or **H** and is determined by the product of the **moment of inertia** and the **angular velocity** (measured in radians/s) of a body or segment. It is expressed in the units of **kg.m²/s** (kilogramme metre squared per second).

Angular momentum = moment of inertia (kg.m²) × angular velocity (rad/s)

$L = I\ \omega\ (kg.m^2/s)$

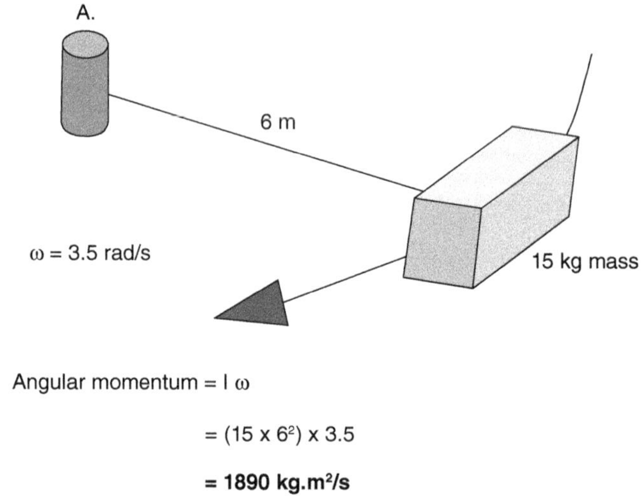


**Figure C6.6** Angular momentum

Consider *Figure C6.6*, which represents the 15 kg mass that was used previously as an example to calculate the moment of inertia values. In *Figure C6.6*, the mass is now given an angular velocity of 3.5 rads/s. The angular momentum is the product of moment of inertia ($15 \times 6^2$) and angular velocity (3.5) and is expressed as **1890 kg.m²/s.** If there were zero angular velocity then there would also be zero angular momentum. The radius of rotation also determines the angular momentum and is a very important factor because of the mathematical squared component. As we have seen before with moment of inertia, if we double the distance (radius of rotation = 12 m) we quadruple the angular momentum to **7560 kg.m²/s** (which is 4 × the previous 1890 kg.m²/s value).

Angular momentum (as with the moment of inertia) must be expressed with reference to an **axis of rotation** and the calculation of the whole body angular momentum is the sum of all its individual body segment momenta. For human body segments that rotate about an axis other than their centre of gravity the parallel axis theorem is used and this is also applied to angular momentum calculations. We have seen previously that for rotation about an arbitrary axis A, the moment of inertia is:

$$I_A = I_{CoG} + md^2$$

Hence, angular momentum for the segment about its own centre of gravity and for the segment about a parallel axis can also be expressed as:

$$L_A = (I_{CoG} + md^2)\ \omega$$

Multiply out the brackets:

$$L_A = I_{CoG}\ \omega + md^2\ \omega$$

where:

$L_A$ = Angular momentum (kg.m²/s) about axis of rotation A

$I_{CoG}$ = Moment of inertia about an axis through the centre of gravity of the segment m = the mass of the body or segment

d = the distance between an axis through point A and a parallel axis through the centre of gravity of the segment

ω = angular velocity (rad/s)

The calculation of whole limb (e.g. the leg complex) angular momentum about the hip axis of rotation is determined by summating the angular momentum of individual segments about an axis through their own centre of gravity and the angular momentum of the segment about the axis of rotation (i.e. the hip axis). *Figure C6.7* helps to illustrate this in more detail. It is important to reiterate that this calculation is axis and plane specific.

## CONSERVATION AND TRANSFER OF ANGULAR MOMENTUM

During movement of the body the total angular momentum possessed by the body (in the absence of any external force) remains constant. When the body is in flight, gravity is not considered to be an external force that affects angular momentum because it does not cause any rotational component (i.e. it will act through the centre of gravity and there is no moment arm). This conservation of angular momentum principle is derived essentially from Newton's first law of motion, which in angular terms can be expressed as follows:

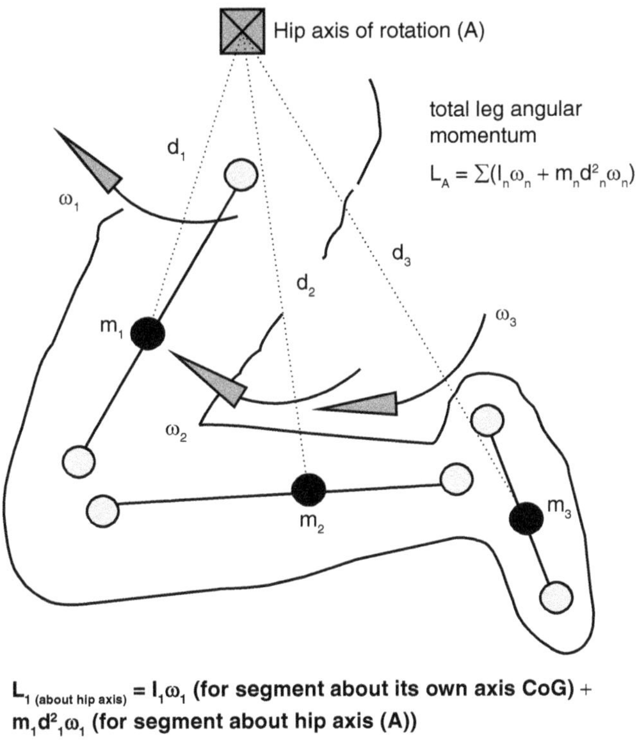


**Figure C6.7** Total leg angular momentum

> Every object will remain in a non-rotating state or will continue to rotate about a particular axis with uniform angular velocity unless acted upon by an unbalanced eccentric force or couple.

As an athlete prepares to jump into the air (such as in the case of a diver on a springboard) the athlete, while in contact with the ground (or board), will generate angular momentum (creating rotation – by applying forces at perpendicular distances to axes of rotation – thus creating reaction torques or moments). Once he/she is in the air (during flight as in the case of the diver) and in the absence of external forces (neglecting gravity effects) the athlete's angular momentum will remain constant. In this respect it is possible for the athlete to change his/her **moment of inertia** in order to increase or decrease his/her **angular velocity**. The diver will form a tight tuck around the centre of gravity rotation axis (transverse axis – sagittal plane) and the angular rotation (causing somersaults) will increase because the athlete has decreased the moment of inertia of the body. Similarly, if the athlete extends the body the moment of inertia will increase and the angular rotation (angular velocity) will be reduced. This is one of the reasons why the diver will extend suddenly at the end of the dive (i.e. to reduce the angular rotation and enter the water with minimal rotation in a straighter, more controlled and aesthetic position).

Similarly, it is possible to transfer the angular momentum possessed by the body in one axis to another different axis of rotation within the body. This is how divers initiate twist and tilt manoeuvres in the air when they only seem initially to have rotation about one axis and in one plane of movement. In addition, it is also possible to see that mechanically divers can initiate angular momentum for one part of the body while in the air (remembering that the overall angular momentum of the body must remain constant). For example, a diver who leaves the board with zero angular momentum can vigorously move or rotate an arm in a particular axis of rotation. The corresponding effect will be that the athlete will generate angular momentum (for this segment) about this axis and plane of motion. This angular momentum must therefore (because of the conservation of angular momentum principle) be balanced by an angular momentum in an opposite direction by another part or segment (i.e. zero or fixed whole body angular momentum). Hence, it may be the case that the athlete is seen to have to counteract this new angular momentum created in the arms by a simultaneous rotation and movement in the legs. In this context it is important to reiterate the effects of moment of inertia. Segments or body parts (such as the legs) that have large moments of inertia will thus have smaller angular rotations. Therefore, the high angular rotation created by a diver vigorously dropping an arm in mid-flight may be balanced by what appears only a small rotation of the legs (because the legs can have a much larger moment of inertia than the arms and the amount of angular momentum must remain constant).

*Figure C6.8* shows a diagram that helps to explain this principle of conservation of angular momentum. It is important to remember when considering this diagram that the angular momentum is determined from the product of moment of inertia and angular velocity ($\mathbf{L} = \mathbf{I}\,\omega$) and it is constant in the absence of any external force or torque. Thus, it can be seen that if **I increases**, then **ω must decrease**. Similarly, if **I decreases**, then **ω must increase**.

There are many examples in sport and exercise where moment of inertia and angular momentum are important and throughout this section we have seen examples such as soccer kicking, the leg drive in sprinting, the golf swing and diving. However, the following provides more examples of these principles in a sport application.

- The ice skater pirouetting on the ice with the arms out to the side would have a high moment of inertia and small angular velocity. Bringing the arms in will reduce the moment of inertia and increase angular velocity (rotation speed).
- The gymnast performing several somersaults and twists in the air would need to reduce the moment of inertia about different axes in order to increase rotation velocity and have any chance of performing the number of twists and somersaults before landing.

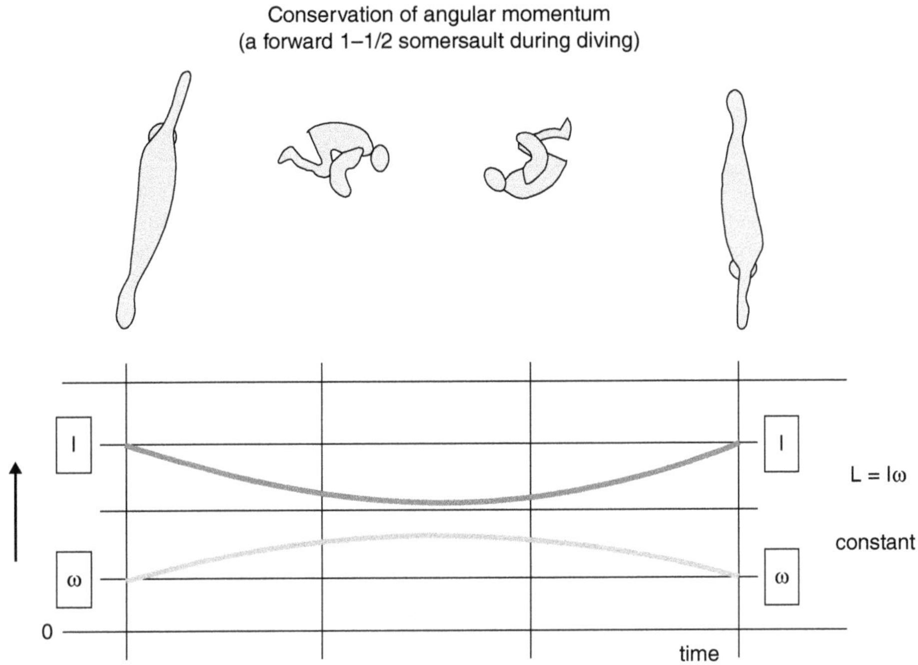


**Figure C6.8** Conservation of angular momentum

- The downhill racing skier using long skis (having a large moment of inertia) will find it difficult to initiate turns easily (neglecting friction effects), whereas the slalom skier with shorter skis (smaller moment of inertia) will find turning much easier.
- The young child using a set of shorter-shafted golf clubs or a tennis racket with a shorter handle and lighter head will find it easier to play golf or tennis.
- The basketball player performing a "slam dunk" shot is forced to counteract the rotation initiated in the arms by movement in the legs.
- The elite tennis player serving with a bent arm action in order to reduce the moment of inertia of the arm and racket and increase rotational velocity, which results in a faster serve.

| *Key notes* | |
|---|---|
| **Moment of inertia** | This refers to the reluctance of an object or body to start rotating or change its state of rotation (i.e. speed up or slow down). It is measured as the product of the mass of the body and the distance of the mass from the axis of rotation squared ($I = mr^2$). Double the axis of rotation of the mass and you will quadruple the moment of inertia value. The moment of inertia value can be changed by redistributing the body mass about the axis of rotation (such as in the case of "tucking" or extending the body during a somersault). |
| **Parallel axis theorem** | This is a method of calculation that is used to determine whole body or body part (such as the legs) moment of inertia. The parallel axis theorem states that the moment of inertia about a parallel axis is equal to the moment of inertia of the segment in its original position plus the product of the mass and the square of the distance between the parallel axis. |

*(continued)*

*(continued)*

| | |
|---|---|
| **Angular momentum** | This is the product of the moment of inertia and angular velocity. It remains constant in the absence of any external force/torque and the principle of conservation of angular momentum is derived from the angular analogue of Newton's first law of motion. In flight the angular momentum possessed by a body is constant and gravity is not considered to be an external force affecting the amount of angular momentum present in a system. In flight, gravity will act through the centre of gravity and there is no moment arm created – hence, no external torque or moment is present. Athletes can transfer angular momentum to different body parts and different axes and planes of movement/rotation. Athletes can also initiate angular momentum of some body part from a state of having no whole body angular momentum (such as dropping an arm during the flight phase in diving). However, this angular momentum must be counterbalanced by angular momentum of some other body part in the opposite direction (because of the principle of conservation where angular momentum was zero to begin with and must be zero at the end). Individual segment angular momentum is determined by calculating the angular momentum of the segment about its own centre of gravity plus the angular momentum of the segment about the whole body's centre of gravity. Whole body angular momentum is the summation of all the individual segment momenta. Angular momentum is affected by the mass of the body, the distribution of this mass and its angular velocity, and it is plane and axis specific. In addition, it has significant implications for performance in sport. |

## FURTHER READING

The following resources provide additional reading around the relevance of the moment of inertia and angular momentum to two different sports situations.

1 Hiley, M. J., & Yeadon, M. R. (2003). Optimum technique for generating angular momentum in accelerated backward giant circles prior to a dismount. *Journal of Applied Biomechanics, 19*(2), 119–130.
2 Bahamonde, R. E. (2000). Changes in angular momentum during the tennis serve. *Journal of Sports Sciences, 18*(8), 579–592.

# NEWTON'S LAWS OF MOTION

## Angular motion

*Paul Grimshaw*



As we have already seen, Newton's first law of motion relates to situations where forces are balanced and the net effect of external forces acting on an object is zero. As a reminder, Newton's first law, which related to linear motion, states:

> Every object will remain at rest or continue with uniform motion unless acted upon by an unbalanced force.

This law can also be applied to the linear momentum of a body in that it is also true that the momentum (mass multiplied by velocity) possessed by a body is constant in the absence of any external force. This law is saying that a body will either stay at rest (with no momentum) or keep moving (with a constant momentum) unless acted upon by an external force. An object in motion that is not affected by a net external force will have a constant linear momentum because it has a constant linear velocity and a fixed mass. In angular terms this law can be reworded as follows and can be made applicable to angular motion:

> The angular momentum of a body remains constant unless a net external torque is exerted upon the body.

In this application we are dealing with rotation and rotational effects and the term **force** has been replaced with the term **torque** (a twisting or turning moment). In this context, it is worth identifying that in the application of this law to rotation it has purposefully **not** been stated in the context of **constant angular velocity** (as in the case of the linear analogue) because as we will see in the rotational application for the human body (which is not a rigid body) this is not necessarily the case.

The **angular momentum** (usually represented by the **symbol L**) of a body or object can be expressed by the **moment of inertia** of the body multiplied by its **angular velocity**.

**angular momentum = moment of inertia × angular velocity**

$$\mathbf{L = I \times \omega \text{ or } L = I\omega \ (kg.m^2/s)}$$

Angular momentum is measured in the units of **kilogramme metre squared per second** ($kg.m^2/s$). The moment of inertia of an object is the reluctance of the object to start or stop rotating, or to change its state of rotation. **Moment of inertia** is measured in the units of **kilogramme metre squared** ($kg.m^2$).

The moment of inertia of an object refers to the object's ability to resist rotation. The larger the moment of inertia, the more the object will resist rotation. Similarly, the smaller the moment of inertia of an object, the less will be its resistance to start, stop or change its rotational state (*Figure C7.1*). The **moment of inertia** is calculated from the distribution of **mass** (m) about an axis of **rotation** (r). It can be expressed mathematically as $\mathbf{I = m \times r^2}$. The moment

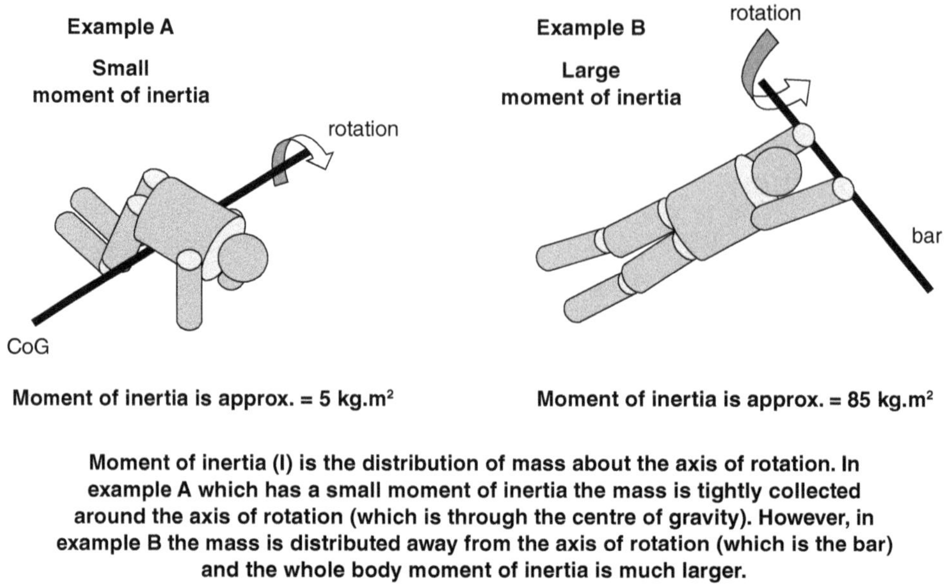


**Figure C7.1** Moment of inertia in sport

of inertia of a body is related to a specific axis of rotation and there will be different moment of inertia values for each axis that the body is rotating about. For example, there may be a moment of inertia of the whole body about a longitudinal axis or about an anterior–posterior axis. Also, moment of inertia can be expressed for individual parts or individual segments of a body (e.g. the upper leg can have a moment of inertia about the hip joint axis of rotation or the lower leg a moment of inertia about the knee joint axis of rotation).

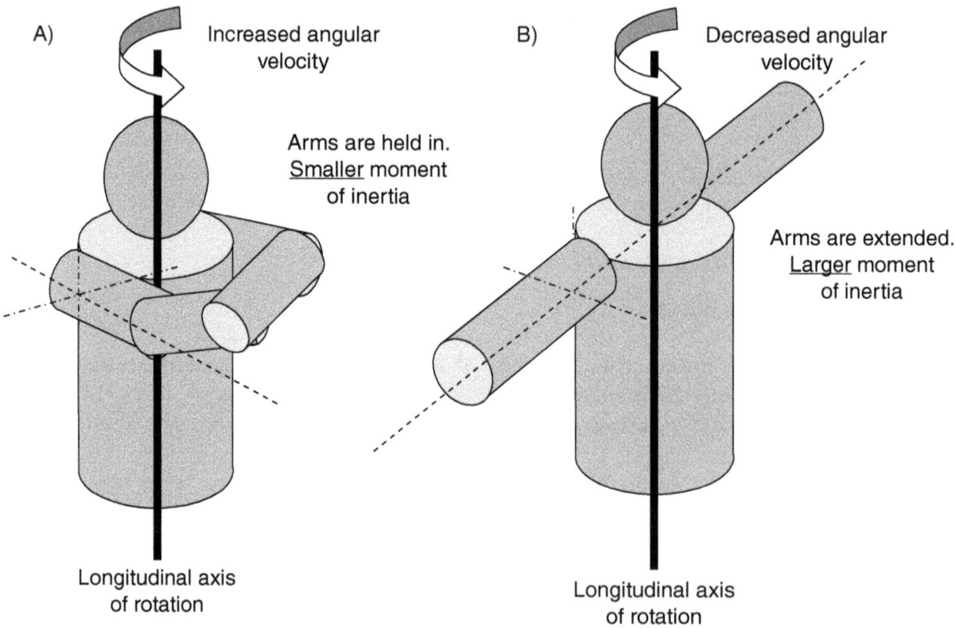


**Figure C7.2** Changes in moment of inertia

In *Figure C7.1* and *Figure C7.2*, it can be seen that when the mass is distributed much closer to the axis of rotation the moment of inertia is reduced. In *Figure* C7.2 (Example A), the arms are held close in tightly around the body. This has the effect of distributing the mass (i.e. the mass of the arms) much closer to the axis of rotation (the longitudinal axis). This reduces the moment of inertia of the body about this axis of rotation. In the case of Example B, the arms are held outwards and the moment of inertia is increased.

The angular analogue of Newton's first law states that the **angular momentum** of a body will remain **constant** unless the body is acted upon by an **external torque.** At this point it is particularly important to stress that angular momentum is related to a particular axis of rotation. If the body that is being considered is made up of several parts, then the total angular momentum is the sum of all the individual momenta of each body part (that is acting about the same axis of rotation).

In angular terms the **angular momentum** can also be expressed by multiplying the **square of the distance** of the object from the axis of rotation ($\mathbf{r^2}$) by the **mass** of the object (**m**) and its **angular velocity** ($\omega$).

$$L = mr^2 \times \omega$$

Since $\mathbf{mr^2}$ = the **moment of inertia** of an object then the **angular momentum** is the **moment of inertia (I)** multiplied by the **angular velocity** of the object (refer to the equation for angular momentum ($L = I\omega$) shown previously). This angular momentum will also occur about a particular axis of rotation.

Imagine a diver taking off from the springboard in an attempt to perform a double somersault before he/she enters the water. In order to perform this effectively, the diver will create angular momentum while still in contact with the springboard. For example, he/she will have applied a force to the board, which will create a torque or twisting moment on the body (because of its position in relation to the centre of gravity or in this case the axis of rotation). This will enable the diver to leave the board with angular rotation and hence angular momentum. *Figure C7.3* shows this in more detail.

Once the diver is airborne and in the absence of any external torque they will have a constant angular momentum (Newton's first law of angular motion) that will be maintained

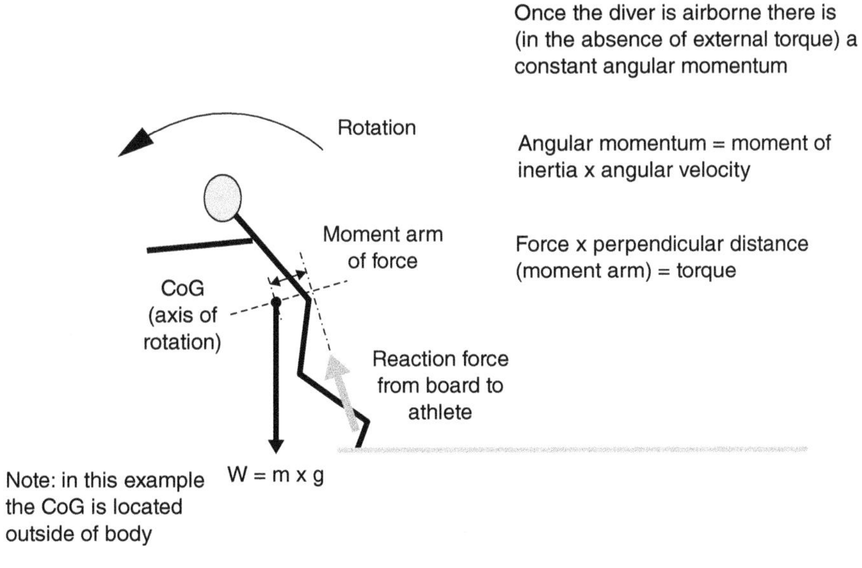


**Figure C7.3** Angular momentum in sport

throughout the dive. Hence, in order to perform the necessary somersaults, the diver will need to adjust his/her moment of inertia. The basic somersault takes place about the transverse axis (and in the sagittal plane) and the angular momentum at board contact will be created about this axis of rotation (although it could also be present about other axes of rotation). Since the amount of somersaulting angular momentum is constant and the diver is able to change his/her moment of inertia (by tucking up like a ball or by extending the limbs), the angular velocity must change in order to maintain this constant angular momentum principle described by Newton's first law. If the diver reduces his/her moment of inertia (tucks up like a ball) the angular velocity will increase, and he/she will be able to perform more somersaults in a short amount of time (i.e. higher angular velocity). Conversely, if the diver wanted to enter the water in a straight and controlled position (with limited rotation) he/she would straighten the body and increase the moment of inertia. This would slow down their rotation (i.e. reduce the angular velocity). During all this activity the angular momentum of the diver will remain constant (*Figure C7.4* helps to show this diagrammatically).

$L = I\,\omega$ (constant angular momentum in the absence of any external torques)

$L = \uparrow I$ and $\downarrow \omega$ (increased moment of inertia and decreased angular velocity)

$L = \downarrow I$ and $\uparrow \omega$ (decreased moment of inertia and increased angular velocity)

Newton's first law of motion states that in the absence of angular momentum and in the absence of any external torque the amount of angular momentum will remain constant; that is **zero angular momentum at take-off = zero angular momentum after take-off.** If this is true, how is it that a diver without any angular momentum at take-off can still initiate twists and turns in the air during the dive?

Consider the case of the diver in *Figure C7.3* and in particular what would happen if the athlete had zero angular momentum at take-off and yet the athlete was still required to twist and turn during the dive. This zero angular momentum would be where the reaction force (from the board) was applied directly through the centre of gravity of the athlete (i.e. not at a distance

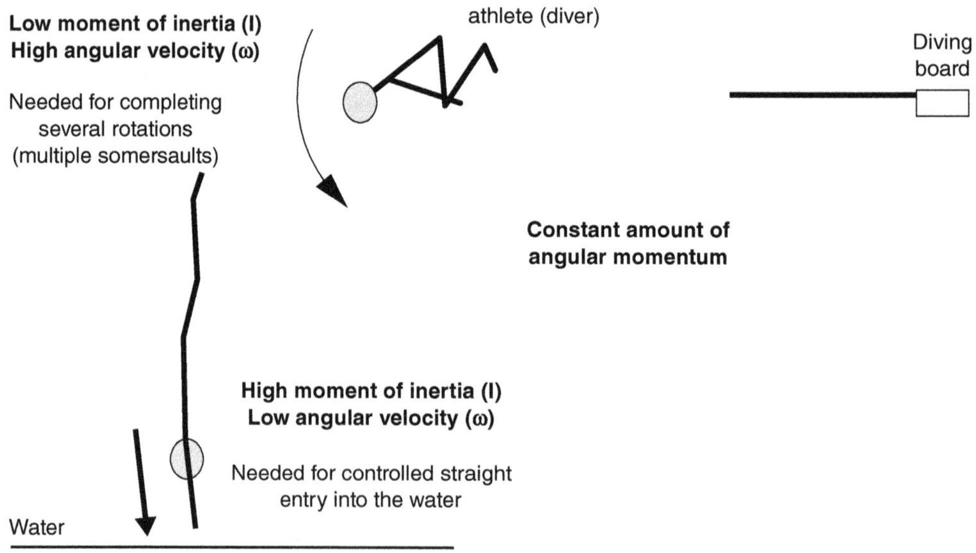


**Figure C7.4** Constant angular momentum in sport

(moment arm) from the axis of rotation). In this example the diver would leave the board with no angular velocity and hence no angular momentum (L = Iω). The athlete could initiate twists and turns by using asymmetrical or symmetrical movements of the arms or legs. For example, the diver could pull one of his/her arms across the chest in a rotational movement about the longitudinal axis. This would create a certain angular momentum for part of the body and an angular momentum in the opposite direction for the remainder of the body (i.e. summation of zero) about a certain axis of rotation (e.g. the longitudinal axis) and in a certain direction (i.e. clockwise or anti-clockwise). The arm would have a small moment of inertia and would be moving with a large angular velocity. Hence, the body would have some angular momentum about this axis (because as we have seen before the equation for angular momentum is L = Iω). However, since the diver had no initial angular momentum (at take-off) about the longitudinal axis, Newton's first law says that this directional angular momentum must be counteracted by angular momentum in an opposite direction (i.e. making the total angular momentum about this axis equal zero (the net angular momentum)). In diving the twist that is initiated by the asymmetrical or symmetrical arm movements is often counteracted by the legs in changing from a rotational twist to a pike and then back to a straight controlled extended position for entry into the water. In this application, there would be potentially more control of the rotation in the air and at entry to the water because the athlete controlled the amount of angular momentum he/she created by moving the arm. Hence, since this must be balanced by angular momentum in an opposite direction the athlete will experience opposite controlled angular momentum to balance out the equation (because they had zero angular momentum to begin with). Therefore, just before contact with the water (which would be an external force) the net angular momentum would be equal to zero, which is consistent with the angular analogue of Newton's first law of motion.

Perhaps an easy way to remember and apply this is to imagine a cat falling out of a tree backwards. Even if the cat had zero angular momentum at the time that it fell, it is still somehow able to land on its feet (by twisting and turning in the air). As the cat falls backwards it brings its front legs close to the head, which reduces the moment of inertia about the longitudinal axis. This causes a large rotation (angular velocity) of the upper body clockwise (approximately 180 degrees). At the same time the cat also extends the lower legs out away from the body, creating a large moment of inertia (about the longitudinal axis). This results in only a small amount of angular rotation of the lower body in an opposite anti-clockwise direction (approximately 5 degrees). This balances the angular momentum equation for both clockwise and anti-clockwise rotation about this specific axis. The upper body is now facing the ground and is ready for landing. However, in order to get the lower part of the body facing the ground the cat does exactly the same but opposite (it reduces the moment of inertia of the lower body and increases moment of inertia of upper body). This gives the lower body a large clockwise rotation and the upper body only a small amount of anti-clockwise rotation. The final result is that the cat lands on both its front and back feet, and it does so with a net angular momentum of zero. However, like the diver, it was clearly able to initiate a twist and turn in mid-air without any initial angular momentum. Although this is not directly applicable to human movement, it demonstrates that Newton's first law applies to angular motion.

As a reminder, **Newton's second law** for objects in linear motion was concerned with bodies that are subjected to unbalanced forces and it is stated as follows:

> When a force acts on an object, the change in motion (momentum) experienced by the object takes place in the direction of the force and this is proportional to the size of the force and inversely proportional to the mass of the object.

In the angular analogue of this law we replace the term **force** with **torque**, the term **mass** with **moment of inertia** and the term **momentum** with **angular momentum**. This law can now be reworded as follows:

When a torque acts on an object, the change in angular motion (angular momentum) experienced by the object takes place in the direction of the torque, and this is proportional to the size of the torque and inversely proportional to the moment of inertia of the object.

This can be expressed algebraically by the equation:

$$T = I \times \alpha$$

where:

T = net torque

I = moment of inertia

$\alpha$ = angular acceleration

Remembering that Newton's second law is concerned with unbalanced forces or torques we can see that if we apply a torque to an object, and the result of the **net torque** acting on the object is **not zero**, the object will have **angular acceleration** (Newton's first law: where we apply an external torque and cause a rate of change of angular velocity, i.e. an angular acceleration). The torque created a twisting or turning moment and the object moved with an angular velocity (i.e. it rotated). In the application of a net torque that is not zero the object would accelerate with angular acceleration (because it is rotating). The angular acceleration of the object will take place in the same direction as the applied torque. The amount of **angular acceleration** will be dependent on the amount of **applied torque** and the **moment of inertia** of the object. As we have seen, the human body is not a rigid body and as such it does not have a constant moment of inertia. The larger the moment of inertia of the object the less angular acceleration it will

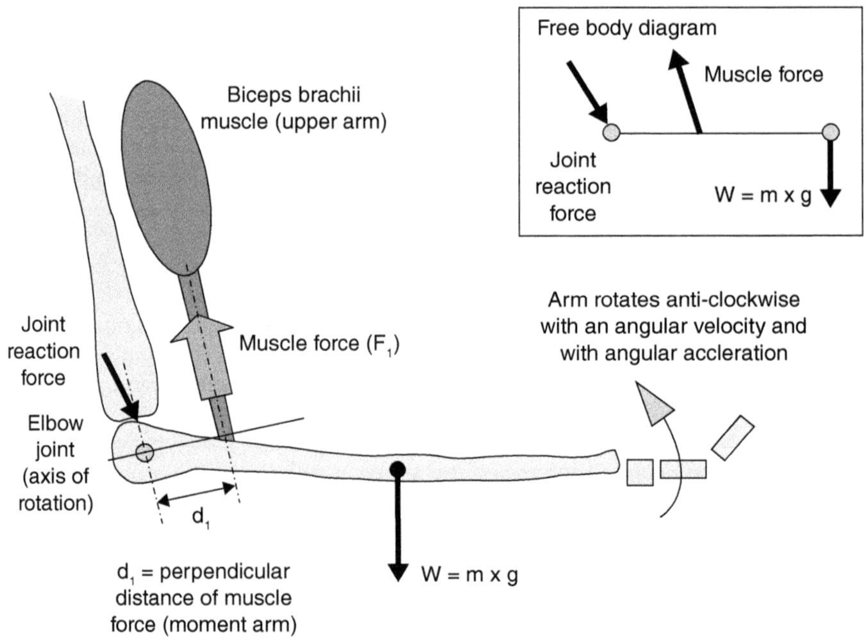


**Figure C7.5** Application of torque and moments

have for a given applied torque. Conversely, for a given applied torque (or net torque of greater than zero), the smaller the moment of inertia of the object the greater the angular acceleration it will have. *Figure C7.5* helps to illustrate this in more detail.

Considering *Figure C7.5*, we can see that the biceps brachii muscle in the upper arm exerts a force at a perpendicular distance to the axis of rotation (the elbow joint). This creates a torque (a twisting and turning moment at the elbow joint) in an anti-clockwise direction. This torque causes the elbow joint to flex and the lower arm to rotate anti-clockwise (the same direction as the applied torque). As the arm rotates it will have a certain angular acceleration and, hence, angular velocity. The arm was initially held stationary and the net torque applied to the arm in this case is not zero (i.e. the muscle will exert a torque on the arm). Hence, the arm will accelerate with an angular acceleration anti-clockwise. The rate of change of angular velocity of the arm (or the angular acceleration) will be dependent upon how much torque is applied to the arm. The amount of angular acceleration will also be dependent on the moment of inertia of the arm. As the arm is rotated in an anti-clockwise direction, the moment of inertia of the arm will resist the rotation, which is created or initiated by the torque. The greater the moment of inertia of the arm, the smaller will be the angular acceleration of the arm. Conversely, the smaller the moment of inertia, the larger will be its angular acceleration. Hence, for a given applied torque there would be different angular accelerations for different moment of inertia values for the arm (like in the case of different-sized and different-shaped arms). This has important implications when exercising using weights and using weight-training machines.

In this context it is important to point out that, as we have seen, **angular momentum** (which is the object's **moment of inertia** multiplied by its **angular velocity**) is a **vector** quantity (i.e. it has **magnitude** and **direction**). Usually, we are concerned with rotation about one axis of rotation and it is applicable, therefore, to refer to angular momentum about a single origin or a single axis of rotation. In addition, it is worth repeating that the total angular momentum of a body about any axis of rotation is made up by adding all the angular momenta of the various parts or segments of the body which are rotating about that axis. Within biomechanics this has important implications for explaining human movement and hence requires more complex analyses in the study of angular momenta about multiple axes of rotation.

As we have seen, a net torque (that is not zero) that acts on an object will cause an **angular acceleration** of the object in the **direction of the net torque.** The amount of net external torque will equal the rate of change of angular momentum (i.e. from the angular analogue of Newton's second law).

$$\text{Torque}_{\text{net}} = \frac{\text{Change in angular momentum}}{\text{Change in time}}$$

$$\text{Torque}_{\text{net}} = \text{Rate of change in angular momentum}$$

The change in angular momentum of an object can be determined by examining the initial and final angular momentum possessed by the object:

$$\text{Change in angular momentum} = \text{angular momentum}_{\text{final}} - \text{angular momentum}_{\text{initial}}$$

Thus, we can now include this in the equation for torque:

$$\text{Torque}_{\text{net}} = \frac{\text{angular momentum}_{\text{final}} - \text{angular momentum}_{\text{initial}}}{\text{Change in time}}$$

Mathematically this can be expressed as:

$$T_{net} = \frac{L_{final} - L_{initial}}{t_{final} - t_{initial}}$$

Rearranging this equation produces the following:

$$T_{net} \times (t_2 - t_1) = L_{final} - L_{initial}$$

This can now be expressed as the equation for angular impulse:

**$T_{(net)} \times t$ = Angular impulse (torque multiplied by time)**

**$L_{(f)} - L_{(i)}$ = Change in angular momentum (final – initial)**

**Angular impulse = Change in angular momentum**

This equation has important implications for the effective execution of rotational movements with human motion. For example, considering the diver in *Figure C7.3*, it is possible to see that the diver creates angular momentum by applying a force at a distance from an axis of rotation. Angular momentum is moment of inertia multiplied by angular velocity. As the diver creates torque he/she will create rotation (angular acceleration and, hence, angular velocity). The amount of rotation will depend on the torque created and the moment of inertia of the body. As we have seen, a net external torque that is not zero causes angular acceleration. Angular acceleration is the rate of change of angular velocity. The greater the angular acceleration of the body, the greater the rate of change of angular velocity (i.e. the more rotations we can create in a shorter space of time). As the diver creates torque he/she applies a force on the board for a certain length of time. In angular terms, since the reaction force is applied at a distance from the axis of rotation (the moment arm), a torque will be created. Now, as we have seen, the angular acceleration possessed by the object (the rate of change of angular velocity) is related to the torque and the moment of inertia. If the athlete has a large moment of inertia as he/she creates the torque, the rotational component will be small. As the rotational component of the action is small (i.e. reduced angular velocity), the athlete will be able to create a torque for a longer period of time. This is achieved because they will have less rotation effect and would be able to stay in contact with the board for longer before the rotation would cause them to have to leave it into the dive. This application of torque for a longer period of time will create a greater change in angular momentum (i.e. they will have more angular momentum). Consequently, the more angular momentum they have when they leave the take-off board into the dive, the more they are able to rotate in the air. For example, because they have a large angular momentum they can reduce their moment of inertia and perform more rotational somersaults in a given time (i.e. they will be rotating faster because once they leave the board and in the absence of external torque the amount of angular momentum they have will be constant).

This ability to create a torque in a controlled manner and to be able to apply this torque for a long period of time results in a **greater change** in **angular momentum**. If they had zero angular momentum before they started to apply the torque it follows that they will have more angular momentum the longer time that they can apply the torque (**angular impulse = change in angular momentum**). This has very important implications within human movement and is applicable in many examples within sport where the athletes use angular momentum and rotational movements to generate both linear and angular velocities. For example, the rotational component used by modern javelin throwers, the rotational running across the circle technique used by discus throwers, the golfer using rotational movements of hips and shoulders to generate torque that is transferred to the club to accelerate it quicker, and the tennis player serving with rotational movements about the longitudinal axis in order to impart large amounts of spin and velocity to the ball. All these examples and more utilise the angular analogue of Newton's second law of motion.

In linear terms, **Newton's third law of motion** states:

> Whenever an object exerts a force on another there will be an equal and opposite force exerted by the second object on the first.

In the angular analogue of Newton's third law of motion, the term **force** is replaced with the term **torque** and can be written as:

> Whenever an object exerts a torque on another there will be an equal and opposite torque exerted by the second object on the first.

Torque, as we know, is a turning or twisting moment that causes an angular acceleration of an object. In the context of this third law of motion, it is important to remember (as with the linear version) that it is the forces or torques that are equal and opposite and not the net effect of the forces or torques. The equal and opposite torques will act on each body differently (because the two bodies are different). As with the linear analogue of this law of motion, it is important to remember the consideration of **external** and **internal** force or torque. If the body is in equilibrium (when the algebraic sum of all the torques acting is zero) under the action of external torque, both the external and internal torque systems are **separately** in equilibrium. **In considering the net effect of external torques or forces acting on a body we would summate only the external torques that are acting on that body.** Hence, if body A exerts an external force or torque on body B, body B will exert an external reaction force or torque on body A. However, in considering the net effect on body B we would only consider the external forces or torques acting on body B (which is the external force or torque provided from body A – this is of course ignoring any other external forces such as gravity).

In the case of the arm in *Figure C7.5,* which is undergoing a flexion movement (an anti-clockwise rotation of the lower arm (the forearm)), the torque created by the biceps brachii muscle exerts a torque on the lower arm (via the elbow). This torque causes the lower arm to rotate in an anti-clockwise direction. The reaction torque to this will be a torque in the opposite direction created on the upper arm. As the torque is applied to the arm (to accelerate it anti-clockwise) there will be an equal and opposite torque acting on the upper arm. This is why when you conduct an arm curl during weightlifting you can feel the stress/strain in the upper arm: because the torques are equal and opposite and act independently on two different bodies; that is, the torque of the biceps acts on the lower arm (anti-clockwise) and the reaction torque acts on the upper arm and the net result is zero (equal and opposite torques). In this scenario, movement takes place because the torque acting on the lower arm is considered as an external torque acting on that body (Newton's second law). As we have seen this movement is dependent upon the moment of inertia of the object. The torque created is dependent upon the force applied and the moment arm at which the force is applied (the perpendicular distance from the axis of rotation). The angular analogue of Newton's third law has important implications with human movement and in particular for consideration of injury. *Figure C7.6* shows some further example of the action–reaction torque within sport and exercise.

Considering *Figure C7.6*, it is possible to see three examples of the action–reaction torque situation: the tennis forehand, the soccer kick and the golf swing. In the tennis forehand example, the torque provided by the muscles to the forearm will create an equal and opposite torque acting on the upper arm. This has important implications for the development of medial or lateral epicondylitis (the "tennis elbow" injury). In the soccer kick example, the torque is created on the upper leg as it swings through in an anti-clockwise direction (created on the upper leg by the hip flexors). This causes a reaction torque at the pelvis. This has important implications for common hamstring injury problems experienced in soccer. Finally, in the golf swing, the torque

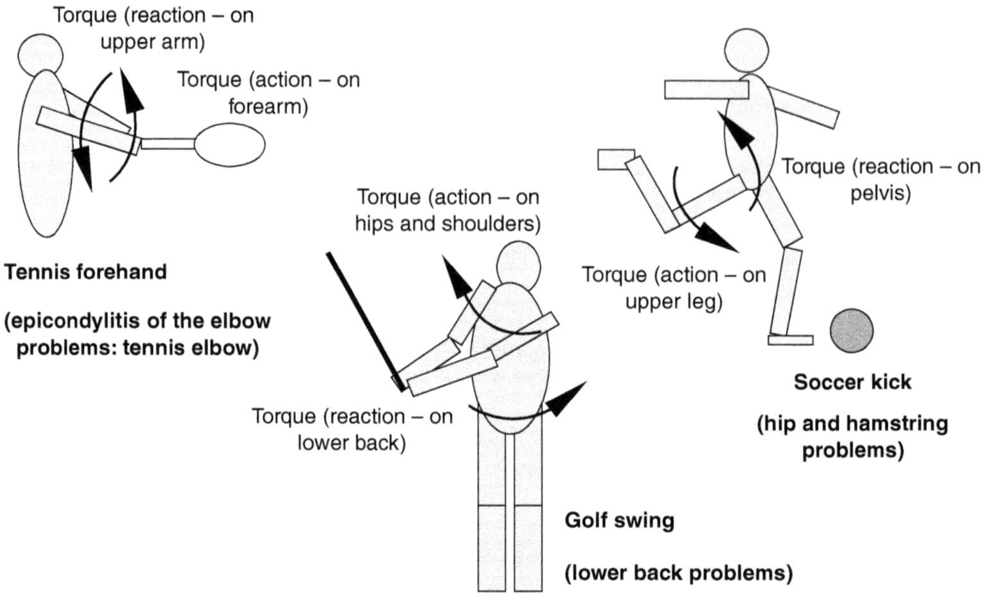


**Figure C7.6** Newton's third law of angular motion in sport

that is created by rotating the hips and shoulders through the backswing and downswing (in both clockwise and anti-clockwise directions) causes a reaction torque on the lower back, which has potential for the development of lower back pain.

All of the angular analogues of Newton's laws of motion are critical in the effective execution of human movement and performance. In addition, they serve as a strong basis for more complex analyses of multi-axial and rotational movements that are common to many sporting actions.

*Key notes*

| | |
|---|---|
| **Newton's first law of motion (angular analogue)** | The angular momentum of a body remains constant unless a net external torque is exerted on the body. The angular momentum (L) of a body can be determined by the moment of inertia (I) multiplied by its angular velocity (ω). The moment of inertia of a body is described as the reluctance of an object to start or stop rotating or change its state of motion. Moment of inertia of a body is calculated from the distribution of mass (m) about an axis of rotation (r). Moment of inertia = mass (m) × radius$^2$ ($r^2$). The further away from the axis of rotation a mass is distributed, the larger will be the moment of inertia. |
| **Application** | The ice skater, in a jump, holds the arms close to the body during a pirouette move (rotation about the longitudinal axis), which reduces the moment of inertia of the body about this axis. This offers less resistance to a change in its state of rotational motion (about this axis). If the skater had created an amount of angular momentum before he/she left the ice, this angular momentum (in the absence of an external torque or force) would remain constant. Since angular momentum = moment of inertia × angular velocity, a reduced moment of inertia would result in an increased angular velocity (i.e. more rotations in a given time). |

| | |
|---|---|
| **Newton's second law of motion (angular analogue)** | When a torque acts on an object, the change in angular motion (angular momentum) experienced by the object takes place in the direction of the torque and this is proportional to the size of the torque and inversely proportional to the moment of inertia of the object. Algebraically this is expressed as $T = I \times \alpha$ (where T = torque, I = moment of inertia and $\alpha$ = angular acceleration). |
| **Application** | In the case of arm curling with weights, the biceps brachii muscle applies a torque to the lower arm (the forearm). Depending on the moment of inertia of the arm and the weights, this torque may cause an acceleration of the arm (anti-clockwise). The amount of this acceleration is dependent on the moment of inertia offered in resistance to this movement. The smaller the moment of inertia, the greater will be the acceleration for a given applied net torque. |
| **Newton's third law of motion (angular analogue)** | Whenever an object exerts a torque on another there will be an equal and opposite torque exerted by the second object on the first. |
| **Application** | The torque created on the upper leg by the hip flexors during the kicking action in soccer will create an equal and opposite torque that is exerted on the pelvis. This has important implications for hamstring injury. The torque that is created on the body by the shoulders and hips during the backswing in golf will create a reaction torque acting in the lower back (lower back injury implications). |

## FURTHER READING

The following resources provide additional reading on the concept of Newton's laws (angular motion) in sport.

1 Salo, A., Grimshaw P. N., & Marar, L. (1997). 3-D biomechanical analysis of sprint hurdles at different competitive levels. *Medicine and Science in Sports and Exercise*, *29*(2), 231–237.
2 Marshall, R. N., & Elliot, B. C. (2000). Long-axis rotation: The missing link in proximal-to-distal segmental sequencing. *Journal of Sports Sciences*, *18*(4), 247–254.
3 Schache, A. G., Blanch, P., Rath, D., Wrigley, T., & Bennell, K. (2002). Three-dimensional angular kinematics of the lumbar spine and pelvis during running. *Human Movement Science*, *21*(2), 273–293.

# SECTION D

# Muscle and joint kinematics and kinetics

# MUSCLE AND JOINT FORCES

## Statics and dynamics

*Paul Grimshaw*

**Static calculations** of muscle and joint forces are useful in that they form the basis of forward and inverse dynamics (*Figure D1.1*), which are currently used in many modelling and computer simulation programs. Understanding the basic calculations of the forces that act on a joint during various forms of human movement is essential if we are to have any detailed understanding of injury prevention mechanisms and/or performance characteristics. Although the calculations in this text are based on a two-dimensional approach and are primarily presented in the sagittal plane, they are nevertheless valuable for the student of biomechanics in providing an understanding for more complex three-dimensional problems. In some applications they are valid, as in the case of estimating the loads on the joint during squatting in weightlifting or, alternatively, estimating the forces on the elbow during the holding of the arm in the 90-degree position (like when doing an arm curl in weight lifting). The methods used are from standard mechanics, the application of trigonometry and the use of free body diagrams and they provide the reader with skills that can be used to model the loads and forces that could be responsible for injury. Furthermore, by knowing the muscle forces involved in an activity it may also be possible to assess the effects of exercise programmes.

The movement that is generated by a muscle force is determined by several factors, namely the degrees of freedom of the joint (i.e. a hinge joint, such as the elbow, or a ball and socket joint, such as the hip), the geometry of the articulating surfaces and the passive restraints to the

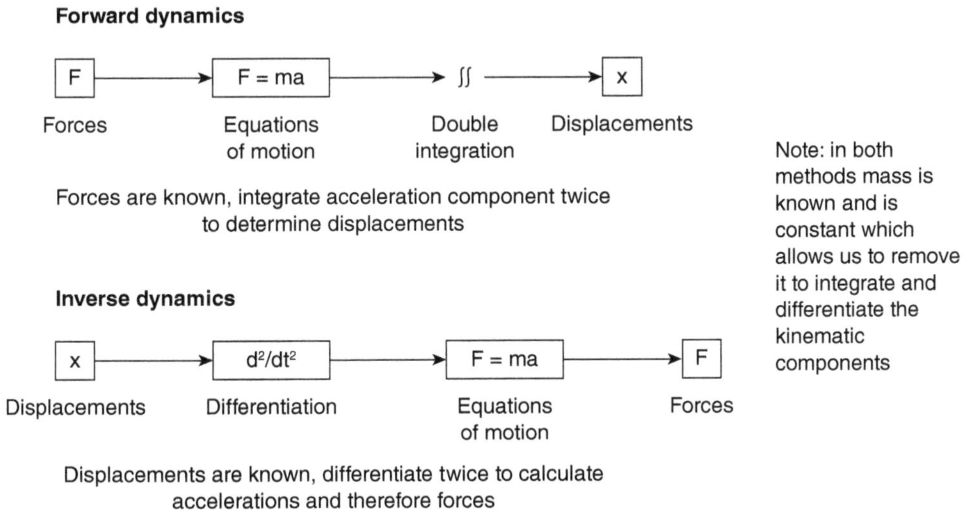


**Figure D1.1** Forward and inverse dynamics

movement, such as those provided by the ligaments, capsule and joint soft-tissue structures. In static calculations of joint motion and muscle forces we assume that the forces are in equilibrium (i.e. are balanced and with no movement). Hence, we can use the equation derived from Newton's second law, which states that the sum (net affect) of the forces acting on the system is zero (i.e. equilibrium):

$$\Sigma F = 0$$

Next, we use the method of free body diagrams (see the Free Body Diagrams section elsewhere in this text) in order to replace the muscle and joint reaction forces and include the weight of bones and objects. We measure muscle insertion points and generally (in this two-dimensional example) ignore ligament and other soft-tissue forces. Finally, we assume that the joints are frictionless. An example of an applied problem and a free body diagram for the lifting/supporting of a weight (ball) in the hand during an arm curl is shown in *Figure D1.2* and *Figure D1.3*.

For the example given in *Figure D1.2*, we have the following information: solve the problem to find the muscle force in the biceps brachii muscle (F1) for the static (holding the arm still in this position with no movement) condition, where the sum of the forces would be equal to zero. In addition, calculate the joint reaction force that would be present at the elbow joint. **Assume that, in this case, the force in the brachioradialis muscle is zero.**

- Arm flexed to 90 degrees with the forearm and hand in the horizontal position
- Mass of weight (ball) in hand = 2 kg
- Mass of forearm and hand (derived from anthropometric tables) = 1.5 kg
- $d_{1H}$ = distance (horizontal) from elbow joint to biceps brachii force (F1) = 3 cm

  *Note: the* $d_{1H}$ *distance is not the perpendicular distance to the muscle force*
- $d_{2H}$ = distance (horizontal) from elbow joint to brachioradialis force (F2) = 20 cm

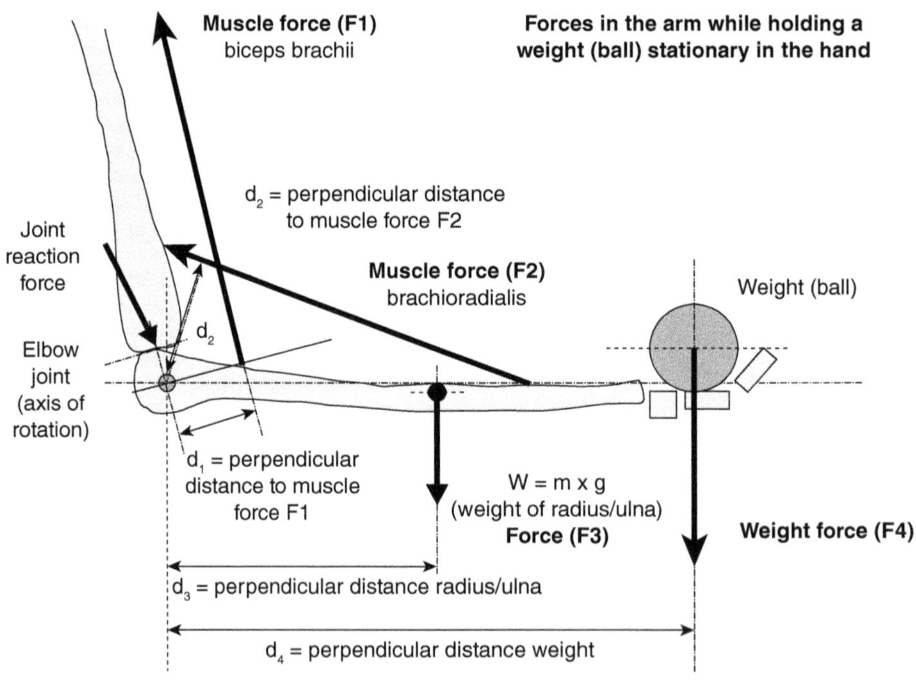


**Figure D1.2** Forces in the arm while holding a weight (ball) in the hand (not drawn to scale)

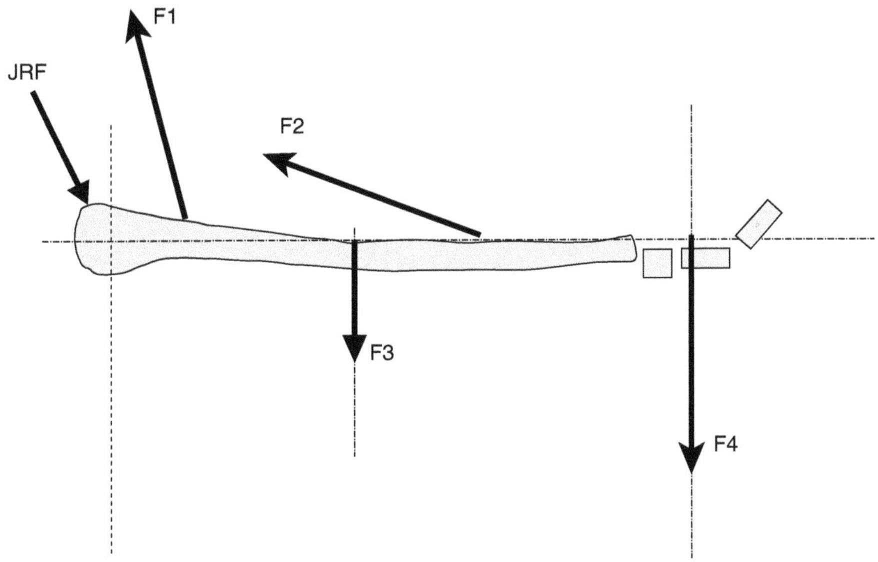


**Figure D1.3** Free body diagram (not drawn to scale)

*Note: the $d_{2H}$ distance is not the perpendicular distance to the muscle force, but this is not needed for this example. It is shown for any future calculations when brachioradialis force is not zero.*

- $d_3$ = distance (horizontal) from elbow joint to centre of gravity of radius/ulna = 15 cm

  *Note: In this case, the $d_3$ distance is the perpendicular distance*

- $d_4$ = distance (horizontal) from elbow joint to weight (ball) in hand = 35 cm

  *Note: In this example, the $d_4$ distance is the perpendicular distance*

- The angle of F1 muscle force (biceps brachii) to the arm is 75 degrees
- The angle of F2 muscle force (brachioradialis) to the arm is 20 degrees

  *Note: In this example, the arm is horizontal*

As stated, the distances given to the muscle forces in these examples are horizontal distances and hence are **not** always the perpendicular distances. Therefore, the perpendicular distances may need to be calculated before the solution can be attempted.

## TASK

1. Write the equations of static equilibrium and show that there are three unknowns.
2. By assuming the force in the brachioradialis muscle is zero:

   a Calculate the force in the biceps brachii muscle and the reaction force (both direction and magnitude) acting at the elbow joint.

   b If the arm is extended a further 30 degrees (i.e. below the horizontal) the angle of the biceps brachii muscle with the forearm is reduced to 45 degrees. What happens to the magnitude of the joint reaction force at the elbow (**homework problem**)?

## Solution

In this example we are dealing with a two-dimensional problem in a static condition (i.e. no movement) and it is important to show both the horizontal (Fx) and vertical (Fy) components of force. Hence, the equations of static equilibrium that we are concerned with are as follows.

$$\Sigma \text{Fx} = 0$$

$$\Sigma \text{Fy} = 0$$

$$\Sigma \text{M} = 0$$

This indicates that the sum of the forces in the x and in the y directions are equal to zero and the sum of the moments (clockwise and anti-clockwise (sometimes termed counter-clockwise)) is also equal to zero. Next, draw the force and moment diagram from *Figure D1.2* and *Figure D1.3* that will help us solve this problem (this is shown in *Figure D1.4*).

Now, determine the force in the biceps brachii muscle that is needed to hold the arm still. Using the equations below, we get the following:

$$\Sigma \text{Fx} = 0$$

$$\Sigma \text{Fy} = 0$$

$$\Sigma \text{M} = 0$$

Sum of the clockwise and anti-clockwise moments = 0

$$((1.5 \times 9.81 \times 0.15) + (2.0 \times 9.81 \times 0.35)) + (-\text{F1} \times \text{Sin } 75° \times 0.03) = 0$$

$$((2.21) + (6.87)) + (-\text{F1} \times 0.029) = 0$$

$$9.08 = \text{F1} \times 0.029$$

$$9.08/0.029 = \text{F1}$$

$$\mathbf{313.10\ N = F1}$$

Note: The Sine of 75° is used from trigonometry to determine the perpendicular distance of the muscle force from the elbow joint (see *Figure D1.4*).

In order to determine the effect on the joint reaction force acting on the elbow, we now need to resolve this muscle force in the biceps brachii into its horizontal (Fx) and vertical (Fy) components. *Figure D1.5* helps to illustrate how this is achieved in more detail.

Resolve forces horizontally (Fx) and vertically (Fy) using the following:

$$\Sigma \text{Fx} = 0$$

$$\Sigma \text{Fy} = 0$$

Using the following notations, we can **resolve the forces horizontally:**

Fmx = horizontal component of the muscle force

Fjx = horizontal component of the joint reaction force

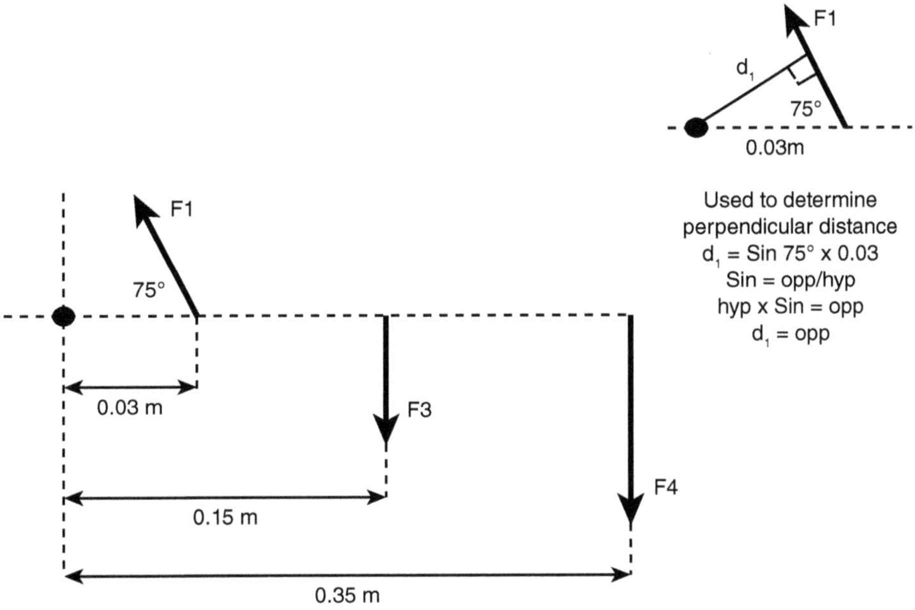


**Figure D1.4** Forces and their perpendicular distances of application (not drawn to scale)

Note that the other two weight forces do not have a horizontal component.

Therefore:

$$\sum Fx = Fmx + Fjx = 0$$

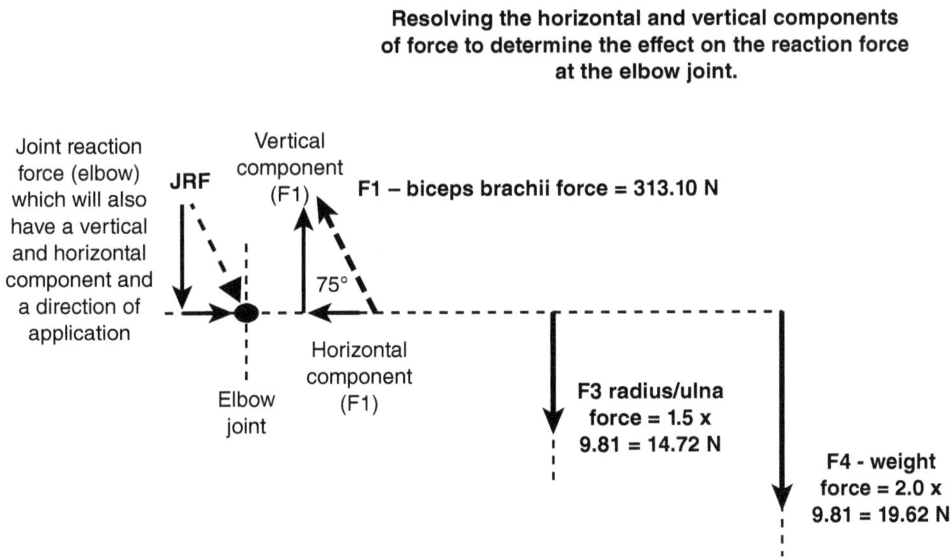


**Figure D1.5** Resolving the components of force (not drawn to scale)

The horizontal component of the muscle force F1 can be calculated using trigonometry:

$$F \text{ Cos } \theta = \text{Fmx}$$

$$313.10 \times \text{Cos } 75° = \text{Fmx}$$

$$313.10 \times 0.259 = \text{Fmx}$$

$$\mathbf{81.09\ N = Fmx}$$

Substitute into the equation above for the sum of the forces in the x direction

$$\Sigma \text{Fx} = 81.09 + \text{Fjx} = 0$$

Therefore:

$$\mathbf{Fjx} = -81.09\mathbf{N}$$

This indicates that the elbow joint reaction force is equal and opposite in direction to the horizontal component of the biceps brachii force. This can also be seen in *Figure D1.5*, where the joint reaction force is directed to the right whereas the biceps brachii muscle horizontal component of force is directed to the left.

Using the following notations, we can now **resolve the forces vertically:**

Fmy = vertical component of the biceps brachii muscle force

Fjy = vertical component of the joint reaction force

$Fw_1y$ = vertical component of the weight 1 force (the radius/ulna)

$Fw_2y$ = vertical component of the weight 2 force (the ball)

Again, using the equation from above for the vertical direction:

$$\Sigma \text{Fy} = 0$$

We can express this as follows:

$$\Sigma \text{Fy} = \text{Fmy} + \text{Fjy} + \text{Fw}_1\text{y} + \text{Fw}_2\text{y} = 0$$

Note that it is important to express the direction of the forces that we know, hence we know that the vertical component of both the weight forces is downwards (negative) and the vertical component of the muscle force is upwards (positive). At this point we do not know the direction of the vertical component of the joint force:

$$\Sigma \text{Fy} = \text{Fmy} + \text{Fjy} - \text{Fw}_1\text{y} - \text{Fw}_2\text{y} = 0$$

The vertical component of the biceps brachii muscle force is derived using trigonometry:

$$\text{Fmy} = F \text{ Sin } \theta$$

$$\text{Fmy} = 313.10 \times \text{Sin } 75°$$

$$\text{Fmy} = 313.10 \times 0.966$$

$$\mathbf{Fmy = 302.45\ N}$$

Given the nature of the biceps brachii force (see *Figure D1.5* for direction), this would be a vertically upwards directed component of force. Substitute all the known values into the equation for the vertical component of force shown above.

$$\Sigma Fy = 302.45 + Fjy + (-1.5 \times 9.81) + (-2.0 \times 9.81) = 0$$

$$\Sigma Fy = 302.45 + Fjy - 14.72 - 19.62 = 0$$

$$-Fjy = 302.45 - 14.72 - 19.62$$

$$-Fjy = 268.11$$

$$\mathbf{Fjy = -268.11\,N}$$

This negative value indicates to us that the vertical component of the joint reaction force at the elbow is going in a downwards direction. This can be confirmed by looking at the direction of the joint reaction force shown in *Figure D1.5*.

Now these horizontal and vertical components of the elbow joint reaction force need to be resolved into one force and an angle (direction) of application. Using trigonometry and the following values for the horizontal and vertical components of the joint reaction force (Fjx and Fjy), we can now calculate the elbow joint reaction force (Fj).

$$\mathbf{Fjx = -81.09\,N}$$

$$\mathbf{Fjy = -268.11\,N}$$

$$Fj = \sqrt{Fjy^2 + Fjx^2}$$

$$Fj = \sqrt{(-268.11^2 + -81.09^2)}$$

$$Fj = \sqrt{(71882.9 + 6575.6)}$$

$$Fj = \sqrt{78458.5}$$

$$\mathbf{Fj = 280.10\,N}$$

Now, determine the angle of orientation of the joint reaction force.

$$\text{Tan}\,\theta = \frac{Fjy}{Fjx}$$

$$\text{Tan}\,\theta = \frac{-268.11}{-81.09}$$

$$\text{Tan}\,\theta = 3.31$$

$$\boldsymbol{\theta = 73.19^\circ}$$

These values represent the angle of application (73.19° expressed to the left horizontal) and magnitude (280.10 N) for the joint reaction force that is applied to the elbow in this example. For clarity, this is depicted in more detail in *Figure D1.6*. Finally, *Figure D1.7* shows the diagram that will be useful in solving the second part of this problem (i.e. for homework). Please note that the angle for the calculation of the perpendicular distance for the muscle force (biceps brachii) and

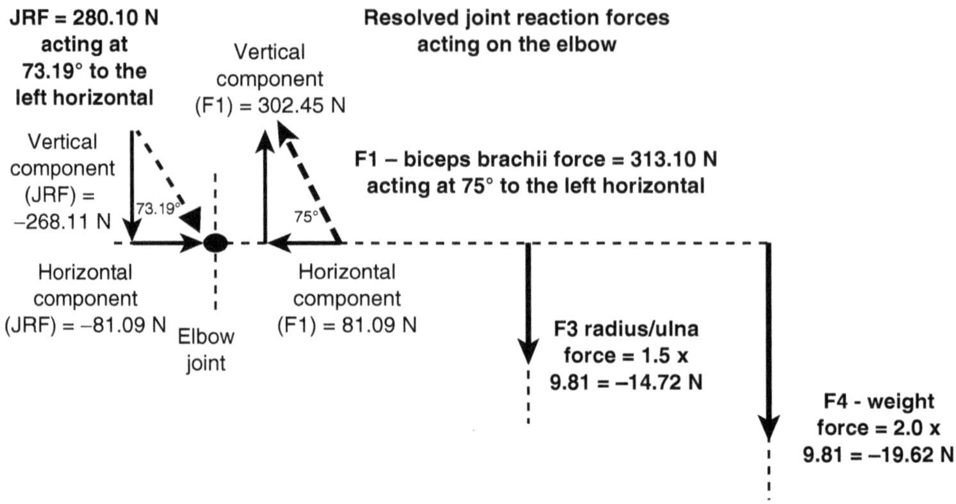


**Figure D1.6** Muscle and joint reaction forces (not drawn to scale)

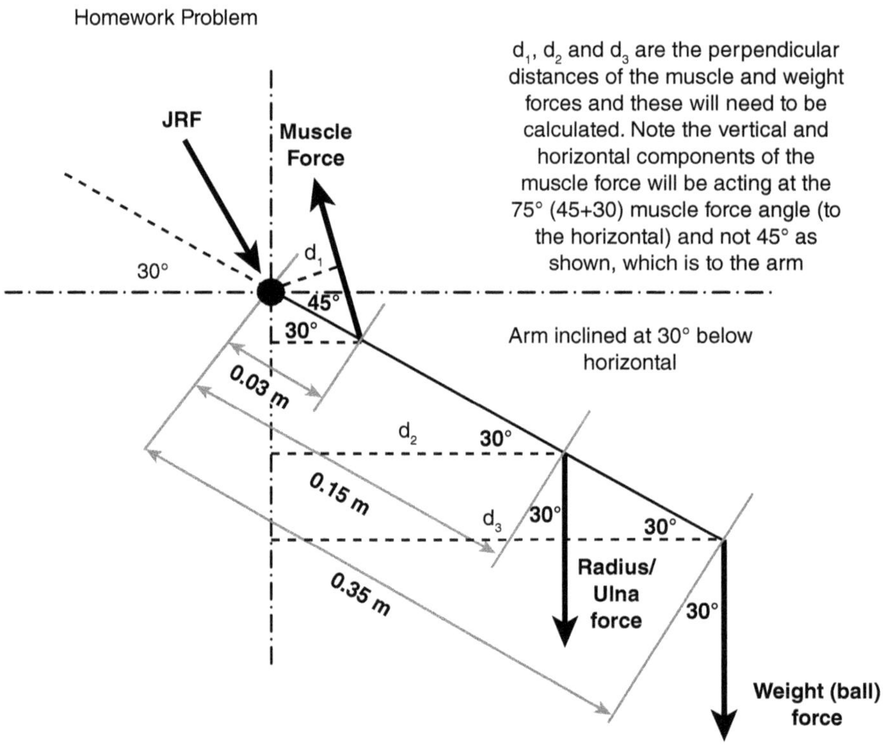


**Figure D1.7** Applied problem (homework) (not drawn to scale)

the weight forces will be different to that seen in the previous example, because the arm is inclined at a different position. Also, while the working solution is not given, the following homework solution guide provides details of the approximate values that you should obtain:

- Muscle force in biceps brachii around 370 N
- Joint reaction force (elbow) around 340 N at an angle of 73 degrees to the left horizontal

  *Note: both values are increased when compared with the values calculated when the arm was held horizontal.*

Despite the limitations of these static calculations, which include their 2D application and the ignoring of soft-tissue passive structures and friction, they have still shown us the forces that could be responsible for injury mechanisms. The vertical force (**–268.11 N**) acting on the elbow joint would essentially be a compressive force trying to push the joint surfaces together, whereas the horizontal force (**–81.09 N**) would be a type of shear force that is trying to push the joint apart. This type of analysis is essential (both in its fundamental knowledge and its application in modelling) if the biomechanics specialist is to be able to help define and prevent injury mechanisms in sport and human movement.

**Dynamics calculations** are applicable in biomechanics since most forms of human movement are not static. The acceleration approach is used to solve kinetic problems and determine the effect of dynamics on muscle and joint forces. Angular acceleration is considered to be a result of a torque, which is defined as a force acting at a perpendicular distance from an axis of rotation (like in the case of the biceps brachii muscle force acting about the elbow joint). In static applications, the clockwise moments or torques are balanced by the anti-clockwise moments resulting in an angular acceleration of zero (i.e. no movement or equilibrium). In dynamic applications, the torques are not equal to zero and an angular acceleration is produced. The resultant torque has the same causal relationship to angular acceleration that resultant force has to linear acceleration. The resistance to changes in angular velocity is termed the moment of inertia and is an important consideration in dynamics calculations.

In *Figure D1.8*, dynamics calculations are applied to the same application (but not the same values) as presented in *Figure D1.2* for an arm holding a mass in the hand and then rotating the mass anti-clockwise (flexing the elbow) with an angular acceleration. In this example, a 75 kg athlete is holding a 10 kg mass in the hand and he/she is accelerating this mass anti-clockwise at an angular acceleration of 80 rad/s$^2$.

## TASK

Determine the muscle and joint forces that are involved in this movement. Again, in this example, we are making the same assumptions as in the static analysis with regards to frictionless joints and soft-tissue structures and we are also considering the brachioradialis force to be zero. The free body diagram of this example is shown in *Figure D1.9*.

In dynamics we utilise the acceleration approach to solve problems. Also, remember from *Figure D1.1* that inverse dynamics is the computation of muscle and joint forces from kinematics (displacement, velocity and acceleration). Hence, the first and second conditions of equilibrium, which we used for the statics calculations, are now represented as the following dynamic equations.

$$\Sigma Fx = ma_x$$

$$\Sigma Fy = ma_y$$

$$\Sigma M = I\alpha$$

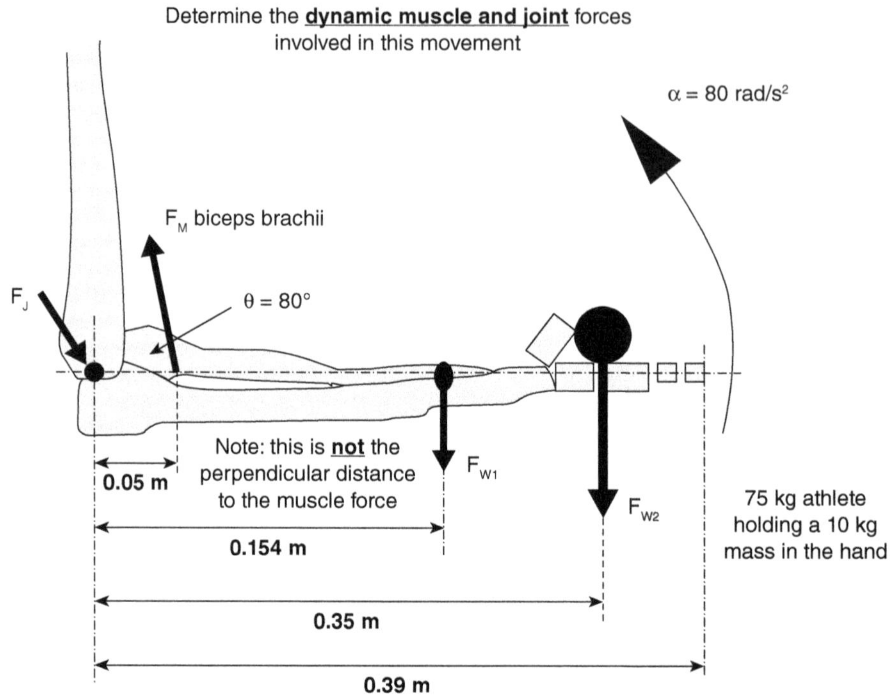


**Figure D1.8** Dynamics calculation problem (not drawn to scale)

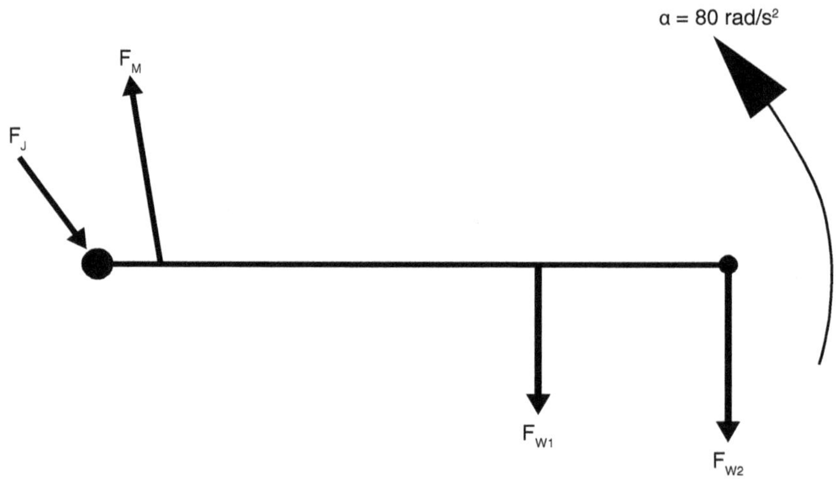


**Figure D1.9** Free body diagram of dynamics calculation problem (not drawn to scale)

where:

$\Sigma Fx$ = sum of forces in the x (horizontal) direction

$\Sigma Fy$ = sum of forces in the y (vertical) direction

$\Sigma M$ = sum of the moments (clockwise and anti-clockwise)

m = mass of the object or segment (kg)

$a_x$ = linear acceleration of the object/segment in the x (horizontal) direction ($m/s^2$)

$a_y$ = linear acceleration of the object/segment in the y (vertical) direction ($m/s^2$)

I = moment of inertia of the object/segment ($kg.m^2$)

$\alpha$ = angular acceleration of the object/segment ($rad/s^2$)

As we know, the moment of inertia of an object is defined as the resistance of the object to start, stop or change its state of rotation. For bodies with a concentrated mass (regular shapes), it is determined by the following:

$$I = \sum mr^2$$

where:

I = moment of inertia ($kg.m^2$)

m = mass of the object (kg)

r = radius from the axis of rotation (m)

However, as we also know, the human body is not a regularly shaped body; hence, in order to calculate the human body moment of inertia values, we use the following formula:

$$I_A = I_{CoG} + md^2$$

where:

$I_A$ = moment of inertia of the object about the axis A ($kg.m^2$)

$I_{CoG}$ = centre of gravity constant to account for different limb moment of inertia values ($kg.m^2$)

m = mass of segment (kg)

d = distance from axis of rotation (m)

Often in human movement, segments are continually rotating about different axes of rotation, such that the body or segment may rotate about either its proximal (i.e. the end nearest the point of attachment) or distal (i.e. the end furthest from the point of attachment) ends. For example, in the case of doing an arm curl with "weights", the rotation of the lower arm (forearm and hand) would be about the elbow (i.e. the proximal end of the lower arm). However, in the case of a subject doing a cartwheel, the rotation of the body/segments (such as the forearm) would be about the wrist (i.e. the distal end of the forearm, but not including the hand in this example).

Therefore, in order to simplify calculations from constantly having to use multiple values from the formula given above (essentially the parallel axis theory), we introduce what is termed the **radii of gyration** measurement. This then allows us to modify the equation and use one measure for the calculation of the moment of inertia of specific segments about their axes of rotation. This new equation is as follows:

$$I_A = mk^2$$

where:

$I_A$ = moment of inertia of the object about the axis A ($kg.m^2$)

m = mass of segment (kg)

k = radius of gyration, specific to which axis the segment is rotating about (m)

*Table D1.1* identifies the position of the respective radii of gyration for when a body is either rotating about a proximal or distal axis for each segment.

The moment of inertia formula can now be revised to include the centre of gravity constant to account for the irregular and changing shape of the body segments.

$$I_A = I_{CoG} + mk^2$$

where:

$I_A$ = moment of inertia of the body or segment about an axis through a point A ($kg.m^2$)

$I_{CoG}$ = moment of inertia about a parallel axis through the centre of gravity (accounts for irregular shapes) ($kg.m^2$)

m = mass (kg)

k = radius of gyration (m) – determined from anthropometric tables (*see the example in Table D1.1*)

We now have a system that we can use to determine the moment of inertia of a segment when it is rotating about a particular axis, such as in the case of the elbow, shoulder or knee joint.

**Table D1.1** Radii of gyration as percentages of segment length

| *Segment* | *From the proximal end %* | *From the distal end %* |
|---|---|---|
| Head, neck and trunk | 83.0 | 60.7 |
| Arm (upper) | 54.2 | 64.5 |
| Forearm | 52.6 | 64.7 |
| Hand | 58.7 | 57.7 |
| Upper limb | 64.5 | 59.6 |
| Forearm and hand | 82.7 | 56.5 |
| Thigh | 54.0 | 65.3 |
| Leg | 52.8 | 64.3 |
| Foot | 69.0 | 69.0 |
| Lower limb | 56.0 | 65.0 |
| Leg and foot | 73.5 | 57.2 |

Proximal is nearest to the point of attachment (to the body) and distal is the furthest away from the point of attachment to the body. For example, if the forearm and hand were rotating about the elbow joint they would be rotating about their proximal attachment.

The values in the table are derived from data presented in *Winter, D. A. (1990). Biomechanics and Motor Control of Human Movement (2nd ed.). New York: Wiley-Interscience Publishers, pp. 56–57 (3rd edition published 2004).*

It is important to note that some of the values used in this section and the theory behind some of these concepts is outlined in the Moment of Inertia and Angular Momentum section of this text. For example, the mass of the forearm (radius/ulna) and hand is determined from the anthropometric tables (*Table C6.2* in the Moment of Inertia and Angular Momentum section of this text), where the lower arm (forearm and hand) accounts for 0.022 × the mass of the body.

Mass of forearm and hand = 0.022 × 75 kg = **1.65 kg**

Weight of forearm and hand = 1.65 kg × 9.81 m/s² = **16.19 N** = $F_{w1}$

Mass of the "weight" held in the hand = 10 kg

Weight of the "weight" held in the hand = 10 × 9.81 m/s² = **98.1 N** = $F_{w2}$

Next, determine the moments that are acting on the system. *Figure D1.10* helps to illustrate this in more detail.

Clockwise moments are negative and anti-clockwise moments are positive.

## *Clockwise moments*

$F_{w1} \times 0.154$ m = 16.19 N × 0.154 m = **−2.49 Nm**

$F_{w2} \times 0.35$ m = 98.1 N × 0.35 m = **−34.34 Nm**

## *Anti-clockwise moments*

$F_M \times \sin 80° \times 0.05$ m = $F_M \times 0.985 \times 0.05$ = **+$F_M$ × 0.49 m**

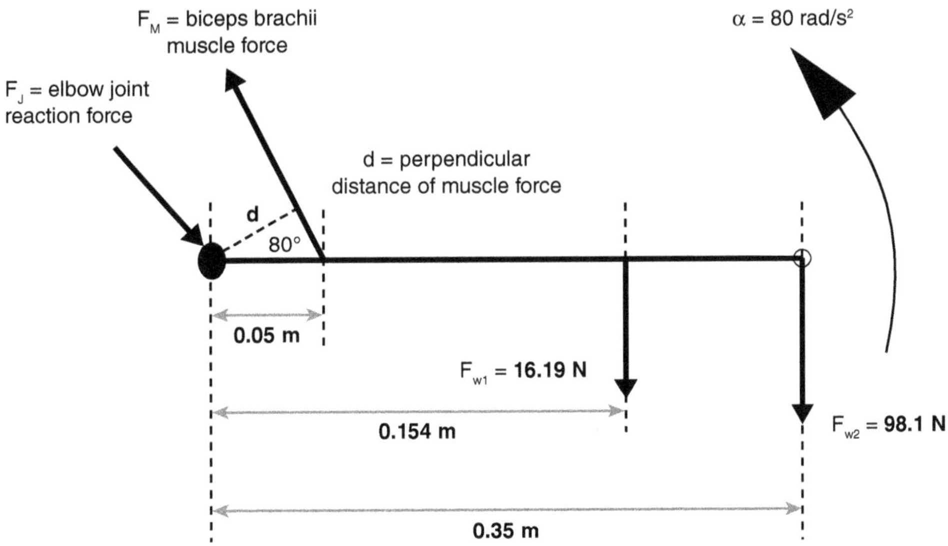


**Figure D1.10** Moment calculations (not drawn to scale)

Determine the moment of inertia values:

$$I = mr^2,\ I = mk^2 \text{ and } I_{CoG}$$

For the 10 kg weight (i.e. the regular shape that is held in the hand at 0.35 m) the moment of inertia is given as:

$$F_{w2} = 10.0\ \text{kg} \times 0.35^2 = \mathbf{1.23\ kg.m^2}$$

The radius of gyration for the forearm and hand is determined from *Table D1.1*. In this example, the forearm segment and hand are rotating about the elbow joint or proximal axis of rotation. Hence, the radius of gyration is located 82.7% from the elbow joint along the length of the arm and hand, which is given as 0.39 m (see *Figure D1.8*). Hence, the radius of gyration for lower arm is:

$$\frac{(82.7 \times 0.39)}{100} = 0.323\,\text{m}$$

Now we can work out the moment of inertia of the forearm and hand as follows:

$$I = mk^2$$

$$I = 1.65\ \text{kg} \times 0.323^2 = \mathbf{0.172\ kg.m^2}$$

Also, we determine the $I_{CoG}$ constant for the lower arm (i.e. forearm and hand), taken from *Table C6.1* in the Moment of Inertia and Angular Momentum section of this text, to be as follows:

$$I_{CoG}\ \text{constant} = \mathbf{0.007\ kg.m^2} \text{ for the forearm}$$

$$I_{CoG}\ \text{constant} = \mathbf{0.0005\ kg.m^2} \text{ for the hand}$$

In order to determine the effects of the moment of inertia for the arm, hand and "weight" in this example, we now add these values together.

$$I = I_{CoG} + mk^2 + mr^2$$

$$I = 0.007 + 0.0005 + 0.172 + 1.23 = \mathbf{1.41\ kg.m^2}$$

Now, we can use the formula for the sum of the moments in a dynamic application to solve the problem for the muscle force ($F_M$) in the biceps brachii muscle needed to accelerate this mass at 80 rad/s².

$$\Sigma M = I\alpha$$

$$+(F_M \times 0.049) - 2.49 - 34.34 = 1.41 \times 80$$

$$F_M \times 0.049 = 112.8 + 2.49 + 34.34$$

$$F_M = 149.63/0.049$$

$$\mathbf{F_M = 3053.67\ N}$$

Note that, in comparison to the static example where a similar arm was held stationary (with a 2 kg "weight" or ball), this is considerably more than the 313.10 N that was found in that example. This shows the importance and value of the dynamic calculations when movement and acceleration is involved (as in most cases in human movement).

Next, we need to translate this biceps brachii muscle force into the joint reaction forces that are acting at the elbow joint. These joint forces can now be calculated in a similar manner to the static calculation methods but taking into account the linear acceleration of the centre of mass of the system (arm, hand and "weight" in the hand) in both the horizontal and vertical directions. For example, we have to combine the first condition of equilibrium with the acceleration approach using the following formula (as shown earlier):

$$\Sigma Fx = ma_x$$

$$\Sigma Fy = ma_y$$

In this case, the accelerations $a_x$ and $a_y$ are the linear horizontal (x) and linear vertical (y) accelerations of the arm, hand and "weight". The angular acceleration of the arm at any point (collection of points) along the arm will be the same, but the linear accelerations along the arm will differ depending on the position on the arm that we are considering these linear accelerations (point mass) to be acting. The Linear–Angular Motion section in this text helps to explain this in more detail. As the limb rotates the linear accelerations will vary depending upon the position (angle) of the limb. Using the linear and angular relationships these accelerations are given as:

$$a_x = -r \times \alpha \times \text{Sin } \theta$$

$$a_y = r \times \alpha \times \text{Cos } \theta$$

where:

$a_x$ = horizontal acceleration (m/s²)

$a_y$ = vertical acceleration (m/s²)

$r$ = radius that the point of consideration is acting at from the axis of rotation (usually in a simple system without added 'weights' this would be the centre of mass position of the forearm and hand) (m)

$\alpha$ = angular acceleration (rad/s²)

$\theta$ = the angle of the arm to the horizontal at the point of consideration

However, it is now necessary for us to determine the point that we are considering for the linear accelerations to be acting in this system. This can be achieved by taking moments about the system. *Figure D1.11* helps to illustrate this in more detail:

This time we use the second condition of equilibrium where the sum of the moments is equal to zero as we are trying to calculate a position on the limb and weight system that accounts for the effect of the two other masses acting on the arm (i.e. the mass of the radius/ulna and hand and the mass of the "weight" in the hand).

$$D \times F = 16.19 \times 0.154 + 98.1 \times 0.35$$

The combined F in this example would be 16.19 + 98.1 = 114.29 N

$$D \times 114.29 = 2.49 + 34.34$$

$$D = 36.83/114.29$$

$$\mathbf{D = 0.32\ m}$$

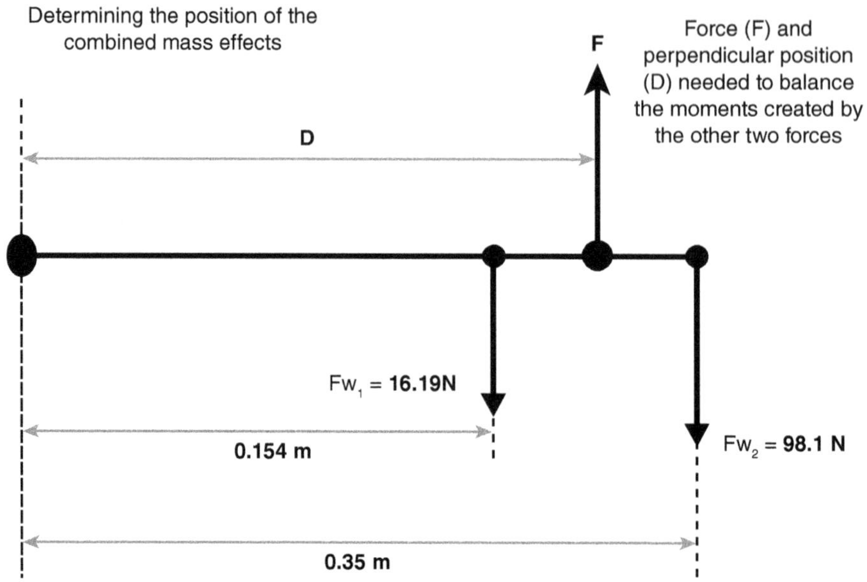


**Figure D1.11** Determining combined mass position (not drawn to scale)

This value (distance from the elbow joint) for D represents a position along the limb that we can consider the point that the combined effect of the two masses (radius/ulna and hand and the "weight" in the hand) to be acting. This is the position of the point mass that we will now use in our acceleration calculations to determine the joint reaction forces at the elbow. *Figure D1.12* helps to illustrate this in a diagrammatic form, with the elbow joint forces moved for clarity.

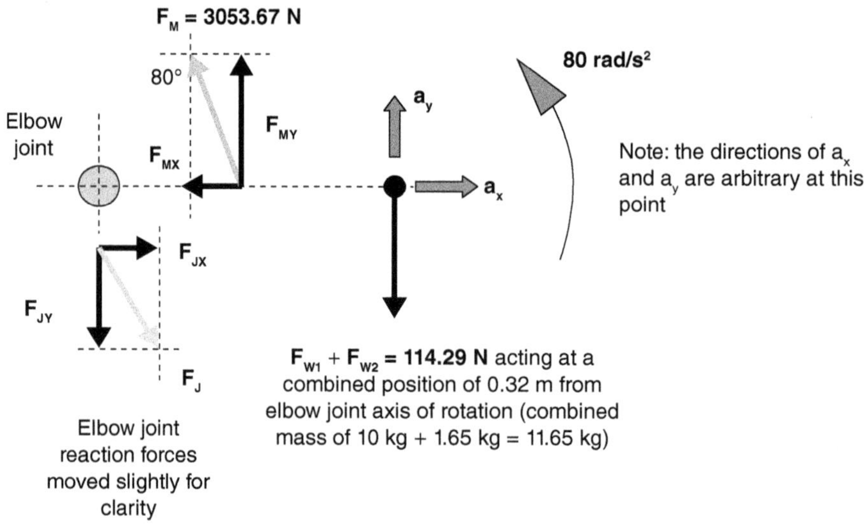


**Figure D1.12** Muscle and joint forces (not drawn to scale)

First, determine the linear accelerations that are acting at the point mass centre that we are considering to be acting at 0.32 m from the elbow axis of rotation (note that this distance also happens to be the same (or very close to the same) distance as the radius of gyration point (0.323 m) for the arm and hand. However, in this case it is a coincidence). Using the two linear acceleration formulae shown above, we know that the arm is at a horizontal position (i.e. zero degrees) and that the arm is being accelerated at 80 rad/s$^2$.

$a_x = -r \times \alpha \times \text{Sin}\ \theta$

$a_y = r \times \alpha \times \text{Cos}\ \theta$

$a_x = -0.32 \times 80 \times \text{Sin}\ 0° = -0.32 \times 80 \times 0 = \mathbf{0}$ **(zero because the Sine of 0°** = 0)

$a_y = 0.32 \times 80 \times \text{Cos}\ 0° = 0.32 \times 80 \times 1 = \mathbf{25.6\ m/s^2}$ (vertical acceleration)

Now, we need to resolve the horizontal and vertical components of the biceps brachii muscle force as we did for the static example shown previously.

Vertical component of the muscle force = F Sin θ

$3053.67 \times \text{Sin}\ 80° = 3053.67 \times 0.985 = \mathbf{3007.86\ N} = \text{Fmy}$

Horizontal component of the muscle force = F Cos θ

$3053.67 \times \text{Cos}\ 80° = 3053.67 \times 0.174 = \mathbf{531.33\ N} = \text{Fmx}$

We now have all the components in order to use the previous equations to determine the joint forces acting at the elbow.

$$\Sigma \text{Fx} = \text{ma}_x$$

$$\Sigma \text{Fy} = \text{ma}_y$$

Vertical force acting at the elbow:

$$\Sigma \text{Fy} = \text{Fmy} + \text{Fjy} + \text{Fw}_1\text{y} + \text{Fw}_2\text{y} = \text{ma}_y$$

$$\Sigma\ \text{Fy} = 3007.86 + \text{Fjy} + 16.19 + 98.1 = 11.65 \times 25.6$$

$$\text{Fjy} = (11.65 \times 25.6) - 3007.86 - 16.19 - 98.1$$

$$\text{Fjy} = 298.24 - 3122.15$$

$$\mathbf{Fjy} = -2823.91\ \mathbf{N}$$

Horizontal force acting at the elbow:

$$\Sigma \text{Fx} = \text{Fmx} + \text{Fjx} = \text{ma}_x$$

$$\Sigma \text{Fx} = 531.33 + \text{Fjx} = 11.65 \times 0$$

$$\mathbf{Fjx} = -531.33\ \mathbf{N}$$

Now, resolve these forces for the resultant force acting at the elbow and the angle of application.

Resultant elbow joint force:

$$Fj = \sqrt{Fjy^2 + Fjx^2}$$

$$Fj = \sqrt{(-2823.91^2 + -531.33^2)}$$

$$Fj = \sqrt{(7974467.7 + 282311.6)}$$

$$Fj = \sqrt{8256779.3}$$

$$\mathbf{Fj} = 2873.46\ \mathbf{N}$$

Now, determine the angle of orientation of the elbow joint reaction force.

$$\text{Tan}\,\theta = \frac{Fjy}{Fjx}$$

$$\text{Tan}\,\theta = \frac{-2823.91}{-531.33}$$

$$\text{Tan}\,\theta = 5.31$$

$$\theta = \mathbf{79.33°}$$

These values represent the angle of application (**79.33°** expressed to the left horizontal) and magnitude (**2873.46 N**) for the joint reaction force that is applied to the elbow in this example. This is depicted in more detail in *Figure D1.13* (with the elbow forces moved slightly for clarity) and as can be seen these are considerably greater than the forces that occur in the static situation (280.10 N) by a factor of more than ten times. Hence, as we can now appreciate, these values have very important implications for injury situations, such as lateral epicondylitis or "tennis elbow", which as the name suggests is often seen in the sport of tennis. In addition, by knowing the muscle forces involved in making the arm move in this way we can also potentially determine the effects of exercise and training programmes.

Finally, *Figure D1.14* provides a homework example for this type of problem. The full solution is not provided but the following values are approximately what should be expected (again ignoring the brachioradialis force).

Body weight (BW) = 86 × 9.81 = 843.66 N

Static forces in biceps brachii muscle = 2.6 × body weight

Static elbow joint force = 2.5 × body weight

Dynamic muscle force = 10.97 × body weight

Dynamic elbow joint force = 10.4 × body weight

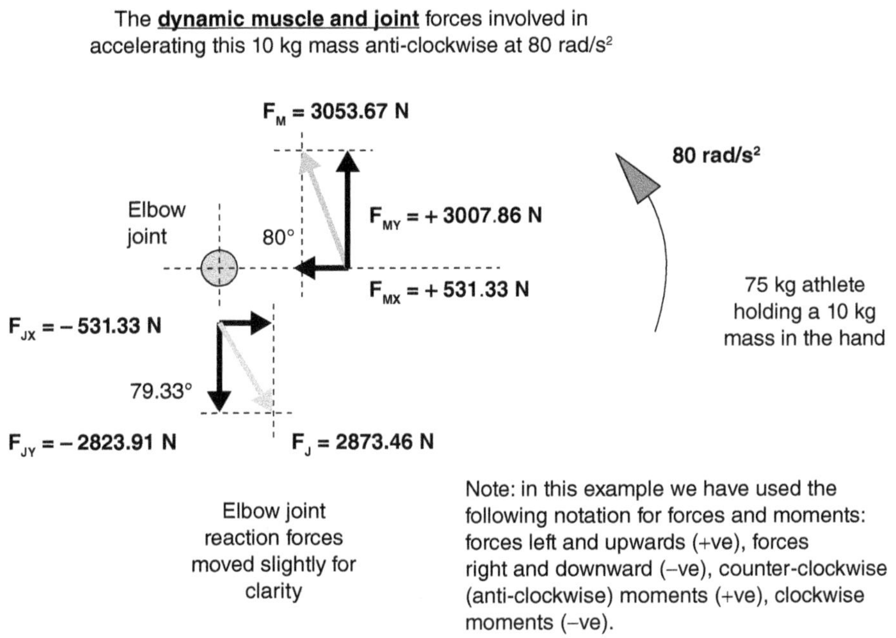


**Figure D1.13** Solution to dynamics problem (not drawn to scale)

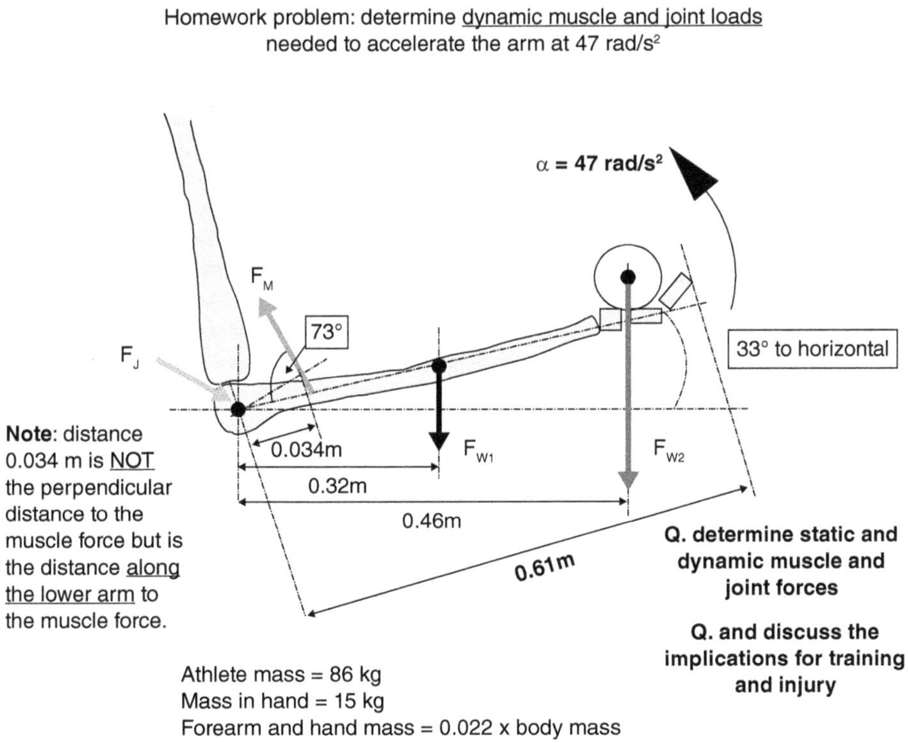


**Figure D1.14** Homework dynamics problem (not drawn to scale)

*Key notes*

| | |
|---|---|
| **Statics** | Static calculations utilise the first and second condition of equilibrium and form the basis of the complex inverse dynamics calculations that are used extensively in biomechanics modelling of joint and muscle forces. They are valuable in providing an understanding of injury potential and performance characteristics and could even be used for determining the effects of an exercise or training programme. |
| **Muscle forces** | Muscle forces are determined using the first and second conditions of equilibrium where the sum of the forces and the clockwise and anti-clockwise (or counter-clockwise) moments are zero. |
| **Joint forces** | Joint forces are determined using free body diagrams and trigonometry. |
| **Dynamics** | Dynamics is applicable since no form of human movement is purely static. Statics form the basis of dynamic calculations. Dynamics utilise the acceleration approach to solving problems and allow biomechanists to determine realistic muscle and joint forces that could be responsible for injury. They are used in most biomechanical modelling programmes and by providing information on the specific muscle forces involved in movement they can potentially help to assess the effects of complex exercise and training programmes. |
| **Moment of inertia** | Moment of inertia is required for calculation of dynamic forces and is determined by: $\mathbf{I = \Sigma\, m\, r^2}$ To account for irregular shapes of the human body, it is important to add the centre of gravity constant. In addition, the radii of gyration measures are used to account for the movement of segments about different axes of rotation. |
| **Radii of gyration** | This is required for calculation of dynamic forces. It is important to determine about what axis the limb or joint is rotating (proximal or distal). |
| **Dynamic solution** | $\Sigma F = ma$ <br> $\Sigma CWM + \Sigma ACWM = I\alpha$ |
| **SUMMARY** | |
| **Injury** | Such muscle and joint forces in a single arm curl can be large, even with the smallest of "weights" held in the hand, and if excessive they can result in the potential for injury to the joint or soft-tissue structures. |
| **Performance** | It is important to understand the muscle and joint forces in static calculations as the training effect in activities such as weightlifting can be assessed. In addition, the design and development of exercise equipment is based on such a working knowledge of these forces and loads. |
| **Dynamics** | Dynamics show that joint forces during movement increase considerably when the limb is accelerated in a specific direction. This increases again (as does the potential for injury) when the athlete has to move more weight than just that of the limb. This information and biomechanical knowledge can be used in the assessment of exercises designed to develop muscle strength. |

## FURTHER READING

The following resources provide some additional reading on the topics associated with static and dynamic calculations in sport and exercise. The third source is a useful reference for the more recent measurements for anthropometric parameters of the human body.

1 Feltner, M. E., & Dapena, J. (1989). Three-dimensional interactions in a two-segment kinetic chain. Part I: General model. *International Journal of Sport Biomechanics, 5*(4), 403–419.
2 Koehle, M. J., & Hull, M. L. (2008). A method of calculating physiologically relevant joint reaction forces during forward dynamic simulations of movement from an existing knee model. *Journal of Biomechanics, 41*(5), 1143–1146.
3 Kikolova, G. S., & Toshev, Y. E. (2007). Estimation of male and female body segment parameters of the Bulgarian population using a 16-segmental mathematical model. *Journal of Biomechanics 40*(16), 3700–3707.
4 Dugan, E. L., Doyle, T. L. A., Humphries, B., Hasson, C. J., & Newton, R. U. (2004). Determining the optimal load for jump squats: A review of methods and calculations. *Journal of Strength and Conditioning Research 18*(3), 668–674.
5 Kornecki, S., & Jaszczak, M. (2010). Dynamic analysis of rowing on concept II type C ergometer. *Biology of Sport, 27*(3), 187–194.

# SECTION E

# Special topics

# WORK, POWER AND ENERGY

*Michael Cole*

The term **work** is commonly used in everyday language to refer to the effort expended in performing a task, but when used in a scientific context the term refers to the movement of an object by the application of a force. The **mechanical work** ($W$) done on an object is expressed in units of **joules** (J) and is defined as the product of the **force** ($F$) applied to an object and its **displacement** ($d$) in the direction of the force. This relationship can be written as:

Mechanical Work = Force × Displacement

$$W = F \times d$$ **(Equation E1.1)**

A question that students often ask is: "if work is being done where does it go?" To appreciate the answer to this question it should be noted that while work is not a quantity that can be seen, its effects can. Usually the effects of doing work are seen as a change in an object's position (e.g. moving it upwards against gravity; *Figure E1.1*), velocity or shape (e.g. *Figure E1.2*).

## TASK

A person is bench pressing a load of 60 kg and, with great effort, is able to slowly raise the barbell to a position that is 40 cm above their chest. How much work is done and where does this work go?

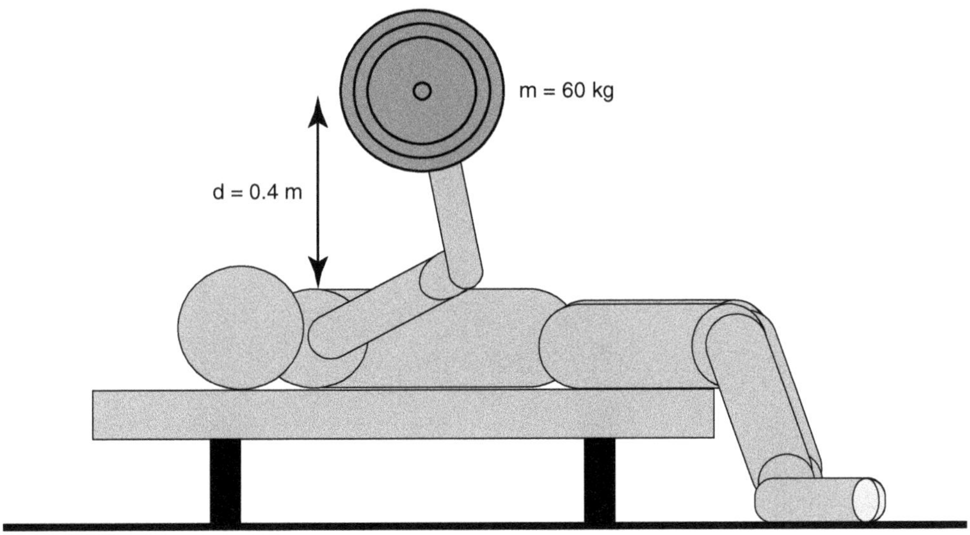


**Figure E1.1** Work example 1: bench press

## Solution

The gravitational load of the weights is derived using the formula:

$F = m \times a$

$F = 60 \text{ kg} \times -9.81 \text{ m/s}^2$

$\mathbf{F = -588.6\ N}$

Note that force is a vector quantity, hence the negative sign represents the downward direction of the force exerted by the weights on the person's hands. To overcome this force and move the barbell, the person must apply an equally large (or greater) force in the opposite direction (i.e. ≥ **+588.6 N**). As the displacement caused by the force is given as the displacement that the barbell has moved from the chest (i.e. 40 cm or **0.4 m**), the work done is given by:

$W = 588.6 \text{ N} \times 0.4 \text{ m} = 235.4$ **Joules**

Where does the work performed during this activity go? It goes into overcoming the gravitational load of the barbell and raising it against gravity.

## TASK

A force of 1000 N is applied to a soccer ball during a kick and this force deforms the ball by 10 cm. How much work has been done on the ball and where does this work go?

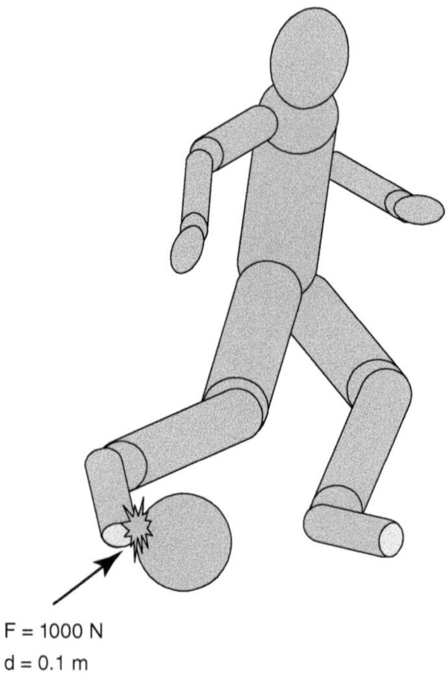


**Figure E1.2** Work example 2: soccer kick

## Solution

The force applied to the ball is given as 1000 N and the displacement caused by the force is given by the deformation of the ball (i.e. 10 cm or 0.1 m). Therefore the work done is given by:

$$W = 1000\ \text{N} \times 0.1\ \text{m} = 100.0\ \textbf{Joules}$$

In this example of a soccer kick, the work performed on the ball goes into deforming the ball. An important implication of the above definition of work is that if the force acting on an object is not large enough to move it, then the work done equals zero. For example, if the person attempting to lift the barbell was unable to move it, no work is done, despite the effort expended by the muscles.

When describing work, it is important to recognise that it can be classified as either **positive work** or **negative work** depending on whether the force is applied in the same direction that the object is moving or in the opposite direction to its movement. In the case of the bench press example, an upward force is exerted by the arm/hand onto the barbell to raise it above the chest; hence positive work is being done on the barbell. In this component of the exercise, the work performed on the barbell goes into changing its position against gravity. During the lowering phase of the movement, the hands continue to apply an upward force, but as the barbell is moving downwards, negative work is being done on the barbell. A percentage of this negative work goes into controlling the descent of the barbell, which is being accelerated downwards by gravity, while some is lost as heat from the muscles. Negative work is preceded by a negative sign and can be interpreted as meaning that the barbell is doing work on the person, rather than the person doing work on the barbell. Of interest in sports performance is the fact that some of this negative work can deform the body's structures (most notably the muscles and tendons), which, given the right technique, can enhance the efficacy of the muscle's shortening capacity (Wilson & Flanagan, 2008). This phenomenon, known as the stretch–shorten cycle, is commonly used to explain why performing a counter-movement prior to a vertical jump can increase jump height and why it is more efficient to run at certain speeds than to walk.

The term **power** (P) is used to describe the rate at which the work is being done. For example, lifting a barbell slowly is different from lifting it rapidly, even though the final outcome (e.g. height lifted) is the same. A second example would be a cyclist who could climb a mountain more quickly by exerting more effort or ascend it more slowly; irrespective of the choice, the outcome would be the same. The difference between the two movements in each case is that the faster movements (i.e. movements performed over shorter time periods) require greater power. Given this relationship, power is defined as the rate at which work is being done and is given by the following equation:

$$\text{Power} = \frac{\text{Work}}{\text{time}} \quad \text{or} \quad P = \frac{W}{t} \qquad \textbf{(Equation E1.2)}$$

Power is measured in **watts** (W) and while these units use the same symbol as work (i.e. W), they are distinguished by the context in which they are used and hence are rarely confused. Given that work (W) is the product of the force applied to an object (F) and the subsequent displacement of that object (d), and that power (P) is derived by dividing the work done by the time taken to perform the work (t), the equation for power can be more clearly written as:

$$P = \frac{F \times d}{t}$$

As $\frac{displacement}{time} = velocity$ :

$$P = F \times v \quad \textbf{(Equation E1.3)}$$

## TASK

What is the power produced in the bench press example given in *Figure E1.1* if it is completed in 2 seconds?

### Solution

The work done on lifting the barbell previously was calculated as 235.4 **Joules** and this work has now been completed in 2 seconds. Given that $P = \frac{W}{t}$, the power required to perform this movement was:

$$P = \frac{235.4}{2}$$

$$\mathbf{P = 117.7\ Watts}$$

## TASK

A cyclist climbs a vertical displacement of 200 m on a mountain road in 3 min 45 s (*Figure E1.3*). If the combined mass of the cyclist and the bicycle is 100 kg, what power has the cyclist produced (ignoring the effects of rolling resistance and aerodynamic drag forces)?

### Solution

The work completed while climbing the mountain road is calculated as:

$$W = F \times d$$

$$W = (100 \text{ kg} \times 9.81 \text{ m/s}^2) \times 200 \text{ m}$$

$$\boldsymbol{W} = 196200 \ \textbf{Joules}$$

The time taken to complete the climb was 3 min 45s or **225 s.** Given that $P = \frac{W}{t}$, the power required to climb the mountain road was:

$$P = \frac{196200}{225}$$

$$\boldsymbol{P} = 872.0 \ \textbf{Watts}$$

**Energy** is defined as the capacity to perform work and is typically considered something that is "stored" or "possessed". Like work, energy is measured in the units of **joules,** which highlights

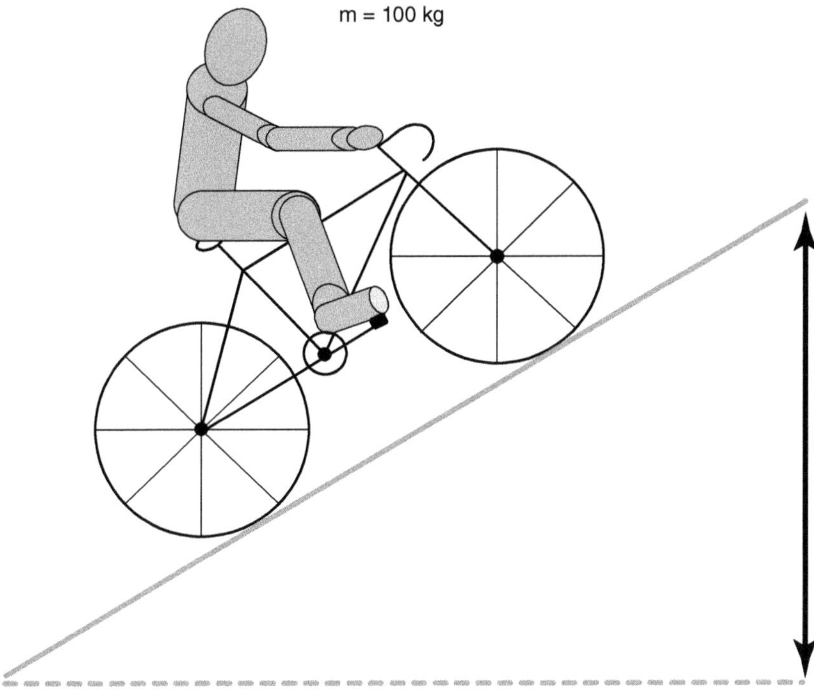


**Figure E1.3** Power example – The climbing cyclist

the important link between these two quantities. In addition to chemical energy that enables muscles to contract, there are two **mechanical** forms of energy that are relevant to the study of biomechanics: i) **potential energy**, which relates to the energy associated with the position or deformation of an object; and ii) **kinetic energy**, which relates to the energy associated with motion.

Potential energy can be sub-divided into two main forms: i) gravitational potential energy; and ii) strain energy. **Gravitational potential energy ($E_{GPE}$)** is the energy that is stored as a result of an object's position in a gravitational field. If an object is at a height (h) above the ground then its gravitational potential energy is given by the equation:

$$E_{GPE} = mass \times gravity \times height \quad \text{or} \quad E_{GPE} = m \times g \times h$$ **(Equation E1.4)**

In the example depicted in *Figure E1.1*, the 60 kg barbell was raised 0.4 m from its resting position, hence it now has a potential energy of $60 \times 9.81 \times 0.4 = 235.4$ **Joules** more than it had in its resting position. This is the same as the work done in moving the barbell to this position and illustrates the close relationship between the work done and the energy stored. It is worth noting that the energy that enabled the work to be done in the first place came from the chemical energy available in the muscles. This example also illustrates the way that energy can be converted from one form (chemical) to another (gravitational potential energy). **Strain potential energy ($E_{SPE}$)** is the energy stored in response to the deformation of a material and is directly proportional to the amount of deformation (*d*) and the material's stiffness (*k*):

$$E_{SPE} = \frac{1}{2}k \times d^2$$ **(Equation E1.5)**

Stiffness is one of the mechanical properties of a material and is determined by an object's physical size, material composition and construction. In the example depicted in *Figure E1.2*, the stiffness of the ball is determined by its pressure, where a lower pressure would allow it to be easily deformed (i.e. it appears soft) and a higher pressure would make it more difficult to deform. The squaring of deformation ($d$) in Equation E1.5 indicates that the relationship between strain energy and deformation is non-linear. That is, a greater increase in deformation results in greater storage of strain energy, but ultimately requires a greater amount of work to be done. A good example to help understand this concept is spring-loaded exercise equipment. Springs are initially quite easy to extend from their resting position, but become progressively harder to stretch as their length increases. Lengthening the spring increases its deformation and ultimately results in the need for an increased amount of strain energy to be stored to further deform the spring.

Like potential energy, there is also two sub-types of kinetic energy: i) **linear kinetic energy ($E_{LKE}$)**, which relates to an object's linear motion; and ii) **angular (or rotational) kinetic energy ($E_{AKE}$)**, which relates to an object's angular motion. Linear kinetic energy is derived using the linear velocity ($v$) and the mass (m) of an object and is described by the equation:

$$E_{LKE} = \frac{1}{2} m \times v^2$$

**(Equation E1.6)**

Linear kinetic energy expresses the work that has to be done for an object to move and reflects the energy stored in the object due to its motion. For example, the sprinter in *Figure E1.4* is required to expend a specific amount of muscle energy during each stride of the race to increase the body's velocity. While the amount of energy expended may be similar for each stride, the non-linear $v^2$ component indicates that the effect of the expended energy on increasing the sprinter's velocity diminishes as their running speed increases. As running speed increases, it becomes increasingly difficult for the sprinter to increase their running speed further. In fact, it requires four times the energy to increase the running speed from 2.3 to 4.6 m/s than from 0 to 2.3 m/s, even though the change in speed is equal. Another issue that is worth considering is that an object that has some linear kinetic energy must have that energy dissipated in order to stop (i.e. the linear kinetic energy must be reduced to zero). For example, in running we extend our leg to brake our forward motion while in jumping we absorb the energy of a counter-movement jump by flexing the ankle, knee and hip joints upon landing. Each movement employs a different strategy, but they share the common goal of reducing the linear kinetic energy in a controlled manner. As these actions usually rely on the expenditure of muscular chemical energy, it becomes apparent that both the increase and dissipation of kinetic energy require the use of muscular energy stores.

## TASK

Calculate the velocity of a 75 kg sprinter over the first five strides and the change in velocity per stride if the energy delivered during each stride is 306 joules.

## Solution

Given that linear kinetic energy is given by $E_{LKE} = \frac{1}{2} m \times v^2$, the sprinter's velocity can be determined by rearranging the formula:

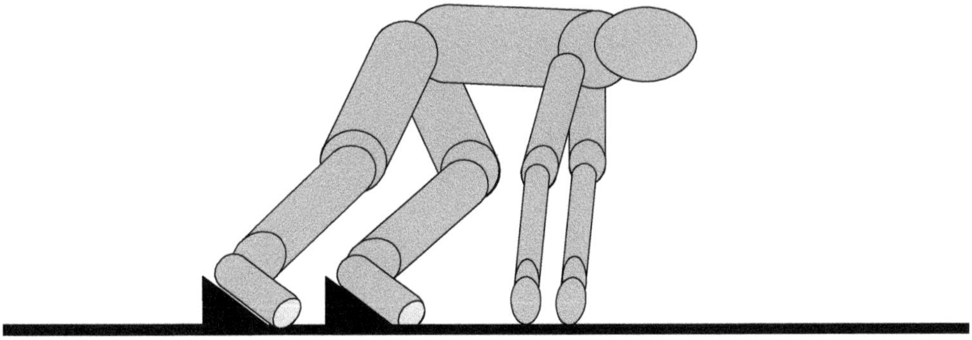


**Figure E1.4** Linear kinetic energy example: 100 m sprint

$$E_{LKE} = \frac{1}{2} \times \frac{m}{1} \times \frac{v^2}{1}$$

$$2 \times E_{LKE} = \frac{m}{1} \times \frac{v^2}{1}$$

$$\frac{2 \times E_{LKE}}{m} = v^2$$

$$v = \sqrt{\frac{2 \times E_{LKE}}{m}}$$

For this example, we will assume that the sprinter is required to expend 3.4 joules of linear kinetic energy per kilogram of body mass per metre of distance covered to accelerate the body (Minetti, Moia, Roi, Susta, & Ferretti, 2002). If each stride taken by the sprinter was 1.2 metres in length, the linear kinetic energy required to complete each stride can be calculated as:

75 kg × 3.4 Joules × 1.2 metres = 306 Joules

However, as the body accelerates during the sprint, it becomes increasingly difficult to further increase running velocity, as the limbs are required to move increasingly fast and there is an obvious need to exert additional external force. Both of these factors contribute to an increased energy expenditure for each successive stride. As such, the change in the sprinter's velocity over the first five strides is calculated as:

$$\text{First stride}: v = \sqrt{\frac{2 \times 306}{75}} = 2.86\ \text{m/s}$$

$$\text{Second stride}: v = \sqrt{\frac{2 \times 612}{75}} = 4.04\ \text{m/s}$$

$$\text{Third stride}: v = \sqrt{\frac{2 \times 918}{75}} = 4.95\ \text{m/s}$$

$$\text{Fourth stride}: v = \sqrt{\frac{2 \times 1224}{75}} = 5.71\ \text{m/s}$$

$$\text{Fifth stride}: v = \sqrt{\frac{2 \times 1530}{75}} = 6.39\ \text{m/s}$$

These data are shown in tabulated form in *Table E1.1*.

Even though energy expenditure increases by the same amount for each stride, the greatest change in velocity is on the first stride with the benefit (indicated by the change in velocity) reducing as velocity increases. It should be noted that, although this example is based on an energy expenditure of 3.4 joules per kilogram per metre travelled, the amount of energy expended per stride can be influenced by many factors, including gradient of running surface, differences in foot landing patterns, running surface and muscle fatigue (Minetti et al., 2002).

**Angular kinetic energy** ($E_{AKE}$) is determined by an object's angular velocity ($\omega$) and moment of inertia ($I$) and is given by the equation:

$$E_{AKE} = \frac{1}{2} I \times \omega^2$$

**(Equation E1.7)**

As most sporting actions involve rotation of the limbs about a joint, energy is contained in the rotation of the limbs. For example, during walking, the knee joint is constantly changing direction as it flexes and extends during each gait cycle. Muscular chemical energy is required to increase the angular velocity of the limbs and to slow them down and change their direction. Thus, actions that require a lot of limb movement (e.g. sprinting) require high levels of energy expenditure.

## WORK AND ENERGY

It has already been noted that work and energy are closely related and that they are measured in the same units (i.e. joules), but unlike work, energy can be stored. In essence, work is the process of changing one form of energy to another (e.g. chemical to linear kinetic), hence the basic relationship between work done ($W$) and energy ($E$) is given by the equation:

**Table E1.1** Relationship between energy expenditure and change in running velocity for a 75kg sprinter

| *Stride* | *Energy (J)* | *Velocity (m/s)* | *Change in velocity* |
|---|---|---|---|
| 0 | 0 | 0 | – |
| 1 | 306 | 2.86 | 2.86 |
| 2 | 612 | 4.04 | 1.18 |
| 3 | 918 | 4.95 | 0.91 |
| 4 | 1224 | 5.71 | 0.76 |
| 5 | 1530 | 6.39 | 0.68 |

$$W = \Delta E \qquad \text{or} \qquad W = E_{final} - E_{initial} \qquad \textbf{(Equation E1.8)}$$

In this formula, the change in energy represents the difference between the energy value at the start of when work is being done ( $E_{initial}$ ) until its end ( $E_{final}$ ). For example, in the sprinting scenario described earlier, the energy change between strides was 306 joules, which resulted from the work being done during each stride. Another example is provided in *Figure E1.5*.

## TASK

A 70 kg high jumper applies an average force of 2000 N over a distance of 0.4 m. Calculate the jumper's velocity at take-off.

## Solution

The high jumper has a zero vertical velocity at the lowest point of the jump ($v_{initial}$) and maximum vertical velocity at the moment of take-off ($v_{final}$). Given that the work done is equal to the change in kinetic energy (Equation E1.8), the high jumper's velocity at take-off can be calculated as:

$$W = \left(\frac{1}{2}m \times v_{final}^2\right) - \left(\frac{1}{2}m \times v_{initial}^2\right)$$

or

$$F \times d = \left(\frac{1}{2}m \times v_{final}^2\right) - \left(\frac{1}{2}m \times v_{initial}^2\right)$$

As the initial vertical velocity was zero, we can disregard this part of the equation to give:

$$F \times d = \left(\frac{1}{2}m \times v_{final}^2\right)$$

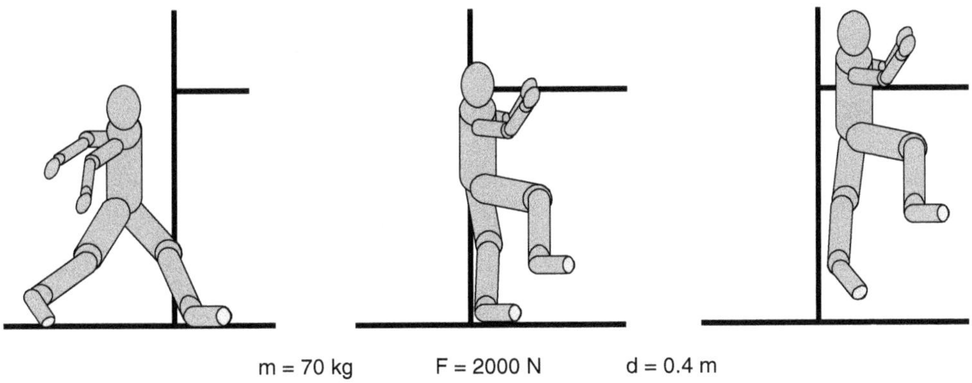


**Figure E1.5** Linear kinetic energy example: high jump

$$2000 \times 0.4 = \left(\left(\frac{1}{2} \times 70\right) \times v_{final}^2\right)$$

$$800 = \left(35 \times v_{final}^2\right)$$

$$\sqrt{\frac{800}{35}} = v_{final}$$

$$v_{final} = 4.78 \text{ m/s}$$

*Key notes*

| | |
|---|---|
| **Work** | Work is defined as the transfer of energy to an object via the application of a force and is evidenced by movement of the object in the same direction as the applied force. Thus, work (W) is expressed in joules (J) and is the product of the force (F) applied to an object and the amount that the object is displaced (d) (i.e. W = F × d). |
| **Positive and negative work** | Positive work is being done when the direction of motion is the same as the direction of the force (e.g. raising a barbell during bench press), while negative work is being done when the direction of motion is opposite to the direction of the force (e.g. lowering a barbell during bench press). Positive and negative work both require energy to activate skeletal muscles, but the energy expended during negative work is often less, due to the utilisation of strain energy stored in the body's tendons. |
| **Power** | Power (P) describes the rate at which work is being done and, hence is calculated as $P = \frac{\text{Work}}{\text{time}}$. For example, lifting a barbell rapidly requires more power than lifting it slowly, even though the final outcome is often the same. Power is expressed in watts (W) and while it shares the same symbol as work (W), it should be noted that the former is a unit of measurement while the latter is a mechanical concept. |
| **Energy** | Energy describes the capacity to do work or perform an action and is expressed in Joules (i.e. the same as work). There are different forms of energy, but the most important to sport and exercise biomechanics are potential energy (gravitational and strain) and kinetic energy (linear and angular). |
| **Work and energy** | Work and energy are closely related, but energy can be stored while work cannot. Energy changes from one form to another by the process of doing work. |

## REFERENCES AND FURTHER READING

The following resources provide additional reading on mechanical efficiency during gait and the role of elastic energy in high-power and high-force activities.

1 Minetti, A. E., Moia, C., Roi, G. S., Susta, D., & Ferretti, G. (2002). Energy cost of walking and running at extreme uphill and downhill slopes. *Journal of Applied Physiology, 93*(3), 1039–1046.
2 Wilson, J. M., & Flanagan, E. P. (2008). The role of elastic energy in activities with high force and power requirements: A brief review. *Journal of Strength & Conditioning Research, 22*(5), 1705–1715.

# CONSERVATION OF ENERGY

*Michael Cole*

The **law of conservation of energy** states that energy can be neither created nor destroyed and expresses the fact that the total amount of energy remains constant as it changes from one form (e.g. chemical) to another (e.g. kinetic). This law is one of the cornerstones of science and while it applies to the energy exchanges that occur in sports and exercise, its practical application to this discipline can be limited. For this reason, a more restricted form of the law has been defined to provide more valuable information for the exercise and sports science discipline.

Several forms of **mechanical energy** have been described elsewhere in this text; principally **gravitational potential energy ($E_{GPE}$), strain potential energy ($E_{SPE}$), linear kinetic energy ($E_{LKE}$) and angular kinetic energy ($E_{AKE}$)**. In the examples provided in the section concerning Work, Power and Energy, the body's **muscles** can be considered structures that convert **chemical energy** to **mechanical energy**. The process of converting chemical energy to mechanical energy inevitably results in a certain amount of energy being converted to **heat**. While this heat may have some biological value in maintaining body temperature, it does not generally contribute directly to the outcome or performance. For example, a ball that is dropped from a height onto a concrete surface compresses and rebounds to a height that is never quite as high as the original drop height, as some of its energy is converted to heat as a result of its compression. The amount of energy converted to heat during this phase provides an indication of the efficiency of energy conversion, which will always be less than 100 per cent when heat is generated. While many energy conversion processes generate heat, the conversion of gravitational potential energy ($E_{GPE}$) to linear ($E_{LKE}$) or angular kinetic energy ($E_{AKE}$) produces no heat and hence is 100 per cent efficient. Given this point, the relationship between these types of energy is referred to as the **conservation of mechanical energy** and is given by the following equation, where total mechanical energy is constant:

$$\text{Total mechanical energy} = E_{GPE} + E_{LKE} + E_{AKE} \qquad \textbf{(Equation E2.1)}$$

It is important to note that Equation E2.1 **does not apply** to situations involving strain potential energy, as the process of deformation causes molecules to rub against each other, which leads to energy loss due to friction. Furthermore, it does not apply to other situations where energy is lost to friction (e.g. when a person or an object slides down a surface) and cannot be used if the influence of air resistance is important (i.e. if relative air speeds exceed 5 to 6 m/s). Rather, the conservation of mechanical energy mainly applies to tasks that involve the body or a projectile moving slowly through the air, such as during athletics jumping, gymnastics, diving, jumping on a trampoline or throwing. It can also be used to understand the techniques used in racket sports where gravitational force can be used to help generate racket head speed (e.g. in a looped forehand or backhand drive). The principle can also explain the manner in which limbs are used in walking and running. In short, the conservation of mechanical energy has widespread application to sport and exercise situations.

## APPLICATION

Consider a trampolinist who is just about to leave the net (i.e. the bouncing surface) to perform a straight bounce (i.e. no rotation); their vertical velocity ($v_{take\text{-}off}$) at this point is at its greatest (*Figure E2.1*). From the trampolinist's last point of contact with the net, their vertical velocity progressively declines until it reaches zero at the highest point of the bounce. As the trampolinist descends, their vertical velocity becomes increasingly negative (remember that velocity is a vector quantity) until they, once again, make contact with the trampoline net.

In this example, the energy according to equation E2.1 remains constant so we can equate the energy conditions at two points (the take-off and the top of flight) to give:

$$E_{GPE_{take-off}} + E_{LKE_{take-off}} = E_{GPE_{top\ of\ flight}} + E_{LKE_{top\ of\ flight}}$$

As $E_{GPE_{take-off}}$ and $E_{LKE_{top\ of\ flight}}$ are both zero, we can simplify the equation to:

$$E_{LKE_{take-off}} = E_{GPE_{top\ of\ flight}} \quad \textbf{(Equation E2.2)}$$

The equation can also be expressed by using the full expressions for each energy term:

$$\frac{1}{2} m \times v^2_{take-off} = m \times g \times h_{top\ of\ flight}$$

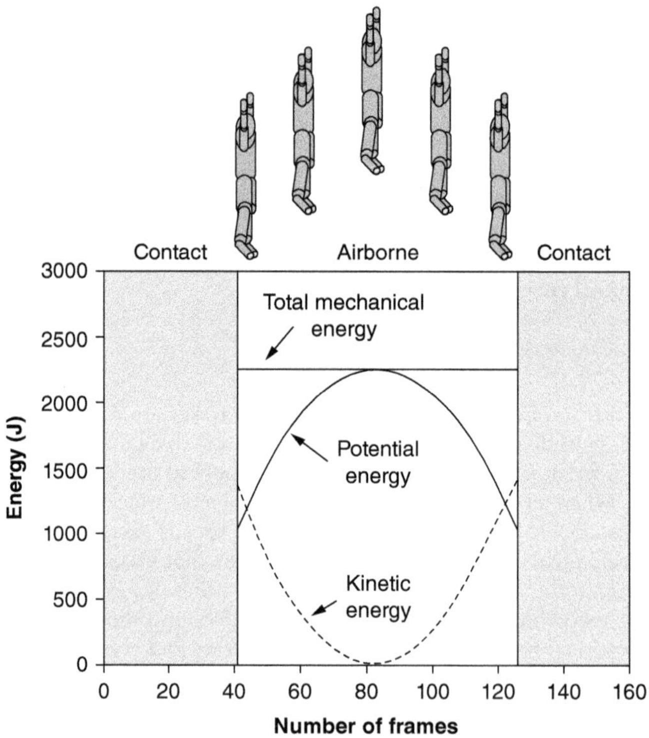


**Figure E2.1** Energy changes during a trampoline bounce

By cancelling the term "m" and re-arranging the equation we arrive at the general expression that links the vertical velocity and height for any projectile motion:

$$v_{take-off} = \sqrt{2 \times g \times h} \qquad \textbf{(Equation E2.3)}$$

## TASK

Considering that the 70 kg high jumper from the example provided in the section on Work, Power and Energy (see *Figure E1.5*) has a vertical take-off velocity of 4.78 m/s, how high would they raise their centre of mass (i.e. how high would they jump)?

## Solution

$$v_{take-off} = \sqrt{2 \times g \times h}$$

$$4.78 = \sqrt{2 \times 9.81 \times h}$$

$$4.78^2 = 2 \times 9.81 \times h$$

$$\frac{22.85}{2 \times 9.81} = h$$

$$h = 1.16\,m$$

To this, of course, we must add the starting height of the centre of mass, which for an adult of average height would be about 1 m. Hence, the maximum possible jump height for this athlete would be around 2.16 m.

## TASK

If a diver executes a dive from the 10 m board, what would be their vertical velocity when they hit the water?

## Solution

Using Equation E2.3, the diver's vertical velocity ($v_v$) at the time of hitting the water can be easily calculated:

$$v_v = \sqrt{2 \times g \times h}$$

$$v_v = \sqrt{2 \times 9.81 \times 10}$$

$$v_v = \sqrt{196.2}$$

$$v_v = 14.01\ m/s$$

Up to now the influence of any rotation involved in the movement has been ignored, but what if the trampolinist was to rotate as they bounced vertically? In this situation, some of the energy associated with the movement would be invested into the rotational component of the task, which would reduce the available linear kinetic energy and overall jump height. In fact, trampolinists start their routines with a series of straight bounces to gain sufficient height to perform their rotational stunts, as these movements ultimately result in a progressive decline in their overall jump height.

## TASK

If a person was to remain rigid and simply fell over by rotating about the feet, what would their velocity be when they hit the ground?

### Solution

To solve this problem we need to know a few things about the person, so for this example we will assume the following

Mass ($m$) = 70 kg

Centre of mass location / radius of gyration ($r$) = 1.0 m

Moment of inertia ($I$) = 80 kg/m$^2$

During the rotational fall the person has only angular kinetic energy, so Equation E2.1 can be rewritten in the same way as Equation E2.2 by substituting angular kinetic energy for linear kinetic energy. Using ground level as the reference zero level, the relationship between gravitational potential energy and rotational kinetic energy can be expressed as:

$$\frac{1}{2}I \times \omega^2 = m \times g \times h$$

Hence:

$$\left(\frac{1}{2} \times 80\right) \times \omega^2 = 70 \times 9.81 \times 1$$

$$\omega^2 = \frac{70 \times 9.81 \times 1}{40}$$

$$\omega = \sqrt{\frac{70 \times 9.81 \times 1}{40}}$$

$$\omega = 4.14 \text{ rad/s}$$

To derive the linear velocity of the centre of mass, which is rotating 1 m above the axis of rotation (i.e. the ankles) at an angular velocity of 4.14 rad/s, the following formula can be used:

$$v = r \times \omega$$

Hence:

$$v = 1.0 \times 4.14$$

$$v = 4.14\,m/s$$

So the person would make contact with the ground at 4.14 m/s at the location of their centre of mass, but as their head is located further from the axis of rotation (i.e. the ankles) this segment will make contact with the ground at an even faster velocity (e.g. 6 to 7 m/s).

It is important to note that in each of these examples, air resistance has been ignored. As velocity increases, it becomes increasingly less reasonable to ignore the influence of air resistance (i.e. motion occurring at velocities greater than 5 to 6 m/s are likely to be influenced by air resistance). The specific influence of air resistance is considered in more detail elsewhere in this text, but for the purposes of using the equations expressed in this section, it is sufficient to be aware of the limitation. For example, the diver's velocity calculated above represents the maximum velocity that would be achieved, but in reality, air resistance would reduce the diver's velocity slightly. Nevertheless, it should be considered that air resistance would not alter the fact that the diver's impact with the water is fast!

*Key notes*

| | |
|---|---|
| **Law of conservation of energy** | The law of conservation of energy is one of the cornerstones of science and states that energy can be neither created nor destroyed. It expresses that the total amount of energy remains constant as it changes from one form to another. Although this law applies to the energy exchanges that occur in sport and exercise, the large number of possible energy combinations can limit its practical application. Nonetheless, a more restricted form of the law that has particular value to this discipline can be identified. |
| **Conservation of mechanical energy** | The conservation of mechanical energy refers to the specific form of the law of conservation of energy which is of value in sport and exercise science as it uses only mechanical forms of energy. It refers to exchanges between just two types of energy: the gravitational potential energy, and linear and angular kinetic energy. In general the conservation of mechanical energy applies to projectile flight where air resistance can be neglected. It cannot be applied where there are obvious energy losses due to friction or other resistances. |

## FURTHER READING

The following resource provides additional reading about the conservation of energy during walking.

1 Arch, E. S., & Fylstra, B. L. (2016). Combined ankle-foot energetics are conserved when distal foot energy absorption is minimized. *Journal of Applied Biomechanics, 32*(6), 571–577.

# E3 MECHANICS OF MATERIALS

*Adrian Burden*

It is important for us to understand how materials behave during sport and exercise as they are fundamental to both performance and the risk of us becoming injured. For example, if subjected to forces that are either too high or are repeated too frequently the **biological materials** in our body (e.g. tendon and bone) may become damaged, resulting in injury. Considering a **non-biological material**, such as a pole vault pole, the amount that it bends when an athlete plants it in the box at the end of the run-up is crucial to how high they will jump. This section will describe how both biological and non-biological materials behave when they experience forces (i.e. are loaded) and how this can relate to performance and the risk of being injured.

During sport and exercise, as well as everyday living, our bodies experience external forces that are often referred to as ground reaction forces (GRF). As outlined elsewhere in this text, these are often expressed in relation to the magnitude of our body weight (BW), and can be one to two times our BW for walking and three to four times BW for faster running. During the same activities, the forces acting at our joints, known as joint reaction forces, can be much higher than GRFs, resulting in biological structures, such as bones and ligaments, being subjected to high forces/loads.

The mechanical properties of a material are determined by the way it reacts to **loading.** The applied load can be categorised as a **force** or a **torque** (or twisting moment) or a combination of these. Loading can either be **gradual** (such as when lifting a barbell) or **impulsive** (such as heel strike impact in running). Loads can either be applied once (**acute loading**) or several times (**repetitive loading**). These latter two characteristics are useful when considering the injury effects of loading, as an acute load can lead to a fracture of the bones or a ruptured tendon, while a repetitive load can lead to an overuse injury, such as tendinitis.

When a load acts on a material, it causes the material to **deform**, and the nature of this deformation can be described and related to its function. When the forces applied to the two ends of a material are directed away from each other, the material is said to be in **tension** (*Figure E3.1a*), and has a tendency to stretch or extend. When the forces are directed towards each other, the material is said to be in **compression** (*Figure E3.1b*), and has a tendency to shorten. When the forces are directed along different lines of action (*Figure E3.1c*), then **shear** is created. When torques (or twisting moments) act at each end of the material in opposite directions, then **torsion** is created (*Figure E3.1d*), causing the material to twist. Combinations of forces and torques lead to more complex types of deformation, such as bending (*Figure E3.2*), when one side of the material is in compression and the other side in tension.

Examples of common load deformation conditions occur in the snatch event in weightlifting. As the weight is lifted from the ground, the bones in the arms (humerus, radius and ulna) experience tension. When the weight is supported above the head the same bones are under compression. Considering the weights and the bar, the weight of the discs acts downward so a shear force is produced on the bar. At the same time the hands produce an upwards force on the bar, causing bending; which can often be seen when heavy weights are on the bar.

As previously mentioned, external loading on the body is measured in newtons (N), and can be normalised to body weight (BW) when we wish to compare forces acting on, for example,

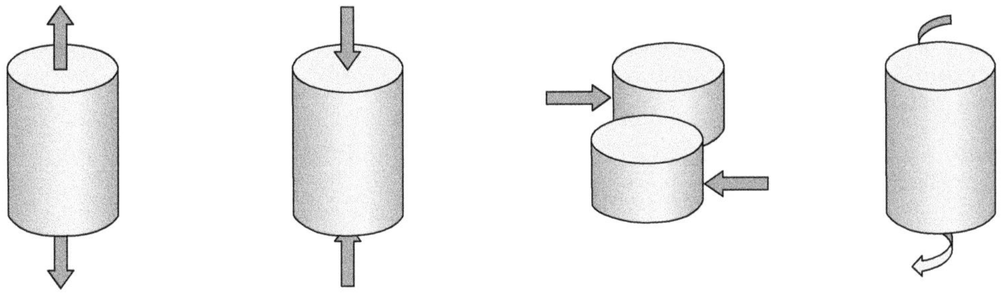


**Figure E3.1** Types of material deformation

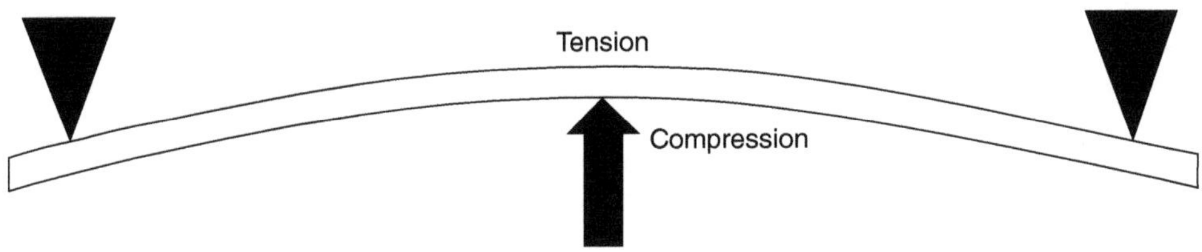


**Figure E3.2** A material experiencing bending

different people. In a similar way, the load acting on materials is normalised in relation to the cross-sectional area of the material being loaded. This is termed **stress**, which is represented by the small Greek letter sigma (σ), and is calculated using the following formula:

$$\text{Stress}(\sigma) = \frac{\text{Force (F)}}{\text{Cross}-\text{sectional area (CSA)}}$$

Thus, let us consider a tibia (CSA = 9 $cm^2$) and a fibula (CSA = 2 $cm^2$) that both experience a load of 2500 N. First we need to convert the two cross-sectional areas into their base SI units, i.e. from $cm^2$ to $m^2$. There are 100 cm in one metre, thus there are 100 x 100 = 10,000 $cm^2$ in one $m^2$. Thus, to convert from $cm^2$ to $m^2$ we need to divide by 10,000. As force is already in SI units, the mechanical stress experienced by the tibia and fibula are 2500/0.0009 = 2,777,778 $N/m^2$ and 16,666,667 $N/m^2$, respectively. While both bones experience the same load, the tibia experiences the lower load stress or load per unit area as it has a greater cross-sectional area. While the units of stress are $N/m^2$, these are more commonly referred to as pascals, or Pa for short (i.e. 1 $N/m^2$ = 1 Pa).

As in the above examples, stresses experienced by materials in the human body are often quite large, so we often present them with a prefix. The most common ones used when presenting stress are kilo (k = 1000), mega (M = 1,000,000) and giga (G = 1,000,000,000). The examples above are in the millions (i.e. 1,000,000), so it is most appropriate to present them as mega pascals (MPa). Thus, the stress experienced by the tibia and fibula would be most appropriately presented as 2.78 MPa (i.e. 2,777,778/1,000,000) and 16.7 MPa (i.e. 16,666,667/1,000,000), respectively. In both cases the answer has been rounded up to three significant figures.

While load is normalised to stress, the deformation of a material is normalised to strain. This is done by dividing the change in dimensions as, or after, the load is applied by the original dimensions of the material. Strain is denoted by the small Greek letter epsilon (ε) and is shown in the following equation:

$$\text{Strain}\ (\varepsilon) = \frac{\text{Change in dimension}\ (\Delta L)}{\text{Original dimension}\ (L_0)}$$

Thus, we can consider an Achilles tendon ($L_0$ = 17 cm) and an anterior cruciate ligament (ACL; $L_0$ = 4 cm) that are both elongated by 1 cm when experiencing tension. As the units of the numerator (top) and denominator (bottom) of the above equation for strain are the same (i.e. cm), we don't need to convert to SI units (we could, but we would get the same answer). The mechanical strain experienced by the Achilles tendon and the ACL are, therefore, 0.0588 and 0.25. Strain is most often expressed as a percentage, simply by multiplying by 100. Thus, while both structures are stretched by the same amount (i.e. 1 cm), they experience very different strains (5.9% for the Achilles tendon and 25% for the ACL).

Stress and strain are usually measured by a machine that applies a particular load to a sample of the material, while measuring that load and the deformation experienced. The load is applied in a particular direction, for example tension, and at a predetermined rate controlled by the machine until the material ruptures or fractures. Obviously this cannot be done for living structures, such as human bone or ligament, so it is done for samples taken from dead bodies, or cadavers, usually from animals.

As both the load and the deformation (e.g. tension) are measured by the machine the resulting load *vs* deformation curve over the duration when the load was applied can be plotted. Using the stress and strain equations, this curve can easily be converted into a stress *vs* strain curve. While the example curve shown in *Figure E3.3* is for cortical bone during tension, many materials display a similar relationship between stress and strain. Initially, when the sample is placed under tension, it will experience **elastic deformation**; that is, if the load were removed the sample would return to its original dimensions. This elastic, or linear, region is shown in *Figure E3.3*. For many materials, stress is linearly related to strain, and this relationship is known as **Hooke's law.** The relationship holds until a material reaches its **elastic limit** or **yield point** (*Figure E3.3*), where the material begins to deform more easily for a smaller increase in stress. Hooke's law implies that as the force (or stress) increases the deformation (or strain) increases in the same proportion and so the force-to-deformation ratio (or stress-to-strain ratio) is constant. This constant is known as the **stiffness** of the material when force and deformation are used to describe its behaviour, and the **modulus of elasticity** when the stress and strain are used. Calculation of the modulus of elasticity for the stress vs strain curve previously shown in *Figure E3.3* is shown in *Figure E3.4*.

As the material experiences further tension after the yield point it begins to exhibit **plastic deformation**; that is, it will begin to experience micro-fractures and, if the load were removed, it would not return to its original dimensions. This is what happens, for example, when a tendon or ligament experiences a minor sprain. Under further tension the material will eventually fracture or rupture completely and experience its ultimate stress and strain (see *Figure E3.3*). The strength of the material is often defined by the ultimate stress that it can withstand. The energy which the material is able to absorb during loading is also often used as a measure of strength and is measured as the area under the stress *vs* strain curve (see *Figure E3.4*).

Thus far, we have discussed applying loads to materials and only made brief reference to unloading them. If a material is elastic, such as rubber, when the load is removed the material quickly returns to its original dimensions, as in *Figure E3.5a*. For other materials, such as structures found in the human body, during unloading the force provided by the material is lower,

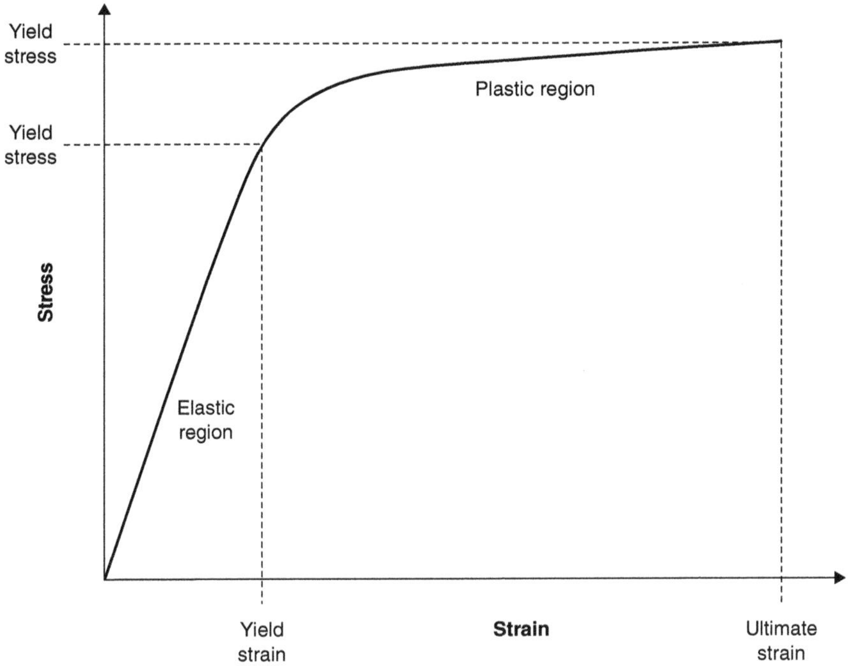


**Figure E3.3** Idealised stress *vs* strain curve of cortical bone loaded under tension

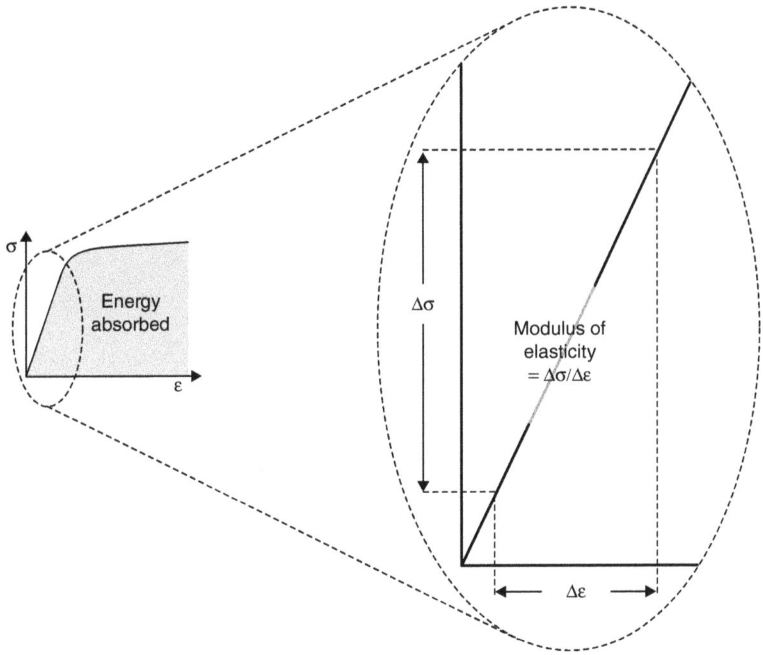


**Figure E3.4** Determination of energy absorbed by a material and its stiffness (from the stress *vs* strain curve in Figure E3.3)

for the same deformation, than it was during loading (*Figure E3.5b*). This occurs because such materials are **viscoelastic**, in that they display both viscous and elastic behaviour. Viscosity is a material's resistance to flow or deformation (e.g. treacle has a higher viscosity than oil, which is more viscous than water). In other words, in viscoelastic materials, the force lags behind during unloading. This is termed **hysteresis**, which is a characteristic of all viscoelastic materials.

The darker shaded area in *Figure E3.5b* represents the energy returned during unloading and is a measure of the **resilience** of a material. The lighter shaded area represents the energy lost during unloading, and is termed the **loop of hysteresis**. Thus, the more elastic a material is, the less energy it will lose during unloading or recoil.

A good example of hysteresis energy loss is when a ball bounces on the floor. If the ball is dropped from a certain height it will never rebound to exactly the same height. The energy stored during compression allows the ball to bounce back. Some of this energy is lost due to friction between the molecules that develop heat during the compression, and so the recoil energy is never quite enough to get the ball back to the height from which it was dropped. The same principle applies to the contact between a tennis racket and ball. Although the racket and ball are poor devices for returning energy, the strings are excellent and hence compensate for these deficiencies.

A special note should be given to sports surfaces. In sports like gymnastics and tumbling the surfaces are described as **area elastic**; that is, they deform over a large area when jumped on and have good elasticity to aid the performer. Wooden gymnasium floors that are "sprung" are also area elastic. Surfaces like real or artificial turf are considered **point elastic**, as they deform in a localised region when jumped on (*Figure E3.6*) and generally have poor elasticity.

Permanent deformations are referred to as **set**, and describe the **plastic** behaviour of materials. Set can be important in some sport materials, for example those used in the midsoles of running shoes. The expanded foam material that is used to provide cushioning as the foot makes contact with the ground gradually permanently deforms through use. This happens because the normally closed cells which make up the foam material gradually fracture and release their internal pressure, and ultimately collapse. When this happens the running shoe midsole becomes thinner and harder. A worn shoe is a known injury risk factor, as it is likely to increase the impact force on heel strike. Shoes that show any sign of this type of wear should be replaced.

There are other properties of materials relevant to sport and exercise. The term **hardness** is used to describe how much resistance a material has to penetration, stretching and wear. A material that is difficult to penetrate is very hard (e.g. the ball bearings in the bearing race of

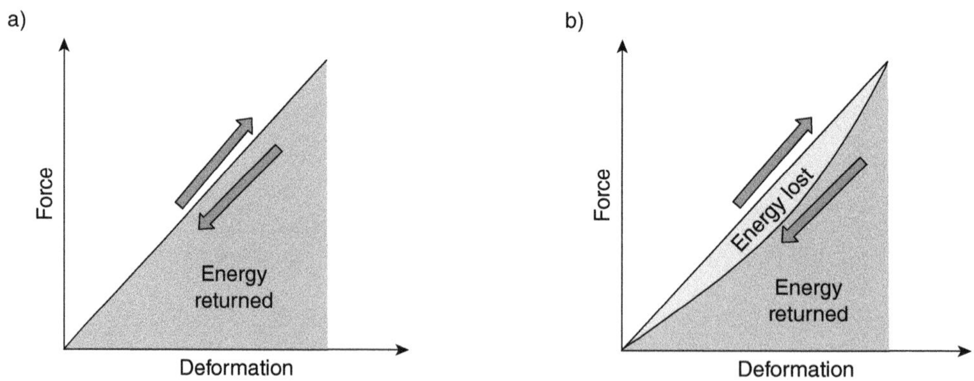


**Figure E3.5** Loading and unloading for (a) an elastic and (b) a viscoelastic material

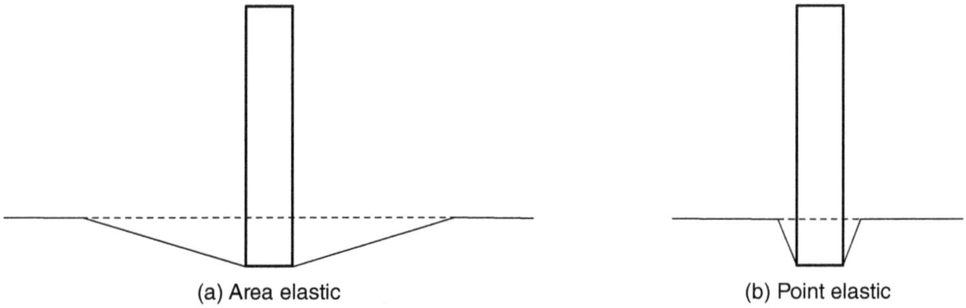


**Figure E3.6** Point and area elastic surfaces

a bicycle). Conversely, a "crash" mat used in gymnastics is very soft. As there is such a wide range of hardnesses in common materials, there are various scales used to measure these. For intermediate materials, such as the midsole material of a running shoe, the Shore "A" scale is used. A measure is taken by a device that has a spring-loaded blunt pointer that is impressed into the surface (*Figure E3.6b*); allowing the amount of penetration to be recorded and used. Clearly, the strength of the spring has to match the general deformability of the surface tested. For very hard surfaces a much stronger spring is used. Running shoe midsole materials range from a Shore "A" value of 20 (hard) to 50 (soft) with 35 being a commonly found value.

*Key notes*

| | |
|---|---|
| **Load and deformation** | Material solids can sustain applied loads, but they have a tendency to deform. Depending on the load applied, their state will be one of tension and have a tendency to extend; compression and tend to shorten; shear and tend to slide; or torsion and tend to twist. |
| **Stress and strain** | Stress is defined as the force per unit area and describes the way the force is distributed through the material. Strain is defined as the increase in length divided by the original length and is often expressed as a percentage. |
| **Elastic and plastic deformation** | For many materials, when a load is applied, they deform elastically, so that when the load is removed they return to their original dimensions. When loaded, stress is linearly related to strain, and this relationship is known as Hooke's law. The relationship holds until a material reaches its elastic limit or yield point where the material begins to deform plastically. Under this type of deformation, the material no longer returns to its original dimension when the load is removed. |
| **Viscoelasticity and hysteresis** | Some materials, such as structures found in the human body, are viscoelastic, in that they display both viscous and elastic behaviour. Viscosity refers to a material's resistance to deformation or flow. In such materials, the force lags behind during unloading, which is termed hysteresis, and is characteristic of all viscoelastic materials. |
| **Area elastic and point elastic surfaces** | Surfaces in sport can be characterised as area elastic or point elastic depending on how they deform. Area elastic surfaces deform over a large area and floors which are designed to be area elastic are generally referred to as sprung floors. These have advantages in terms of energy return to the athlete and are generally more comfortable to play on. Point elastic surfaces deform locally and typify playing fields and artificial playing surfaces. These are generally less comfortable to perform on. |

## FURTHER READING

The following resource provides additional reading around the importance of the material used to make the midsole of a running shoe with respect to controlling ground reaction forces.

1 Kersting, U. G., & Brüggeman, G. P. (2006). Midsole material-related force control during heel-toe running. *Research in Sports Medicine, 14*(1), 1–17.

# IMPACTS

*Michael Cole*

**Impacts** are characterised by **large forces** that are applied over **short time periods.** Sporting examples of impacts include a soccer header, hitting a ball with a tennis racket or striking a golf ball with a driver. In each of these cases, large forces ranging from about 1000 N (soccer header) to approximately 10000 N (golf swing) are applied to the ball over contact times ranging from as little as 0.5 µs (0.0005 seconds) for the golf swing to 20 ms (0.02 seconds) for the soccer header. Given the range of forces and contact times provided in these examples, it is evident that the nature of an impact is dependent on both objects involved in the collision. To understand the nature of an impact, it is necessary to consider the **conservation of linear momentum** and the **coefficient of restitution** relationships, which lead to two equations known as the **impact equations.**

The **conservation of linear momentum** is an important scientific principle that defines how objects behave when they interact. While the principle applies to systems that contain many objects (e.g. a rugby scrum), this text will consider the interaction between just two objects (e.g. a golf ball and driver). The **conservation of linear momentum** represents the **first impact equation** and states that the total momentum possessed by a collection of objects prior to their interaction (in this case a collision) is the same as their collective momentum after the collision. This is expressed in the following equation, where $m_A$ and $m_B$ represent the masses of two objects, $v_A$ and $v_B$ denote the objects' initial velocities and $v'_A$ and $v'_B$ represent the objects' velocities after the collision:

$$m_A \times v_A + m_B \times v_B = m_A \times v'_A + m_B \times v'_B$$

Total momentum of the system before = Total momentum of the system after

**(Equation E4.1)**

The **coefficient of restitution** (denoted by the symbol "*e*") is a numeric value that reflects **elasticity** and describes how much of an object's energy is lost to its deformation during a collision and what proportion remains to facilitate its rebound. Using the simple example of dropping a ball from a height $H_{drop}$ and observing it to rebound to a height $H_{rebound}$, the coefficient of restitution can be determined as:

$$e = \sqrt{\frac{h_{rebound}}{h_{drop}}}$$

**(Equation E4.2)**

Given this example, it is evident that the smallest and largest values for "*e*" are zero (0) and one (1), which are respectively recorded when the ball does not rebound (i.e. $H_{rebound}$ is zero) or when the ball rebounds to the height from which it was dropped. While a super ball can come close to achieving a rebound height equal to the drop height, this latter situation never happens in practice. Recall the expression linking height of drop and velocity of impact from the Conservation of Energy section of this text:

$$v = \sqrt{2 \times g \times h}$$

**(Equation E4.3)**

By substituting Equation E4.3 into Equation E4.2, the expression for "e" can be simplified to:

$$v_{drop} = \sqrt{2 \times g \times h_{drop}} \quad or \quad v_{drop}^2 = 2 \times g \times h_{drop}$$

$$\frac{v_{drop}^2}{2g} = h_{drop}$$

$$v_{rebound} = \sqrt{2 \times g \times h_{rebound}} \quad or \quad v_{rebound}^2 = 2 \times g \times h_{rebound}$$

$$\frac{v_{rebound}^2}{2g} = h_{rebound}$$

Hence, Equation E4.2 can be rewritten as:

$$e = \sqrt{\frac{v_{rebound}^2 / 2g}{v_{drop}^2 / 2g}} \quad or \quad e^2 = \frac{v_{rebound}^2 / 2g}{v_{drop}^2 / 2g}$$

As both the numerator and denominator are divided by 2g, the equation can be further simplified by cancelling this component out:

$$e^2 = \frac{v_{rebound}^2}{v_{drop}^2}$$

Similarly, given that the components on both sides of the equation are squared, they effectively cancel each other out, leaving us with the equation:

$$e = \frac{v_{rebound}}{v_{drop}}$$

In this equation, it should be remembered that velocity is a vector quantity and hence has both magnitude and direction. Therefore, the velocity of a falling ball will be oppositely directed to its velocity while rebounding, hence the direction of its motion must be considered in the calculations. While these examples describe one moving object (e.g. a ball) colliding with a stationary object (e.g. the ground), it is important to note that the coefficient of restitution is also relevant to more complex situations in which two or more moving objects collide. When such an impact occurs, momentum is conserved and hence there is a relationship between the masses and velocities before and after impact. Substituting the velocity terms from Equation E4.1 into the expression for "*e*" gives the more general equation of the coefficient of restitution:

$$e = \frac{v'_A - v'_B}{v_B - v_A}$$

**(Equation E4.4)**

Equation E4.4 depicts the **second impact equation,** which takes into account the relative velocities of the bodies before impact ($v_B - v_A$) and the relative velocities after impact ($v'_A - v'_B$). The reversal of the $v_A$ and $v_B$ terms for the denominator takes into account the change in direction that occurs when the objects rebound following impact (i.e. positive velocity becomes negative velocity). If the rebound situation described above is applied to equation E4.4, the first mass would represent the ball ($m_A$) and the second mass would be the floor ($m_B$). Given that the floor's velocity is zero both prior to ($v_B = 0$) and after ($v'_B = 0$) impact, the equation can be simplified to:

$$e = \frac{v'_A}{-v_A}$$

Experiments show that the coefficient of restitution for various sports balls dropped onto a concrete floor range from 0.75 for a basketball or soccer ball to 0.67 and 0.32 for tennis and cricket balls, respectively. However, it should be noted that the coefficient of restitution is not fixed and that dropping the same sports balls onto a softer surface, such as grass, reduces this value. Similarly, sports balls that are softer or have a lower pressure will also have a lower coefficient of restitution, indicating that the coefficient of restitution is influenced by the characteristics of all elements involved in the collision.

Collisions between two objects can typically be classified into one of two main categories: i) central impacts; or ii) oblique impacts. For **central impacts**, the velocities of the centre of mass of each object are directed towards each other, along the **line of impact** (*Figure E4.1*), as is the case during **head-on collisions**. Given some information about the objects involved in the collision, it is possible to use the two impact equations (equations E4.1 and E4.4) to derive further information about the impact. These equations include seven terms ($m_A, m_B, v_A, v_B, v'_A, v'_B$ and $e$) and can be used to find a maximum of two unknowns. In other words, five of these seven terms must be known to determine a complete solution to the problem. An example is given in *Figure E4.2*.

## TASK

A 60 kg rugby wing is moving at a speed of +8 m/s and collides with a 100 kg forward who is at rest (i.e. velocity equals 0 m/s) (*Figure E4.2*). After the collision, the forward is observed to move at a speed of +3.6 m/s. Determine the velocity of the wing after the collision and the coefficient of restitution between the two players. Assume no interaction with the ground.

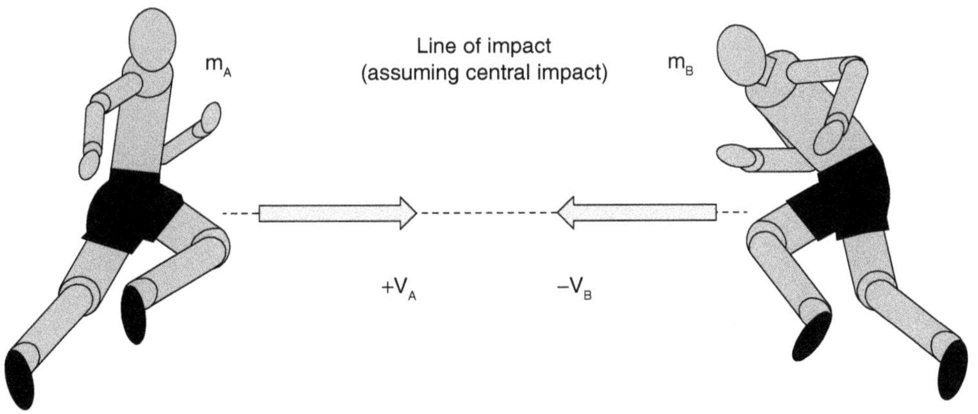


**Figure E4.1** Central impacts occur when two objects collide along the line of impact

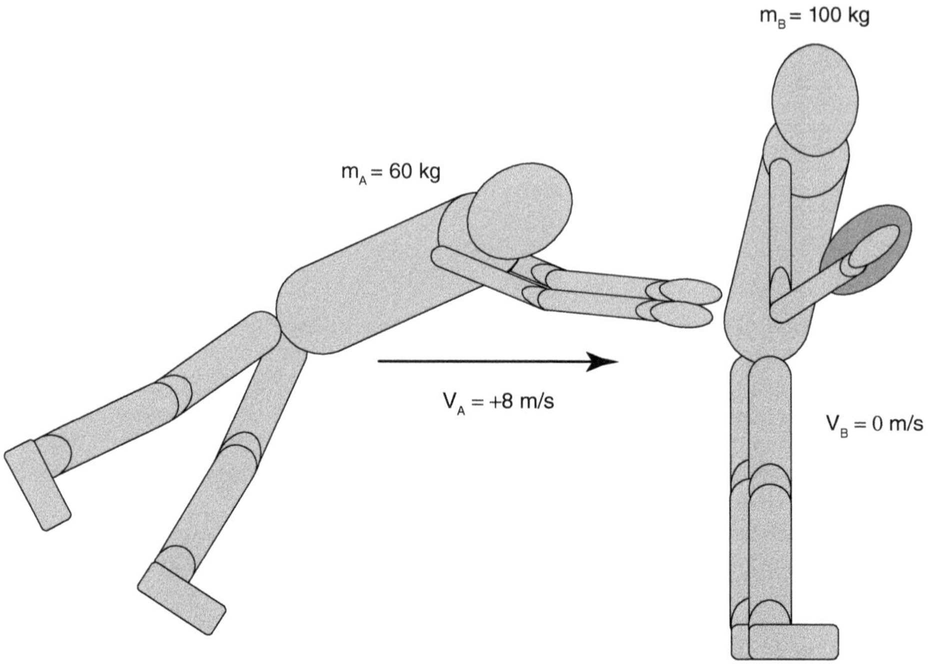


**Figure E4.2** A central impact between two rugby players

## Solution

Known variables:

$$m_A = 60 \text{ kg}$$

$$v_A = +8 \text{ m/s}$$
$$v'_A = ?$$

$$m_B = 100 \text{ kg}$$
$$v_B = 0 \text{ m/s}$$
$$v'_B = +3.6 \text{ m/s}$$

Using the conservation of linear momentum equation:

$$m_A \times v_A + m_B \times v_B = m_A \times v'_A + m_B \times v'_B$$

$$(60 \times 8) + (100 \times 0) = (60 \times v'_A) + (100 \times 3.6)$$

$$480 + 0 = (60 \times v'_A) + 360$$

$$480 - 360 = 60 \times v'_A$$

$$v'_A = \frac{120}{60}$$

$$v'_A = +\mathbf{2.0\ m/s}$$

Using the coefficient of restitution equation:

$$e = \frac{v'_A - v'_B}{v_B - v_A} = \frac{2.0 - (+3.6)}{0 - (+8.0)} = \frac{-1.6}{-8} = 0.2$$

**Note:** the coefficient of restitution ranges from 0 to 1, but is neither positive nor negative.

An interesting fact about some impacts is that the mass being hit can have a post-impact velocity that is greater than the hitting mass (i.e. there is an apparent velocity gain). For example, in soccer the foot has an impact velocity of around 20 m/s, while the ball can be propelled at ≥25 m/s. **Why does this occur and is it possible to get more velocity out than you put in?** By rearranging the two impact equations, they can be expressed in terms of foot and ball velocity. In the example below, it is assumed that the leg has no influence on the foot and that $v_{ball}$ is equal to zero (i.e. the ball is initially stationary). These equations now become:

$$m_{foot} \times v_{foot} = m_{foot} \times v'_{foot} + m_{ball} \times v'_{ball}$$

and

$$e = \frac{v'_{foot} - v'_{ball}}{-v_{foot}}$$

By restructuring this equation, it is possible to isolate the velocity of the ball after impact $(v'_{ball})$ and express this in terms of the velocity of the foot before impact $(v'_{foot})$, as:

$$v'_{ball} = v_{foot} \times \left[ \frac{m_{foot}}{m_{foot} + m_{ball}} \times (1+e) \right]$$

The term $\frac{m_{foot}}{m_{foot} + m_{ball}}$ represents the mass of the foot as a proportion of the foot and ball's combined masses and typical values for an adult male foot and a soccer ball are around 0.8. The term (1+*e*) represents the effectiveness of the impact, which is influenced by the hardness of the ball (as a result of its pressure) and the extent to which the foot deforms at the ankle and metatarsophalangeal joints during impact. A typical value for this term is around 1.5. It should be noted that the deformation and flexing of the foot results in a lower coefficient of restitution than would be typical of an impact between a soccer ball and concrete. Substituting these values into the above equation:

$$v'_{ball} = v_{foot} \times (0.8 \times 1.5) = 1.2 \times v_{foot}$$

This relationship suggests that in a typical kick (maximal instep kick by a competent player) the ball should travel about 20 per cent faster than the foot travels. This speed gain is the result of the greater mass of the foot relative to the ball. If a player can increase the mass of the foot (e.g. by wearing a heavier boot) or increase the quality of the impact (e.g. by having a more rigid foot) then the initial velocity of the ball following impact would be further increased.

In **oblique impacts,** the velocities of each object are directed towards each other at an angle to the line of impact $(\theta_A \text{ and } \theta_B)$ as illustrated in *Figure E4.3*. After impact, the objects move away from each other again but with different velocities and at different angles $(\theta'_A \text{ and } \theta'_B)$. These four angles are added to the seven terms that characterise central impacts (i.e. $m_A$, $m_B$, $v_A$, $v_B$, $v'_A$, $v'_B$ and *e*) to yield **eleven terms** for these situations.

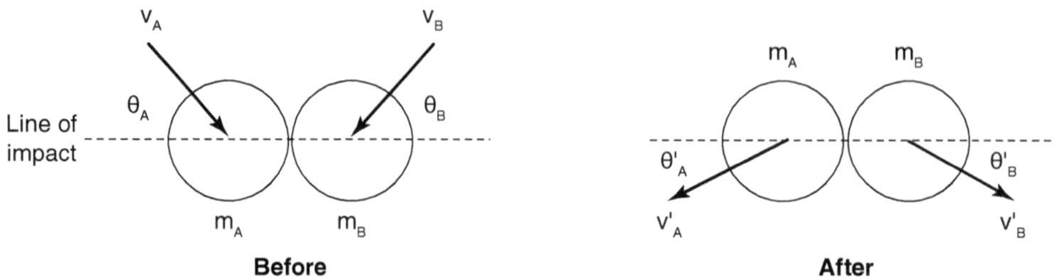


**Figure E4.3** An illustration of an oblique impact in the instant before and after contact

Solving an oblique impact problem requires the velocities to be divided into two components: one directed **along the line of impact** (horizontal in this example) and one that occurs **perpendicular to the line of impact** (vertical in this example) (*Figure E4.4*). The component directed **along the line of impact** is a central impact and can be resolved using the conservation of linear momentum (Equation E4.1). The velocity component terms in *Figure E4.4* are substituted to give:

$$m_A \times (v_A \times \cos\theta_A) + m_B \times (v_B \times \cos\theta_B) = m_A \times (v'_A \times \cos\theta'_A) + m_B \times (v'_B \times \cos\theta'_B) \quad \textbf{(Equation E4.5)}$$

Similarly, using the coefficient of restitution (Equation E4.4) and substituting the velocity components of Equation E4.5 gives:

$$e = \frac{(v'_A \times \cos\theta'_A) - (v'_B \times \cos\theta'_B)}{(v_B \times \cos\theta_B) - (v_A \times \cos\theta_A)} \quad \textbf{(Equation E4.6)}$$

The velocities that occur **perpendicular to the line of impact** (vertical in this example) are not affected by the collision, as the interaction between the objects is considered frictionless in this direction. The absence of an interacting force means that the following can be written:

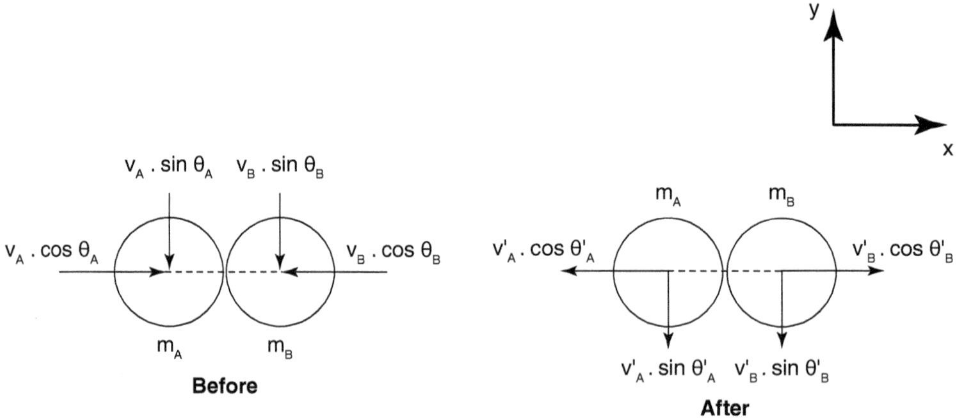


**Figure E4.4** Velocity components for an oblique impact in the instant before and after contact

$$v_A \times \sin\theta_A = v'_A \times \sin\theta'_A \qquad \text{(Equation E4.7)}$$

$$v_B \times \sin\theta_B = v'_B \times \sin\theta'_B \qquad \text{(Equation E4.8)}$$

If seven of the eleven terms are known, equations E4.5 to E4.8 will allow the calculation of four unknowns. An example is given below (*Figure E4.5*).

## TASK

Two identical smooth balls with the same mass collide with the velocities and direction shown in *Figure E4.5*. If the coefficient of restitution is 0.9, determine the velocity and direction of each ball after impact.

## Solution

First obtain the component velocities **before impact.** It is important here to consider that the balls are travelling in opposing directions when calculating the central component of the impact (denoted by the negative $v_{B_x}$ value):

$$v_{A_x} = v_A \times \cos(30) = +26.0 \text{ m/s}$$

$$v_{B_x} = -v_B \times \cos(60) = -20.0 \text{ m/s}$$

$$v_{A_y} = v_A \times \sin(30) = +15.0 \text{ m/s}$$

$$v_{B_y} = v_B \times \sin(60) = +34.6 \text{ m/s}$$

In this example, given their smooth surface and neglecting the effects of other forces (e.g. spin), the balls' velocities **perpendicular to the line of impact** (i.e. the y component) remain unchanged after the impact. In other words:

$$v'_{A_y} = \mathbf{15.0 \text{ m/s}}$$

$$v'_{B_y} = \mathbf{+34.6 \text{ m/s}}$$

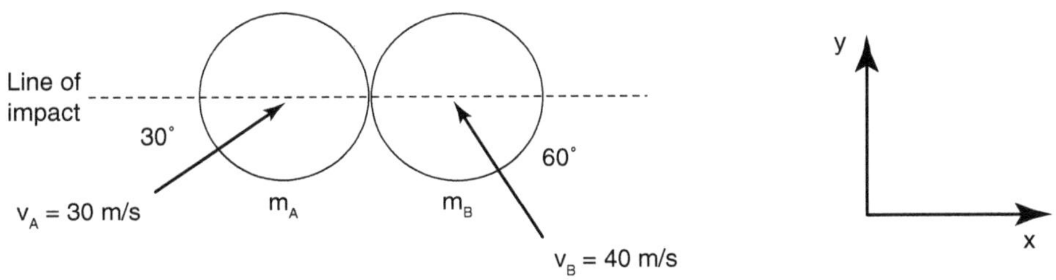


**Figure E4.5** Oblique impact between two balls with identical physical characteristics

The balls' velocities **parallel to the line of impact** (i.e. the x components) are governed by the conservation of linear momentum and coefficient of restitution equations. Calculation of the conservation of linear momentum gives:

$$m_A \times v_{A_x} + m_B \times v_{B_x} = m_A \times v'_{A_x} + m_B \times v'_{B_x}$$

However, as the balls are identical, $m_A$ and $m_B$ are equal, hence the substitution of the values for $v_{A_x}$ and $v_{B_x}$ yields:

$$v'_{A_x} + v'_{B_x} = 26.0 + (-20.0) = +6.0 \text{ m/s} \quad (1)$$

The coefficient of restitution equation yields:

$$e = \frac{v'_{A_x} - v'_{B_x}}{v_{B_x} - v_{A_x}}$$

Therefore:

$$0.9 = \frac{v'_{A_x} - v'_{B_x}}{-20 - (+26)}$$

$$0.9 \times -46 = v'_{A_x} - v'_{B_x}$$

$$-41.4 \text{ m/s} = v'_{A_x} - v'_{B_x} \quad (2)$$

As the balls' masses are the same and are unaffected by the collision, the conservation of linear momentum formula can be simplified to determine the balls' velocities parallel to the line of impact after the collision (i.e. $v'_{A_x}$ and $v'_{B_x}$):

$$v_{A_x} + v_{B_x} = v'_{A_x} + v'_{B_x}$$

Using (1) above to rearrange this equation yields:

$$v'_{A_x} = +6.0 - v'_{B_x}$$

Substitution of this information into (2) above gives:

$$+6.0 - v'_{B_x} - v'_{B_x} = -41.4 \text{ m/s}$$

$$+6.0 + 41.4 = +v'_{B_x} + v'_{B_x}$$

$$+47.4 = 2 \times v'_{B_x}$$

$$\frac{+47.4}{2} = v'_{B_x}$$

$$v'_{B_x} = \mathbf{+23.7 \text{ m/s}}$$

Given $v'_{A_x} - v'_{B_x} = -41.4\,\text{m/s}$, substitution of $v'_{B_x}$ yields:

$$v'_{A_x} - 23.7 = -41.4 \text{ m/s}$$

$$v'_{A_x} = -41.4 + 23.7$$

$$v'_{A_x} = \mathbf{-17.7\ m/s}$$

By incorporating the velocities **perpendicular to the line of impact**, the vector velocities after impact are calculated as:

$$v'_A = \sqrt{v'^2_{A_x} + v'^2_{A_y}} = \sqrt{-17.7^2 + 15^2} = \mathbf{23.2\ m/s}$$

$$v'_B = \sqrt{v'^2_{B_x} + v'^2_{B_y}} = \sqrt{23.7^2 + 34.6^2} = \mathbf{41.9\ m/s}$$

Unlike velocities, which are vector quantities, it is not practical to express angles as positive or negative values. As such, the absolute velocities are used in the following equations to determine the angles:

$$\theta'_A = \tan^{-1}(v'_{A_y} / v'_{A_x}) = \tan^{-1}(15/17.7) = \mathbf{40.3°} \text{ to the left horizontal}$$

$$\theta'_B = \tan^{-1}(v'_{B_y} / v'_{B_x}) = \tan^{-1}(34.6/23.7) = \mathbf{55.6°} \text{ to the right horizontal}$$

It is interesting to note that when an object, such as a ball, makes **contact with a surface** its rebound angle will be reduced relative to its incidence angle (*Figure E4.6*). This change is caused by a loss of energy during the collision, which reduces the vertical component of the ball's velocity, while (in an ideal situation) leaving the horizontal component largely unchanged. The same scientific principles outlined in this section are used to solve these types of problems and an example of a squash ball impacting a vertical wall is given in *Figure E4.7*.

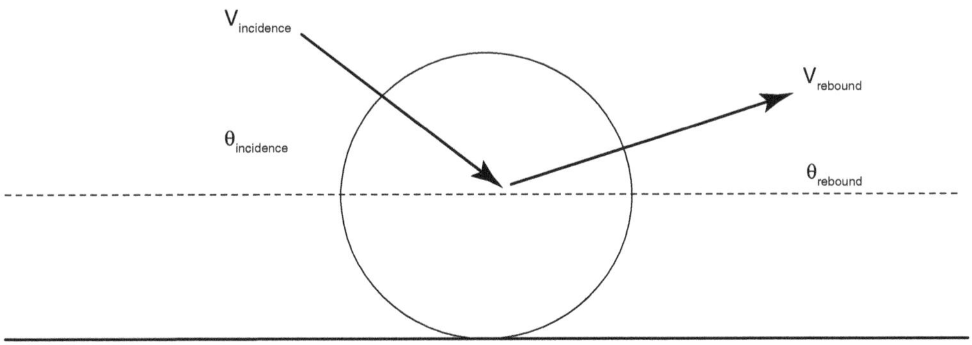


**Figure E4.6** Oblique impact of a ball with a surface

## TASK

A squash ball is hit against a smooth vertical wall with a velocity of 20 m/s at an angle of 30 degrees to the lower vertical (*Figure E4.7*). If the coefficient of restitution is 0.4 determine the magnitude and direction of the ball's velocity after impact.

## Solution

First resolve the component velocities of the ball before impact.

$$v_{x_{ball}} = 20 \times \sin(30) = \mathbf{+10.0\ m/s}$$

$$v_{y_{ball}} = 20 \times \cos(30) = \mathbf{+17.3\ m/s}$$

As the wall in this example is smooth, the ball's velocity **perpendicular to the impact line** (i.e. the x component) is unchanged after impact, but occurs in the opposite direction. Hence:

$$v'_{x_{ball}} = \mathbf{-10.0\ m/s}$$

The ball's velocity **parallel to the impact line** (i.e. y component) after impact is covered by the coefficient of restitution. Note that the wall is stationary (i.e. $v_{y_{wall}}$ and $v'_{y_{wall}} = 0$ m/s):

$$e = \frac{v'_{y_{ball}} - v'_{y_{wall}}}{v_{y_{wall}} - v_{y_{ball}}}$$

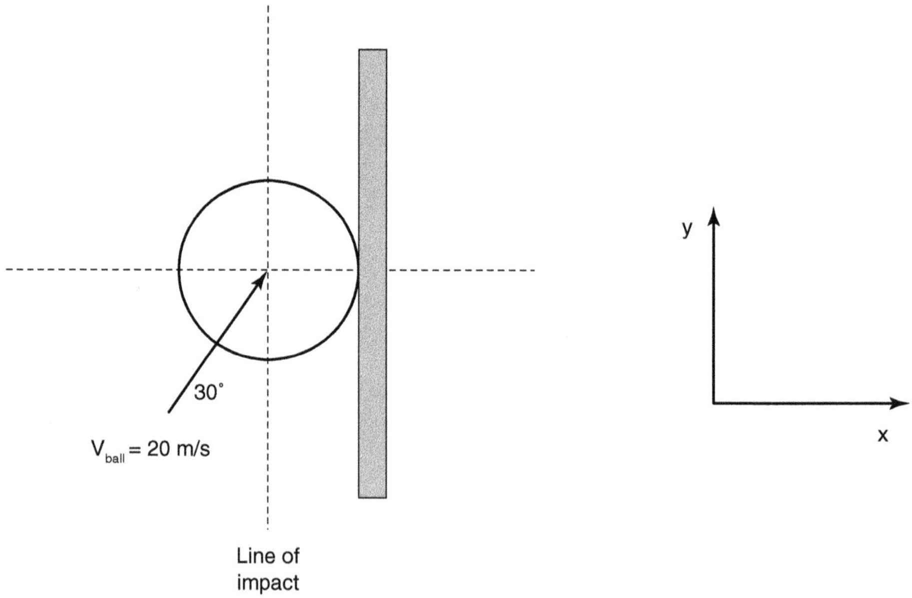


**Figure E4.7** Oblique impact between a squash ball and a smooth vertical wall

$$0.4 = \frac{v'_{y_{ball}} - 0}{0 - (+17.3)} \qquad or \qquad 0.4 = \frac{v'_{y_{ball}} - 0}{0 - 17.3}$$

$$0.4 \times 17.3 = v'_{y_{ball}} - 0$$

$$+6.9\ m/s = v'_{y_{ball}}$$

$$v'_{y_{ball}} = +6.9\ \mathbf{m/s}$$

The resultant velocity is given by resolving the two velocity components:

$$v'_{ball} = \sqrt{v'^2_{x_{ball}} + v'^2_{y_{ball}}} = \sqrt{-10.0^2 + 6.9^2} = \mathbf{12.1\ m/s}$$

Again, given that it is impractical to express angles as positive or negative values, the absolute velocities are used to determine the angle of rebound:

$$\theta = \tan^{-1}(v'_{y_{ball}} \,/\, v'_{x_{ball}}) = \tan^{-1}(10.0 / 6.9) = \mathbf{55.4°}\text{ to the upper vertical}$$

In reality, the condition of a frictionless impact does not apply, as a ball will tend to slide over the surface during the time of contact. As the ball skids, frictional forces act to slow its horizontal velocity and hence decrease its speed in both the perpendicular and parallel directions. While this may lead to the ball rebounding at a higher angle than the incidence angle, the increased angle comes at the cost of a reduced velocity. This tendency for a ball to "sit up" after impact with a surface is a feature of some tennis courts (e.g. clay courts are often described as "slow"), while other surface types, such as grass, allow the ball to slide more easily. As the velocity of the ball is largely unaffected on these surfaces, these surfaces are typically described as "fast". Furthermore, it should be noted that if one or more of the objects involved in a collision are rotating before impact (e.g. top spin placed on a tennis ball), this will influence its interaction with the surface and ultimately alter its final direction of travel. However, the detailed mechanics of these situations are complex and beyond the scope of this text.

*Key notes*

| | |
|---|---|
| **Impacts and collisions** | Impacts are characterised by large forces applied over short contact times and can be examined using the conservation of linear momentum and coefficient of restitution. A collision describes the way in which two objects interact with each other during an impact. |
| **The conservation of linear momentum** | The conservation of linear momentum defines how objects behave when they interact and represents the general situation of a collision in sport. The law states that a system's linear momentum before an interaction is the same as its linear momentum after the interaction (i.e. momentum is conserved). Conservation of linear momentum is the first of two impact equations. |
| **The coefficient of restitution** | The coefficient of restitution (denoted by the symbol "e") quantifies how the energy stored in a deformed material is returned. It provides a numeric value that reflects the elasticity of the materials involved in a collision. The coefficient of restitution can range from 0 when a dropped object does not rebound to 1 when the dropped object rebounds to the height from which it was dropped. The coefficient of restitution is the second impact equation. |

*(continued)*

*(continued)*

| | |
|---|---|
| **Central impacts** | Central impacts, or head-on collisions, occur when the velocities of two objects are directed towards each other along a line of impact. |
| **Oblique impacts** | Oblique impacts occur when the velocities of two objects are not directed along a line of impact. While the two impact equations also apply to oblique impacts, further information is required to solve these problems. |

## FURTHER READING

The following resources provide additional reading about linear and oblique impacts as they relate to soccer and other codes of football.

1 Nunome, H., Ball, K., & Shinkai, H. (2014). Myth and fact of ball impact dynamics in football codes. *Footwear Science, 6*(2), 105–118.
2 Lees, A., & Nolan, L. (1998). The biomechanics of soccer: A review. *Journal of Sports Sciences, 16*(3), 211–234.

# CENTRE OF PERCUSSION

*Paul Grimshaw*

Most striking and throwing actions start at the base of the segments (such as the legs) and travel sequentially up the body through the pelvis, trunk, shoulder, upper arm, forearm and hand. The body has segments (i.e. arms, hands, legs) that are connected to each other in series by joints, and the striking implement (e.g. bat or racket) becomes one additional link in this series of sequential movements. *Figure E5.1* helps to illustrate this sequential segment movement and the summation of speed principle that is directly linked to this action. The striking implement, such as the bat or racket, is the last link in this kinematic and kinetic chain.

This can easily be demonstrated by considering the swinging of a compound pendulum versus the swinging of a rigid connected system. A compound pendulum is composed of a series of masses ranging from 2.2 kg to 40 kg that are connected by links that are hinged at the mass centres. The rigid connected system contains the same masses but these are connected by rigid links that are not hinged. The velocity of the smallest and most distal mass (2.2 kg) is measured, as the two pendulums swing down from a horizontal position under the effect of gravity. The maximum velocity of the rigidly connected system's end point is about 10 m/s, whereas the maximum velocity of the end point in the linked compound system is 18 m/s. *Figure E5.2* helps to illustrate this and shows how it relates to the human arm.

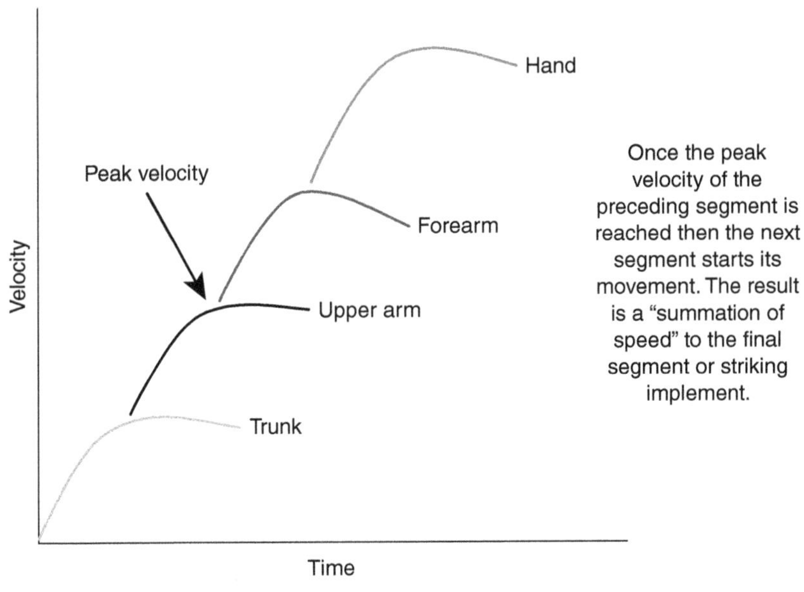


**Figure E5.1** The summation of sequential velocities during a throwing or striking motion

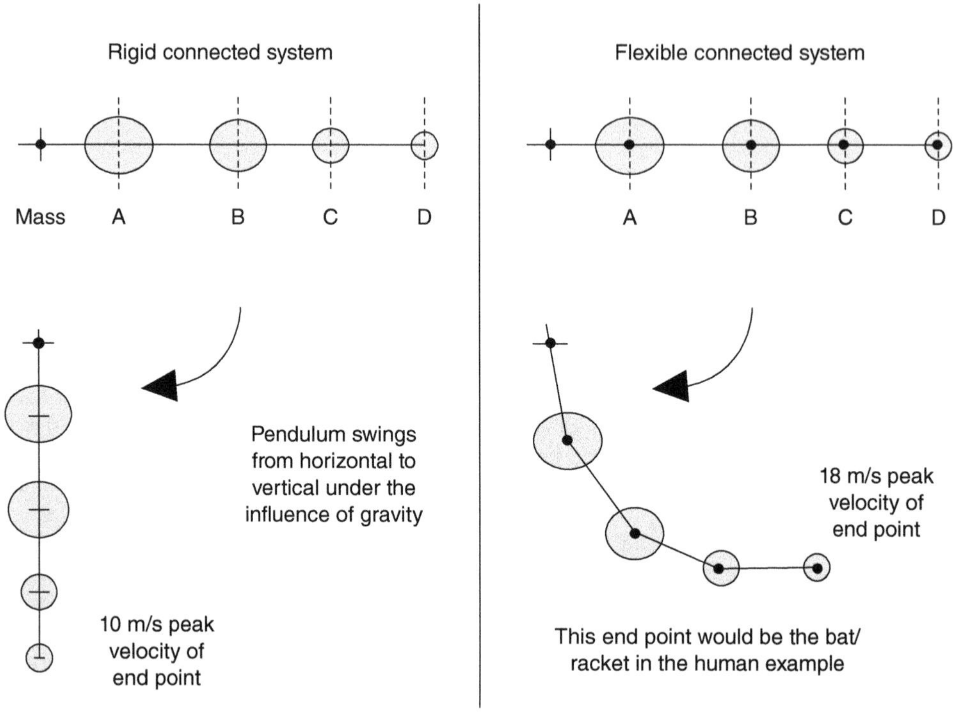


**Figure E5.2** The rigid versus a compound (flexible) pendulum

One of the main objectives of sequential segmental motion in racket sports, such as tennis, cricket and baseball, is to accelerate the end point of the implement (i.e. the bat or racket) so that it is travelling at an optimum speed (velocity) when it makes contact with the ball. As we have seen from the Linear–Angular Motion section of this text, segments of the body travel in angular rotations about proximal (closest to the attachment) or distal (furthest away from the attachment) axes of rotation. The shortest distance between any axis of rotation and the **centre of percussion** of the implement is known as the radius of rotation. The centre of percussion of an implement can be broadly defined as the point (or often area) on the racket or bat that makes the most effective contact with the ball. For example, in tennis a larger centre of percussion area on a racket can minimise the consequences of poorly hit shots. The **centre of percussion** is an important point for both effective performance and prevention of injury.

A simple experiment to explain the centre of percussion is to consider a vertical beam that is suspended from a horizontal wire by a U-bolt. The vertical beam is allowed to move freely along the wire. *Figure E5.3* helps to explain this experiment in more detail.

If, in the example shown in *Figure E5.3*, the force is applied to the beam at exactly the point of the centre of percussion then there will be no movement of the beam in either a rotational or a translational direction, as the two will cancel each other out. This has implications for striking a ball at exactly the centre of percussion of the racket or bat and, in particular, for the prevention of injury to the wrist, elbow and shoulder. By striking the ball at the centre of percussion, there will be limited vibration/movement translated to the rotation or pivot point; that is, the hand, wrist, elbow and/or shoulder, depending on how the shot is hit.

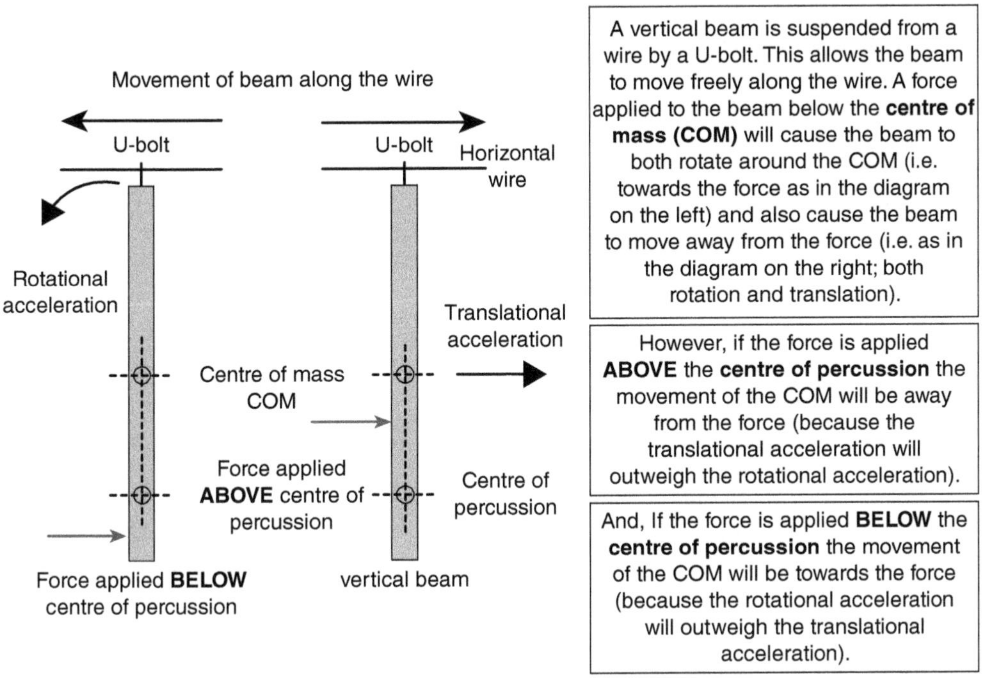


**Figure E5.3** A vertical beam that is suspended from a wire and is free to move (rotate and translate)

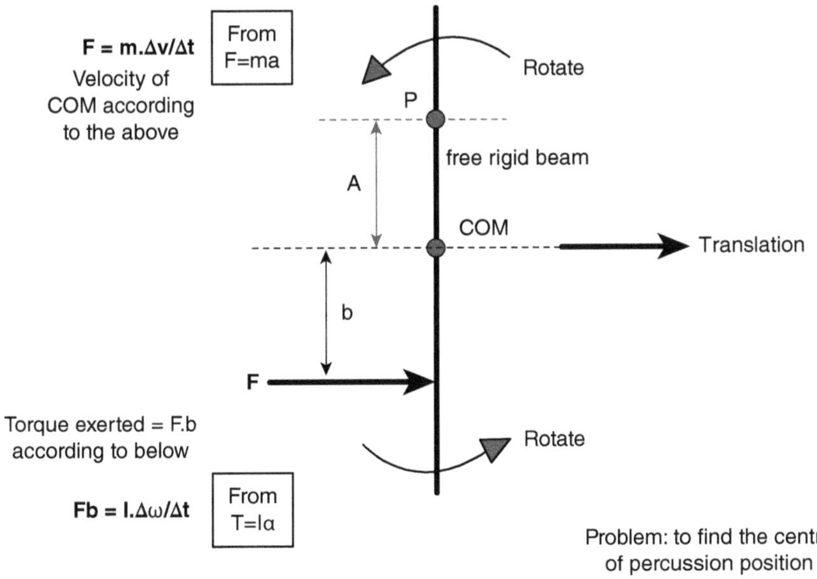


**Figure E5.4** Calculating the centre of percussion

Considering *Figure E5.4*, a force (F) is applied to a rigid beam (not attached to anything in this example and initially at rest), at a right angle (90°) and at a perpendicular distance (b) from the centre of mass (COM). This will result in the COM moving at a velocity (and acceleration) according to the relationship below which is derived from Newton's second law of motion (**F=ma**):

$$F = m\frac{\Delta v}{\Delta t}$$

where:

F = force (N)

m = mass (kg)

$\Delta v/\Delta t$ = linear acceleration or change in velocity over change in time of COM ($m/s^2$)

If "m" is the mass of the rigid beam, the torque (force x perpendicular distance or Fb) exerted on the beam can be determined from the following, which is also derived from Newton's second law of angular motion ($T=I\alpha$):

$$Fb = I\frac{\Delta\omega}{\Delta t}$$

where:

F = force (N)

b = distance (perpendicular) from COM that the force is applied (m)

I = moment of inertia of the beam around the COM axis ($kg.m^2$)

$\Delta\omega/\Delta t$ = angular acceleration or change in angular velocity over change in time ($rad/s^2$)

torque = force × perpendicular distance from the axis of rotation (or the COM in this case)

For any point (P) on the opposite side of the COM from the point of impact or force application (see *Figure E5.4*), the velocity of P is given as follows, which is, in part, derived from the relationship between linear and angular velocity (**v=ωr**). Note that in this example the point (P) could be where the racket or bat is held by the hand, and point of force application (F) is the point of contact or impact between the racket or bat and the ball.

$$v = V - A\omega$$

where:

v = the final (net) velocity of point P (m/s)

V = velocity of the COM (opposite direction to point P, as COM moves forwards as P moves backwards (rotation and translation))

A = the distance of point P from the centre of mass (COM), or r in the $v = \omega r$ example (m)

$\omega$ = angular velocity of point P or technically the segment (rigid beam) that point P is on as we cannot have an angular velocity of a point (rad/s)

Remember, in this example, that velocity is directional and we are considering both linear translation and angular rotation. Therefore, the net (final linear velocity) of point P (v) is the result (sum) from the COM linear velocity and the linear component of the rotational movement of point P (v = ωr). The COM point will move forwards (away from the force) and the point (P) will move backward (or rotate towards the force); hence, will have an opposite directional linear velocity component to that of the COM.

Considering the above equations, we can now rearrange these expressions to determine formulae to find the position of the **centre of percussion.**

$$v = V - A\omega$$

Multiply both sides by Δ/Δt, or express as a change in the variable with respect to a change in time:

$$\frac{\Delta v}{\Delta t} = \frac{\Delta V}{\Delta t} - \frac{A\Delta\omega}{\Delta t}$$

Exchange terms from initial equations shown previously:

$$\frac{\Delta v}{\Delta t} = \frac{F}{m} - \frac{AFb}{I}$$

Combine terms:

$$\frac{\Delta v}{\Delta t} = \left[\frac{1}{m} - \frac{Ab}{I}\right] F$$

If the beam is initially at rest, v is given by the following, which is representative of integrating the force and time curve (to determine the area under the force–time curve) with a constant mass to get velocity ($Ft = m\,(v_2 - v_1)$ = impulse)

$$v = \left[\frac{1}{m} - \frac{Ab}{I}\right] \int F\,dt$$

When the force is applied directly at the centre of percussion, the rotational and linear accelerations (and velocity) will cancel each other out and there will be no (zero) velocity of the rigid beam (v=0).

Hence, when the force is applied at exactly the centre of percussion:

$$v = V - A\omega = 0$$

Now expressing this relationship to force applied, with respect to time or the integral of force and time, we get the following equation. And remember that this position can be calculated for a range of forces acting over a range of time periods. As force is applied to the object for a period of time, there will be an impulse that is applied to the object (see the Impulse and Momentum section in this text). As impulse is derived from the force applied and the time of application, the area that is contained under the force and time graph is the impulse. We know that impulse is defined as the change in momentum possessed by an object (Impulse = Ft = $m(v_2 - v_1)$). In other words, an object or, in this case, the racket or bat, will achieve a greater final velocity ($v_2$) if the impulse applied to the object is increased; assuming that the object starts off from a stationary position. Hence, if we take the integral (see the Kinematic Data: Integration section in this text) of this equation we can express this again as follows:

$$v = \left( \frac{1}{m} - A.\frac{b}{I} \right) \int Ft$$

Since we know that v must equal 0 when the centre of percussion position is determined for F we can finally rearrange this equation to find the **position b**, where the force (F) must be applied for this rigid beam, such that it would be located at the **centre of percussion.**

$$b = \frac{I}{Am}$$

Determined as follows:

$$\frac{1}{m} - \frac{Ab}{I} = 0$$

Move Ab/I to the right-hand side,

$$\frac{1}{m} = \frac{Ab}{I}$$

Multiply both sides by I,

$$I \times \frac{1}{m} = \frac{IAb}{I}$$

Cancel out and work out I,

$$\frac{I}{m} = Ab$$

Divide both sides by A,

$$\frac{I}{Am} = b$$

This value for "b" (from *Figure E5.4*) is now portrayed as the location of the centre of percussion for a force that is applied to the beam when the result is a zero velocity of the beam. However, it is important to note that the centre of percussion is defined as a point for force location or application. Hence, if the specific **grip location** on the racket or bat is changed (i.e. in our case point P), then the centre of percussion point will also change its position on the racket or bat. This is due to the moment of inertia of the rotating system, which is axis-specific, changing.

The term **"sweet spot"**, which is often used in the media/sports literature to describe the centre of percussion point on a racket or bat, is not necessarily the same point as the centre of percussion. Generally, the "sweet spot"' is described by these sources as the point that, when the ball is hit with the implement, the player is not obviously aware that any impact has taken place (i.e. it was a "smooth" or "sweet" shot). It is also inferred that, at this point, there will be minimal vibration transferred to the player's hands/wrist or arm. The "sweet spot", however, is what is technically termed a vibration node (a point of fundamental vibration frequency of the racket strings or frame). The centre of percussion may be a "sweet spot" but the centre of percussion can also change its position and has important implications for both performance and injury prevention. Hitting a ball at the "sweet spot" on a racket or bat is generally considered to be the point at which contact "feels" to be the best. These points can be demonstrated by the tennis racket example shown in *Figure E5.5*.

However, there is some current discussion in the scientific literature that the "sweet spot" and the centre of percussion are indeed different points. For example, it is suggested that some baseball batters believe that the "sweet spot" of a baseball bat is located at a point that is not in the bat at all, but is a point that is actually beyond the end of the bat (which is difficult to imagine if the "sweet spot" is indeed a vibration node). This discussion is unfortunately beyond the scope of this text, but it is an interesting topic for future research.

Whatever the case, the **centre of percussion** is the point where there is a minimal force transference to the hand, wrist and/or arm of the player if the ball was to contact the racket at this point. This point, however, is not fundamentally the point where the ball will travel off with maximum speed or the point at which there is absolutely no vibration on the hands, wrist or arms. As we have seen, the racket (or bat) is moved through the air in a three-dimensional movement pattern and, hence, there will be rotation, translation and vibration about all three

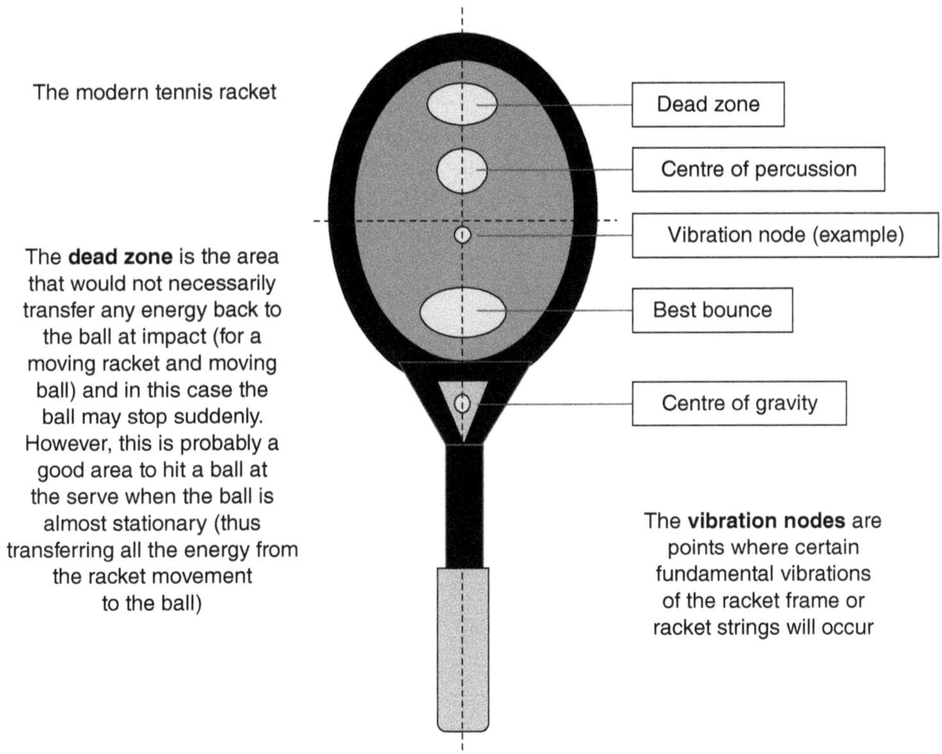


**Figure E5.5** The modern tennis racket

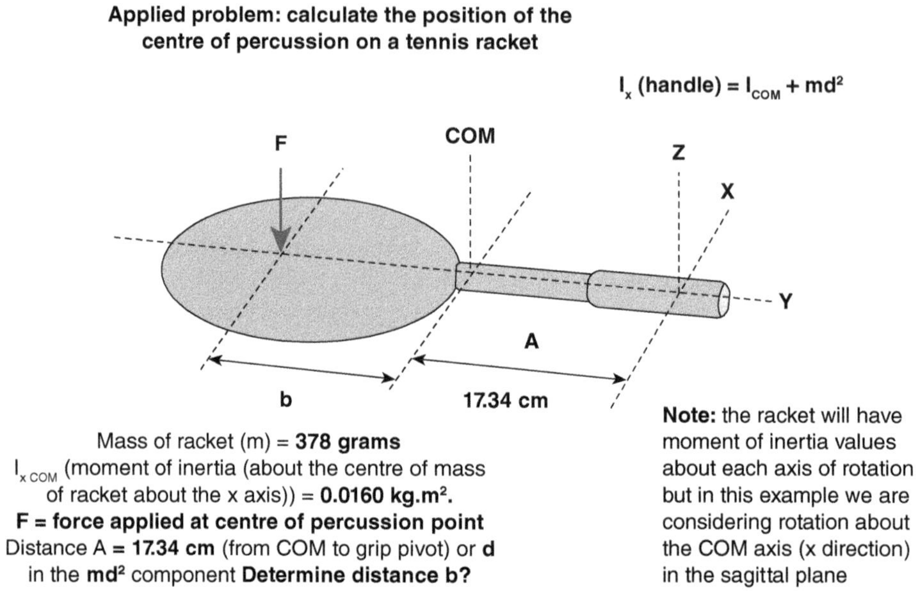


**Figure E5.6** Applied task: determining the centre of percussion of a tennis racket

axes and planes of motion. So, for example, a centre of percussion point for one direction of motion may not be the same centre of percussion point for another direction of motion; simply because the racket will have a different moment of inertia value for each axis that it is rotated about.

*Figure E5.6* shows an applied example of determining the centre of percussion point of a tennis racket and is an experiment that can easily be conducted in a laboratory class by measuring location points and forces acting on a racket or bat. In this example, we have all the information we need to determine the distance (b) along the Y-axis, for example, where the centre of percussion is located.

## Solution

Using the equation below, we can now determine this position:

$$b = \frac{I}{Am}$$

First, determine the moment of inertia of the racket about this axis of rotation.

$$I_x \text{ (handle)} = I_{COM} + md^2$$

$$I = 0.0160 + 0.378 \times 0.1734^2$$

In units of metres (m) and kilograms (kg).

$$I = 0.0160 + 0.378 \times 0.0300 \text{ kg.m}^2$$

$$\mathbf{I = 0.0160 + 0.0114 \text{ kg.m}^2}$$

Now, substitute this into the equation for b.

$b = I/Am$

$b = 0.0274/Am$

**$A = 0.1734$ m**

**$m = 0.378$ kg**

$b = 0.0274/0.1734 \times 0.378$

$b = 0.0274/0.0655$

**$b = 0.418$ m (or 41.83 cm from the COM position along the racket to the tip).**

Considering a modern tennis racket must be less than **73.66** cm (29 inches) long (ITF ruling), this would put the centre of percussion in this example near to the tip of the racket; or **59.17 cm** from the grip position (**41.83 + 17.34 cm**).

The centre of percussion is an important point in the design and testing of a tennis racket and also other striking implements, such as cricket and baseball bats. The point has important applications for hitting effective shots and also hitting shots that minimise the vibrations that are transferred to the wrists, hands, elbows and/or shoulders, which can cause injury (especially through overuse). A common tennis injury that can result from excessive vibrations at the elbow is termed lateral epicondylitis ("tennis elbow") and minimising the vibrations experienced at this joint while playing can be an advantage to the prevention of this injury. Many tennis racket companies are now seen incorporating anti-vibration devices on the strings and on the racket frame that aim to reduce these unwanted vibrations. However, it is clear that further research is needed to determine the exact relationship between the centre of percussion and the so-called "sweet spot", as it is so often called in sporting terminology.

| *Key notes* | |
|---|---|
| **Centre of percussion** | The **centre of percussion** of a striking implement can be broadly defined as the point (or often area) on the racket or bat that makes the most effective contact with the ball. For example, in tennis, a larger centre of percussion area on a racket can minimise the consequences of poorly hit shots. The centre of percussion is an important point for both effective performance and prevention of injury. It is important to note that the centre of percussion is defined as a point for force location or application. Hence, if the specific **grip location on the racket** or bat is changed, then the centre of percussion point will also change its position on the racket or bat (because the moment of inertia (axis-specific) of the rotating system will change). |
| **The "sweet spot"** | The term **"sweet spot"** is often used in the media/sports literature to describe the centre of percussion on a racket or bat. However, it is not necessarily the same point as the centre of percussion. Generally, the "sweet spot", as described by these sources, is the point at which the player is not obviously aware that any impact has taken place when the ball makes contact with this spot on the striking implement (i.e. it was a "smooth" or "sweet" shot). It is also inferred that, at this point of contact, there will be minimal vibration transferred to the player's hands, wrists and/or arms. The "sweet spot", however, is what is technically termed a vibration node (a point of fundamental vibration frequency of the racket strings or frame). The centre of percussion may be a "sweet spot", but the centre of percussion can also change its position and has important implications for both performance and injury prevention. |

## FURTHER READING

The following resources provide additional reading on the topic of the centre of percussion.

1 Miller, S. (2006). Modern tennis rackets, balls and surfaces. *British Journal of Sports Medicine, 40*(5), 401–405.
2 Smith, L. V. (2002). Evaluating baseball bat performance. *Sports Engineering, 4*(2), 205–214.
3 Cross, R. (2008). Customising a tennis racket by adding weights. *Sports Engineering, 4*(1), 1–14.
4 Cross, R. (2004). Center of percussion of hand-held implements. *American Journal of Physics, 72*(5), 622–630.

## SECTION F

# Human movement applications

# WALKING

*Neil Fowler*

In the analysis of any skill, it is important to understand the role of the various joint movements and body segments involved. Walking is no exception. The joint and segmental interactions involved in walking are so complex that it takes most humans a year to be able to "toddle" and a further three to four years to walk perfectly. Gait is defined according to the sequence of **swing** and **support** phases of the legs when the foot is either in the air (swing) or in contact with the floor (support or stance). Walking gait is characterised by periods of double support, where both feet are in contact with the ground, separated by periods of single support when one foot is on the ground while the other leg is swung forwards to make the next step. There is no time at which both feet break contact with the ground at the same time; in other words, there is no flight phase.

One complete gait cycle, for example, from right heel strike (when the heel of the right foot contacts the floor) to the next right heel strike, is known as a **stride** (*Figure F1.1*). Each stride is made up of two **steps**, each step covering the period from one heel strike to the heel strike of the contralateral limb (*Figure F1.2*). The distance covered in a stride is known as the stride length; the rate at which strides are taken is referred to as the stride frequency or cadence, measured as the number of strides per unit of time (e.g. strides per second). Based on the stride length and frequency it is possible to calculate the velocity at which someone is walking (Equation F1.1):

$$\text{Stride length} \times \text{Stride frequency} = \text{Velocity} \qquad \textbf{(Equation F1.1)}$$

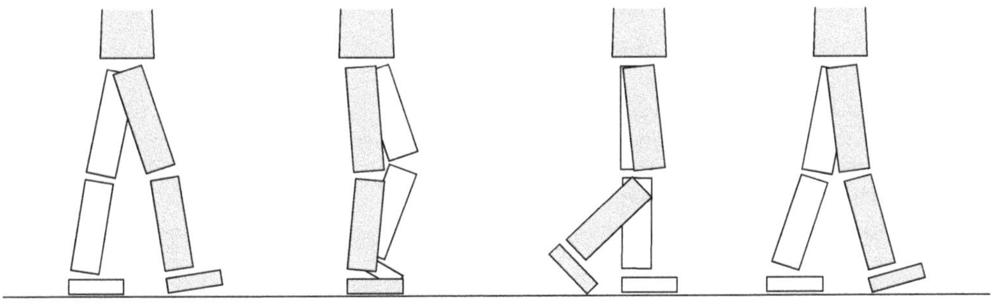

**Figure F1.1** The phases of the walking gait cycle. One full stride of the right leg is shown from heel strike to heel strike

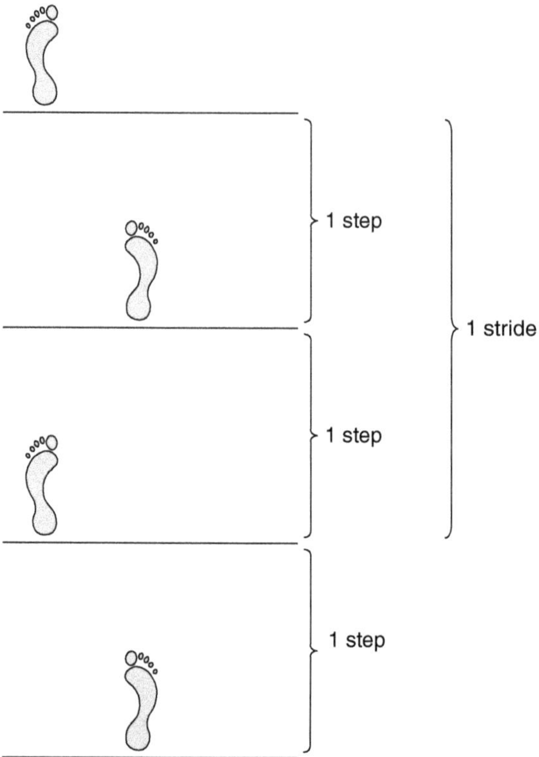


**Figure F1.2** The decomposition of the gait cycle into steps and strides. One stride is made up of two consecutive steps

For example, if an individual's stride length was 1.2 m and their stride frequency was 1.5 strides per second (1.5 Hz or three strides every two seconds) then their velocity could be found by:

Stride length × Stride frequency = Velocity

$1.2 \times 1.5 = 1.8$ m/s or 6.48 km/h

To increase the speed of walking, one must increase the stride length and/or the stride frequency. The relationship between stride length and frequency is complex and relates to the energy cost of moving the limbs through a greater range and at different rates. Therefore, there are limits to how far you can increase walking speed before it becomes either too difficult or too inefficient to continue (i.e. the amount of energy it costs per stride becomes too great).

## TASK

The data in *Table F1.1* represent the average stride length and frequency for a range of walking speeds. Plot graphs of stride length and frequency against walking speed and reflect upon how the strategy employed to increase speed changes.

To understand human gait, it is important to perform a kinesiological analysis of the movements. This involves the description of movement in terms of the sequencing and range of joint

**Table F1.1** Average stride length (m) and stride frequency (Hz) at a range of walking speeds

| | | | | | | |
|---|---|---|---|---|---|---|
| **Stride length (m)** | 0.50 | 0.65 | 0.77 | 0.91 | 1.04 | 1.20 |
| **Stride frequency (Hz)** | 1.00 | 1.55 | 1.95 | 2.20 | 2.40 | 2.50 |
| **Speed (m/s)** | 0.50 | 1.00 | 1.50 | 2.00 | 2.50 | 3.00 |

movements and the muscle actions involved. To assist this form of analysis it is normal to break the action down into phases and consider it joint by joint. In general, limb movements and muscle activation patterns are considered within a **gait cycle**, which is defined as the time period between two successive heel contacts made with the same limb (i.e. a stride).

To examine the movements that occur within a gait cycle, this phase is generally broken down into smaller segments that have a clear start and end point and fit together to give a continuous sequence of movement. In gait, the cycle is first divided into **swing** and **stance** phases; in other words, the periods when the foot is either in the air or in contact with the ground. The ratio of stance to swing times is a useful measure for quantifying normal and abnormal gaits. The normal ratio of stance to swing during walking is generally considered to be 60 per cent stance / 40 per cent swing.

Quite often, however, sub-division of the gait cycle into stance and swing phases still yields sections that are too long to allow us to perform a useful analysis of the movement, hence these phases are further divided into sub-phases. A common division of the gait cycle is into five phases (*Table F1.2*), and these can be used to describe both walking and running actions (*Figure F1.3*).

**Table F1.2** The defined start and end points for the five sub-phases of the gait cycle

| | *Start point* | *End point* |
|---|---|---|
| **Swing phase** | Toe-off | Foot strike |
| *Early swing* | Toe-off | Start of knee extension |
| *Late swing* | Start of knee extension | Foot strike |
| **Stance phase** | Foot strike | Toe-off |
| *Loading phase* | Foot strike | Foot flat |
| *Mid-stance* | Foot flat | Heel raise |
| *Drive-off* | Heel raise | Toe-off |

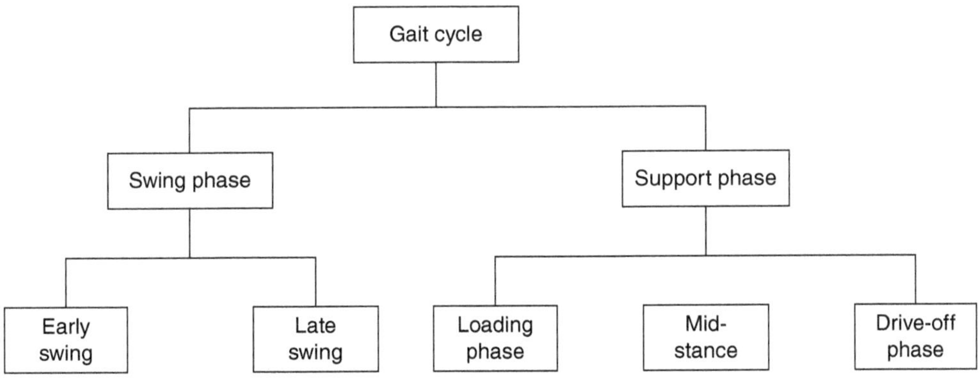


**Figure F1.3** The five phases of the gait cycle

## TASK

Through the use of a video recording or simple observation, consider the sagittal plane movements of the hip, knee and ankle during walking. Record your observations of the direction (flexion/extension) and extent (approximate start and end angles) of movement at each joint and compare your results with the ranges of motion presented in *Table F1.3*.

These movements of the lower limb are representative of normal walking. However, it should be noted that, while the general pattern of walking is quite characteristic, there will be notable inter-individual variability in the absolute joint angles moved through. The values in *Table F1.3* are indicative of the pattern of movement and should not be considered a normative data set.

Movements of the limbs are created or controlled by the contractions of the musculature (*Figure F1.4*), which are mainly involved in the initiation and cessation of limb movements. Much of the action for swinging the leg is achieved by the pendulum effect of gravity and does not require a significant muscular effort.

With the average person taking over 1200 steps per kilometre while walking, it is important to understand the processes involved in the contact between the body and the ground. Application of Newton's first law states that a force must be acting in order for the movements associated with walking to take place. The study of the forces associated with gait therefore forms an important part of any gait analysis. The typical ground reaction force trace shows the periods of single and double support that are characteristic of walking and the way that the forces are transferred from one foot to the other (*Figure F1.5*). The double support phase, where the load is transferred from one foot to the other, allows the loading rate (slope of the curve) to be controlled and therefore kept relatively low. The vertical force remains relatively close to body weight throughout the periods of single support, with the peak forces at initial heel strike and drive-off only reaching values a little above body weight. These peaks result from the body decelerating at heel strike and then accelerating at toe-off; during mid-stance the trough results from a net downwards acceleration of the centre of mass as it passes over the foot.

**Table F1.3** Typical lower limb joint movements during walking gait at a moderate speed. The ranges of motion presented describe the start and end points of the joints during each phase.

| | *Joint* | *Movement* | *Range of motion* |
|---|---|---|---|
| **Swing phase** | | | |
| *Early swing* | Hip | Flexion | 0° extension to 30° flexion |
| | Knee | Flexion | 30° flexion to 60° flexion |
| | Ankle | Dorsiflexion | 5° plantarflexion to 0° dorsiflexion |
| *Late swing* | Hip | Extension | 30° flexion to 25° flexion |
| | Knee | Extension | 60° flexion to 10° flexion |
| | Ankle | Dorsiflexion | 0° dorsiflexion to 5° dorsiflexion |
| **Stance phase** | | | |
| *Loading phase* | Hip | Flexion | 25° flexion to 30° flexion |
| | Knee | Flexion | 10° flexion to 20° flexion |
| | Ankle | Plantarflexion | 5° dorsiflexion to 10° plantarflexion |
| *Mid-stance* | Hip | Extension | 30° flexion to 9° extension |
| | Knee | Extension | 20° flexion to 5° flexion |
| | Ankle | Dorsiflexion | 10° plantarflexion to 20° dorsiflexion |
| *Drive-off* | Hip | Flexion | 9° extension to 0° extension |
| | Knee | Flexion | 5° flexion to 30° flexion |
| | Ankle | Plantarflexion | 20° dorsiflexion to 5° plantarflexion |

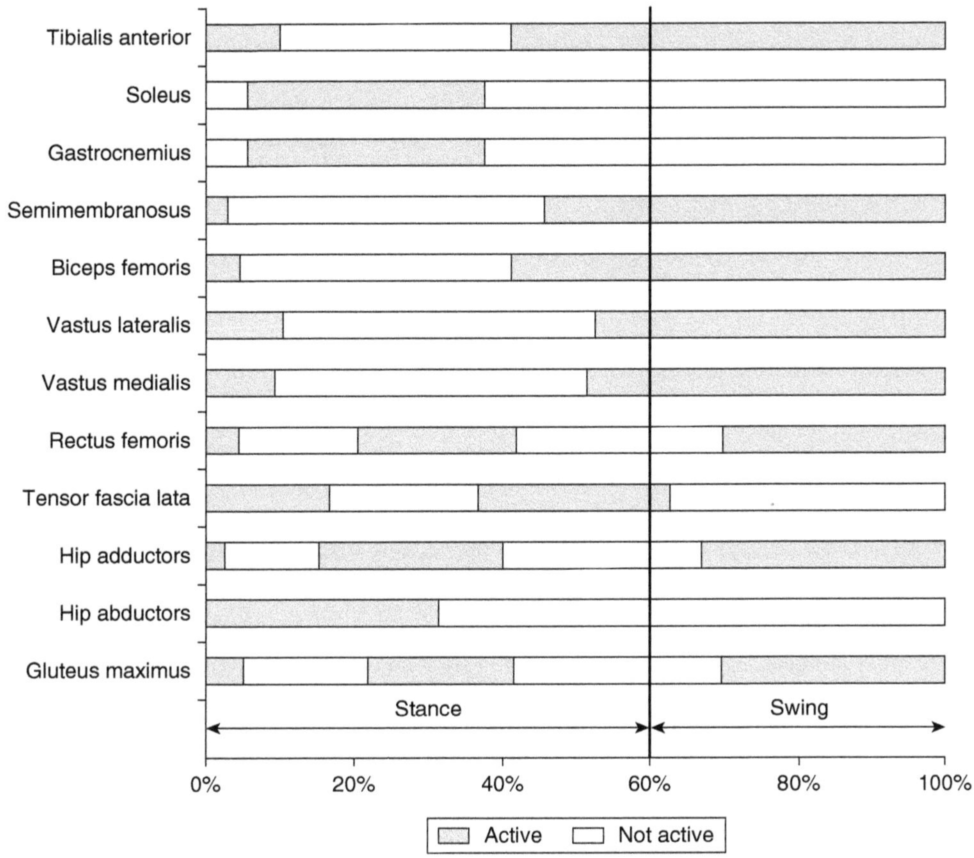


**Figure F1.4** Muscle activity during walking. The horizontal axis is shown as the percentage of a single stride (gait cycle) where zero represents heel strike and 100% the next heel strike for the same limb.

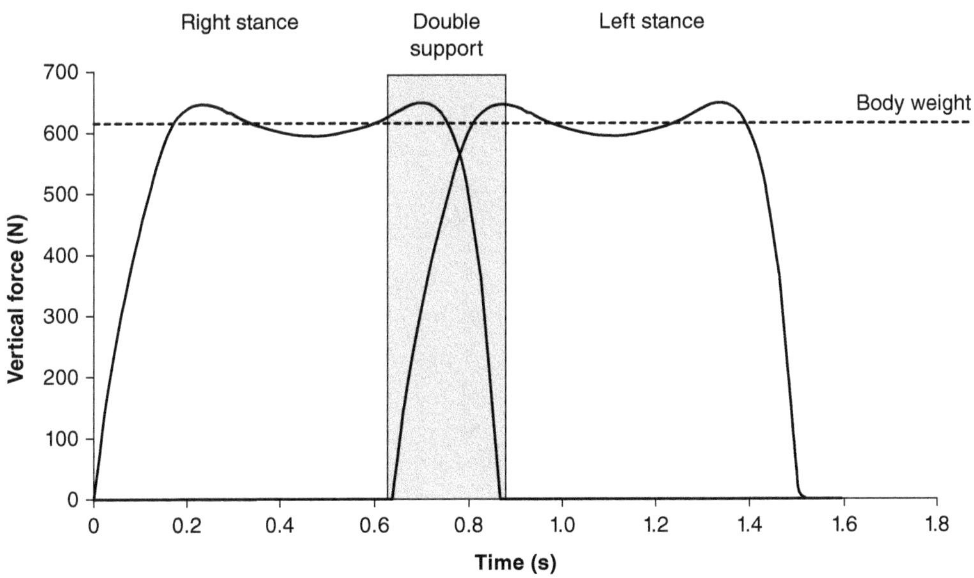


**Figure F1.5** Typical vertical ground reaction forces during normal walking

The anterior–posterior forces (*Figure F1.6*) show the forces acting along the direction of movement; these are known as either braking or driving forces dependent upon their direction. At impact the force is acting in the opposite direction to the movement and is therefore a braking force. The magnitude of the braking force will fluctuate as the gait style changes. As the centre of mass passes over the foot and the forwards drive begins, so the force becomes positive and acts as a driving force. The point at which the force changes from braking to driving is normally between 45 and 50 per cent of the total stance time. Variations from this normal pattern will be suggestive of an abnormal gait.

The role of the upper body during gait is to act as a stabiliser to limit the changes in the angular momentum of the body and to thus conserve energy. If there was no upper body then there would be large positive and negative swings in the angular momentum of the body as the legs rotated forwards and backwards.

The arms swing in a contralateral fashion in exact anti-phase to the legs, in other words the right arm reaches maximum shoulder and elbow flexion when the right leg reaches toe-off and maximum shoulder and elbow extension at right heel strike. Thus, the upper body is being rotated in the opposite direction to the pelvis. This out-of-phase action of the arms generates an opposite angular momentum value to the legs and so reduces the change in the angular momentum of the whole body. It is interesting to note that, despite the greater mass of the legs, the arms are able to generate a momentum almost equal in magnitude to that of the legs. This is possible because the arms are positioned further from the mid-line of the body and so require less mass to achieve the same moment of inertia (Equation F1.2).

$$\text{angular momentum } (L) = \text{moment of inertia } (I) \times \text{angular velocity } (\omega) \qquad \textbf{(Equation F1.2)}$$

In the horizontal plane, the arms have no effect as they act in opposite directions, in other words one forwards and one backwards. Similarly, the arms contribute less than 5 per cent to the total lift of the body in a vertical direction.

The movement of the body during walking has been compared to that of an inverted pendulum. This model represents the lower limb as a single rigid segment with the mass located at

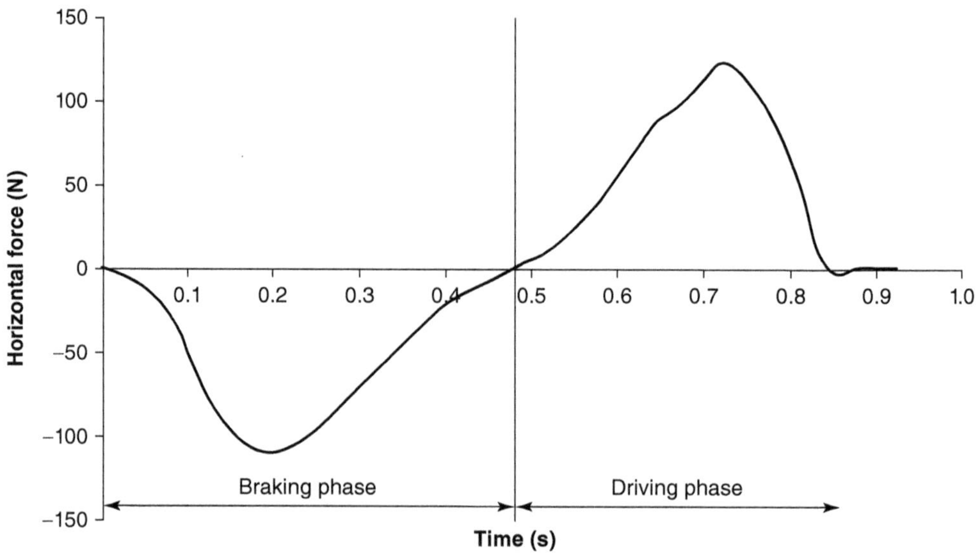


**Figure F1.6** Typical anterior–posterior ground reaction forces during normal walking

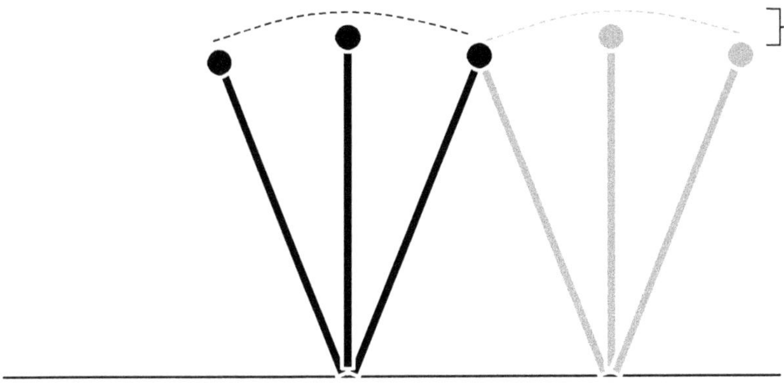


**Figure F1.7** Inverted pendulum model of walking gait

the top (*Figure F1.7*). During single leg stance the centre of mass follows a circular path along the arc created by the leg, during which there is an energy exchange from kinetic to potential energy. This inverted pendulum model of gait helps to reduce the energy cost of locomotion and requires the main energetic input to occur during the periods of double support in which the trajectory of the centre of mass is transferred from one limb to the other.

The inverted pendulum model also offers insight into the theoretical maximum walking speed that can be sustained without a flight phase being required. **Centripetal acceleration** is considered to be the component of acceleration directed towards the centre of an angular path and, in the case of the inverted pendulum, this is the acceleration acting through the limb as a consequence of the linear motion of the centre of mass about the point of contact with the floor (*Figure F1.8*). During this scenario the maximum value for the centripetal acceleration will be equal to "g": the acceleration due to gravity. It is therefore possible to calculate the maximum value for which this is true.

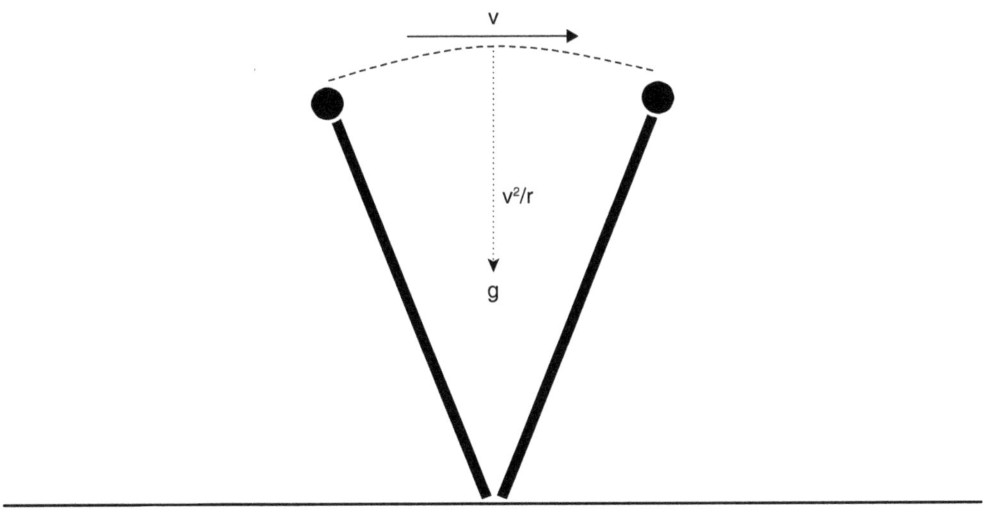


**Figure F1.8** The inverted pendulum model for calculating the theoretical maximum walking speed. Speed is attained when the centripetal acceleration ($v_2/r$) is equal to the acceleration due to gravity.

If the centre of mass illustrated in *Figure F1.8* is assumed to have an average linear velocity of "v" acting perpendicular to the limb (i.e. the inverted pendulum), which has a length (and hence, radius of rotation) of "r", we can use the formula for centripetal acceleration (Equation F1.3) to determine the limb's angular velocity.

$$a_{centripetal} = \text{radius of rotation}(r) \times \text{average angular velocity}^2\,(\omega^2) \qquad \textbf{(Equation F1.3)}$$

Equation F1.4 depicts the relationship between the average linear velocity (v) of a point that is on a rotating arm of "r" length that is rotating at an average angular velocity of "ω".

$$\text{average linear velocity}(v) = \text{average angular velocity}(\omega) \times \text{radius of rotation}(r) \qquad \textbf{(Equation F1.4)}$$

Given this relationship, Equation F1.3 can be rewritten to yield Equation F1.5:

$$a_{centripetal} = \text{radius of rotation}(r) \times \left(\frac{\text{average linear velocity}^2}{\text{radius of rotation}^2}\right)$$

or

$$a_{centripetal} = \frac{\text{average linear velocity}^2}{\text{radius of rotation}} \qquad \textbf{(Equation F1.5)}$$

Given that the maximum value for the centripetal acceleration is equal to one gravitational unit (i.e. the acceleration due to gravity, denoted by the symbol "*g*"):

$$\frac{\text{average linear velocity}^2}{\text{radius of rotation}} \leq g \qquad \textbf{(Equation F1.6)}$$

By rearranging Equation F1.6, we are able to produce an expression that represents the average linear velocity of the point or body located on the rotating arm (Equation F1.7).

$$\text{average linear velocity}(v) = \sqrt{g \times \text{radius of rotation}} \qquad \textbf{(Equation F1.7)}$$

## TASK

For a person with a lower limb length of 0.90 m, calculate their maximum theoretical walking speed using the inverted pendulum model.

*Key notes*

| | |
|---|---|
| **The gait cycle** | Describes the style of locomotion. For walking this is divided into support (when the foot is on the ground) and swing (when the foot is off the ground) phases. The support phase contains periods of single support (one foot on the ground) and double support (both feet on the ground). |
| **Stride** | The movement from when one foot touches the floor to when the same foot next touches the floor. Each stride is made up of two steps. |

| | |
|---|---|
| **Speed** | The speed of walking is found by multiplying the stride frequency (number of strides per second) by the length of each stride. |
| **Forces during walking** | The vertical ground reaction force during walking typically peaks at a little above body weight. The force rises relatively slowly as the load is transferred from one foot to the other during the periods of double support (when both feet are on the floor). The horizontal force is initially negative, indicating that it acts in the opposite direction to the movement and serves as a braking action. During the latter half of the support phase the horizontal force becomes positive to propel the body forwards into the next step. |
| **Upper body movement in walking** | During walking the arms swing in the opposite direction to the legs such that when the left leg is forwards the left arm is back. This movement helps to overcome the angular momentum of the lower body and to reduce the energy cost of walking. |

## FURTHER READING

The following two articles provide additional reading on the biomechanics of walking.

1 Kuo, A. D. (2007). The six determinants of gait and the inverted pendulum analogy: A dynamic walking perspective. *Human Movement Science, 26*(4), 617–656.
2 Farley, C. T., & Ferris, D. P. (1998). Biomechanics of walking and running: Centre of mass movements and muscle action. *Exercise and Sport Sciences Reviews, 26*(1), 253–286.

F2

# RUNNING

*Neil Fowler*

As with walking, the running action is made up of a series of steps in which forward progress is made by sequentially planting the left and right feet on the ground. The style of movement is described as a form of **gait** and, as it is considered a cyclic (repeated) movement pattern, it is referred to as the running **gait cycle.** The running gait cycle is divided according to the sequence of swing (when the foot is in the air) and support/stance (when the foot is on the ground) phases of the legs. During running the periods of single support (where only one foot is on the ground) are separated by a flight phase in which there is no ground contact. Unlike walking gait, there are no double support phases during the running gait pattern (i.e. no periods when both feet are on the floor at the same time). In running, the ratio of stance to swing rises to approximately 40 per cent stance and 60 per cent swing, but the exact ratio depends upon the runner's speed. Specifically, the relative duration of the stance phase decreases with increased running speed, such that stance only constitutes about 20 per cent of the gait cycle during maximal sprinting.

Increases in running speed require an increase in the **stride length** (the distance covered by each stride) and/or an increase in **stride frequency** (the number of strides taken in each unit of time). Up to a speed of approximately 7 m/s most of the increase is achieved through increasing the stride length while maintaining a nearly constant stride frequency. Above this speed the stride frequency increases. It is suggested that the reason why stride length is usually increased first, with changes in stride frequency reserved until higher velocities, is because there is an optimum stride frequency at which the energy cost of running is minimised. If the stride frequency is changed, then the energy cost per unit of distance covered increases, making the athlete less energy efficient. Despite the obvious logic, there is only a low correlation between an athlete's stride length at a given speed and their anthropometric measures. There is a stronger relationship with factors such as strength and flexibility.

## TASK

Plot the data in *Table F2.1* to demonstrate the relationship between stride frequency and running economy ($VO_2$) and identify the most economical stride frequency. If the participant in this experiment is running at 3.33 m/s determine their stride length at this speed.

**Table F2.1** Running economy ($VO_2$ L/min) recorded for an athlete while running at 3.33 m/s using a range of stride frequencies

| **Stride frequency (Hz)** | 1.20 | 1.27 | 1.33 | 1.40 | 1.47 | 1.60 |
|---|---|---|---|---|---|---|
| **$VO_2$ (L/min)** | 3.81 | 3.72 | 3.49 | 3.57 | 3.59 | 3.64 |

To determine the athlete's stride length for a given stride frequency you need to consider the relationship between stride length, stride frequency and running speed:

Running Speed (m/s) = stride length (m) × stride frequency (Hz)

Hence, the following illustrates the calculations for deriving the stride length for the first stride frequency measure.

$$3.33 = \text{stride length} \times 1.20$$

$$\frac{3.33}{1.20} = \text{stride length}$$

$$2.78 = \text{stride length}$$

$$\textbf{Stride Length} = \textbf{2.78 m}$$

Each stride is of course two steps; hence step length can be calculated as:

$$\text{Step Length} = \frac{\text{Stride Length}}{2}$$

$$\textbf{Step Length} = \textbf{1.39 m}$$

Like walking, the running gait cycle is broken into a number of phases to facilitate biomechanical analysis. *Table F2.2* summarises the typical movements seen during running.

**Table F2.2** The typical movement patterns of the lower limb joints during running gait. The ranges of motion presented describe the start and end points of the joints during each phase, with the absolute ranges of motion being dependent upon the athlete's speed (i.e. increased speed generally increases joint range of motion).

| | *Joint* | *Movement* | *Range of motion* |
|---|---|---|---|
| **Swing phase** | | | |
| *Early swing* | Hip | Flexion | 9° extension to 55° flexion |
| | Knee | Flexion | 25° flexion to 90° flexion |
| | Ankle | Dorsiflexion | 20° plantarflexion to 10° dorsiflexion |
| *Late swing* | Hip | Extension | 55° flexion to 45° flexion |
| | Knee | Extension | 90° flexion to 20° flexion |
| | Ankle | Plantarflexion | 10° dorsiflexion to 5° dorsiflexion |
| **Stance phase** | | | |
| *Loading phase* | Hip | Flexion | 45° flexion to 50° flexion |
| | Knee | Flexion | 20° flexion to 40° flexion |
| | Ankle | Dorsiflexion | 5° dorsiflexion to 20° dorsiflexion |
| *Mid-stance* | Hip | Extension | 50° flexion to 15° flexion |
| | Knee | Flexion | 40° flexion to 40° flexion |
| | Ankle | Dorsiflexion | 20° dorsiflexion to 30° dorsiflexion |
| *Drive-off* | Hip | Extension | 15° flexion to 9° extension |
| | Knee | Extension | 40° flexion to 25° flexion |
| | Ankle | Plantarflexion | 30° dorsiflexion to 20° plantarflexion |

## TASK

Using the information presented in this section so far and the equivalent tables provided in this text's section on Walking, describe the key kinematic difference between walking and running. Upon completing this task, you will observe that, despite sharing some similarities, running and walking gait differ in a number of important ways. The major differences are described below.

**Hip:** At foot strike the hip is flexed to approximately 45° (greater than in walking where it is typically 30°), this angle is maintained during early stance by knee flexion. During drive-off the hip extends to approximately 9° at toe-off (the same as walking). Flexion during the swing phase reaches about 55° (compared with only 25° during walking).

**Knee:** At heel strike the knee is flexed to an angle of approximately 25°, compared with 10° during walking (i.e. the knee is never straight at impact). The knee flexes to about 40° by mid-stance (only 20° during walking) and extends from mid-stance through to toe-off. During swing the knee reaches a peak flexion angle of 90°, compared with only 60° during walking.

**Ankle:** The ankle reaches maximum dorsiflexion of about 30° by mid-stance. Plantarflexion at toe-off is significantly greater during running than in walking, reaching 20° compared with only 5° in walking.

As speed increases, the flexion of hip and knee joints during the swing phase increases, which reduces the moment of inertia of the limb and facilitates a faster swing phase. There may also be a slight increase in the degree of knee flexion at initial foot contact.

The phasing of muscle activity in running is shown in *Figure F2.1*. In general, muscles are most active in anticipation for and just after initial contact. Muscle contraction is more important at this time than it is for the preparation for and the act of leaving the ground. As the speed of gait increases, so does the degree of muscle activity. The swing phase becomes more active with a greater muscular contribution to the movement. The increased ranges of joint movement result in a greater period of muscle activity during all phases of the cycle.

So far, only the major movements of the lower limb in the sagittal plane (side view) have been considered. While these are clearly the largest movements that occur, there are other, smaller movements that are of equal or greater importance to potential injury risk. The movements of the subtalar joint have been the focus of much attention in the research literature due to their suspected role in the aetiology of injury. The subtalar joint lies just below the ankle joint and is formed by the talus above and calcaneus (heel bone) below. It is at this joint that the movements of **inversion** (turning the sole of the foot inwards) and **eversion** (turning the sole of the foot outward) mainly occur. During any form of gait, the motions of the subtalar joint and the other plane joints in the foot and ankle serve a shock-absorbing function. Just prior to impact, the foot is positioned in a **supinated** position (inverted, adducted and plantar flexed) such that the outside portion of the heel makes first contact with the ground. Immediately following impact, the foot flattens as the whole of the foot is placed onto the floor and the subtalar joint moves from inversion to eversion. This movement is known as **pronation**. As this happens the plantar fascia (ligaments and tendons of the sole of the foot) becomes stretched and the supporting musculature works eccentrically to resist this flattening. These actions help to reduce the impact force by effectively softening the impact and slowing the descent of the body more gradually.

Once in mid-stance the foot remains in the pronated position as the body weight moves over the foot, before supinating again to form a rigid lever for toe-off. *Figure F2.2* shows a "typical" trace for the rear foot (pronation/supination) angle during a single stance phase. Note that most of the pronation happens during the first 50–100 ms after foot strike. It is therefore a very rapid movement. The normal range of motion is from ~10° supination to ~10° pronation.

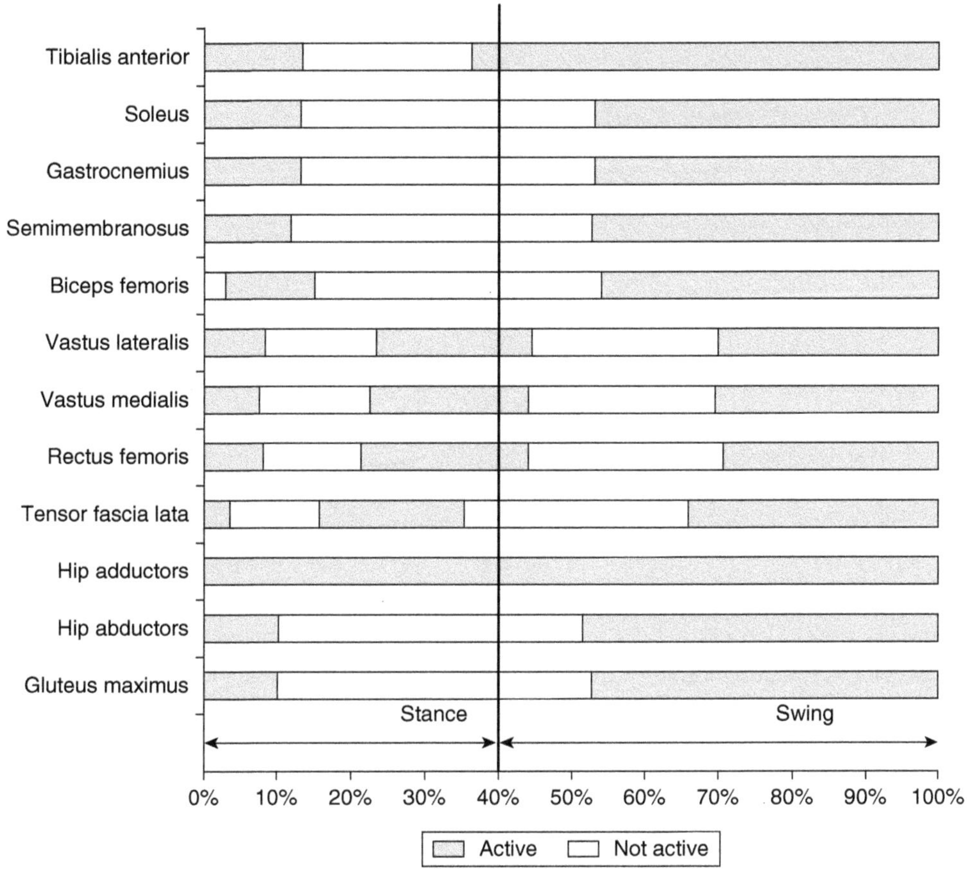


**Figure F2.1** Muscle activity during running. The horizontal axis is shown as the percentage of a single stride (gait cycle) where zero represents heel strike and 100% the next heel strike for the same limb.

If the foot strikes the ground in a pronated position, or if the muscles and plantar fascia are loose, then there will be little resistance to the normal range of pronation and the shock-absorbing function of the foot will be lost. The nature of the subtalar joint means that, if the foot pronates, then there must be an accompanying rotation of the tibia. Pronation leads to an internal rotation of the tibia while supination leads to an external rotation. Excessive pronation >20° will lead to an excessive internal rotation of the tibia. This will disrupt the normal loading of the ankle joint and may lead to misalignment problems at the knee and anterior knee pain.

There are a number of factors that have been suggested to be related to the amount and the rate of pronation during gait. The greater the body weight of an individual, the greater will be the load during stance. Excessive body weight will tend to lead to a more flattened arch position and either greater or more rapid pronation of the foot. As running speed increases, the foot strikes the ground in a more supinated position while pronation ends at more-or-less the same final angle at all speeds. This will give a greater range of motion and, because stance time decreases with running speed, the rate of pronation is greater at faster speeds.

During running, the vertical movement of the body is greater than in walking as a consequence of there being a flight phase. As the body will be falling from a greater height it will have

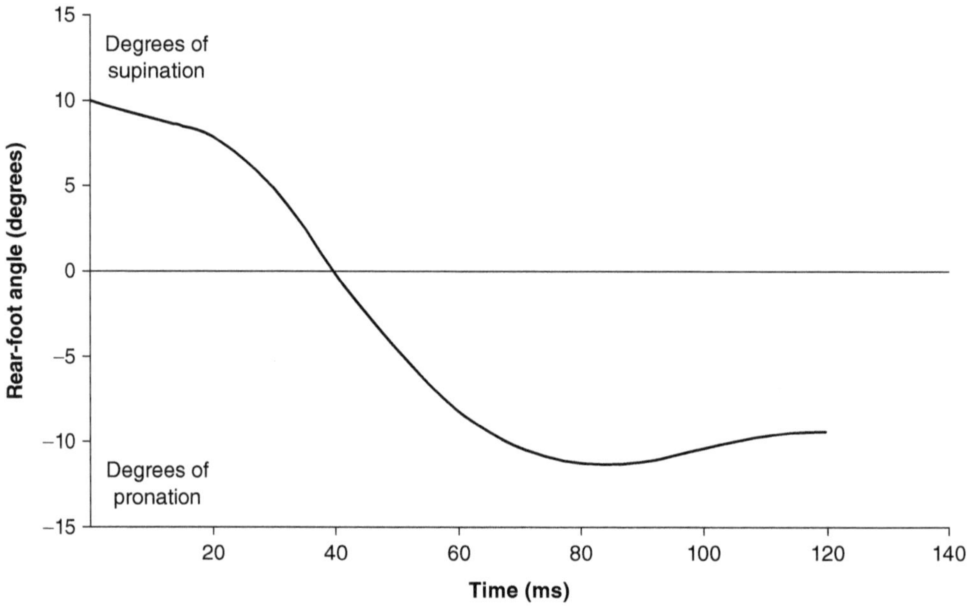


**Figure F2.2** Typical rear-foot movement during running. The data ends during mid-stance, as it is not possible to accurately measure the rear-foot motion two-dimensionally after the heel lifts from the floor; hence, the data does not show the return of the foot to a supinated position for toe-off.

a greater vertical velocity at foot strike. The slope of the force–time curve (loading rate) is also greater, reaching a peak after only about 0.05 s (whereas in walking this peak does not occur until about 0.15 s). This will mean that the tissues of the body are loaded more rapidly and a greater stress is placed upon them. Furthermore, during running the impact is experienced by a single leg, rather than there being a gradual transfer of load from one leg to the other, as seen during the double-support phase of walking.

The vertical ground reaction forces experienced during running are directly related to body weight. This is logical, as Newton's second law states that the force is proportional to the mass of the body. Typically, vertical ground reaction forces in running are of the order 2 to 2.5 times body weight. Similarly, as the speed increases, so does the impact peak force and the rate of force loading. Vertical ground reaction forces increase by about 1 body weight (from ~2 to ~3 times body weight) when running speed increases from 3 to 6 m/s (9-minute mile pace to 4.5-minute mile pace). In contrast, the vertical drive-off force is relatively unaffected by running speed, but as contact times decrease with increased speed the rate at which the force is applied to the body significantly rises.

About 80 per cent of runners adopt a heel-to-toe running pattern, which produces a very typical pattern of ground reaction forces (*Figure F2.3*). Running style is classified according to which part of the foot makes first contact with the ground, such that a rear-foot striker would be someone who makes first contact with the ground with the rear third of the foot. Similarly, runners who make first impact with the middle third of the foot are classified as mid-foot strikers, while those who first contact the ground with the forward third of the foot are classified as forefoot strikers. Runners with a mid- or forefoot impact will tend to experience lower impact forces as loading is cushioned by the eccentric activation of the gastrocnemius and soleus (calf) muscles.

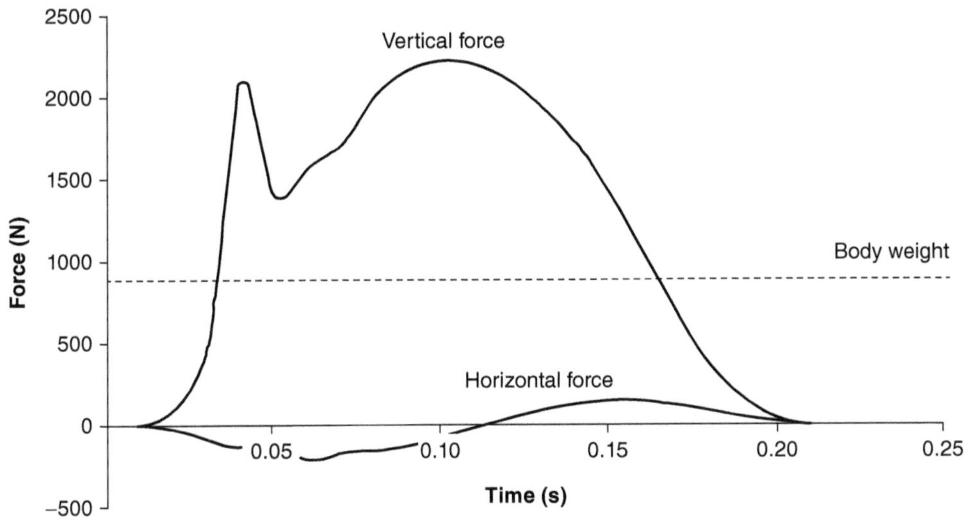


**Figure F2.3** Typical vertical and horizontal ground reaction forces during heel-to-toe running

In the horizontal plane, the force–time profiles demonstrate the existence of a braking and propulsive phase to the gait cycle. During the first period of contact, the ground reaction force is oriented backwards towards the centre of mass, resulting in a reduction to the velocity. From the point of mid-stance, the force reverses and becomes positive, acting to accelerate the body towards toe-off. During steady-state (constant velocity) running, the braking and propulsive impulses will be equal; if these are not in balance then the body will experience either a net increase or decrease in velocity.

While the description provided is considered typical of ground reaction forces during running, it is widely accepted that modern running shoes act to reduce the impact forces seen during walking and running. The various cushioning devices found in the shoes function to prolong the impact and reduce the stress on the musculoskeletal system. The choice of running surface will also have a significant effect on the magnitude of the forces. The harder a surface the greater will be the forces experienced. More compliant surfaces give a greater time to stop the downward motion of the body and so reduce the impact force. However, running on a soft surface uses up more energy and will lead to a slower pace or earlier fatigue.

## TASK

Identify the following variables on the vertical and horizontal force–time graphs in *Figure F2.4* and estimate, as accurately as you can, the magnitude of each of them.

1 Duration of stance phase (i.e. stance time)
2 Impact peak
3 Time to impact peak (from start of stance phase)
4 Active peak
5 Braking peak
6 Duration of braking phase (as percentage of stance time)
7 Propulsive peak
8 Duration of propulsive phase (as percentage of stance time)

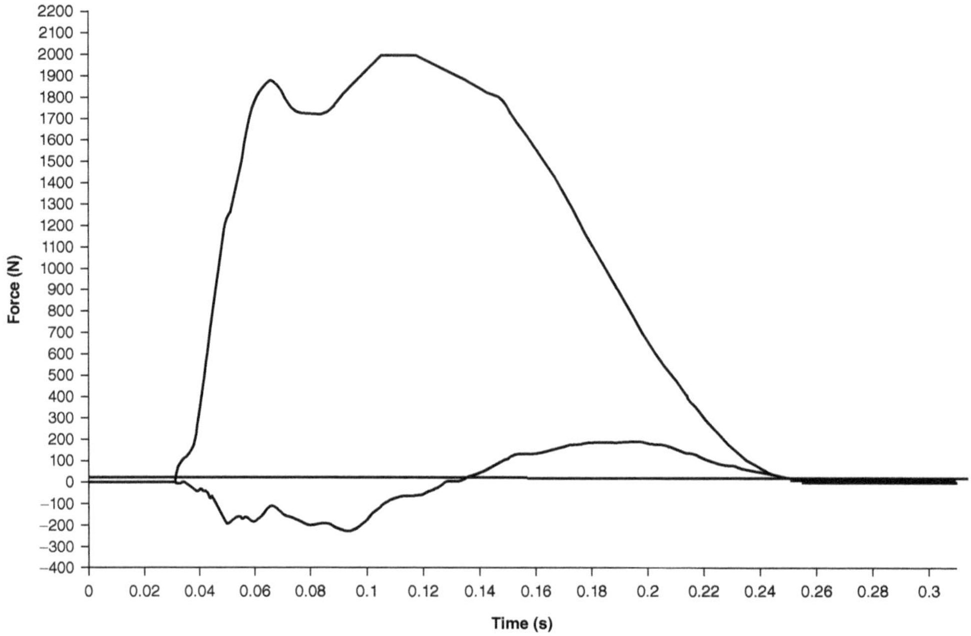


**Figure F2.4** Vertical and horizontal ground reaction forces measured during running

# TASK

Use the horizontal ground reaction force data in *Table F2.3* to calculate the braking and propulsive impulse and thus determine whether there is a net increase or decrease in velocity.

**Table F2.3** Anterior–posterior (horizontal) ground reaction force

| *Time (s)* | *Force (N)* |
|---|---|
| **0.00** | 0 |
| **0.01** | –10 |
| **0.02** | –30 |
| **0.03** | –55 |
| **0.04** | –80 |
| **0.05** | –120 |
| **0.06** | –190 |
| **0.07** | –210 |
| **0.08** | –210 |
| **0.09** | –190 |
| **0.10** | –150 |
| **0.11** | –90 |
| **0.12** | –40 |
| **0.13** | 5 |
| **0.14** | 60 |
| **0.15** | 110 |
| **0.16** | 150 |

| | |
|---|---|
| **0.17** | 180 |
| **0.18** | 190 |
| **0.19** | 175 |
| **0.20** | 150 |
| **0.21** | 95 |
| **0.22** | 40 |
| **0.23** | 10 |
| **0.24** | 0 |
| **0.25** | 0 |

As the body rises and falls and speeds up and slows down during the gait cycle, there are changes in the **potential** and **kinetic** energies in the body. Potential energy describes the energy due to a body's position and is related to its weight and height above the ground (mass × gravity × height). Kinetic energy is the energy due to the body's motion and is related to the mass and the velocity of the body (1/2 mass × velocity$^2$). In running, the changes in these two energies are in phase, such that when potential energy is high, so is kinetic energy. Running has been likened to an individual on a pogo stick, propelling the body from a low point during the middle portion of stance (stance phase reversal) to a peak during the flight phase. To reduce the total energetic cost of running two mechanisms are used:

1 The storage and later return of elastic potential energy via the stretch of elastic structures;
2 The passive transfer of energy from one body segment to another.

These seek to ensure that the amount of metabolic energy required to run is minimised. During the initial period of ground contact, the body's energy decreases as the velocity slows during the braking phase and the centre of mass is lowered by hip and knee flexion. Some of this energy can be stored in the tissues of the lower limb, for example, the Achilles tendon in the form of elastic potential energy. Later, during the drive-off phase, this energy can be returned to contribute to the increase in centre of mass height and velocity. In doing so, this storage and reuse of energy reduces the requirement for the work to be performed by active muscular contraction.

Given the information provided, it is evident that runners subject their bodies to a considerable amount of loading during the course of their running career. Many athletes run between 50 and 100 miles (approximately 80–160 km) every week, which translates to more than 100,000 weekly impacts with the ground. Consider the fact that much of this running is carried out on concrete surfaces and aspects of injury become evident very quickly. While this cumulative loading effect can be responsible for a variety of musculoskeletal and joint problems in runners, sub-optimal running mechanics, which may include excessive or insufficient amounts of both pronation and supination, can increase the risk of injury to the shins, knees, hips and back. The amounts of pronation and supination that a runner experiences can be affected by the type of running shoe that they wear, the surface they are running on, their running style and the type and intensity of training they undertake. Furthermore, excessive and insufficient amounts of pronation and supination may emerge in response to another injury that requires the athlete to excessively pronate or supinate to continue running without pain.

The relationship with pronation and supination and rotation of other structures is shown in *Table F2.4*. It is important to note that this is a ratio representation and that, for example, for every degree of pronation that occurs there would be 2.5 degrees of internal rotation of the tibia and fibula complex (1:2.5 ratio). By comparison, for every degree of supination there is only 0.5 degrees of tibia and fibula external rotation (2:1 ratio). This is one of the many

**Table F2.4** Pronation and supination relationship with other structures of the leg

| | |
|---|---|
| SUPINATION<br>(External rotation) | 2:1 ratio with fibula and tibia rotation<br>1:1.5 ratio with femoral rotation<br>1:1 ratio with pelvic rotation |
| PRONATION<br>(Internal rotation) | 1:2.5 ratio with fibula and tibia rotation<br>1:1.5 ratio with femoral rotation<br>2:1 ratio with pelvic rotation |

reasons why over-pronation is more problematic for running injuries than over-supination. For example, for 10 degrees of pronation the tibia and fibula complex would internally rotate approximately 25 degrees.

Attempting to correct the degree of over-pronation in a runner may not be the correct solution for the medical practitioner. In fact, treatment of such injury-promoting issues requires a careful balance and interpretation of the exact cause of the excessive or insufficient pronation and supination. For example, an excessive pronator may land on the ground in a rolled-over or pronated position (i.e. on the medial (inside) edge of the heel or mid-foot) and then continue to pronate too much and for far too long into the stance phase. On the other hand, an excessive supinator may land on the lateral edge of the heel and then not pronate at all. This athlete may roll outwards on the outer edge of the heel from heel strike all the way through to toe-off. These are two extreme cases of over-pronation and over-supination. Injuries such as patella tendinitis, plantar fasciitis, shin splints, iliotibial band friction syndrome and patellofemoral pain syndrome are just a few of the many that can manifest from pronation and supination concerns. Of these common running-related injuries, **patellofemoral pain syndrome** is considered one of the more problematic injuries and it typically presents as an anterior knee joint pain.

When the foot moves from heel strike to mid-stance the foot normally undergoes a pronation movement. The **ankle dorsiflexes**, the **calcaneus everts** and the **forefoot abducts**, causing the tibia and fibula complex (lower leg) to **rotate internally** (*Figure F2.5*). When the foot

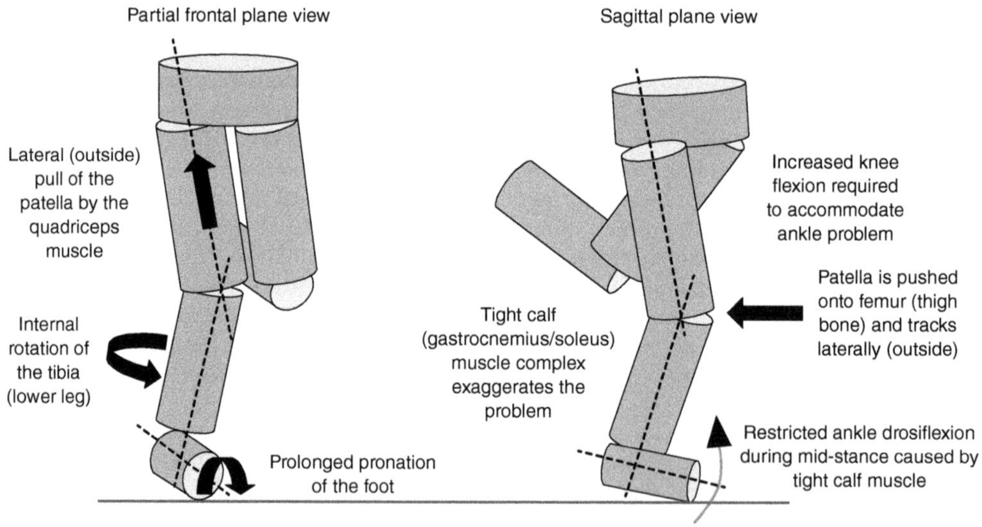


**Figure F2.5** The mechanics of patellofemoral pain syndrome

pronates past the point of mid-stance and, indeed, when the foot pronates too much (usually measured by the amount of eversion of the calcaneus), the lower leg internally rotates excessively and for too long. This pronation continues into the stance phase and past the point of mid-stance. The leg (knee) reaches a point of maximum knee flexion and the quadriceps cause a pull on the patella that attempts to move this bone laterally (away from the body's mid-line). However, because the lower leg is still internally rotated and the foot is still pronated, this lateral pull causes the patella to track over the lateral femoral condyle. Normally the lower leg would be externally rotating and the foot supinating at this point, which allows the patella to be pulled naturally within the groove between the femoral condyles.

To add to this problem, as the knee is flexed during mid-stance, the ankle normally undergoes a degree of dorsiflexion. This movement is decelerated by the gastrocnemius and soleus muscle complex. If the athlete has a tight gastrocnemius–soleus muscle complex, then normal amounts of ankle dorsiflexion are prevented. This has the result of making the athlete increase the amount of knee flexion, which further forces the patella down onto the femur and causes additional aggravated knee pain (*Figure F2.5*).

Symptoms of patellofemoral pain syndrome are generally described as pain in and around the knee joint. This pain is increased when the athlete climbs stairs, walks or even sits. Often an audible "click" may be heard as the knee moves, and chronic cases can lead to an inflammation of the bursa underneath the patella and painful knee swelling. Ultimately, if left untreated, patellofemoral pain syndrome can lead to irreversible degenerative changes to the patella bone. Finally, it is important to remember that patellofemoral pain syndrome may result from a problem that is evident in another part of the body (e.g. the back) and indeed may be a symptom caused by another injury.

The treatment rationale for this problem consists of recommending changes to training programmes and/or training surfaces, while anti-inflammatory medications may also be prescribed. In more severe cases, it may be necessary to consider surgery, particularly if the patella is constantly being pulled laterally and is in need of reattachment in a more biomechanically optimum position. Other non-invasive methods include the use of recommended running shoes and/or the prescription of orthotic devices to control the excessive and prolonged amounts of pronation.

Modern running shoes comprise a number of characteristics that help to reduce the onset of patellofemoral pain syndrome and these include: i) an extended medial support that aims to prevent excessive inward rolling (calcaneal eversion); ii) increased mid-sole density to, once again, control the excessive inward roll of the foot; and iii) little or negative heel flare. The heel flare is the angle made by the sole component of the shoe when viewed from the rear. Older shoe models, such as the Brooks Rage for women (which are no longer manufactured), possessed a specific roll bar at the mid-foot of the shoe. This was really effective in controlling prolonged and pronounced pronation.

## TASK

Use the internet to search for specific types of anti-pronating shoes that are available to athletes.

Orthotic devices are another method for controlling excessive pronation and prolonged pronation, although they can also be designed to correct for supination problems. They are a type of insole, usually made by a podiatrist, that is placed inside the shoe. Orthotics can take on many different forms and can range from rigid orthotics with limited shock-absorbing capacity to soft orthotics that are shock absorbing. The type of orthotic device depends very much on the problem and the type of foot strike of the athlete. Currently, there is a scientific debate as to whether orthotic devices are actually a form of management or a form of treatment. For example, if the orthotic devices are taken away from the athlete after a period of use, will the foot continue to function as though the orthotics were still present?

## TASK

Search the Internet for research on orthotic devices to see what information you can find to clarify whether they are a form of management or treatment for patellofemoral pain syndrome.

| *Key notes* | |
|---|---|
| **The gait cycle** | For running, the gait cycle is divided into support (when the foot is on the ground) and swing (when the foot is off the ground) phases. There are also periods of flight, when both feet are off the ground simultaneously. |
| **Running speed** | Running speed is the product of stride length and stride frequency. Increases in speed are normally achieved by increasing stride length up to about 7 m/s, after which changes in stride frequency are required. Changes in stride frequency normally lead to increased energy expenditure per unit of distance covered (economy). |
| **Foot strike** | Running gait is often described according to which part of the foot makes first contact with the ground. In most runners the first contact is in the rear third of the foot and these runners are described as "rear-foot strikers". A "mid-foot striker" makes first contact in the middle third of the foot and a "forefoot striker" in the front third. |
| **Forces during running** | The peak vertical impact force during running is typically about 2–2.5 times body weight, but varies based on running speed (faster speeds result in higher force) and running shoes and surface characteristics (softer surfaces lead to lower force). Early in the support phase, the horizontal force slows the forward progression of the body (braking), but at about 50 per cent of this phase the force becomes positive and accelerates the body into the flight phase (propulsion). |
| **Pronation and supination** | During running, the joints between the talus and the calcaneus (subtalar joints) experience movements known as pronation and supination. Pronation consists of a combination of eversion, adduction and dorsiflexion, which flattens the foot to the floor. Supination raises the foot's arch via inversion, adduction and plantar flexion. Typically, the foot moves between 10 degrees of supination at foot strike and 10 degrees of pronation by mid-stance during running, and excessive and insufficient amounts of pronation and supination can be the cause of injury. |
| **Patellofemoral pain syndrome** | Patellofemoral pain syndrome is a common running injury which develops in the anterior part of the knee joint and can result from over-pronation problems. Athletes often hear an audible "click" both within and around the knee joint. Treatment options include: i) training programme changes; ii) anti-inflammatory medications; iii) running shoe prescription; or iv) surgery (as a last resort). |
| **Orthotic devices** | Orthotic devices can correct excessive and prolonged pronation problems. These devices can be totally controlling or they can be shock absorbing. There is still current debate on their effective use/function. |

## FURTHER READING

The following resources provide additional reading on the biomechanics of running:

1 Novacheck, T. F. (1998). The biomechanics of running. *Gait and Posture, 7*(1), 77–95.
2 Nigg, B. M., Cole, G. K., & Brüggemann, G. P. (1995). Impact forces during heel-toe running. *Journal of Applied Biomechanics, 11*(4), 407–432.
3 Williams, K. R., & Cavanagh, P. R. (1983). A model for the calculation of mechanical power during distance running. *Journal of Biomechanics, 16*(2), 115–128.

# F3 JUMPING

*Neil Fowler*

Jumping is a movement fundamental to a large number of sports, the goal of which may be to try to achieve the greatest possible distance, either horizontally or vertically, as in the long jump or high jump events. Alternatively, the jump may be performed in order to intercept an airborne object, as might be the case when heading a soccer ball. Jumps can be performed either from a standing position or following a run-up and might involve a take-off from one or both feet. As a consequence, there is no single description for the movements involved in jumping. However, there are sufficient similarities between two-footed and single-footed jumps to allow each movement to be broken down into similar phases (*Table F3.1*) and facilitate a similar analytical approach.

During the **counter-movement** phase, the hip, knee and ankle joints all undergo a period of flexion. Generally, the amount of knee and ankle flexion is similar between jumping performances, but the degree of hip flexion will depend upon the demands of the task and the specific situation in which the jump is performed (i.e. for greater jump heights there will be a greater range of hip flexion). The counter-movement serves two purposes: i) to move the body into a better position to start the propulsive phase; and ii) to stimulate the stretch–shorten effect. If an athlete attempts to jump from an upright standing position without performing a counter-movement, it is obvious that there is very limited potential for the performer to generate lift from this position as the joints of the lower limb are at, or close to, the end of their movement range. To jump from this position can only really be achieved by ankle plantarflexion. By performing a counter-movement the joints are initially flexed, thus permitting a greater range of movement for the propulsive phase.

The greater range of motion allows for the creation of a larger impulse (force × time) during the **propulsive phase,** as the force produced by the hip, knee and ankle extensors can be applied over a greater time. Given that this impulse is directly related to the change in velocity of the body, the greater impulse results in a higher take-off velocity and thus a greater jump height.

## TASK

A volleyball player who has a mass of 75 kg performs a two-footed jump to try to block the ball. If they start from a static standing position and create a vertical impulse of 280 Ns, find the take-off velocity and height achieved.

**Table F3.1** The four common phases of jumping.

| *Phase* | *Start point* | *End point* |
|---|---|---|
| **Counter-movement** | Initiation of movement | Maximum knee flexion |
| **Propulsion** | Maximum knee flexion | Take-off |
| **Flight** | Take-off | Ground contact |
| **Landing** | Ground contact | End of movement |

To solve this problem, we must recall that the impulse–momentum relationship tells us that the impulse is equal to the change in momentum. Given momentum is defined as the product of a body's mass and its velocity (i.e. momentum = mass × velocity) and that, in this example, the body has started from a stationary position (i.e. initial velocity and momentum are zero), we can determine take-off velocity as follows:

$$\Delta\text{momentum} = \Delta\text{impulse}$$

$$(\text{mass} \times \text{velocity}_{\text{final}}) - (\text{mass} \times \text{velocity}_{\text{initial}}) = \text{impulse}_{\text{final}} - \text{impulse}_{\text{initial}}$$

$$(75 \times \text{velocity}_{\text{final}}) - (75 \times 0) = 280 - 0$$

$$(75 \times \text{velocity}_{\text{final}}) = 280$$

$$\text{velocity}_{\text{final}} = \frac{280}{75}$$

$$\mathbf{velocity_{final}} = 3.73\ \mathbf{m/s}$$

Once the body leaves the ground, it will begin decelerating due to gravitational acceleration and, at the top of the jump, its vertical velocity will be equal to zero. Using the equations of uniform acceleration, we are able to determine the height jumped ("s" in Equation F3.1):

$$v^2 = u^2 + 2 \times a \times s \qquad \textbf{(Equation F3.1)}$$

For simplicity, Equation F3.1 can be rewritten more specifically for this example:

$$\text{velocity}_{\text{final}}^{\ 2} = \text{velocity}_{\text{initial}}^{\ 2} + 2 \times \text{acceleration}_{\text{gravity}} \times \text{height} \qquad \textbf{(Equation F3.2)}$$

Therefore:

$$3.73^2 = 0^2 + 2 \times 9.81 \times \text{height}$$

$$3.73^2 = 2 \times 9.81 \times \text{height}$$

$$\frac{3.73^2}{2 \times 9.81} = \text{height}$$

$$\textbf{height} = 0.71\ \textbf{m}$$

During the counter-movement phase, the main action of the hip, knee and ankle extensors is to resist or control the flexion that is occurring at these joints as a natural consequence of gravitational force. As the muscles are producing a resistive force while their length is being increased, their action during the counter-movement is eccentric in nature. This eccentric contraction is often referred to as a pre-stretch, as the stretch of the muscles precedes the concentric contraction of these same muscles in the following propulsive phase. Pre-stretching a muscle before it is shortened leads to an increase in the force that the muscle can produce. This increase in force following the pre-stretch is known as the **stretch–shorten cycle**. The faster the stretch and the shorter the delay between the stretch and shortening components, the greater the increase in muscle force. As such, it is important that the counter-movement is performed rapidly, and that there is a minimal delay between the end of the counter-movement and start of the propulsive phase.

If the body is lowered slowly, into a position comparable to that achieved at the end of the counter-movement, and this position is held prior to the upward movement, the jump is described as a squat jump; as the jump begins from a squat position. The jump distance achieved in squat jumps is usually about 10 per cent less than that achieved from a comparable counter-movement jump, suggesting that the counter-movement adds about 10 per cent to the height that can be achieved.

During the propulsive phase, the joints of the lower limb (hip, knee and ankle) undergo extension. The hip is always the first joint to start to extend, accelerating the large, heavy trunk segment. Extension at the knee and ankle joints follows after a short delay, and the initiation of knee and ankle extensions may happen either simultaneously or in sequence (knee then ankle, or ankle then knee). However, there is no clear evidence to suggest that performance is better with any particular sequence for the initiation of knee and ankle movements.

In addition to the movements of the lower limb, **arm swing** plays an important part in the performance of standing jumps. During the counter-movement phase, the arms are swung downward and backward before swinging forwards and upwards in the propulsive phase. Arm swing has been shown to add between 10 and 20 per cent to the distance achieved in a jump. To be effective, the arm swing must be timed appropriately, such that take-off occurs at the point when the arms are at, or close to, maximum velocity.

The exact mechanism through which the arm swing contributes to jumping performance has not been determined. However, it is believed to be related to both a direct contribution to momentum of the centre of mass and/or the creation of greater ground reaction forces. If the body is considered as a series of individual segments, then it is possible to calculate the velocity, and hence momentum (mass × velocity), of each. The velocity of the centre of mass, and hence momentum, is a consequence of the summation of these individual segmental velocities and momenta. If the arms are being swung upwards and forwards then they will contribute to the centre of mass's forwards and upwards motion.

The most accurate method for obtaining measures of the take-off velocity is to use a force platform to record the vertical ground reaction force. From the force–time data it is thus possible to determine the impulse as the start point for the calculation above. Another common method used to determine the height of a vertical jump is to measure the flight time and use this to calculate the height achieved. This calculation also makes use of the equations of uniformly accelerated motion (Equation F3.3).

$$s = u \times t + \frac{1}{2} a \times t^2$$

or

$$\text{displacement} = \text{velocity}_{\text{initial}} \times \text{time} + \frac{1}{2}\text{acceleration} \times \text{time}^2 \qquad \textbf{(Equation F3.3)}$$

If it is assumed that the take-off and landing are performed at the same relative height, then the apex of the jump will occur at exactly half the flight time, at which point the velocity will be zero, and the acceleration will be due to gravity (–g). Therefore, at the apex of the jump, the body will have a zero velocity and, from this point to landing, the body will experience a displacement equal to the height jumped in one half of the flight time. Hence, jump height(s) can be estimated via Equation F3.4:

$u = 0$

$t = \frac{1}{2}$ flight time

$a = g$

**Note:** the minus sign has been omitted for acceleration, as direction is not important in this example.

$$s = 0 \times t + \frac{1}{2} \times g \times (\frac{1}{2} t)^2$$

$$s = \frac{1}{2} \times g \times (\frac{1}{2} t)^2$$ **(Equation F3.4)**

## TASK

If an athlete performs a vertical jump on a jump mat which records their flight time as 0.8 seconds find the actual height jumped.

To solve this problem, we can use Equation F3.4:

$$s = \frac{1}{2} \times g \times (\frac{1}{2} t)^2$$

$$s = \frac{1}{2} \times 9.81 \times (\frac{1}{2} \times 0.8)^2$$

$$s = 0.78\mathbf{m}$$

However, the use of flight time to predict jump height is problematic, as it assumes the take-off and landing are at the same relative height and that the height of the jump is represented by the displacement of the centre of mass during flight. However, in reality, there is normally a difference between take-off and landing heights. For example, at take-off, the arms are generally elevated and the lower limbs are near to full extension, while in preparation for landing, the lower limb is usually in a position of partial flexion at the hip and knee. As such, the centre of mass is typically lower at landing than at take-off. The triple extension of hip, knee and ankle at take-off also means that the centre of mass may be up to 10 cm above its position in normal standing at take-off; displacement that is not accounted for in flight time.

In many athletic events and other sports, jumping actions are performed from a single leg and following a run-up. Performance in running jumps is also determined by the velocity of the body at take-off. Unlike in standing jumps, the body will have an initial velocity as a consequence of the run-up and thus the take-off velocity will be determined by the combination of the run-up velocity and the change in velocity due to the take-off impulse. In the case of horizontal jumps, such as the long jump, the athlete will attempt to generate a large initial velocity during the run-up. Strong positive correlation coefficients have been reported between run-up speed and jump distance, indicating that greater jump distances can be achieved with a faster run-up. Despite the strength of the relationship between run-up speed and distance, it is worthy of note that athletes do not achieve their maximum running speed during their run-up. Each individual will have an optimal run-up speed somewhere below their maximum running speed. Increasing the run-up speed above this leads to a decrease in performance as the athlete is unable to generate sufficient vertical impulse during the take-off phase.

## TASK

*Table F3.2* contains two sets of data obtained from an experiment to assess the effect of run-up speed on long-jump distance. The individual athlete data shows the relationship between running speed and distance jumped for the same athlete, while the population data represents the average performance of a larger sample of athletes who are each jumping at their preferred run-up speed. Plot the data presented in *Table F3.2* and describe the different relationships.

**Table F3.2** Long jump distance achieved from a range of run-up speeds for an individual athlete and population averages from skilled long jumpers.

| *Approach velocity (m/s)* | *Individual athlete distance (m)* | *Population data distance (m)* |
|---|---|---|
| 6.0 | 4.3 | 2.0 |
| 6.5 | 4.8 | 2.9 |
| 7.0 | 5.2 | 3.4 |
| 7.5 | 5.7 | 4.0 |
| 8.0 | 6.2 | 4.7 |
| 8.5 | 6.5 | 5.2 |
| 9.0 | 6.8 | 6.0 |
| 9.5 | 7.2 | 6.6 |
| 10.0 | 7.6 | 7.2 |
| 10.5 | 7.7 | 7.9 |
| 11.0 | 7.5 | 8.7 |

What do you think would happen if the individual athlete attempted to jump after a run-up at a greater speed than those shown?

Similar to two-legged jumps, the actions of the take-off leg during single-legged jumps involve the hip, knee and ankle of the take-off leg experiencing an initial flexion after heel strike, followed by extension. These movements give rise to what are commonly described as the compression and drive-off phases. The compression phase is comparable with the counter-movement and involves similar mechanisms, although the range of joint motion is much less than seen in two-legged jumps.

In addition to the action of the take-off leg, the contralateral (opposite) leg also makes an important contribution to the take-off velocity. The non-take-off leg is often referred to as the "free leg" as it is free to swing in space. The motion of the free leg and the arms contribute to the take-off velocity in a similar way to that described for the arms in standing jumps. The momentum within these segments increases centre of mass velocity by 10 to 15 per cent provided that the movements are timed correctly.

During the take-off phase, the body experiences a combination of horizontal and vertical forces. As the line of action of these does not always pass through the centre of mass, there is a resultant moment about the centre of mass that tends to cause rotation. Generally, the ground reaction forces act to create a forwards or somersaulting moment. Hence, during take-off, the body experiences a torque impulse (torque × time) that causes a change in angular momentum (moment of inertia × angular velocity). Once airborne, the body tends to rotate forwards and the performer has to take action to control the consequences of this rotation on their performance.

In the long and triple jump events, the athlete is required to control the rate of forwards rotation to allow them to achieve an optimal landing position. As the amount of angular momentum is constant during flight, there is no possibility to apply a corrective torque while the body is in the air. As such, there are two methods that an athlete may adopt to **control the rotation** of their body. The first requires a simple consideration of the definition of angular momentum (Equation F3.5):

angular momentum (kg.m²/s = moment of inertia (kg.m²) × angular velocity (rad/s)

or

$$L = I \times \omega \qquad \textbf{(Equation F3.5)}$$

As the angular momentum (L) is constant, the rotation of the body (ω) during flight will depend upon the moment of inertia (I) and, hence, the body's position. If the moment of inertia is maximised by adopting an elongated body position in flight the angular velocity will be minimised and only a small degree of rotation will occur during the flight. This is seen in the long jump by athletes adopting a "hang" technique where the arms are extended above the head and the legs extended.

The other method for controlling angular motion during flight requires an analysis of the segmental contributions to whole-body angular momentum. If the body is considered as a series of individual segments, the angular momentum of each segment can be calculated and used to determine the whole-body angular momentum. If one group of segments are rotated in such a way as to generate an angular momentum equal to the whole-body angular momentum there would be no net rotation about the centre of mass. This is what happens in the "hitch-kick" technique, where the athlete performs what appears to be a running leg action in the air. The angular momentum generated by rotating the legs makes a sufficient contribution to whole-body angular momentum to prevent the forwards rotation of the body.

*Key notes*

**The counter-movement**

Most forms of jumping are initiated with a downward movement of the body referred to as a counter-movement. This action serves to increase the distance that can be achieved in the jump by about 10 per cent. This increased performance is attributed to a greater range of movement during the propulsive phase and the use of the stretch–shorten cycle.

**The stretch–shorten cycle**

The stretch–shorten cycle describes the sequence of movement whereby an active muscle is first stretched (eccentric action) and then shortened (concentric action). The stretching phase results in a more forceful shortening of the muscle than if there had been no pre-stretch.

**Jump height**

In a standing vertical jump, the height achieved is directly proportional to the vertical velocity at take-off such that:

$$\text{Jump height} = \frac{\text{Take-off velocity}^2}{2 \times \text{gravity}}$$

The vertical velocity is related to the vertical impulse during the contact period such that:

$$\text{Take-off velocity} = \frac{\text{Force} \times \text{Time}}{\text{Mass}}$$

The impulse can be determined by integrating the vertical ground reaction force.

**Arm swing**

Arm swing is believed to add 10 to 20 per cent to the height or distance achieved in a jump. This is related to both a direct contribution to centre of mass momentum and to the creation of greater ground reaction forces. The centre of mass momentum is a summation of the individual segments' momenta. If the arms are swung upward and forwards they contribute to the body's forwards and upwards motion.

*(continued)*

*(continued)*

| | |
|---|---|
| **Control of angular momentum** | During the flight phase of a jump, the body will tend to rotate (somersault) in accordance with the body's angular momentum at take-off.<br><br>Angular momentum = Moment of inertia × Angular velocity<br><br>In order to control the rate of rotation, jumpers may change the orientation of their limbs to elongate their body position and increase their moment of inertia; ultimately slowing the rate of rotation (i.e. as in the "hang" technique). Alternatively, the athlete may elect to rotate their arms and legs in a forwards direction to slow the rate of forwards body rotation by transferring momentum (i.e. as in the "hitch-kick" technique). |

## FURTHER READING

The following two resources provide additional reading on the biomechanics of jumping in sport.


1 Hay, J. G. (1993). Citius, altius, longius (faster, higher, longer): The biomechanics of jumping for distance. *Journal of Biomechanics*, *26*(suppl. 1), 7–21.
2 Bobbert, M. F., & van Ingen Schenau, G. J. (1988). Coordination in vertical jumping. *Journal of Biomechanics*, *21*(3), 249–262.

# THROWING



*Neil Fowler*

Although many sports involve the use of some form of overarm throwing or striking action, the specific kinematic attributes of each may vary depending on the unique requirements of the skill being performed. With this in mind, it would be unrealistic to try to cover all of the specific variants of the throwing action; hence, this section will focus upon the general movement patterns and highlight how these can be varied to achieve different goals.

Throwing is considered to be an **open chain** movement, which means that the distal end of the moving segment (i.e. the hand) is free to move in space. This is in contrast to **closed chain** movements, where the distal segment is constrained such that it is not free to move relative to the other parts of the body (e.g. the foot during weight-bearing tasks). The differentiation of movement into open and closed chains is somewhat contrived and the terms are more commonly used in the context of physiotherapy than biomechanics.

In general, the throwing action can be broken down into three fundamental phases, which include: i) the preparation phase; ii) the pulling/driving phase; and iii) the follow-through phase. However, as the throwing action is rather complex, it is common to further divide the pulling phase into early and late parts, yielding four phases (*Table F4.1*).

The **preparation phase** provides a longer pulling path for accelerating the arm and serves to pre-stretch the musculature. The movements of the preparation phase are initiated by a forceful contraction of the prime movers for each action (concentric contraction). If a run-up is used before the throw, this will help to create momentum within the body and make the preparatory movements faster and require less muscular effort to initiate. As the body reaches the end position of the preparation phase, the rotator muscles of the trunk become stretched and the stretch reflex stimulates an eccentric contraction. Similarly, the muscles of the anterior shoulder region become stretched by the abduction and horizontal extension of the shoulder. These eccentric contractions facilitate the use of the stretch–shortening cycle to enhance the force of the early preparation phase and thus increase the velocity of the movement.

**Table F4.1** The start and end points for the four common sub-phases of the throwing action

| | | *Description* |
|---|---|---|
| Preparation | Start point | First backward movement of the hand |
| | End point | Maximum horizontal extension of the shoulder |
| Early pulling phase | Start point | Maximum horizontal extension of the shoulder |
| | End point | Maximum external rotation of the shoulder |
| Late pulling phase | Start point | Maximum external rotation of the shoulder |
| | End point | Release of object |
| Follow-through | Start point | Release of object |
| | End point | Maximum shoulder extension |

The **early pulling phase** is the primary phase for accelerating the motion of the upper limb. This involves a sequential movement of the trunk and upper limb, which serves to transfer momentum from the proximal to the distal segments. During the early pulling phase, those muscles stretched in the preparation phase overcome the external force and begin to contract concentrically to rotate and flex the trunk. The initial rotation of the pelvis is closely followed by trunk rotation, which collectively serve to accelerate the shoulder axis in the horizontal plane. The muscles of the anterior shoulder region continue to work eccentrically as the forward rotation of the trunk tends to leave the arm lagging behind, which induces greater external rotation of the shoulder. Many coaches teach that the thrower should "lead with the elbow". This in reality does not happen. The elbow remains behind the shoulder axis throughout much of the throw and it is the initial trunk rotation that generates the external rotation of the shoulder that leaves the lower arm lagging behind. The degree of external rotation at the shoulder is also related to the elbow angle. If the elbow is allowed to flex beyond 90 degrees, the moment of inertia of the limb will decrease and a smaller rotational torque will be exerted on the shoulder. To preserve elbow flexion, the triceps reach their peak activity during the early pull phase and are considered to be working in an isometric/eccentric fashion due to the lack of significant elbow movement evident at this time.

During the **late pulling phase**, the elbow extends rapidly as the radius of the arm is increased to generate maximum linear velocity in the distal segment. This motion does not involve a muscular action of the triceps; the elbow would extend at the same time and with the same velocity even if the triceps had not been functioning. The movement is performed by the transfer of momentum from the trunk and upper arm to the lower arm segment. If the muscle were to be active it would be unable to contribute significant force to the movement, due to its high-speed nature.

The rapid arm movements are gradually slowed via the eccentric contractions of antagonistic muscles during the **follow-through phase**. If the movements are brought to an abrupt halt then the tension developed in the antagonists will be great and the risk of injury will be increased. A long and gradual follow-through is the most desirable, but is not always practical within the sporting context.

**Table F4.2** Typical movements of the trunk and upper limb joints during the four phases of a right-handed throw

| *Phase* | *Joint* | *Description* |
|---|---|---|
| Preparation | Trunk | Left lateral flexion; clockwise rotation; hyper-extension |
| | Shoulder | Horizontal extension; abduction |
| | Elbow | Flexion |
| | Wrist | Extension |
| Early pulling phase | Trunk | Counter-clockwise rotation; flexion |
| | Shoulder | Horizontal flexion; external rotation |
| | Elbow | No movement |
| | Wrist | No movement |
| Late pulling phase | Trunk | Counter-clockwise rotation; flexion |
| | Shoulder | Internal rotation |
| | Elbow | Extension |
| | Wrist | Flexion |
| Follow-through | Trunk | Counter-clockwise rotation; flexion |
| | Shoulder | Adduction; internal rotation; extension |
| | Elbow | Flexion |
| | Wrist | Flexion; pronation |

As with most open chain exercises, the freedom of the distal segment allows substantial variation in the specific orientation of the various segments and the relative length of each phase during the performance of the skill. These variations give rise to various techniques that may include: i) overarm throwing; ii) round-arm throwing; iii) bowling; and iv) overhead striking, serving or smashing. Within each of these different techniques there are many more variations, which makes it impossible to define all of them. Essentially the techniques differ by the orientation of the trunk and the degree of shoulder abduction. These differences in limb orientations will lead to some differences in the prime mover muscles involved in the action. However, the general phasing and the nature of the muscle actions will be consistent across all of the variants of the throwing action.

Nevertheless, a general description of the movement patterns of the trunk and upper limb joints is provided in *Table F4.2* to assist with understanding the ways in which these joints interact to facilitate the throwing action.

## TASK

Use a suitable video camera to conduct a movement analysis of two contrasting styles of throwing (e.g. a javelin throw for maximum distance and a baseball pitch). Using the movement descriptors above, compare the ranges of movement and relative phasing of the different styles of throwing and consider why these differences occur.

## TASK

If a javelin of mass 0.6 kg is released at a speed of 30 m/s following a run-up that produced an initial velocity of 7 m/s and a pulling phase that lasted 0.1 s, how much force is acting on the javelin?

## Solution

To calculate the average force acting on the object (javelin) during the throw, it is necessary to apply the impulse-momentum relationship, which gives:

$$\text{Force} \times \text{Time (impulse)} = \text{Mass} \times (\text{release velocity} - \text{initial velocity})$$

So, in this example:

$$\text{Force} \times 0.1 = 0.6 \times (30 - 7)$$

$$\text{Force} = \frac{(0.6 \times 23)}{0.1}$$

$$\text{Force} = 138\ \mathbf{N}$$

It is possible to determine the relative importance of each of the joint movements involved in a throw by calculating the degree to which they lead to the development of the velocity of the hand and, by extension, the object thrown. This can be achieved by analysing both the angular velocity at the joint and the perpendicular distance between the axis of rotation and the ball. The product of angular velocity and radius gives the linear velocity:

$$\text{linear velocity}\left(v\right) = \text{angular velocity}\left(\omega\right) \times \text{radius (r)}$$

Hence, summing the linear velocity contributions from each joint of the upper limb would yield the final velocity of the ball, such that:

$$v_{release} = v_{shoulder} + v_{humerus} + v_{forearm} + v_{hand}$$

In this example, it should be noted that $v_{shoulder}$ is considered to be the velocity of shoulder segment, which is generated during the run-up and via the movements of the lower body and trunk. For the remaining segments, their contribution to the release velocity ($v_{release}$) will be related to the angular velocity (ω) and radius (r), such that:

$$v_{humerus} = (\omega_{int/ext\,rotation} \times r) + (\omega_{flex/ext} \times r) + (\omega_{abd/add} \times r)$$

$$v_{forearm} = (\omega_{pro/sup} \times r) + (\omega_{flex/ext} \times r)$$

$$v_{hand} = (\omega_{rad/ulna\,deviation} \times r) + (\omega_{flex/ext} \times r)$$

When applied to an overarm throw that resulted in a release speed of about 28 m/s, the following contributions have been reported:

Internal rotation of humerus = 8.0 m/s

Wrist flexion = 7.0 m/s

Horizontal flexion of humerus = 6.5 m/s

Forearm pronation = 4.0 m/s

Forward motion of shoulder = 2.5 m/s

While shoulder flexion/extension, ulna deviation and elbow extension do not make a significant contribution to release speed, they may be important to ensuring that the release of the object is optimal in terms of angle, orientation or spin. In general, this relationship demonstrates the importance of maintaining as large a radius as possible between the distal segment and axis of rotation in order to maximise the release velocity of the object. However, this will inevitably increase the moment of inertia of the limb and increase the work necessary to achieve the same angular velocity.

The optimal release speed and angle of a throw is primarily determined by the objective of the throwing task. For example, a basketball player shooting a free throw does not need to generate a maximum velocity at release, but does need to find an optimal combination of height, angle and release speed to ensure that the ball intercepts the basket. In contrast, a shot-putter will want to find a combination of release characteristics that maximise the flight distance of the shot. Due to its relatively high mass the shot will be largely unaffected by aerodynamic forces during its flight, but in other throwing sports aerodynamic forces can be important. For example, while a javelin thrower is also seeking to maximise the release conditions to increase their throwing distance, the shape and weight of the javelin mean that they must also pay attention to factors such as the angle of attack (the angle between the long axis of the javelin and its velocity vector) and the degree of pitch/pitching moment (the tendency to rotate upwards or downwards) due to the more significant aerodynamic forces.

## TASK

At the point of release, a shot-putter creates a height of 2.2 m, a velocity of 13 m/s, and a release angle of 35 degrees. Calculate the range of the throw.

## Solution

To calculate the throw's range (distance), we can use the projectile motion equation below:

$$\text{Range} = \left(\frac{(v^2 \times \sin(2 \times \theta))}{2 \times g}\right) \times \left(1 + \sqrt{1 + \frac{(2 \times g \times h)}{v^2 \times \sin^2(\theta)}}\right)$$

Hence,

$$\text{Range} = \left(\frac{(13^2 \times \sin(2 \times 35))}{2 \times 9.81}\right) \times \left(1 + \sqrt{1 + \frac{(2 \times 9.81 \times 2.2)}{13^2 \times \sin^2(35)}}\right)$$

$$\text{Range} = \left(\frac{169 \times 0.94}{19.62}\right) \times \left(1 + \sqrt{1 + \frac{43.16}{169 \times 0.33}}\right)$$

$$\text{Range} = 8.10 \times \left(1 + \sqrt{1.78}\right)$$

$$\text{Range} = 8.10 \times 2.33$$

$$\textbf{Range} = \mathbf{18.87\,m}$$

*Key notes*

| | |
|---|---|
| **The phases of a throw** | The throwing action is divided into preparation, early pulling/driving, late pulling/driving and follow-through phases. The duration and range of motion in these phases will vary according to the purpose of the throw. |
| **The preparation phase** | The preparation phase provides a longer pulling path for accelerating the arm and serves to pre-stretch the musculature. Specifically, the anterior shoulder muscles are stretched by the abduction and horizontal extension of the shoulder. These eccentric contractions facilitate the use of the stretch–shortening cycle to enhance the force of the early preparation phase and thus increase the velocity of the movement during the pulling phase. |
| **The pulling phase** | The pulling phase is where the velocity of the throw is developed and is the primary phase for accelerating the motion of the upper limb. During the early pull phase, the shoulder axis is accelerated in the horizontal plane by pelvic and then trunk rotations, causing the flexed lower arm to lag behind due to external rotation of the shoulder. During this phase, those muscles stretched during the preparation phase also overcome the external force and begin to contract concentrically, contributing to trunk rotation and flexion. During the late pulling phase, the shoulder internally rotates and the elbow extends. Collectively, these movements sequentially accelerate the joints and transfer momentum from proximal to distal segments. |
| **The follow-through** | The follow-through brings the throwing arm to a controlled stop. Muscle actions in this phase are mainly eccentric. A shorter follow-through requires more forceful muscle contractions. |

## FURTHER READING

The following resources provide additional reading on the biomechanics of throwing in sport.

1 Fleisig, G. S., Barrentine, S. W., Escamilla, R. F., & Andrews, J. R. (1996). Biomechanics of overhand throwing with implications for injuries. *Sports Medicine, 21*(6), 421–437.
2 Morriss, C., & Bartlett, R. (1996). Biomechanical factors critical for performance in men's javelin throw. *Sports Medicine, 21*(6), 438–446.

# KICKING

*Michael Cole*

Kicking is a fundamental skill in many sports but can vary from kicking a ball from the ground, as in performing a penalty kick in soccer, or kicking a ball that has been dropped from the hands, as in the drop punt in Australian Rules football. Given the wide variety of ways that kicking is used in different sports, it is not surprising that these different kicking actions rely on different parts of the foot to strike the ball (e.g. the dorsal (top) aspect, the toes, the lateral or medial border of the foot). Furthermore, these different kicking actions also differ with respect to the relative position of the ball at the point of contact (e.g. in the air *vs* on the ground) and its relative proximity to the body at the point of contact. Although each of these variables will affect the magnitude of the joint movements seen during a performance, the gross movements of the lower limb generally follow a similar pattern from the point where the supporting leg is planted on the ground in all forms of kicking. It is this general movement pattern that will be described in this section.

As with all sporting actions, the biomechanical analysis of kicking technique typically begins with the movement being broken down into phases, for which the specific movements of the segments are described. For the kicking action, these phases are defined in *Table F5.1* and are specifically referred to as the i) preparation; ii) driving; and iii) follow-through phases.

Not surprisingly, each of these phases can also be described by the unique combination of lower limb joint movements (*Table F5.2*), which ultimately contribute to the coordinated movement sequence that facilitates the acceleration of the foot towards the ball.

Of the three phases outlined above, the driving phase is often further sub-divided into two parts, with the first characterised by hip flexion with little to no knee extension and the second representing the component where the hip has achieved its peak angular velocity and the knee begins to extend to ball contact. The relative duration of each phase will vary depending on the type of kick being executed and on the circumstances under which the athlete is performing. For example, some kicks might have very short preparation phases to maximise the likelihood of surprising the opposition, while others may have a deliberately long preparation

**Table F5.1** The start and end points for the three common phases of the kicking action. Unless otherwise noted (*), all start/end points relate to the kicking leg.

| *Phase* | *Start/End* | *Description* |
|---|---|---|
| *Preparation* | Start point | Supporting foot contact* |
| | End point | Maximum hip extension |
| *Driving* | Start point | Maximum hip extension |
| | End point | Ball contact |
| *Follow-through* | Start point | Ball contact |
| | End point | Maximum hip flexion |

**Table F5.2** Joint movements of the kicking leg that describe the three common phases of the kicking action, as performed with the right foot

| *Phase* | *Joint* | *Movement* |
|---|---|---|
| *Preparation* | Pelvis | Clockwise rotation (i.e. away from ball) |
| | Hip | Extension; abduction |
| | Knee | Flexion |
| | Ankle | Neutral |
| *Driving* | Pelvis | Anti-clockwise rotation |
| | Hip | Flexion; adduction |
| | Knee | Extension |
| | Ankle | Plantarflexion |
| *Follow-through* | Pelvis | Anti-clockwise rotation |
| | Hip | Flexion |
| | Knee | Extension; hyperextension |
| | Ankle | Dorsiflexion |

phase to maximise accuracy and force production. Given this situation, it is often not possible to prescribe set ratios for the duration of kicking phases.

When the aim of the kick is to maximise the velocity of the ball following contact, the athlete aims to accelerate their kicking foot as much as possible throughout the movement to ensure that it is travelling as fast as possible when it strikes the ball. By considering the specific mechanics of each of the three phases in isolation, it is possible to see how these movements contribute to this end goal (*Figure F5.1*).

During the preparation phase, extension and abduction of the hip and flexion of the knee require coordinated muscle activation to raise the kicking limb against gravity. Ultimately, these actions move the kicking foot away from the eventual direction of the kick, which serves two functions: i) to increase the angle through which the limb must move before it makes contact with the ball; and ii) to allow a greater opportunity for foot velocity to be developed. In the context of the laws of motion, understanding the importance of these functions is very straightforward. In particular, it is well known that the greater the time (or distance) over which a body (or segment of a body) is accelerated, the greater the final foot velocity. As the hip of the kicking leg continues to extend and the knee continues to flex throughout the preparation phase, the length of the hip flexors and knee extensors is rapidly increased (i.e. they are stretched). During the latter stages of this phase, the increased length (stretch) of the hip flexors

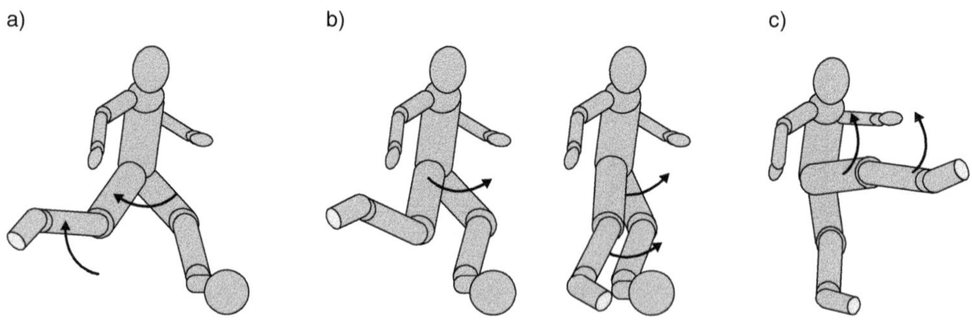


**Figure F5.1** The three primary phases of the kicking action, including: a) preparation phase; b) driving phase; and c) follow-through phase

and knee extensors initiates stretch reflexes, which activate these muscles and causes them to start contracting eccentrically. These eccentric contractions contribute to preventing the hip flexors and knee extensors from being over-stretched and potentially injured, but also slow the movement of the kicking leg. When sufficient tension is developed, the hip flexors overcome the backwards motion of the limb and begin to contract concentrically to flex the hip. At this point, it should be acknowledged that there are a number of other important neurophysiological mechanisms involved in this process (e.g. the response of Golgi tendon organs to increased muscle tension); however, an in-depth discussion of these topics is beyond the scope of this text.

During the driving phase, movements of the kicking limb include rapid flexion and adduction of the hip and extension of the knee; however, these two movements do not start at the same time. A temporal analysis of the movement would reveal that the hip begins flexing first, while the knee is still flexed. During the latter stages of the driving phase, a forceful concentric contraction of the knee extensors causes the knee to extend. As the knee extends, the moment of inertia of the leg increases and the velocity of hip flexion is reduced. As this happens, momentum is transferred from the thigh to the shank and the movement continues with minimal muscle action. This movement sequence conforms to the summation of velocities principle (otherwise known as the summation of speed principle) and is important for facilitating the rapid increase in the angular and linear velocities of the kicking leg (*Figure F5.2a*). This increased capacity to angularly accelerate the lower limb is also facilitated by the reduced moment of inertia of the leg (about the hip joint) when the knee is flexed relative to when the knee is extended. Given that Newton's second law of angular motion states that the angular acceleration of a limb is directly influenced by its moment of inertia and the magnitude of the torque produced, it is clear that it will be easier to angularly accelerate the hip joint when the knee is flexed. Once the hip joint has achieved its maximum angular velocity, the knee joint begins to extend to increase the length of the limb prior to the point of impact (*Figure F5.2b*). By ensuring that, at ball contact, the shank and foot are further away from the axis of rotation (i.e. the hip joint) and that the lower limb is acting as a single rigid lever, an athlete is able to maximise the linear velocity of the foot and transfer more energy to the ball. The peak linear velocity of the ball during the maximal instep kick is reported to range from 18 to 35 m/s for adult male players and mathematical descriptions indicate that these velocities are 1.2 times the velocity of the foot at the point of ball contact. The increased velocity of the ball relative to the foot is facilitated by the impulse–momentum relationship and the material properties of objects involved in the contact, including the foot, boot and ball. The linear velocity of moving segment is found by multiplying the angular velocity of the limb (ω) by its distance (radius) from the axis of rotation (r).

$$v\ (\text{m/s}) = \omega\ (\text{rad/s}) \times r\ (\text{m})$$

It should be noted that the movements of the ankle joint during the driving phase are primarily directed towards aligning the desired point of contact of the foot with the target area on the ball's surface. For example, if it is the intention of the athlete to kick the ball with their instep, rather than the dorsal aspect of the foot, they will exhibit different foot kinematics during the movement to alter the orientation of the foot.

Towards the end of the driving phase, the knee approaches full extension, which stimulates a stretch reflex response from the knee flexors that are also being stretched by action of the hip flexors. This evokes an eccentric contraction of the muscles that flex the knee, which ultimately initiates the slowing of the limb to protect the knee joint from being damaged due to hyperextension. After ball contact, the kicking action enters the follow-through phase, which involves almost exclusive eccentric muscle function aimed at gradually reducing the high angular and linear velocities of the kicking limb. Understandably, the movements that occur during the follow-through can have no influence on the outcome of the kick, but this phase is important

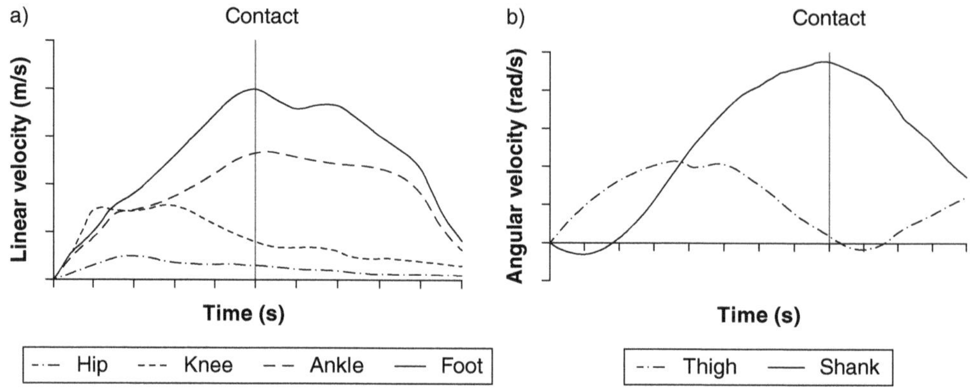


**Figure F5.2** Graphical representation of the linear and angular velocities of the swinging leg during the kicking action. Specifically, the graphs show: a) linear velocities of the hip, knee, ankle and foot; and b) angular velocities of the thigh and shank. Note: The vertical line on each graph represents the point of contact between the foot and the ball.

for protecting the muscles and joints involved in the movement by allowing the mechanical loads to be dissipated over a longer time period. Again, in alignment with Newton's second law of angular motion, the longer the time allowed to slow the moving limb, the smaller the force required to achieve the outcome, as the impulse (force × time) created to move the limb will be the same as the impulse required to stop the limb.

Thus far, we have only considered movements of the lower leg, but the upper body is also important during the kicking action to maximise distance and balance the body. During preparation for a right-footed kick, the contralateral arm projects laterally via shoulder abduction to balance the body, while the pelvis rotates clockwise to swing the right leg away from the ball. During the driving phase, the abducted contralateral arm is swung backwards (away from the intended direction of the kick), which complements the anti-clockwise rotation of the shoulders, trunk and pelvis. During the follow-through, the pelvis and kicking leg continue to rotate anti-clockwise, but the contralateral arm and upper torso begin to exhibit clockwise rotation. The opposing rotations of these segments serve to control the change in angular momentum and improve the stability of the performance.

The mechanical description outlined previously only considers the kicking action from the point at which the supporting limb makes initial contact with the ground. However, in many situations the kick will be preceded by a run-up, which provides the body with some initial momentum and which can result in maximum ball velocities of up to 7 m/s faster than a similar kick performed without a run-up. In situations where it is feasible to perform a run-up, such as when taking a penalty kick in soccer, an approach angle of between 30 and 45 degrees is often adopted. By adopting an angled approach to the ball, an athlete is able to produce greater pelvic rotation, which increases the acceleration path of their kicking limb and allows them to tilt their leg in the frontal plane to make better contact with the ball.

Like most sporting activities, injuries in kicking-related sports are generally caused by deficits in a player's fitness, training errors, environmental factors or mechanisms related to sport-specific equipment. Under controlled conditions, the forces that are transmitted across the knee joint while performing the kicking action are considerable; however, in competition, players are frequently required to perform the task while being challenged by an opponent. For example, in soccer, a tackle made by an opposing player has the potential to place the knee in a vulnerable position and can lead to the ligaments of this joint being easily injured. Under these situations, a rupture of the anterior cruciate ligament (ACL), with or without a concomitant tear of the medial collateral ligament (MCL), can be quite common. The ACL is one of the main supporting

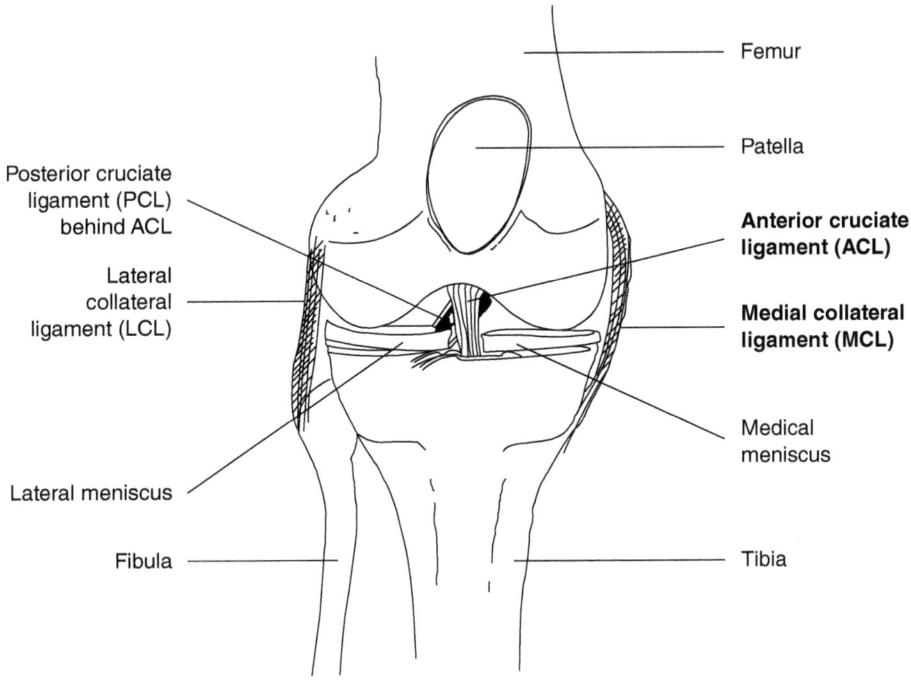


**Figure F5.3** An anterior (frontal plane) view of the major skeletal, cartilaginous and ligamentous structures of the right knee

ligaments of the knee (*Figure F5.3*) and is responsible for supporting the knee in a movement known as anterior tibial translation, where the tibia is moved anteriorly (forward) with respect to the femur. In addition, the ligament also contributes to the stability of the joint during rotation of the tibia on the femur. The ligaments, together with the muscles, provide joint support and stability, and injury to these ligaments of the knee can seriously affect an athlete's career.

The typical mechanism of ACL injury in soccer is where an athlete's knee is forcefully struck on the lateral side (often in a contact tackle situation), causing the knee to buckle medially (otherwise known as a valgus knee position). In this situation, the knee is usually flexed and there is a degree of internal rotation of the femur on an externally rotated tibia, which is secured to the ground by the soccer boot (*Figure F5.4a*). In addition, the ACL can easily be torn when the leg is positioned in severe hyperextension and the force of another player causes the hyperextension to go beyond that normally allowed by the knee joint (*Figure 5.4b*). This causes excessive anterior translation of the tibia with respect to the femur that, when combined with sudden deceleration and any degree of internal or external rotation on a fixed foot (i.e. due to the studs/bars of the soccer boot), means the ligament is susceptible to partial or complete rupture.

In situations where the ACL is completely ruptured, it is typically necessary to surgically repair the damaged structures using either a graft derived from the athlete's semi-membranous or patella tendon (autograft), from other human donors (allograft) or from a non-biological artificial source. Reconstruction using a semi-membranous or patella tendon autograft is generally the preferred option and despite each graft type having its own disadvantages, both have been shown to be capable of restoring knee stability to a degree that allows an athlete to return to play. If a surgeon does not consider the ACL to be partially or completely ruptured, non-operative treatments such as knee braces, anti-inflammatory injections and/or physiotherapy aimed at strengthening the quadriceps and hamstring muscle groups may be prescribed. However, in

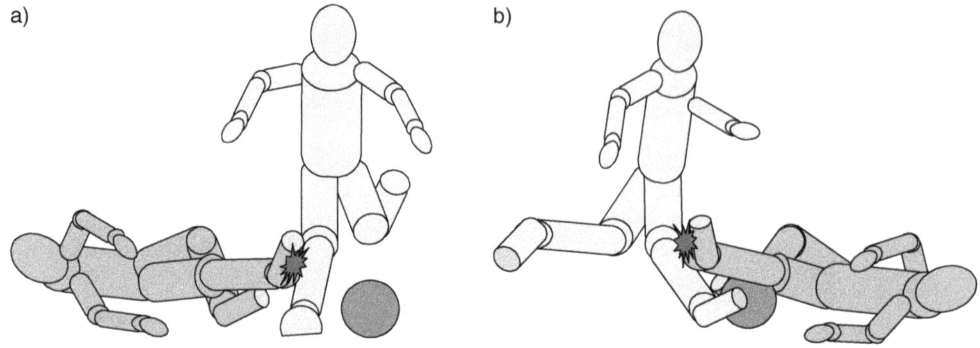


**Figure F5.4** A diagrammatical depiction of common football-related knee injuries resulting from: a) external tibial rotation and valgus knee motion; and b) knee hyperextension

many cases, those who initially receive non-operative treatments become candidates for surgical reconstruction or repair, which has led to considerable debate about the effectiveness of non-operative therapies.

*Key notes*

| | |
|---|---|
| **Phases of the kicking action** | Biomechanical analysis of kicking involves the movement being broken down into three phases: i) preparation phase; ii) driving phase; and iii) follow-through. |
| **Preparation phase** | During the preparation phase, the kicking leg moves away from the eventual direction of the kick to: i) increase the range through which the limb moves before ball contact; and ii) allow a greater opportunity for foot velocity to be developed. |
| **Driving phase** | The driving phase is initiated by the rapid rotation of the shoulders, trunk and pelvis, which are closely followed by powerful activation of the hip flexors and, later, the knee extensors of the kicking limb. Temporal analysis reveals that the hip begins flexing first, but during the latter stages of the driving phase, a forceful concentric contraction of the knee extensors causes the knee to extend towards ball contact. These two components of the driving phase (i.e. commencement of hip flexion, commencement of knee extension) are often used to further sub-divide this phase into two separate components. |
| **Follow-through** | After ball contact, the kicking action enters the follow-through phase, which almost exclusively involves eccentric muscle function aimed at gradually reducing the high velocities of the kicking limb. Although this phase of the kicking action has no influence on the outcome of the performance, it is vital for protecting the muscles and joints involved in the movement, as it allows the mechanical loads to be dissipated over a longer period of time. |
| **Knee ligament injuries** | Anterior cruciate ligament (ACL) or combined ACL/medial collateral ligament (MCL) injuries are the most common knee injuries in soccer. They are often caused by a forced valgus of the knee with an internally rotated femur and/or externally rotated tibia on a planted foot (often with the force of contact from another player). |

## FURTHER READING

The following resources provide additional reading on the biomechanics of kicking in sport:

1 Lees, A., Asai, T., Andersen, T. B., Nunome, H., & Sterzing, T. (2010). The biomechanics of kicking in soccer: A review. *Journal of Sports Sciences, 28*(8), 805–817.
2 Kellis, E., & Katis, A. (2007). Biomechanical characteristics and determinants of instep soccer kick. *Journal of Sports Science and Medicine, 6*(2), 154–165.
3 Lees, A., & Nolan, L. (1998). The biomechanics of soccer: A review. *Journal of Sports Sciences, 16*(3), 211–234.

# F6 SWIMMING

*Paul Grimshaw*

**Swimming velocity** can be defined as the speed that a swimmer is able to achieve through the water by movements of the body. This velocity is dependent upon two factors: **stroke rate** (stroke frequency) and **distance per stroke** (stroke length).

Swimming velocity (v) = Stroke rate (SR) × Distance per stroke (DPS)

**v = SR × DPS**

The **stroke rate** of a swimmer is determined from the time it takes to complete both the pulling and the recovery phase of the stroke, and **distance per stroke** is governed by the **propulsive** and **resistive forces** that act on the swimmer as they move through the water.

In order to increase swimming velocity, an athlete can increase either or both of these components (stroke rate or distance per stroke). However, an increase in one component should not be achieved at a loss or detriment in the other.

**Stroke rate** (frequency) can be improved by increasing the number of strokes for each length (lap) of the pool (cadence). However, one of the main drawbacks to increasing swimming velocity by this method is that the more hurriedly the swimmer tries to swim the more likely there will be deterioration in the swimmer's technique. **Distance per stroke** can be achieved by increasing the **propulsive forces** while reducing the effect of the **resistive forces**. Hence, it is critical to understand these two types of force in more detail.

**Resistive forces** are a direct consequence of drag, which can be classified into three types: **form, wave and surface drag.**

**Form drag** is concerned with the cross-sectional area of the body that is exposed to the oncoming flow of water, the shape of the body, and the relative velocity of the fluid flow (*Figure F6.1*). **Wave drag** is involved when the swimmer moves at the interface between the air and the water (*Figure F6.2*). During wave drag, some of the energy of the swimmer is transformed into wave motion, which also acts to drag the swimmer backwards. **Surface drag** (*Figure F6.3*) is concerned with the amount of body surface area, the smoothness of the body's surface, and the relative velocity of the oncoming flow (frictional drag).

**Form drag** can be reduced by adopting a more streamlined shape, as in the case of the swimmer in Example A in *Figure F6.1*. In Example A, the athlete is lying almost flat in the water (level with the surface of the water). This would create less frontal resistance to the oncoming flow of water (note: the athlete actually moves forwards and the oncoming water is stationary). Form drag is probably the **largest resistive force** in swimming, and the most effective way to reduce form drag is to try to adopt a more streamlined (hydrodynamic) body position. The ability to be more streamlined is, however, closely related to the amount of buoyancy (ability to float) possessed by the swimmer. The more buoyant a swimmer the easier it is to maintain a more streamlined body shape. In the context of form drag it may be important to point out that some elite athletes do not swim in the most streamlined position. Elite swimmers who have

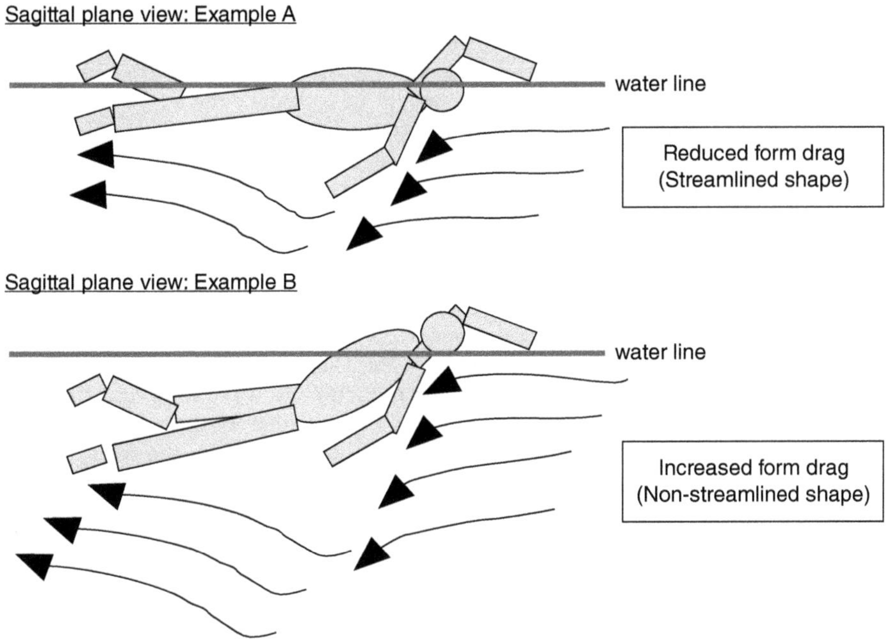


**Figure F6.1** Form drag in swimming

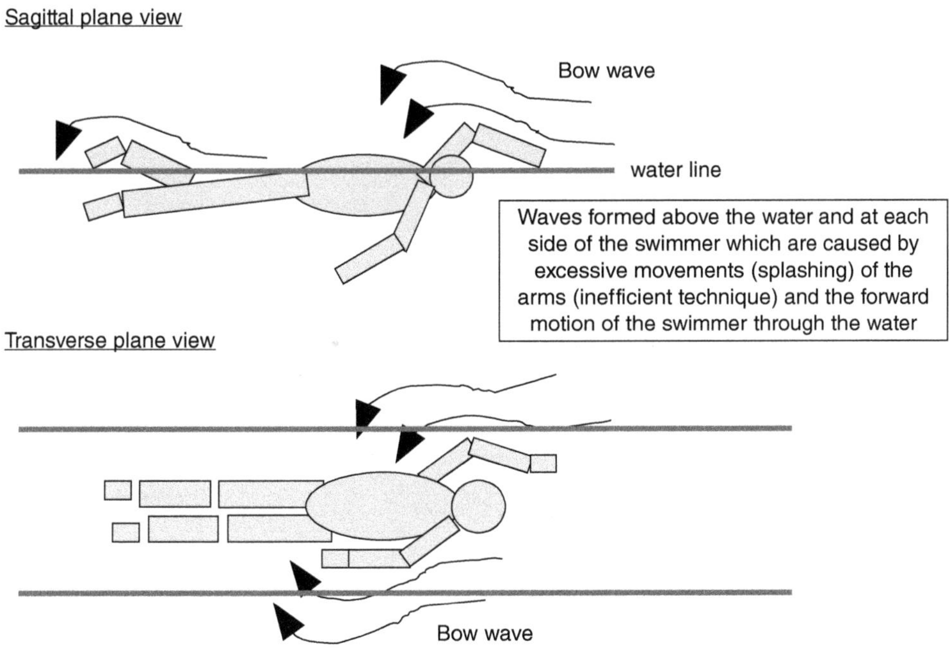


**Figure F6.2** Wave drag in swimming

a very powerful leg kick will adopt a more angled swimming position (higher upper body and lower leg position) in order to utilise the leg kick more effectively.

**Wave drag** probably accounts for the next most significant resistive force in swimming. The method used to reduce this type of drag would be to reduce the size of the bow wave (a v-shaped wave created by an object moving across a liquid surface) created by the swimmer. Any fast movements of the body, such as arm recovery, should be performed in the air rather than while in contact with the water. Large bow waves in swimming result from excessive vertical and lateral movements of the body. Crashing the arms and hands into the water during entry can create large bow waves. The hands should slip and glide into the water. In freestyle (which is the most common swimming stroke), the hand should be initially placed side-on and the arm should glide into the stroke through the same hole (path) made in the water by the hand entry. Often, large amounts of wave drag are created by the side-to-side movements through the water (like a snake-type movement), which is a common fault seen in adult swimmers. The reduction in wave drag and the effect on performance can be seen by observing that many elite swimmers now spend as much of the race as possible, within the defined rules, under the water (from the start and at each turn). Indeed, many modern competitive strokes now utilise more underwater actions, for example the modern breastroke, which now uses undulating underwater movements that resemble the butterfly stroke.

**Surface drag** is created by the amount of surface area actually in contact with the water during swimming (*Figure F6.3*). In this context, very little can be done to reduce surface drag during swimming apart from the wearing of the friction-reducing "shark suits" developed by sports companies such as Speedo and Adidas (full-length versions that covered the arms, body, neck and legs were banned by the international body for swimming (FINA) in 2010).

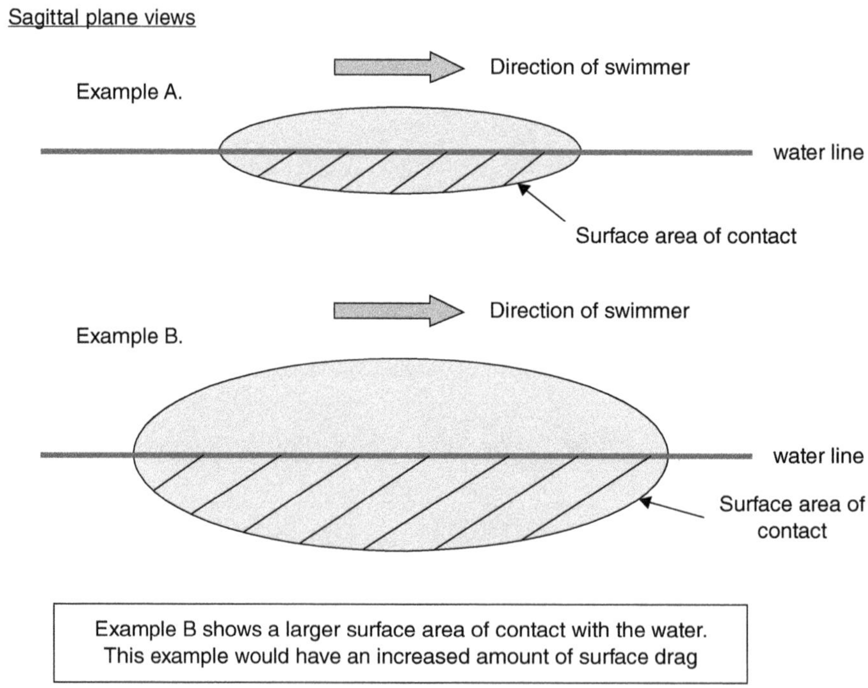


**Figure F6.3** Surface drag in swimming

Other methods, such as wearing swimming caps and body shaving, are also used to reduce surface drag. However, these techniques are probably only really helpful at the elite level of the sport. In swimming, 90 per cent of the drag comes from the shape of the swimmer as he/she moves through the water (form drag) and only 10 per cent is attributed to the friction caused between the skin, the costume and the water; but at the elite level this 10 per cent could mean the difference between breaking a world record or not. In 1875 Captain Matthew Webb, while swimming the English Channel, wore a swimming costume that weighed (mass) 4.5 kg (10 lb). The now banned one-piece body suits weighed (mass) only a few ounces (0.09 kg), which saved over 98 per cent in weight (mass) compared with the original swimming costume used in 1875. It was speculated that there was an 8 per cent reduction in drag resistance while wearing these suits, and that they were even better than swimming with no suit on at all. Given the number of world records broken while wearing the body suits, it is not surprising that FINA decided to ban them from the sport in 2010.

All these resistive forces can be calculated providing that certain aspects of the swimming conditions are known. *Figures F6.4*, *F6.5*, *F6.6* and *F6.7* show how these resistive forces can be calculated and the variables that must be known in order to do this.

Prior to the 1970s, **propulsive forces** in swimming were thought to be due entirely to the action–reaction method (i.e. push backwards in the water and you moved forwards – Newton's third law). This was termed **drag propulsion**, as it relied on the large surface area of the hand to push the water backwards (like the paddle-wheel propulsion used in small boats). However, in the years that followed the 1970s, the term **lift propulsion** was introduced. This was primarily attributed to the work of James Counsilman in the USA, and it involved both lateral and vertical movements of the hand through the water. This technique is still used by many elite-level swimmers today.

**Form or pressure drag**
The flow of water around a swimmer separates from the swimmer at certain points depending on the speed (velocity), size and shape of the swimmer

Once this separation point is reached the flow reverses and creates distinct eddy currents (vortices of water or a whirling mass of fluid)

This causes a pressure differential between the front and rear of a swimmer

Next, a pressure drag results (effectively pulling the swimmer backwards) which is proportional to the pressure difference times the cross-sectional area of the swimmer

Pressure drag (Dp) depends on the square of swimming speed (v), a dimensionless drag coefficient $C_D$ (e.g. bluff body (separated flow over much of its surface i.e. larger $C_D$) versus torpedo-shape (less separation i.e. smaller $C_D$) of the swimmer), the cross-sectional area of the swimmer and the density of the liquid (water)

$$Dp = 1/2\, p\, Ap\, v^2\, C_D$$

Dp = pressure drag
$p$ = density of water
Ap = cross-sectional area
v = swimmer velocity
$C_D$ = drag coefficient

**Figure F6.4** Calculation of form or pressure drag

**Wave drag**
Water in front of the swimmer piles up and forms a hollow behind the swimmer. This creates a wave system. As the velocity of swimming increases, the wave length (crest to crest) and wave amplitude increase

When certain speeds are reached the wave length equals the water line length of the swimmer (proportional to the height of the swimmer). This speed is often termed **hull speed** and is the speed at which the swimmer is trapped in the hollow between crests of waves (constantly climbing out of and falling into the hollow)

More effort by the swimmer creates a higher wave amplitude and hence more energy is needed to climb out of the hollow. In practice it is impossible to swim faster than the **hull speed**

The magnitude of the wave drag is defined as the **Froude-number (Fr),** which is a dimensionless number

$\mathbf{Fr = v/\sqrt{gL}}$

v is the speed of the swimmer
L is the length of the swimmer
g is acceleration due to gravity

Swimmers with different heights but equal body form will create the same wave system (relative to body dimensions). This means equal Froude numbers which for the shorter swimmer is a lower (less) swimming velocity

**Figure F6.5** Calculation of wave drag

**Surface or friction drag**
The water directly in contact with the body (skin, suit) is held to the swimmer by adherent (sticky, clinging) forces. This layer of water moves with the same speed as the swimmer. The water speed (layers) reduces as the distance away from the swimmer increases. For example, the water far in front, at the back and at the far sides of the swimmer is usually at rest

The water closest to the body is slowed down by the layer next to it and this layer is slowed down by the next layer and this continues until there is still water (i.e. no movement). The **friction force** created is proportional to the total surface area of the swimmer (i.e. that in contact with the water) and the speed of the swimmer

If the water around the swimmer is arranged in orderly layers (i.e. laminar or streamlined flow) the drag force exerted by the water equals the friction force. However, at certain speeds of swimming the flow becomes **turbulent** (i.e. choppy or rough water), which depends on the shape and size of the swimmer

This turbulence usually happens very suddenly and it occurs at a critical value called the **Reynolds number** (Re – a dimensionless scaling number)

**Figure F6.6** Calculation of surface or friction drag

$$Re = vL\,p/\eta$$

v is the velocity of the swimmer
L is the length of the swimmer
*p* is the density of water
η is the viscosity of water ($0.897 \times 10^{-3}$ N.s.$m^{-2}$)

Depending on the shape of the object the critical Reynolds number will often be in the order of 500,000

For a swimmer travelling at a speed of 2 m/s, with a body length of 2 m, $p$ = 1000 kg.$m^{-3}$ and η of $0.897 \times 10^{-3}$ the Re number would be 4,459,308 which shows that turbulence will always be present in competitive swimming

**Figure F6.7** Reynolds number: surface or friction drag. Swimmer Cate Campbell (AUS) is shown wearing the one-piece swimsuit that is currently allowed in the female competition.



The term **lift** gives a slightly false impression as to how the principle works in swimming. It implies that the force is always directed upward (i.e. to lift the body). In swimming this is not necessarily the case, as the lift force can act in almost any direction and, more accurately, acts at a right angle (i.e. at 90 degrees) to the direction of movement of the object (or fluid flow) that is causing the lift force to be created. Since it is the hand that would cause the lift force to be created in swimming, it is clear that this force can occur in any direction.

Lift is based around **Bernoulli's principle** of fluid dynamics, which is more commonly seen applied to aerodynamics (movement through the air) and the movement or flight of aeroplanes. *Figure F6.8* shows the cross-sectional view of an aeroplane wing. When the wing moves

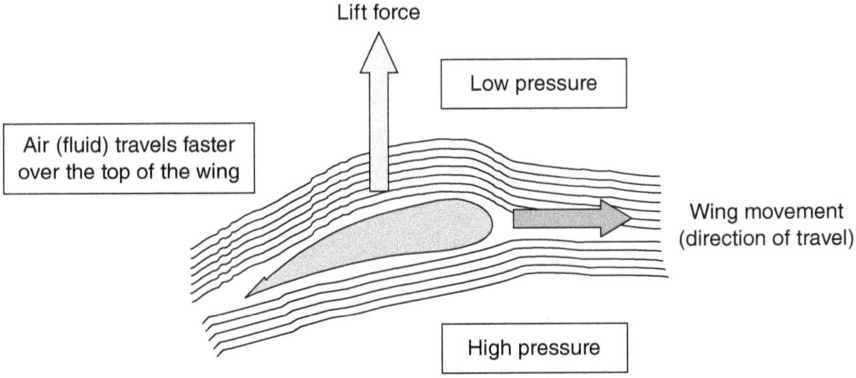


**Figure F6.8** Lift force on an aeroplane wing

forwards (propelled by the jet engines of the plane) the layers of air that are oncoming to the wing separate. Some travel over the top and others travel below the wing. Due to the shape of the wing the path over the top of the wing is longer than the path underneath the wing. The shape of the wing and its inclination causes the air over the top of the wing to travel faster than the air underneath the wing (it also has a greater distance to travel). The result is that this difference in the speed of air travel causes a pressure differential to occur. A high pressure is created below the wing and a low pressure is created above the wing. According to Bernoulli's principle, which relates to the pressure differential and the association of lower pressure with a faster fluid velocity, the result is a "lift" force that acts at right angles to the direction of fluid flow or, in this case, the direction of the aeroplane wing (since it is travelling forwards). As we have seen, the wing is travelling forwards and here the lift force is acting upwards (at 90 degrees to the forwards motion). It is this lift force that enables the aeroplane to take off the ground and fly through the air (i.e. the lifting of the wings). This continues to occur in flight as the wing is still being propelled through the air (fluid) by the jet engines of the aeroplane.

In swimming, the hand performs the same function as the aeroplane wing and it can also create a lift force if it is moved through a fluid (water) in a certain manner (i.e. shape and angle). As an example, the next time you are a passenger in a moving car try carefully holding your hand out of the car window in a shape like an aeroplane wing (cup your hand to create the shape seen in *Figure F6.8*). The result will be that your hand will lift upwards as the car travels forwards. In this case the hand is stationary and the car is travelling forwards, which provides the flow of air over your hand. If you angle your hand at different positions to the oncoming air flow you will see the effect of different lift forces. At some positions and angles of the hand you can even create a lift force that is directed downwards and not upwards. This downwardly directed lift force is often used on racing cars by having the spoiler on the back of the car angled and positioned in a certain way. This creates a downward lift force that helps to keep the car firmly on the ground, offering better traction at high speeds.

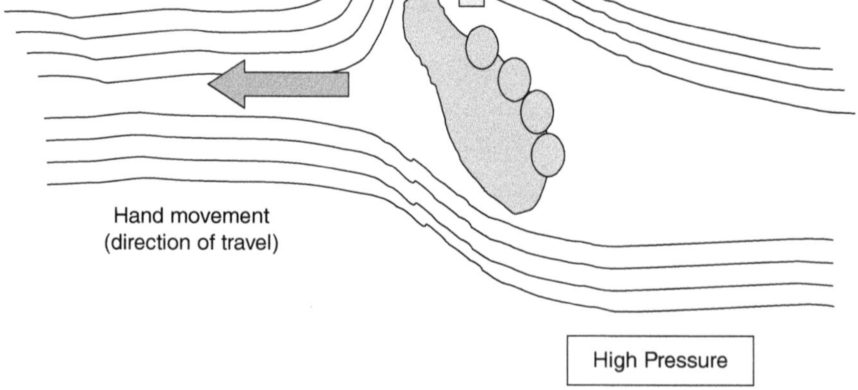


**Figure F6.9** Lift force on the hand during swimming

In the case of the swimmer, although the hand is moved through the water at much slower speeds it can still create enough lift force to propel the body forwards through the water. As a practical example, position yourself vertically in the water and use the horizontal sculling action of your hands to keep you afloat. This method of flotation is primarily dependent upon the lift forces that you create while your hands are sculling horizontally under the water. In the freestyle stroke, the hand is angled and moved in an elliptical pull pattern throughout the arm-pull phase of the stroke. Depending on the hand position as you pull it through the water it will create different amounts and directions of lift (or force). If the hand movement is in the appropriate direction, the lift forces that are created can be directed either to help keep you in a flat horizontal position or indeed propel you forwards through the water. *Figure F6.9* shows the lift forces created by the positioning of the hand as it is moved through the water (which is analogous to the aeroplane wing).

During swimming, the advantages of this application are significant. In the old action–reaction method of propulsion (drag propulsion) half of the stroke was classified as recovery because you cannot push directly backwards at all points throughout the pull phase. However, by combining the lift (lift propulsion) and drag method of propulsion (drag propulsion) the whole of the pulling phase can be utilised to propel the body forwards through the water. In the modern swimming stroke, athletes use lift and drag propulsion to swim more effectively.

The direction in which the hand is inclined in the water is termed the **angle of attack**, and increasing the angle of attack will increase the lift forces. However, with **increasing lift forces** there will also be **increased drag forces** that will act against the movement of the hand through the water. The angle of attack position can reach a limit before there would be less lift force created. In aerodynamics the optimum lift to drag ratio can be achieved with an angle of attack of between 4 and 15 degrees. However, in swimming the angle of attack is at an optimum between 30 and 50 degrees. This occurs because the swimmer is prepared to generate greater lift at the cost of trying to overcome the larger drag forces. Distance swimmers trade off some propulsive lift by having a smaller angle of attack of the hand (i.e. less lift force). This has the benefit of reducing the energy cost of the stroke because the swimmer does not have to overcome so much

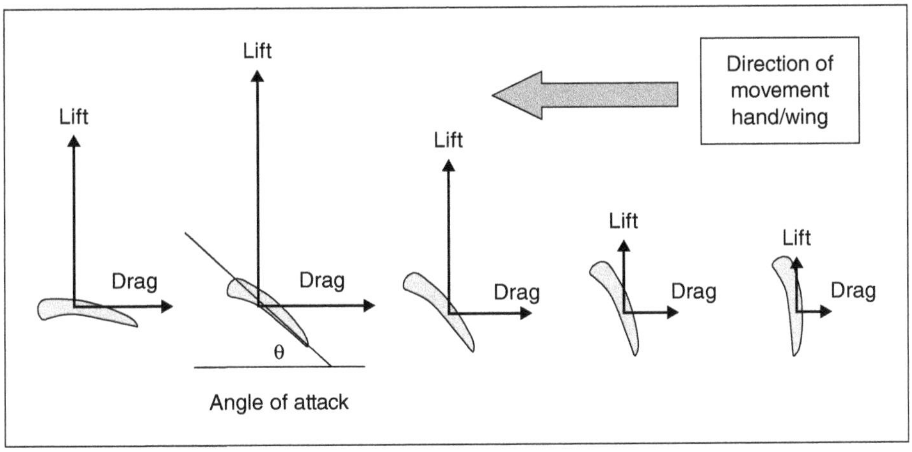


As the "angle of attack" is increased the amount of lift increases. However this increased lift force is only apparent up to a point and beyond this point the lift force reduces as the angle of attack increases. Note the associated drag force with each lift force component

**Figure F6.10** Lift-to-drag ratios during hand positions in swimming

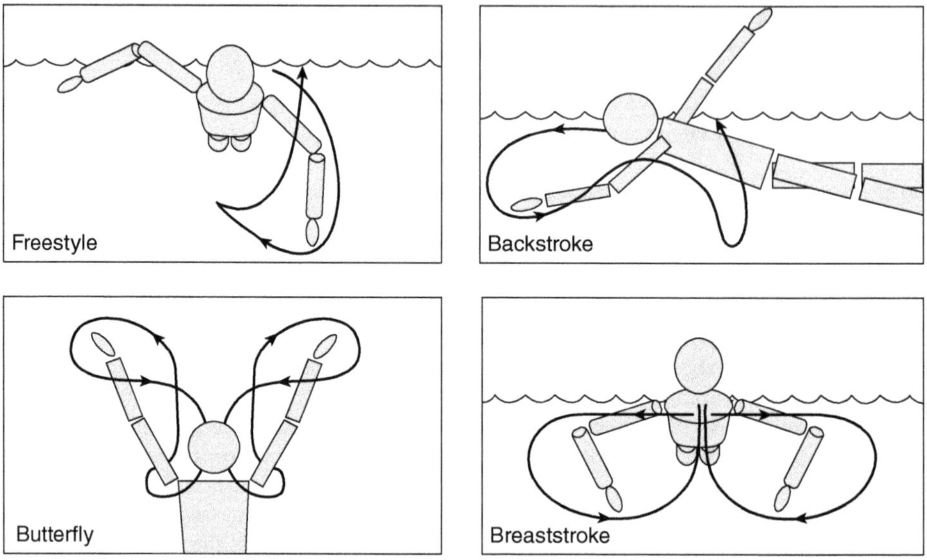


**Figure F6.11** The modern pull patterns of the four competitive strokes

drag force during the movements of the arms/hands. Sprinters, on the other hand, use a larger angle of attack position because the race is much shorter and the increased lift is imperative for greater propulsion (i.e. faster progression). Each swimmer has to develop a "feel" for the water, and it is important to note that the drag force does not act against the direction of the swimmer but acts against the direction of the movement of the hand or object that is creating it. *Figure F6.10* helps to show the lift and drag force ratios during the different pitches of the hand or aeroplane wing.

In modern swimming, the lift and drag propulsion method is used throughout all the four competitive strokes. Athletes adopt complex pull patterns under the water to utilise these principles more effectively (*Figure F6.11*). Today, in freestyle, many elite swimmers will use a pronounced bent arm (i.e. at the elbow) pattern throughout the arm-pulling phase. This is a similar action to that used in climbing a ladder. Distance swimmers in particular will be seen utilising the bent-arm-pull pattern. However, most strokes will still have common features for effective propulsion through water. These are summarised as follows: curved elliptical pull patterns; an extend – "catch the water" – flex – extend pull pattern; a high elbow position (bent arm); hand entry to create minimum splash; utilisation of both lift and drag propulsion; and streamlined body alignments (with the exception of the strong, powerful, leg-kicking athletes). Two objectives are apparent for effective propulsion in swimming:

1) To propel the body forwards with respect to the hands (using an optimum combination of lift and drag propulsion); and
2) To minimise resistance to the propulsion of the body (reduce drag and maintain optimum body alignment).

In swimming, it is speculated that 85 per cent of the propulsion comes from the arm movement through the water. However, there are many who would argue that the legs are a far more significant contributor to propulsion than only 15 per cent. As already mentioned, many athletes with powerful leg kicks will angle their body downward to utilise this propulsive force provided by the legs. Throughout the leg kick, the same principle for lift and drag propulsion also applies (depending on the angle, position and direction of movement of the foot). Arm action in swimming should not be totally classified as either lift or drag propulsive. When the hand moves predominantly backwards, it is likely that the majority of propulsion would be drag propulsion. Conversely, when the hand is moved laterally and vertically, lift propulsion would be more prominent. An effective coach would be tolerant of different techniques and should always be prepared to change a swimmer's technique if inefficiencies are detected. However, this can only be achieved with a good working knowledge of the biomechanics of effective propulsion through a fluid, such as water.

Considering that competitive swimmers can easily complete more than a million shoulder rotations per year (up to 10,000 m training per day with between 15 and 25 strokes per 25 m distance) it is inevitable that this will potentially result in injury. **Shoulder injury** to swimmers is often described by the medical professions as one or more of the following: shoulder joint (glenohumeral joint) instability; inflammation of the supraspinatus tendon and often also the biceps tendon and sub-acromial impingement (impingement of the soft tissue structures lying below the acromion), which are more commonly known or classified as **"swimmer's shoulder"** (*Figure F6.12*).

The modern arm action in most swimming strokes is a precise sequence that first involves arm extension upon water entry, then a "catch" of the water, next arm "pull" through the water in a path that allows elbow bend and inward and outward sweeping movement of the hand, and then finally recovery where the arm is prepared for the next re-entry into the water.

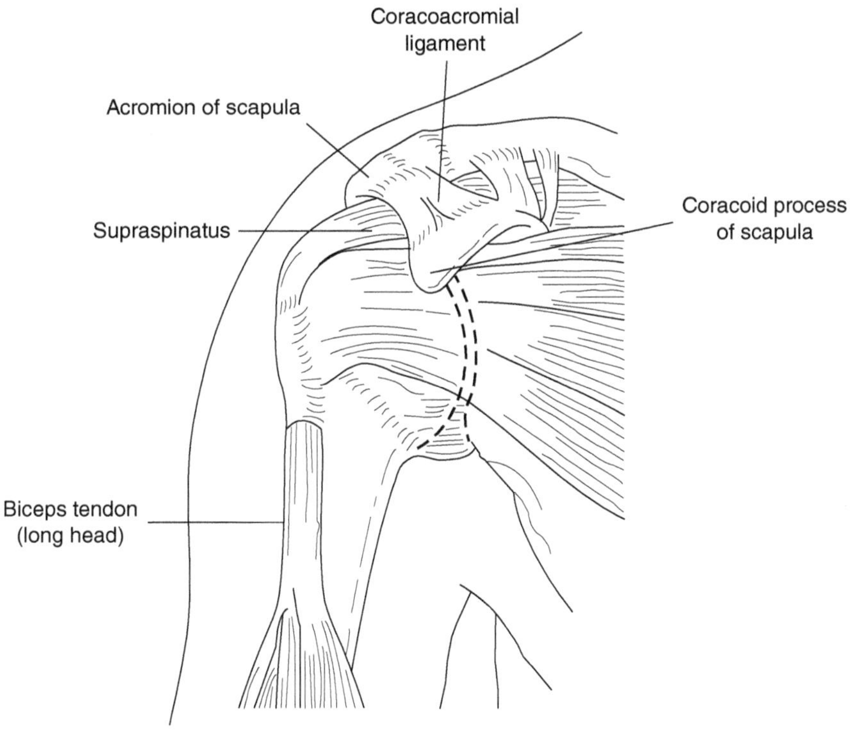


**Figure F6.12** Anatomical structures of the shoulder

## MECHANISM OF INJURY

### The "catch" phase (arm entry to the water)

During the initial extension of the arm and "catch" phase of the freestyle swimming stroke, the shoulder is required to internally rotate and abduct. The large moment experienced during extension and catch tends to cause the arm to be forcibly elevated.

### The "catch" and pull phase

During the pull phase, the shoulder is adducted and internally rotated and often the arm follows an inverted question mark pattern under the water. This allows the application of a force for a longer period of time, utilising the lift principle of propulsion. This action causes the head of the humerus (the upper arm bone) to move under the coracoacromial arch, which can lead to a potential impingement situation. One of the most common technical faults during this phase is the "dropped elbow". The dropping of the elbow during the pull phase causes increased unwanted external shoulder rotation. However, the "high elbow" technique, which provides the muscles with a mechanical advantage, may also present an associated impingement risk. From initial "catch", the hand sweeps down and slightly

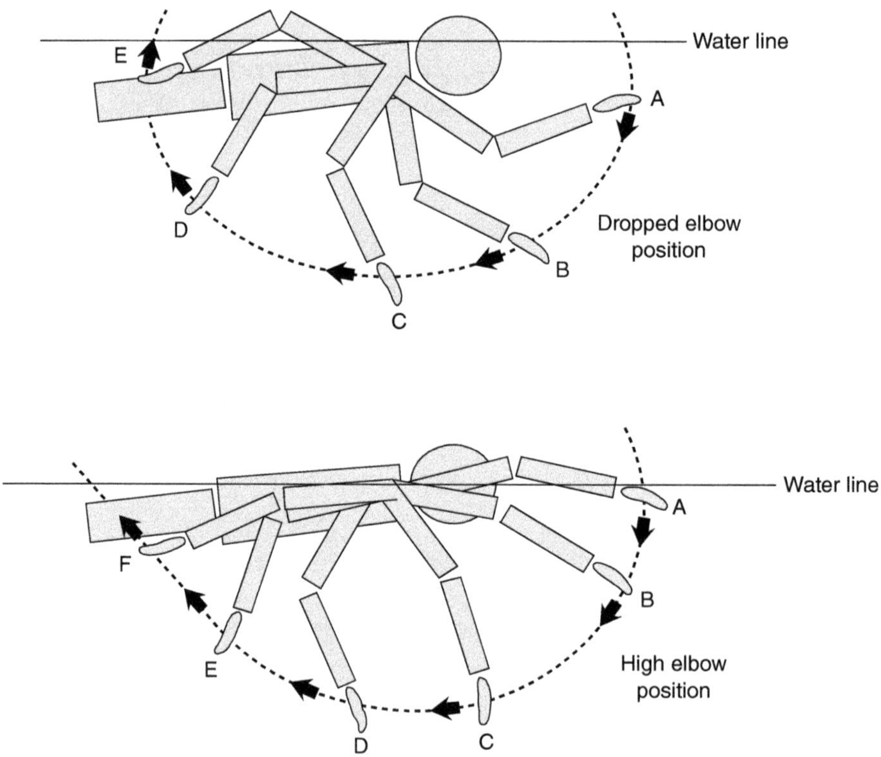


**Figure F6.13** Dropped and high elbow positions during freestyle

outwards while ideally maintaining a high elbow position to the deepest point of the stroke. Next, the hand sweeps inwards and upwards until an angle of approximately 90 degrees of the upper arm and forearm is reached. The in-sweep phase may take the hand past the mid-line of the body or under the outside edge of the body. From the mid-stroke position, the hand is first swept outwards and then backwards, finishing at the end of the pull past the hips as it exits the water. An increased acceleration of the hand towards the end of the pull and internal rotation and adduction of the shoulder may present a potential impingement problem (*Figure F6.13* shows both the dropped and high elbow position during the modern front crawl swimming pull phase).

## The recovery phase

Following the pull phase, the arm leaves the water, usually elbow first (elbow lift), and is required to undergo a recovery phase. During this time, the shoulder is required to abduct and externally rotate as the arm is prepared for re-entry in the water. However, the arm at the beginning of this phase is often still internally rotated. The recovery phase is one of the more important phases in terms of risk for possible shoulder impingement. The shoulder is required to rotate externally and abduct to clear the arm over the water and then quickly to prepare for extension and entry to the "catch" phase position. The timing of the external rotation and the magnitude of this action during the recovery are critical components of risk associated with shoulder impingement and it has been shown that swimmers who execute the recovery phase with a large amount of initial internal rotation of the shoulder are susceptible to potentially increased shoulder impingement. External rotation of the shoulder is needed to allow for complete abduction and, thus, to prevent the greater tuberosity of the humerus making contact with the acromion.

## Prevention and rehabilitation

Excessive internal rotation of the shoulder during the pull phase, combined with late external rotation during the recovery phase, is said to lead to increased risk of shoulder impingement. In addition, reaching across the mid-line on entry, insufficient body rolling, one-sided breathing and asymmetrical muscle balance also contribute to the problem.

Possible solutions to a shoulder impingement problem include the following: during the entry and pull phase of the stroke, the swimmer should try to avoid a large elevation angle at entry and rather increase the tilt angle of the arm to achieve the optimum position. Similarly, the swimmer should avoid a fully extended elbow on entry. The swimmer could also help to resist the forcible elevation caused by entry and catch by developing the shoulder extensor muscles: namely latissimus dorsi, pectoralis major, teres major and triceps brachii. Also, a streamlined hand entry position is advisable. During recovery, the swimmer should try to achieve external rotation of the shoulder early in the recovery phase in order to have time to prepare the hand and arm for re-entry to the water.

Other factors which are also recommended for the prevention of shoulder impingement in swimming include not using hand paddles, if there is a current problem, and possibly changing from distance swimming training to sprint training in order to reduce the number of stroke cycles in a training session. Finally, the swimmer could adopt a bilateral (both sides) breathing technique, assume as horizontal a position as possible in the water (i.e. feet up) and allow some degree of body roll (although there are still current issues of debate as to how much body roll is required for optimum performance and injury prevention).

| | |
|---|---|
| **Swimming velocity** | This is defined as the velocity that a swimmer is able to achieve through the water. The swimming velocity of a swimmer depends on the stroke rate (SR) and the distance per stroke or stroke length (DPS). Hence, swimming velocity (v) is given as SR × DPS. Stroke rate is determined from the time it takes the swimmer to complete the pulling and recovery phases of the arm stroke. Distance per stroke is governed by the propulsive and resistive forces that act on the swimmer. |
| **Resistive forces** | These are a direct consequence of drag and they can be classified as form, wave and surface drag. Form drag is related to the cross-sectional area of the body that is exposed to the water, the shape of the body and the relative velocity of the fluid flow. Form drag can be reduced by adopting a more streamlined position in the water. This type of drag is probably the most significant in terms of resistive forces offered to a swimmer's progression through the water. Wave drag is related to the waves that are created at the interface between the swimmer and the water. Large bow waves (a v-shaped wave caused by an object moving through a fluid) act to drag the swimmer backwards. Any fast movements, such as arm recovery, need to be performed in the air rather than in the water. Many swimmers now perform as much of the race as they can underwater in order to reduce such bow waves. Surface drag is related to the amount of surface area actually in contact with the water during swimming. |
| **Propulsive forces** | Propulsive forces take the form of drag and lift propulsion. Drag propulsion through the water is achieved by pushing the water directly backwards (i.e. the swimmer moves forwards), whereas lift propulsion utilises the same principle of lift force that is used to cause aeroplanes to fly. Swimmers use a combination of both drag and lift propulsion to propel them through the water. Modern techniques utilise complex underwater pull patterns that optimise the maximum amount of propulsion that can be achieved through these two methods. Many modern elite swimmers in freestyle now adopt a pronounced bent elbow pull pattern that is like the action used to climb a ladder. |
| **Swimming injury** | During training and competition, swimmers can complete over a million shoulder rotations per year. The potential for injury due to poor technique and overuse is high. Shoulder impingement in swimming is a common problem and involves shoulder instability, inflammation of the supraspinatus and/or bicep tendon, and impingement of sub-acromial structures. |
| **Mechanism of injury (freestyle)** | A large extension moment is experienced at the shoulder, which has potential for injury. A dropped elbow position can lead to excessive shoulder rotation and increased risk of injury: a high-risk phase for "swimmer's shoulder", where the timing and magnitude of external shoulder rotation are critical. The effective coach would consider technique modifications and body position corrections/alterations. The use and amount of body roll and unilateral and bilateral breathing are important components in any assessment of injury. Consideration should be made of the possible detrimental use of hand paddles. The possible reduction in training distance together with the conditioning of the shoulder rotator muscles could be helpful. |

## FURTHER READING

The following resources provide additional reading on the biomechanics of swimming and swimming injuries:

1 Johnson, J. N., Gauvin, J., & Fredericson, M. (2003). Swimming biomechanics and injury prevention: New stroke techniques and medical considerations. *The Physician and Sportsmedicine, 31*(1), 41–46.
2 Yanai, T., & Hay, J. G. (2000). Shoulder impingement in front-crawl swimming II: Analysis of stroking technique. *Medicine and Science in Sports and Exercise, 32*(1), 30–40.
3 Toussaint, H. M., & Beek, P. J. (1992). Biomechanics of competitive front crawl swimming. *Sports Medicine, 13*(1), 8–24.
4 McCabe, C. B., & Saunders, R. H. (2012). Kinematic differences between front crawl sprint and distance swimmers at a distance pace. *Journal of Sports Sciences, 30*(6), 601–608.

# F7 GOLF

*Michael Cole*

The golf swing is a complex and asymmetrical movement pattern that requires precise inter-segmental coordination and neuromuscular control to maximise both the distance and direction of the stroke. To achieve these goals, each golfer will generally adopt a unique technique that they have developed over years of training and that is specific to their anthropometrical and biomechanical characteristics when performing the full golf swing (i.e. a long-range tee-shot or a shorter iron shot). However, despite these subtle, yet inherent differences in technique, the movement patterns exhibited by different players generally share a number of important kinematic characteristics, which are broadly described in this section. Historically, the golf swing was characterised by a long backswing that involved the pelvis (essentially the hip area) and thorax (essentially the shoulder area) turning through equally large ranges of motion (*Figure F7.1*). Following impact between the ball and the golf club, the body finished in a position that was characterised by a relaxed and upright position for both the pelvis and thorax, which were aligned perpendicular to the target (i.e. the flag). Within the biomechanics literature, this technique is often referred to as the original or **classic golf swing**. However, over the past 60 years, this smooth and rhythmic technique has been replaced by the **modern golf swing** (*Figure F7.2*), which stresses the muscles and joints to generate a more powerful and effective movement. To simplify the analysis of the golf swing, coaches and scientists generally divide the movement into four phases that include the **address, backswing, downswing** and **follow-through**. This section describes the general patterns of movement during each of these phases, but it should be acknowledged that different players may use different techniques to achieve a desirable outcome.

**The address** or pre-swing phase is an important component of the golf swing, as it is allows the golfer to position themselves to ensure a stable foundation from which to generate power throughout the swing. For the modern golf swing, the address typically involves the golfer standing with their feet approximately shoulder width apart, with their knees bent 20 to 25 degrees,

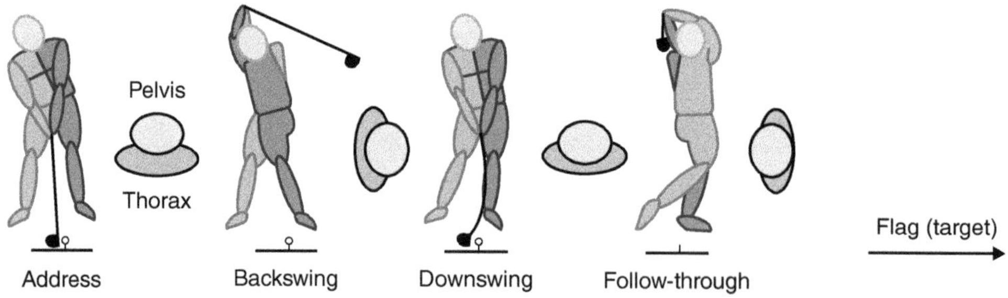


**Figure F7.1** Illustration of the classic golf swing. The darker and lighter ellipses respectively portray the relative orientation of the thorax and pelvis (transverse plane) at each stage of the golf swing

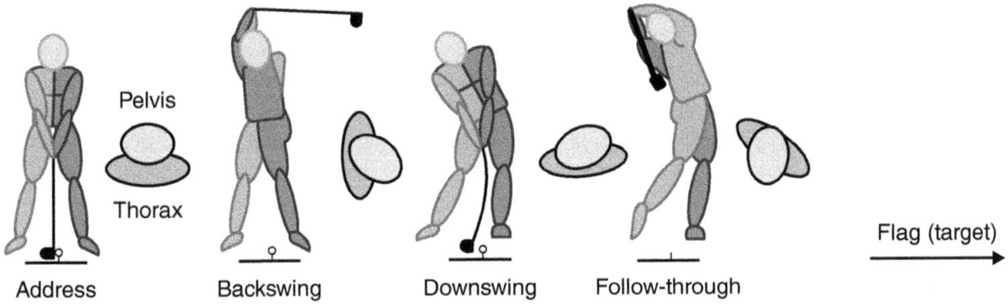


**Figure F7.2** Illustration of the modern golf swing. The darker and lighter ellipses respectively portray the relative orientation of the thorax and pelvis (transverse plane) at each stage of the golf swing

their trunk flexed to approximately 45 degrees and their weight evenly distributed over their feet. The trunk, hip and knee flexion that occurs during this phase brings the centre of gravity closer to the base of support, which ultimately flattens the arc of the swing and improves the overall stability of the golfer throughout the movement. Golfers also tend to angle their lead foot (left in right-handed golfers) about 20 to 30 degrees towards the target during ball address, which is believed to facilitate pelvic rotation about the hip joint and reduce torsional loads on the spine during the downswing.

**The backswing** is initiated by an anterior shear force produced by the trail foot (the right in right-handed golfers) and a posterior shear force produced by the lead foot (the left in right-handed golfers), which collectively produce a clockwise torque that rotates the pelvis and clubhead away from the ball and shifts the body's weight towards the trail foot. During this phase, the trail shoulder goes through a combination of abduction, flexion and external rotation, while the lead shoulder is adducted, flexed and internally rotated. At the top of the backswing, the lead arm is fully extended and the upper thoracic spine is rotated, such that its posterior aspect is facing the target. As the pelvis rotates about its longitudinal axis, the thorax is rotated through a range that is almost double that of the pelvis (40–50° *vs* 90–100°), which serves to eccentrically load the muscles of the hips, trunk and upper extremities before they are shortened at the initiation of the downswing. The eccentric loading of these muscles prior to their shortening is a direct application of the stretch–shorten cycle; a theory that suggests that increasing a muscle's length prior to shortening it produces a more effective and efficient muscle contraction. Collectively, these movements result in the golfer's lead arm being fully extended, their trail arm being flexed at the elbow and their upper thoracic spine and shoulders being rotated through approximately 100 degrees at the completion of this phase.

While the thorax and club continue to rotate away from the target, the **downswing** is initiated by a forceful contraction of the hip and knee extensors on the trail side (the right in right-handed golfers), which causes the pelvis to rotate anti-clockwise and laterally slide, shifting the body's weight back towards the lead leg. As the pelvis rotates towards an open position (facing the target), the thorax, arms, wrists and club unwind in a sequential fashion, adhering to the summation of speed (or summation of velocity) principle. This progression of motion from proximal to distal segments, often referred to as "leading with the hips" in golf, is suggested to further accentuate the stretch of those muscles involved in the stretch–shorten cycle throughout the swing. While the pelvis and thorax continue to rotate and the pelvis laterally slides, the club and lead forearm (the left in right-handed golfers) form an angle of ≤90 degrees at the wrists, which reduces the moment of inertia of the arm/club segment and increases the angular velocity of the golf club. For a proficient golfer, the angular relationship between the forearm and golf club is generally maintained throughout the downswing, until the club reaches an approximately horizontal position (80–100 ms prior to impact with the golf ball). At this point, the lead forearm supinates and the trail forearm pronates, while the wrists move rapidly from radial deviation to ulnar deviation. This movement sequence is commonly referred to as wrist uncocking and the ability to delay the onset of ulnar deviation is reported to be associated with the capacity to achieve greater segmental angular velocities at impact.

The **follow-through** phase begins after impact between the clubhead and the golf ball and is a phase of the golf swing that is characterised by the upper thorax rotating through a range of almost 180 degrees from the point of ball impact. This large degree of rotation is achieved via a combination of pelvic and spinal rotation and scapula protraction/retraction and results in the vertebral column finishing in a hyper-extended position, the trail shoulder (the right in right-handed golfers) pointing towards the target and the hands positioned high above the golfer's head. This posture is commonly referred to as the "reverse C" position, due to the shape that the body's limbs form when viewed from the frontal plane (*Figure F7.3*). While the speed of the movement steadily decreases following impact, it has been argued that the skeletal and soft tissue structures of the spine may not be able to safely accommodate the large range of movement and high spinal loads (Hosea & Gatt, 1996) that occur at impact and throughout the follow-through. Specifically, around impact and during the early follow-through phase, the anteroposterior and compressive loads reach values that are near to the known tolerance limits of some bony and soft tissue structures of the spine; possibly explaining why up to 41 per cent of lower back injuries are sustained around these time points.

**Figure F7.3** Illustration of the finishing posture of the modern golf swing, as viewed from the frontal (coronal) plane. This position is commonly referred to as the "reverse C" position, due to the shape formed by the golfer's limbs at this point (denoted by the dotted line)

Considering the asymmetrical nature of the golf swing, and the excessive joint ranges of motion and spinal loads associated with its execution, it is not surprising that the repeated performance of this skill is often considered one of the primary causes of injury in participants. In the early 2000s, it was estimated that the game of golf was enjoyed on a regular basis by approximately 60 million people worldwide; a popularity that is strengthened by the limited number of restrictions that one's gender or age can pose on involvement. However, the asymmetric movement patterns, extreme ranges of motion and large spinal forces experienced by golfers during the downswing and follow-through reportedly lead to more than 60 per cent of players experiencing an injury directly related to their participation. Golf-related injuries typically result from the repeated performance of the golf swing and/or poor swing mechanics, with injuries to the lower back (36%), shoulder (13.5%) and elbow (11.7%) most commonly reported (Fradkin, Windley, Myers, Sell, & Lephart, 2007). In an effort to better understand the possible mechanisms of **golf-related lower back injuries**, research has highlighted a number of differences in the swing mechanics of golfers with lower back pain. For example, while golfers with lower back pain exhibit similar address positions, trunk flexion angles and lateral flexion angles towards the trail side (the right in right-handed golfers), they have greater lateral flexion towards the lead side (the left in right-handed golfers) and less trunk rotation towards the trail side during the backswing compared with uninjured players. Furthermore, golfers who present with lower back pain have been shown to achieve reduced ranges of trunk rotation during a maximum voluntary effort, yet consistently rotate their trunk through a greater range of motion (i.e. supra-maximally) during the performance of the golf swing. Similarly, there is some evidence that golfers with lower back pain have a significantly reduced capacity to internally and externally rotate the lead hip (the left in right-handed golfers) during the golf swing. This reduced hip joint mobility is argued to result in external loads being transmitted from distal to proximal body segments during the golf swing, leading to larger mechanical loads being placed on the lumbar spine. With respect to segmental speeds during the golf swing, players with lower back pain are reported to achieve similar trunk extension velocities, right-side lateral bending velocities and axial trunk rotational velocities, but demonstrate increased trunk flexion velocities and lateral flexion velocities towards the lead side compared with asymptomatic players. Nevertheless, it should be noted that right-side trunk rotation, lateral trunk flexion angle (and velocity) towards the lead side and trunk flexion velocity all peak during the backswing, a phase that is not typically reported to be a high-risk component of the movement. Interventions that target improved strength and mobility of the trunk and/or altered swing mechanics have been shown to benefit golfers with lower back pain (Grimshaw & Burden, 2000; Grimshaw, Giles, Tong, & Grimmer, 2002), but given the multi-factorial nature of such injuries, it is unlikely that a generalised intervention strategy will benefit all players equally.

*Key notes*

| | |
|---|---|
| **Phases of the golf swing** | To facilitate analysis, the golf swing can be divided into four distinct phases: the address, backswing, downswing and follow-through. |
| **Address** | Address is a relatively static component of the golf swing and allows the player to position themselves appropriately to ensure a balanced and powerful swing. |
| **Backswing** | During the backswing, a heavy emphasis is placed on restricting pelvic rotation, while the thorax rotates through a large range of motion. By restricting pelvic turn relative to thorax turn, the muscles of the hip, trunk and upper limbs are eccentrically loaded; storing energy that can be released during the downswing. |

*(continued)*

*(continued)*

| | |
|---|---|
| **Downswing** | The downswing is initiated by a forceful contraction of the hip and knee extensors on the trail side (the right in right-handed golfers), which changes the direction of pelvic rotation and causes a lateral weight-shift towards the lead side. As the club reaches an approximately horizontal position, the wrists demonstrate rapid ulnar deviation that assists with accelerating the clubhead to the point of impact. |
| **Follow-through** | After impact, speed of movement steadily decreases and concludes with the golfer's spine in a hyper-extended position, their trail shoulder (the right in right-handed golfers) pointing towards the target and their hands raised high above their head. This posture is commonly referred to as the "reverse C" position. |
| **Golf-related lower back injuries** | Although the golf swing begins to slow after impact between the clubhead and the ball, epidemiological research suggests that up to 41 per cent of lower back injuries are sustained from impact through to the end of the follow-through. It is argued that many of these injuries are likely to be because the skeletal and soft tissue structures of the spine may not be able to safely accommodate the large range of movement and high spinal loads that occur during this period. Tailored conditioning of the spine's stabilising muscles and/or technique modification may help to reduce lower back injuries in golfers. |

## REFERENCES AND FURTHER READING

The following references provide additional reading about the incidence and possible risk factors of golf-related injuries.

1 Fradkin, A. J., Windley, T. C., Myers, J. B., Sell, T. C., & Lephart, S. M. (2007). Describing the epidemiology and associated age, gender and handicap comparisons of golfing injuries. *International Journal of Injury Control and Safety Promotion, 14*(4), 264–266.
2 Grimshaw, P. N., & Burden, A. M. (2000). Case report: Reduction of low back pain in a professional golfer. *Medicine and Science in Sports and Exercise, 32*(10), 1667–1673.
3 Grimshaw, P. N., Giles, A., Tong, R., & Grimmer, K. (2002). Low back and elbow injuries in golf. *Sports Medicine, 32*(10), 655–666.
4 Hosea, T. M., & Gatt, C. J. (1996). Back pain in golf. *Clinics in Sports Medicine, 15*(1), 37–53.

## SECTION G

# Measurement and analysis techniques

# OPTOELECTRONIC ANALYSIS

*Michael Cole*

**Optoelectronic motion analysis** is currently one of the major data collection tools used in biomechanics and is used to **collect movement-related information** on objects (usually humans) as they perform a particular task or activity. Unlike more traditional methods of motion analysis, optoelectronic systems attempt to reduce the complexity of data collection and the time associated with data analysis by automatically registering the position of markers (usually located on anatomical landmarks). While these systems offer a clear advantage over more traditional video-based motion analysis systems, major **disadvantages** include their relative lack of portability, the cost of cameras and the need for specialised software to operate the system and process the data.

Optoelectronic motion analysis systems can be broadly sub-divided into two major types based on the type of markers used and the associated procedures adopted by the hardware to register their positions. **Passive** optoelectronic motion analysis systems comprise a series of cameras that project infrared light onto reflective spheres (commonly called **markers** or **targets**) that are situated on the object of interest (i.e. the markers are "passive"). The reflected light is **optically** registered by the cameras and **electronically** converted to information that translates into the location of each marker in three-dimensional space. In contrast, an **active** optoelectronic motion analysis system comprises a series of cameras that register signals (usually infrared) emitted by the markers themselves (i.e. the markers are "active"). To energize the markers in these systems, a power source is required and, hence, typically adds extra weight and complexity to the experimental set-up. As most optoelectronic systems used in biomechanics are passive systems, only these will be considered in this section.

Usually a passive three-dimensional (3D) optoelectronic motion analysis system comprises six to eight cameras (quite often more), which are spread around the volume in which measurements are to be made. The layout depicted in *Figure G1.1* shows an example of a typical optoelectronic video camera and illustrates the location of eight cameras around a measurement volume of approximately 27 $m^3$ (3 × 3 × 3 m). Note that, in the middle of this figure, the reflective joint markers placed on a human research participant have been linked within the software to create a model that represents their upper and lower body segments.

The **cameras** used with optoelectronic motion analysis systems generally include a circular array of infrared light-emitting diodes around the lens. Importantly, infrared light cannot be seen by the human eye so using this type of light does not distract the performer or influence their movements. When the cameras are active, these diodes flash and emit infrared light that is returned from the reflective surface of the markers (in the same way as a camera flash is reflected by a mirror). The reflected light is subsequently detected and recorded by an infrared-sensitive chip that is situated within the lens of each camera, producing an "image" of the reflective objects (in this case, markers) for each burst of infrared light (*Figure G1.2*). The cameras repeat this process at a rate of between 120 and 420 times each second (i.e. 120 to 420 Hz), although sampling rates of up to 1000 Hz are possible with specific equipment.

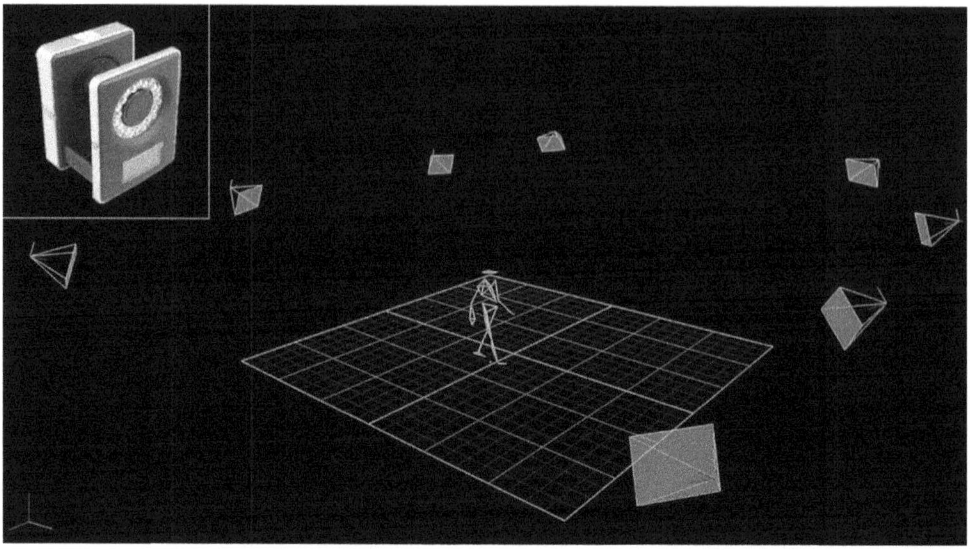

Figure G1.1 An example of a typical optoelectronic camera and a software-generated image depicting the configuration of an eight-camera system used to assess adult walking patterns. Note: the location and relative orientation of each camera is represented by the eight pyramidal shapes

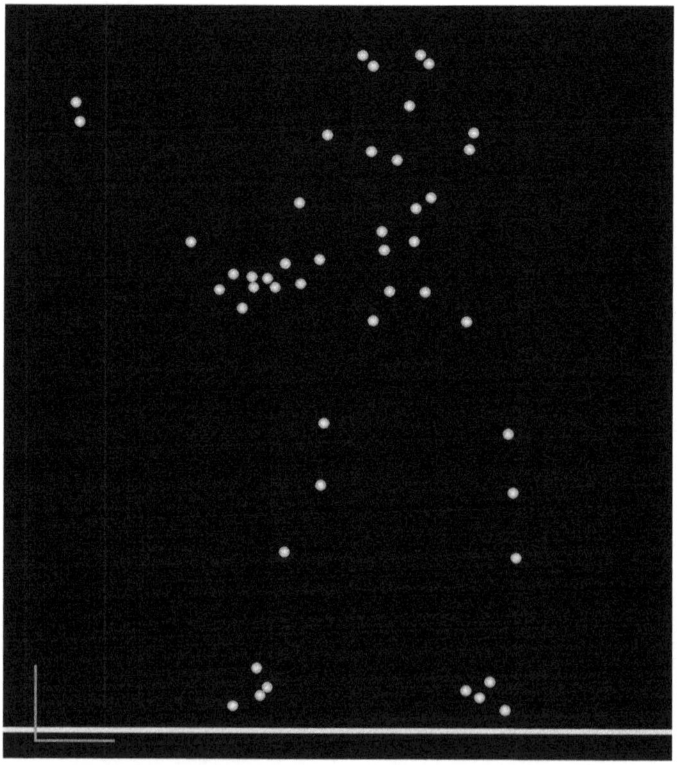

Figure G1.2 A typical image of the markers (grey spheres) as seen by the optoelectronic motion analysis system's infrared cameras during the performance of a golf swing

Each of the two-dimensional images collected by the cameras is transmitted to the host computer where they are merged (or "reconstructed") within the manufacturer's **software** using mathematical methods, such as the direct linear transformation (DLT) procedure (Abdel-Aziz & Karara, 1971). To reconstruct the three-dimensional marker locations, it is essential that the marker be seen by at least two cameras. This means that optoelectronic motion analysis systems are limited to applications where markers can be positioned in such a way that they are not obscured (e.g. by loose-fitting clothing or long hair). After the initial identification of each marker, most modern software is able to track the position of the marker throughout the remainder of the trial. However, this method is best considered a semi-automatic process, as human intervention is often required to solve reconstruction difficulties that originate from markers being poorly identified by the system or becoming temporarily occluded by the body's segments (marker drop-out). Nevertheless, such interventions are typically minimal for a system that is appropriately set up for a particular activity.

As with any motion analysis system, **calibration** of the space within which the performer is going to operate is essential to ensuring that the calculated marker trajectories and subsequent kinematics are accurate. Each manufacturer has developed their own system of calibration but the most common methods involve moving a calibrated hand-held implement that contains two or more markers of known 3D separation (often referred to as a calibration wand) around the entire movement area (*Figure G1.3*). This process is known as **dynamic calibration** and serves to provide essential information for determining the position of each camera relative to all others in the system. When performing the dynamic calibration, it is important to ensure that the calibration wand covers an area that is larger than would be required for the athlete or participant to perform the desired movement to limit the risk of data loss and to ensure that the movement is accurately represented. Following dynamic calibration, the calibration wand or separate calibration frame is positioned in the capture volume at the point where the researcher wishes to set the origin of the global coordinate system (i.e. the 0, 0, 0 point of the X, Y and Z axes). This process is known as **static calibration.** Using the manufacturer's recommended procedure it is possible to obtain reconstruction accuracies of less than 1 mm and reconstruction precisions of around 0.2 mm.

The **spherical markers** used with modern optoelectronic motion analysis systems are not too dissimilar to those used with some older video-based 3D motion analysis systems and are typically comprised of a polystyrene or hard plastic ball that is covered with a reflective tape. Commercially available markers vary considerably in diameter and can range from as small as 3 mm to as large

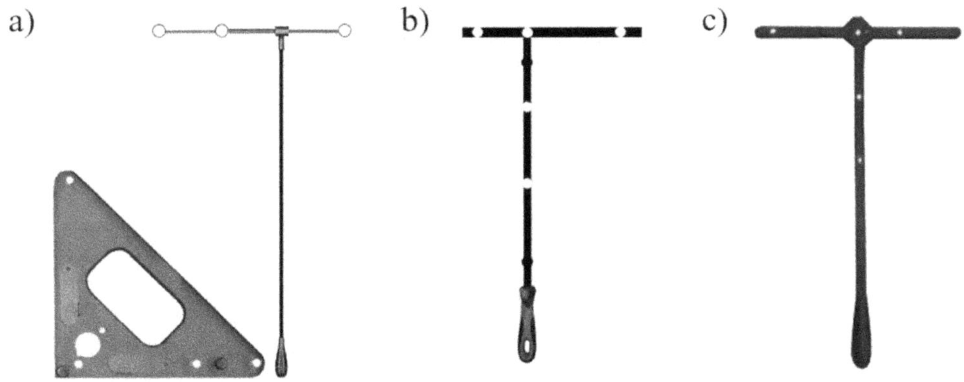


**Figure G1.3** A selection of calibration wands (and calibration frames) that are typically used to calibrate optoelectronic motion analysis systems: a) Vicon 3-marker wand with L-frame; b) Vicon 5-marker wand; and c) Vicon active wand

as 30 mm, depending on the specific application. As a general rule, one should choose a marker that covers about 1/200th of the camera's field of view; hence for a field of view of 3 m (i.e. 3000 mm) the marker needs to be around 15 mm in diameter. The number and placement of markers is usually guided by the **marker set** selected by the researcher at the beginning of the study. For example, the marker set depicted in *Figure G2.4* uses 39 markers placed on the feet, lower limbs, pelvis, trunk, upper limbs, hands and head to assess whole body kinematics during movement. These markers define a 15-segment biomechanical model that consists of the feet (2), shanks (2), thighs (2), pelvis (1), thorax (1), head (1), upper arms (2), forearms (2) and hands (2).

The main purpose of optoelectronic motion analysis systems is to obtain the three-dimensional positions of each marker (i.e. x, y and z) as a function of time. This data is usually output by the system to a data file that can be used to compute a range of linear and/or angular kinematics either within the manufacturer's software or using a third-party application. Additionally, if the opto-electronic system is integrated with other instruments, such as a force platform (which measures ground reaction forces), it is possible to combine each system's outcomes (i.e. ground reaction forces and joint kinematics) to calculate other measures, such as joint moments and/or powers. Collectively, this data forms the basis of **gait analysis**, and integrated optoelectronic systems have a major role to play in this area.

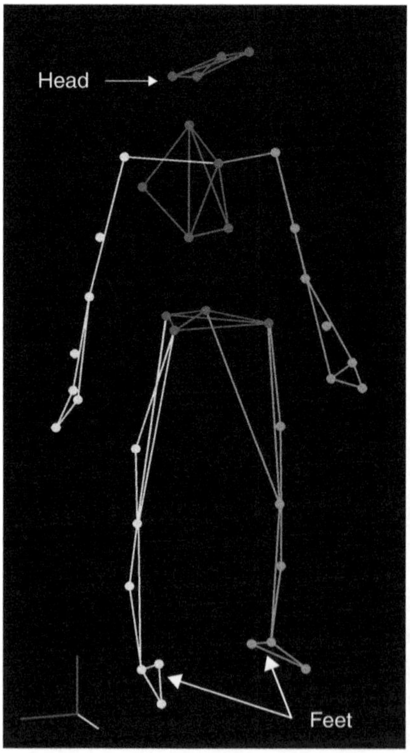


**Figure G2.4** A 39-point marker set for assessing full-body kinematics

*Key notes*

| | |
|---|---|
| **Optoelectronic motion analysis** | Optoelectronic motion analysis uses a series of cameras that project infrared light onto reflective spheres called markers (or targets). The reflected light is optically recorded by the cameras and automatically registers the location of the markers in 3D space. Optoelectronic motion analysis is therefore simpler and less time consuming than older video-based methods. |
| **Optoelectronic cameras** | Optoelectronic cameras are derived from early work involving video technology and have a series of infrared light-emitting diodes around their lens, which flash between 120 and 420 times per second (i.e. 120–420 Hz), although sampling rates of up to 1000 Hz are possible. |
| **Marker sets and biomechanical models** | Numerous biomechanical models are described in the literature. For example, the general patterns of whole-body motion can be assessed with 39 markers that define a 15-segment biomechanical model consisting of the feet (2), shanks (2), thighs (2), pelvis (1), torso (1), head (1), upper arms (2), lower arms (2) and hands (2). More complex models exist for more comprehensive three-dimensional assessments. |
| **Calibration** | Optoelectronic systems require the desired performance area to be calibrated before use to ensure that the cameras accurately identify the markers. This process involves the researcher moving around the desired capture area with a hand-held implement comprising two or more markers of known 3D separation. This process allows the system to determine the relative location of each camera and facilitates accurate 3D reconstruction of the collected data. |

## REFERENCES AND FURTHER READING

The following references describe the methods underpinning the direct linear transformation procedure, which is often used for the 3D reconstruction of multiple two-dimensional images.

1 Abdel-Aziz, Y. I., & Karara, H. M. (1971). Direct linear transformation from comparator coordinates into object space coordinates in close-range photogrammetry. In *Symposium on Close-Range Photogrammetry* (pp. 1–18). Urbana, Illinois: American Society of Photogrammetry.
2 Chen, L., Armstrong, C. W., & Raftopoulos, D. D. (1994). An investigation on the accuracy of three-dimensional space reconstruction using the direct linear transformation technique. *Journal of Biomechanics, 27*(4), 493–500.
3 Wood, G. A., & Marshall, R. N. (1986). The accuracy of DLT extrapolation in three-dimensional film analysis. *Journal of Biomechanics, 19*(9), 781–785.

# G2 WEARABLE SENSORS

*Michael Cole*

The kinematics of human motion are typically described in terms of linear and angular displacement, velocity and acceleration. Displacement and velocity can be measured with reasonable accuracy using video-based and optoelectronic motion analysis systems, but given that these systems measure positional data, some form of differentiation must be used to determine velocity and acceleration. Each time the original data is differentiated, the effect of any small measurement error in the data is multiplied. To determine acceleration from position data, it is necessary to calculate the second differential (i.e. change in displacement/time is the first differential; change in velocity/time is the second differential); hence, the acceleration data is often contaminated with a large amount of error.

Suppose an athlete is filmed side-on (i.e. looking at the sagittal plane) with an older-model conventional video camera and we measured the displacement frame by frame (or at 25 Hz) to obtain the information shown in *Table G2.1*. By calculating the changes in displacement between consecutive frames and dividing the result by the time, it would be possible to determine the velocity (i.e. velocity = displacement/time). The same process can be followed to find the frame-to-frame changes in velocity to allow acceleration to be derived.

In *Table G2.1*, the actual "error-free" displacement data with the subsequent calculations for velocity and acceleration can be seen. However, in *Table G2.2*, a small amount of error (plus or minus 2 cm or 0.02 m) has been added to most of the position data. These errors occur as a consequence of the measurement process. It can be seen that even very small amounts of error in the displacement data lead to large errors in the acceleration data; making it necessary to find an alternative approach to determining accelerations during movement.

**Table G2.1** Velocities and accelerations calculated from "error-free" coordinate data

| *Frame* | *Position (metres)* | *Change in displacement (metres)* | *Change in time (seconds)* | *Velocity (m/s)* | *Change in velocity (m/s)* | *Acceleration ($m/s^2$)* |
|---|---|---|---|---|---|---|
| 1 | 50.00 | | | | | |
| | | 0.40 | 0.04 | 10.00 | | |
| 2 | 50.40 | | | | 0.50 | 12.50 |
| | | 0.42 | 0.04 | 10.50 | | |
| 3 | 50.82 | | | | 0 | 0 |
| | | 0.42 | 0.04 | 10.50 | | |
| 4 | 51.24 | | | | 1.00 | 25.0 |
| | | 0.46 | 0.04 | 11.50 | | |
| 5 | 51.70 | | | | 1.00 | 25.0 |
| | | 0.50 | 0.04 | 12.50 | | |
| 6 | 52.20 | | | | | |

**Table G2.2** Velocities and accelerations calculated from coordinate (position) data containing small errors

| *Frame* | *Position (metres)* | *Change in displacement (metres)* | *Change in time (seconds)* | *Velocity (m/s)* | *Change in velocity (m/s)* | *Acceleration ($m/s^2$)* |
|---|---|---|---|---|---|---|
| 1 | **50.02** | | | | | |
| | | 0.36 | 0.04 | 9.00 | | |
| 2 | **50.38** | | | | 2.50 | 62.5 |
| | | 0.46 | 0.04 | 11.50 | | |
| 3 | **50.84** | | | | 2.00 | –50.0 |
| | | 0.38 | 0.04 | 9.50 | | |
| 4 | **51.22** | | | | 2.50 | 62.5 |
| | | 0.48 | 0.04 | 12.00 | | |
| 5 | 51.70 | | | | 0 | 0 |
| | | 0.48 | 0.04 | 12.00 | | |
| 6 | **52.18** | | | | | |

An alternative method for determining acceleration would be to measure the forces acting upon a body and to use Newton's second law of motion ($\Sigma F = m \times a$) to calculate the resultant acceleration. However, this method is only possible when it is practical to measure the contact forces acting upon the body of interest. There are many applications where this is not possible because the body of interest is not in contact with any surfaces or the movement of interest occurs in a situation where contact forces cannot be easily measured.

Nevertheless, obtaining accurate and reliable acceleration data is essential to many areas of biomechanics. For example, good acceleration data is necessary for the calculation of joint reaction forces (the internal forces acting across joints in the human body obtained through the process of mathematical modelling). There are also many applications of the use of acceleration data to drive control devices in the automotive and aeronautical industries. To solve the difficulties associated with obtaining acceleration indirectly using displacement or force data an alternative method is to measure the acceleration directly using an **accelerometer**.

In applications that involve flight, such as aircraft and satellites, accelerometers are often based on the properties of rotating masses. However, the most common design in human movement combines **Newton's law of mass acceleration** and **Hooke's law of spring action.** Newton's second law states that if a mass (m) experiences an acceleration (a), then there must be a net force (F) acting on the mass; that is $F = m \times a$. Hooke's law states that if a spring of stiffness k is stretched a distance of x from its equilibrium position, there must be a net force acting on the spring given; that is $F = k \times x$. If these two equations are combined it reveals that the displacement of the spring will be proportional to the acceleration (Equation G2.1).

$$F = m \times a = k \times x$$

Therefore:

$$a = \frac{k \times x}{m}$$

**(Equation G2.1)**

*Figure G2.1* shows an accelerometer constructed of a small mass attached to a spring. When there is no acceleration, the spring rests at its natural length ($x_1$) and there is no horizontal force acting upon the mass. If the system is accelerated to the right, the spring's lengthening exerts

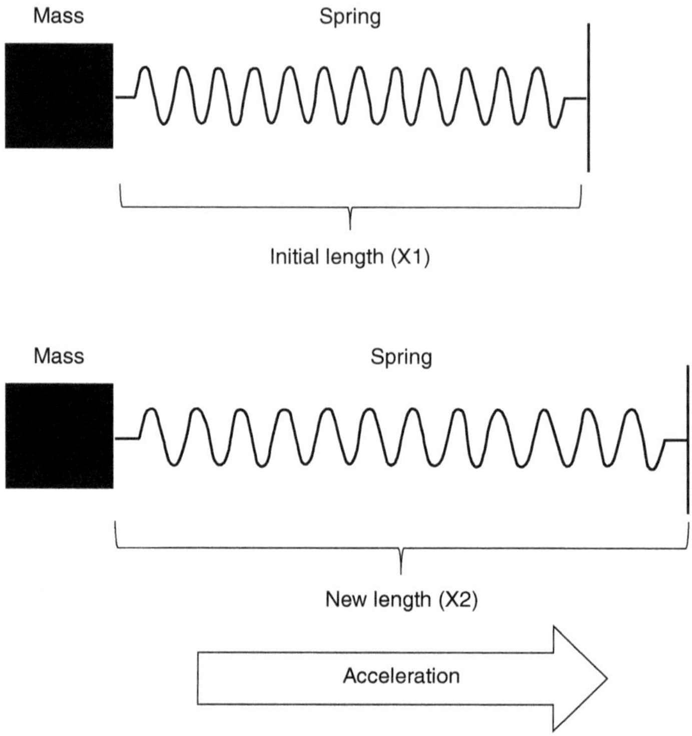


**Figure G2.1** A spring–mass-based accelerometer demonstrating the link between the deformation of the spring ($x_2 - x_1$) and the acceleration ($a = k\ x(x_2 - x_1)/\text{mass}$)

a force on the mass. As the spring's length continues to increase, increasing force is developed until the mass experiences an acceleration equal to the remainder of the system. If the mass's displacement is known, Equation G2.1 can be used to calculate the amount of acceleration.

Many accelerometers are based on this **spring–mass principle**, but accelerometers often differ in how they measure the mass's displacement. Of the many types of commercially available accelerometers (*Table G2.3*), either capacitive or piezoelectric types are the most commonly used in sport and exercise biomechanics. However, accelerometers that include piezoelectric technology are generally more expensive than those using capacitive technology.

It is important to remember that acceleration is a vector quantity and thus has both magnitude and direction. In the example provided in *Figure G2.1*, only the magnitude of the acceleration in the direction that the spring is being stretched can be measured. This means that the accelerometer is only able to measure in one dimension and that the acceleration calculated is only representative of acceleration in that direction. To gain a complete picture of the acceleration of a body it is necessary to have three accelerometers: one aligned with each of the axes of motion. Some accelerometers are produced containing separate sensors in each plane to allow three-dimensional measurements with a single device.

It is important to note the effect that changing the orientation of the accelerometer has on the output. Consider the spring–mass system in *Figure G2.1*, where the spring is oriented to be aligned with the horizontal plane. When the system is at rest, there is no force acting between the spring and the mass. If the system is rotated 90 degrees, so the mass sits above the spring (*Figure G2.2*), the mass would compress the spring and a negative acceleration of 1g would be recorded (where negative 1g represents the acceleration due to gravity or $-9.81\ m/s^2$). If the

**Table G2.3** Summary of the different types of commercially available accelerometers and the key technologies upon which they are based

| *Sensor category* | *Key technologies* |
|---|---|
| **Capacitive** | Comprise two metal planar surfaces (or plates), one or both of which are electrically charged. Acceleration increases/decreases the gap between the plates; changing the system's electrical capacity. The nature of this change allows the acceleration's magnitude and direction to be measured. |
| **Piezoelectric** | Contain piezoelectric crystals attached to a mass and make their measures on the premise that acceleration is proportional to force. When the system is accelerated, the mass deforms the piezoelectric crystals, causing a change in electrical charge that is proportional to the acceleration. |
| **Piezoresistive** | Also known as strain gauge accelerometers. Quantify accelerations by measuring changes in the electrical resistance of a material (e.g. a metal beam) in response to deformation caused by the system's acceleration. |
| **Hall effect** | Measure voltage changes due to variations in magnetic field surrounding a metallic beam. Acceleration of the system affects the beam's position and, in turn, influences the magnetic field. This change is encoded as an electrical signal and allows the nature of acceleration to be determined. |
| **Magnetoresistive** | Work on a similar premise to hall effect sensors, but measure changes in the electrical resistance of a material (e.g. a metallic beam) in response to a change in the surrounding magnetic field. |
| **Heat transfer** | Comprise a single heat source that is centred in a substrate and has equally spaced thermoresistors on four of its sides. When the system is not accelerating, the heat gradient measured by the thermoresistors will be symmetrical. However, under acceleration, the heat gradient becomes asymmetrical; allowing the amplitude of the acceleration to be assessed. |

system were oriented the opposite way around, such that the mass hung below the spring, the spring would lengthen until the force it exerts on the mass is equal to the mass's weight. This force is the result of the acceleration due to gravity, where the baseline acceleration is said to be equal to 1g. Given this understanding, it is important that careful thought is given to the orientation of the accelerometer. Typical values for the acceleration experienced in different situations are provided in *Table G2.4* and are usually reported relative to the acceleration due to gravity (g) or in units of metres per second per second ($m/s^2$).

When using accelerometers, it is critical to ensure that the measuring unit is firmly affixed to the body, as it is important to be confident that the motion of the accelerometer is the same as that of the body being measured. In the case of human movement, it is important to realise that not all parts of the body experience the same motion, and that movement of the soft tissues (skin, fat, muscle) will be different to that of the skeletal system. For this reason, it is typical to measure human movement using accelerometers that are attached to the body over sites that have minimal soft tissue between the skin and the underlying skeleton. The malleoli (ankles), sacrum, spinous processes, occipital protuberance and acromion processes are all common examples of locations where accelerometers are fixed on the body. Even at these bony sites, a firm mounting is necessary and combinations of mounting glues and various adhesive tapes have been used to ensure good fixation.

With respect to the shock-absorbing qualities of the body, comparisons can be made of the different body regions to see how accelerations change as they are absorbed by the body's tissues. During an activity, such as running, the body normally experiences relatively high accelerations (and forces, due to body mass – and essentially weight on Earth – being constant)

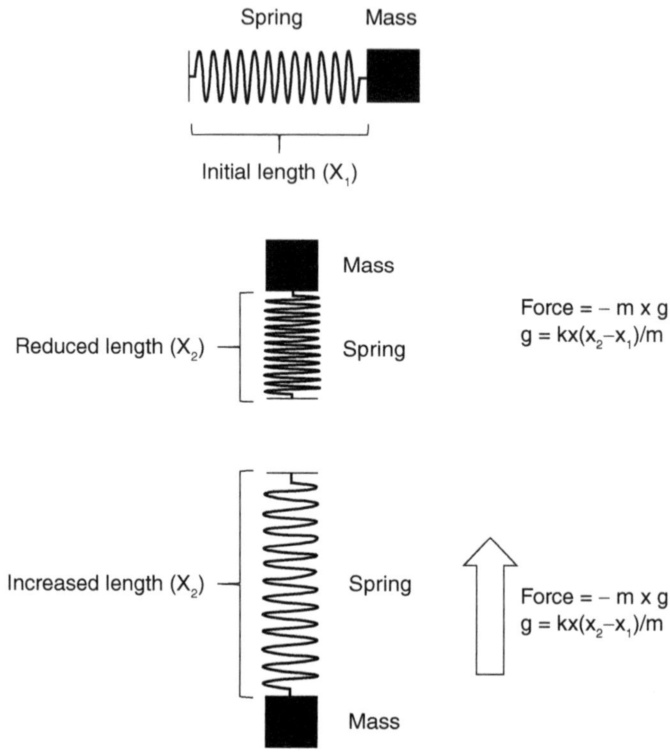


**Figure G2.2** The effect of orientation on the output of an accelerometer

**Table G2.4** Typical accelerations experienced by the human body

| | *Acceleration (g)* | *Acceleration (m/s²)* |
|---|---|---|
| Earth's gravity | 1 | 9.81 |
| Passenger car in corner | 2 | 19.62 |
| Bobsled rider in corner | 5 | 49.05 |
| Average human unconsciousness | 7 | 68.67 |
| Fighter jet maximum (vertical) | 9 | 88.29 |

at impact. These accelerations (and forces) are gradually dissipated by equipment, such as footwear, and/or the skeletal and soft tissue structures of the body. For this reason, lower accelerations are typically recorded when running shod compared with running in unshod conditions. Similarly, lower accelerations are also recorded at segments positioned further up the kinetic chain (e.g. the head) compared with the shank (*Table G2.5*).

Gyroscopes are sensors that are specifically designed to measure the rate of angular displacement (i.e. angular velocity); that is, how fast the sensor is rotating about its own axis or axes. Additionally, using gyroscopic data it is possible to mathematically derive measures of angular position or orientation from the angular velocities that these devices measure. Like accelerometers, gyroscopes may be capable of recording the angular velocity about one, two or three axes of rotation, with each of these units respectively known as uni-axial, bi-axial and tri-axial gyroscopes. While there are numerous types of gyroscopes for application in different settings,

**Table G2.5** Attenuation of running-related accelerations due to specialised equipment (footwear) and skeletal and soft tissue structures

| | *Body location* | *Acceleration (g)* |
|---|---|---|
| **Shod *vs* unshod running** | | |
| Running barefoot | Shank | ~9 g |
| Running in shoes | Shank | ~8 g |
| **Head *vs* shank accelerations** | | |
| Running in shoes | Shank | ~8 g |
| Running in shoes | Head | ~3 g |

one of the more common types is the vibrating structure gyroscope. The measures of angular velocity derived from these devices are based on the behaviour of a small centrally-located mass that is connected to an inner frame via a series of springs. The inner frame is subsequently connected to a fixed electronic circuit board via another series of springs that are oriented perpendicular to the springs attached to the mass (*Figure G2.3*).

Amongst other things, the fixed electronic circuit board provides power to the vibrating structure gyroscope, which causes the centrally located mass to continuously resonate in a sinusoidal manner along the springs that connect it to the inner frame. When a mass or collection of masses (i.e. a system) is rotated, the system will experience both centripetal (centre-seeking) and centrifugal (centre-fleeing) forces. Furthermore, the rotating system will also experience a **Coriolis force** (often called a Coriolis acceleration, given that the system's mass does not

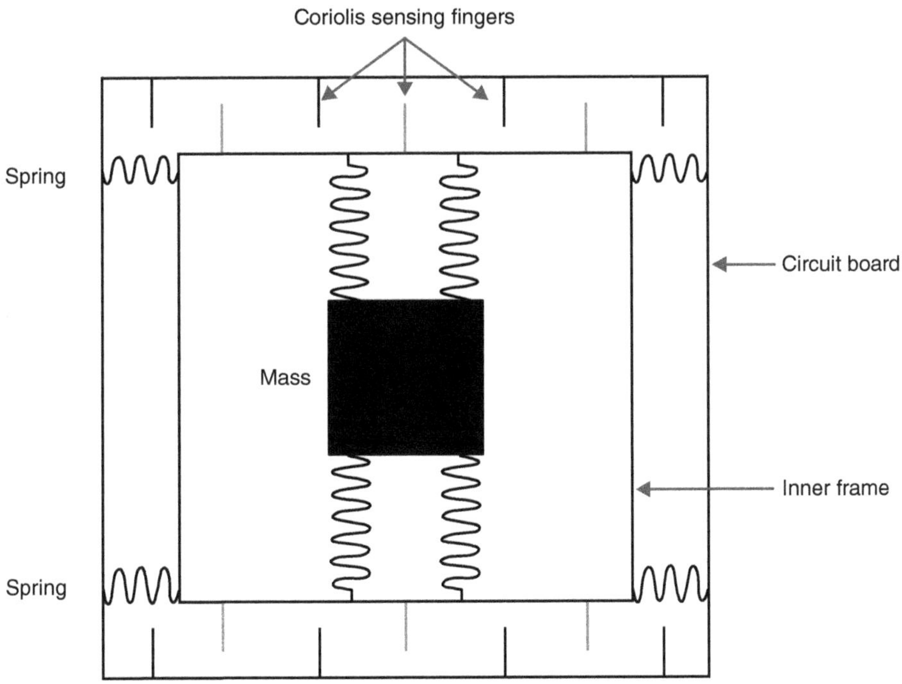


**Figure G2.3** The basic components of a vibrating structure gyroscope

change), which acts perpendicular to the line of action of the centripetal/centrifugal forces and induces lateral movement with respect to the axis of rotation. For example, consider a simple scenario where a person is sitting with a ball in the centre of a rotating circular platform and they are asked to try to roll the ball into a hole that is cut into the platform's upper surface. Assuming that the platform is rotating in a clockwise direction and that the person rolls the ball in much the same way as they would if the platform was not rotating (i.e. they make no allowances for the rotation), the ball would be expected to overshoot the hole and miss to the left (*Figure G2.4a*). Alternatively, if the platform was rotating anti-clockwise (*Figure G2.4b*), the ball would overshoot the target, but it would miss to the right of the hole. In both scenarios, the overshoot component of the performance is attributable to the centrifugal (centre-fleeing) force that the rotating system is experiencing, while the miss to the left or to the right is the result of the Coriolis force. It is the combined effects of the centrifugal and Coriolis forces that contributes to the curvilinear trajectory of the rolling ball.

In a similar way, clockwise rotation of a vibrating structure gyroscope induces a Coriolis force that causes the inner frame of the system to move to the left along the line of action of the springs between the inner frame and the fixed electronic circuit board, while the centrifugal force causes the centrally located mass to shift away from the axis of rotation (*Figure G2.5a*). Similarly, an anti-clockwise rotation of the gyroscope will induce a Coriolis force that causes the inner frame to shift to the right of the axis of rotation, while the centrally-located mass will, again, be drawn away from the centre of rotation by the centrifugal force (*Figure G2.5b*). In both cases, the vibrating structure gyroscope establishes the magnitude of the Coriolis force using a series of capacitance-detecting "fingers" that are located along the outer edge of the inner frame and the inner edge of the electronic circuit board's fixed housing. For example, when a Coriolis force causes the inner frame to move to the left, the distance between the capacitance-detecting fingers on the inner frame and the fixed housing of the circuit board is reduced. Similarly, when the inner frame moves to the right, a different set of capacitance-sensing fingers move closer to each other. The magnitude of the change in capacitance is driven by how much the fingers have moved with respect to one another and, hence, how much the inner frame has been displaced. By measuring the size of the change in capacitance and evaluating which capacitance-detecting fingers have moved closer to each other, the sensor is ultimately able to detect the size and direction of the angular velocity imparted upon the system.

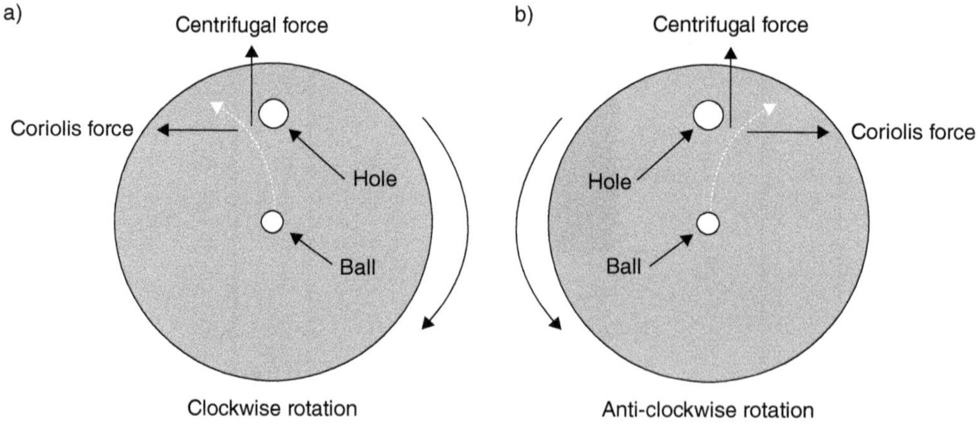


**Figure G2.4** The relationship between centrifugal (centre-fleeing) force and the Coriolis force during a task that involves rolling a ball toward a target (hole) on a platform that is rotating a) clockwise and b) anti-clockwise

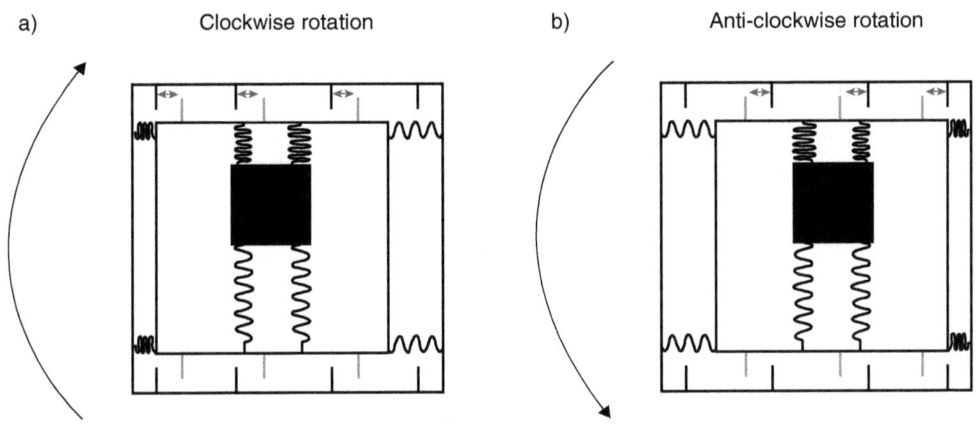


**Figure G2.5** The relative movements of the centrally-located mass and the inner frame of a typical vibrating structure gyroscope during a) clockwise rotation and b) anti-clockwise rotation

Although the technologies described in this section may have begun as large mechanical devices, modern wearable sensors, such as the accelerometers and gyroscopes that are now included in sports equipment and smartphones, are essentially much smaller electronic analogues of their mechanical predecessors. The miniaturisation of these sensors has made it

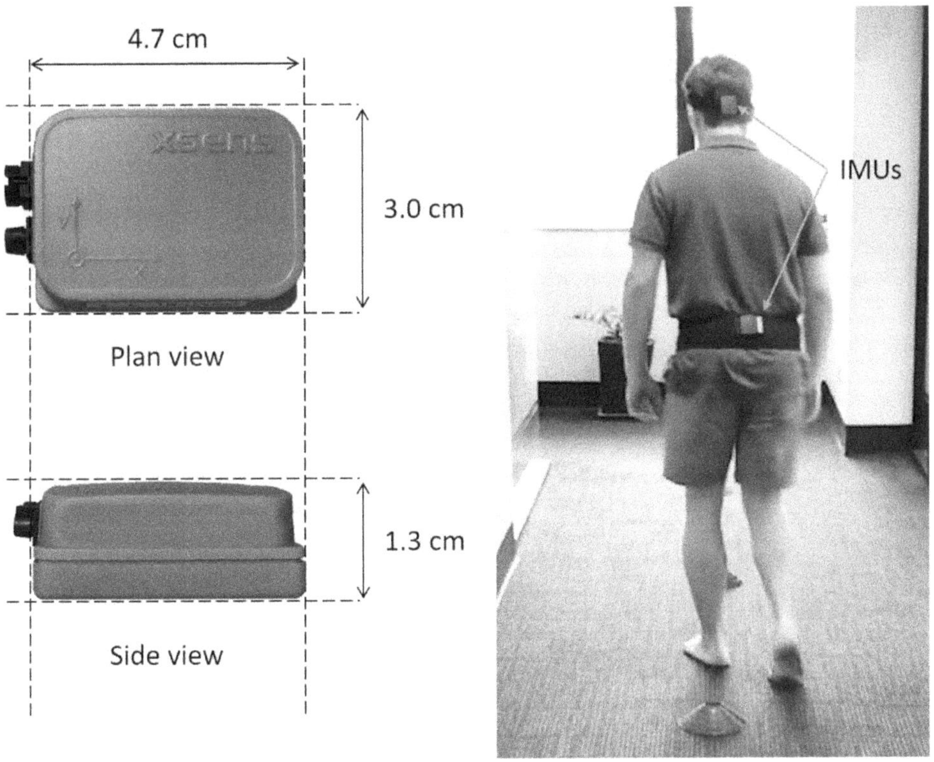


**Figure G2.6** A close-up view of a commercially available inertial measurement unit (left) and the same units affixed to an individual's pelvis and head to assess movement rhythmicity during gait (right)

possible for tri-axial accelerometers, gyroscopes and magnetometers (sensors that detect and measure magnetic fields) to be integrated into a single measuring unit that is generally no larger than a small box of mints. Although these measuring units are marketed under many different names, they are collectively referred to as inertial measurement units or IMUs and are fast becoming an alternate means for assessing the biomechanics of human motion in the laboratory, in the clinic and in the field (*Figure G2.6*). These sophisticated wearable devices use a mathematical process known as **sensor fusion** to combine the data from the individual sensors to develop a dataset that is more reliable than what could be derived from the individual sensors alone. The fusion of data from a tri-axial accelerometer, gyroscope and magnetometer involves complex mathematics and considerable research and development time. For this reason, the specific sensor fusion techniques used by different manufacturers are subtly different and the details underpinning their methods are often labeled proprietary information. A number of manufacturers are now marketing motion analysis systems that incorporate multiple IMUs, which are used to biomechanically model the kinematics of human movement. Compared with optoelectronic motion analysis, such IMU-based systems are known to provide reliable and valid measures of segmental motion and sagittal joint kinematics during gait; however, the validity and reliability of lower limb joint kinematics in the frontal and transverse planes are reportedly much poorer. These results highlight the potential shortcomings of IMU-based motion analysis systems and emphasise the need for biomechanists to carefully evaluate the obvious flexibility of these systems with their likely limitations.

*Key notes*

| | |
|---|---|
| **Acceleration** | Acceleration is a vector quantity and is defined as the rate of change in velocity. It is calculated as the change in velocity divided by the change in time and is the second derivative of position data. Accelerations derived from displacement data are often contaminated with noise (errors) and may be unsuitable for analysis. |
| **Accelerometer** | Accelerometers are devices that directly measure the acceleration of a body. |
| **Newton's second law** | Newton's second law states that for a mass (m) to experience an acceleration (a) of a specific magnitude, there must be a net force (F) acting on the mass. This relationship is mathematically depicted as $F = m \times a$. |
| **Hooke's law** | Hooke's law states that the net force (F) required to deform a spring from its equilibrium position, is proportional to the stiffness of the spring (k) and the extent to which it is deformed (x). Hence, this relationship can be given as $F = k \times x$. |
| **Gyroscope** | Gyroscopes are sensors that measure angular velocity; that is, the rate at which the sensor is rotating about its own axis or axes. Gyroscopes are capable of recording the angular velocity about one, two or three axes of rotation and, if needed, angular position or orientation can be mathematically derived from the angular velocities that they measure. |
| **Magnetometer** | Magnetometers are sensors that are capable of detecting and measuring the direction, strength and relative change of magnetic fields. |
| **Inertial measurement unit (IMU)** | Inertial measurement units or IMUs are small wearable sensors that typically include tri-axial accelerometers, gyroscopes and magnetometers. The combination of data sets from the various sensor types is an intricate process that involves a sophisticated mathematical procedure, known as sensor fusion. By fusing accelerations with angular velocities (or orientations) and magnetometer data, IMUs are capable of providing data that is more reliable than what could be derived from the individual sensors alone. |

## FURTHER READING

The following resources provide additional reading on the use of wearable sensors for assessing segmental motion and joint kinematics during walking for a range of populations.

1 Zhang, J. T., Novak, A. C., Brouwer, B., & Li, Q. (2013). Concurrent validation of Xsens MVN measurement of lower limb joint angular kinematics. *Physiological Measurement, 34*(8), N63–69.
2 Blair, S., Duthie, G., Robertson, S., Hopkins, W., & Ball, K. (2018). Concurrent validation of an inertial measurement system to quantify kicking biomechanics in four football codes. *Journal of Biomechanics, 73*, 2432.
3 Hubble, R. P., Naughton, G. A., Silburn, P. A., & Cole, M. H. (2015). Wearable sensor use for assessing standing balance and walking stability in people with Parkinson's disease: a systematic review. *PLoS One, 10*(4), e0123705.
4 Cole, M. H., van den Hoorn, W., Kavanagh, J., Morrison, S., Hodges, P. W., Smeathers, J. E., & Kerr, G. K. (2014). Concurrent validity of accelerations measured using a tri-axial inertial measurement unit while walking on firm, compliant and uneven surfaces. *PLoS One, 9*(5), e98395.

# G3 FORCE PLATE

*Adrian Burden*

Newton's third law of motion dictates that for every (action) force that is applied by one body to another body, a (reaction) force is exerted by the second body on the first that is equal in magnitude and opposite in direction (*Figure G3.1*). In sport and exercise biomechanics, the reaction force exerted by the ground on an individual is often studied and is termed the ground reaction force (GRF). The force plate, or platform, embedded into the ground in a variety of settings (e.g. a laboratory or athletics track), is used to measure the GRF. Force data can be combined with the velocity of the centre of mass (COM) to obtain the power of the whole body and with kinematic and anthropometric data to determine joint reaction forces.

When a force is applied to the force plate, in-built transducers experience a deformation that is proportional to the magnitude of the force. This deformation subsequently produces a change in the voltage measured from the transducer that is proportional to the amount by which the transducer has deformed. Thus, the change in voltage measured by the transducer is proportional to the magnitude of force that it experiences. Force plates used in sport and exercise biomechanics use either strain gauge or piezoelectric transducers. Although piezoelectric platforms are more sensitive to rapid changes in force, they suffer from a change in output voltage with no change in applied force (i.e. drift). As such, they are more suited to measure forces from relatively short-lived, dynamic activities such as walking, running and jumping. Strain gauge plates are less susceptible to drift, but are not as sensitive as piezoelectric models. As such, they are preferred for recording forces from longer, less dynamic activities, such as archery or shooting.

Regardless of the type of transducer used, there should ideally be a linear relationship between the force applied to the platform and the measured voltage. Assuming linearity, the gradient of the relationship is effectively the calibration coefficient, which is used to convert volts into newtons. In situations where the relationship is non-linear (*Figure G3.2a*), a higher-order polynomial (e.g. quadratic) can be fitted to the data points to provide the calibration coefficient. Force plates should also display minimal hysteresis (*Figure G3.2b*), so that the relationship between force and voltage that is observed when the plate is loaded is the same as when it is unloaded. Transducers are arranged in force plates so that they measure three components of GRF that are parallel to the plate's three orthogonal axes (*Figure G3.3*). There should be minimal cross-talk, which is the detection of force by the transducers in one direction (e.g. parallel to the Z axis) and also by those in one or both of the other directions (i.e. parallel to the X and/or Y axes). In addition, the force plate should have a natural frequency that is much higher than the highest frequency of the signal being measured, and ideally higher than 800 Hz. This is so that the force being applied to the plate does not cause it to vibrate, which would affect the magnitude of the force being detected.

The voltage output from the transducers must be amplified before being recorded and stored, usually on a computer. Sampling of the signal into a computer should also use an analogue-to-digital-converter that has at least 12 bits (ideally 16 bits) to ensure that very small changes in force are able to be detected. To satisfy the Nyquist theorem, which is discussed in the Sampling Theory section of this text, the signal should normally be sampled at a minimum of 500 Hz, and ideally higher if forces are recorded during impacts.

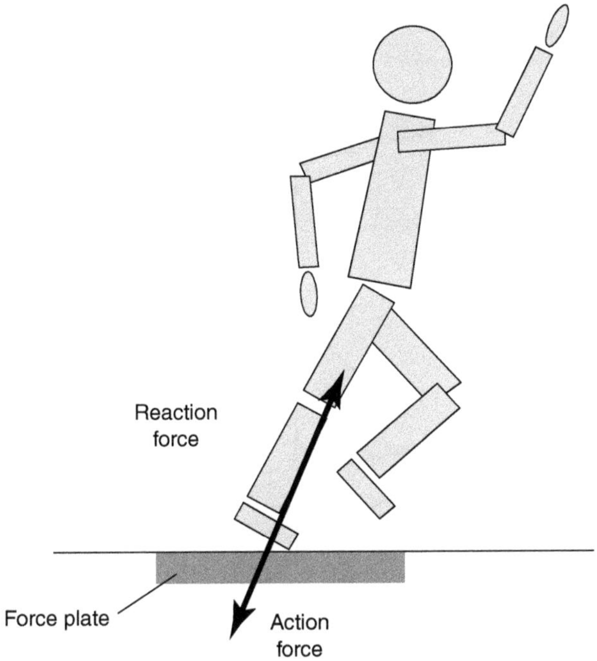


**Figure G3.1** Depiction of Newton's third law of motion

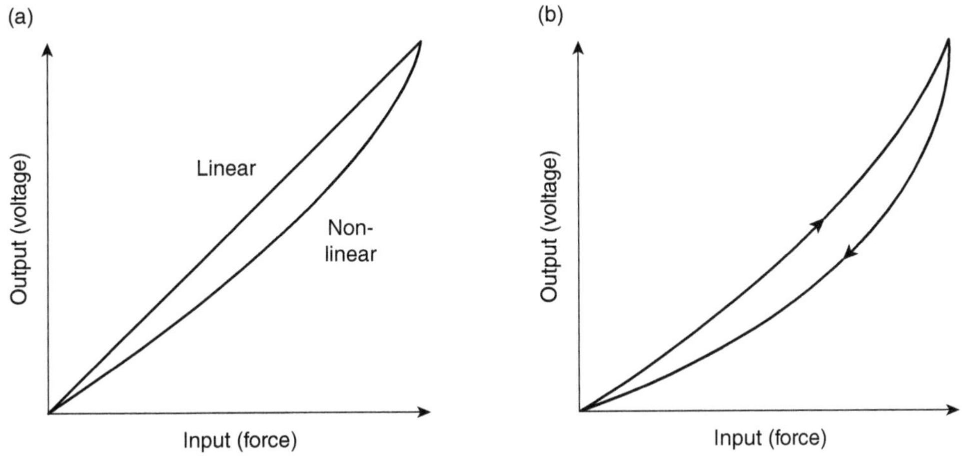


**Figure G3.2** Depiction of (a) linearity and (b) hysteresis of a force platform

Two conventions exist to identify the three components of GRF that force plates measure. The convention shown in *Figure G3.4*, which is commonly used in the United Kingdom, labels positive Fz in the vertical (upwards) direction, normal to the surface of the plate. Positive Fy acts along the forward (horizontal) direction, parallel to the long axis of the plate, and positive

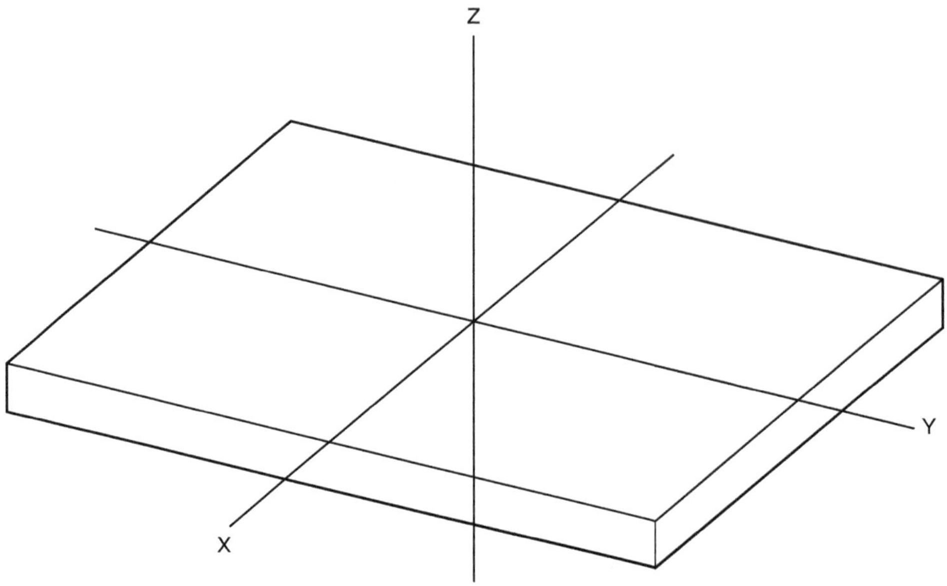


**Figure G3.3** Orthogonal force plate axes

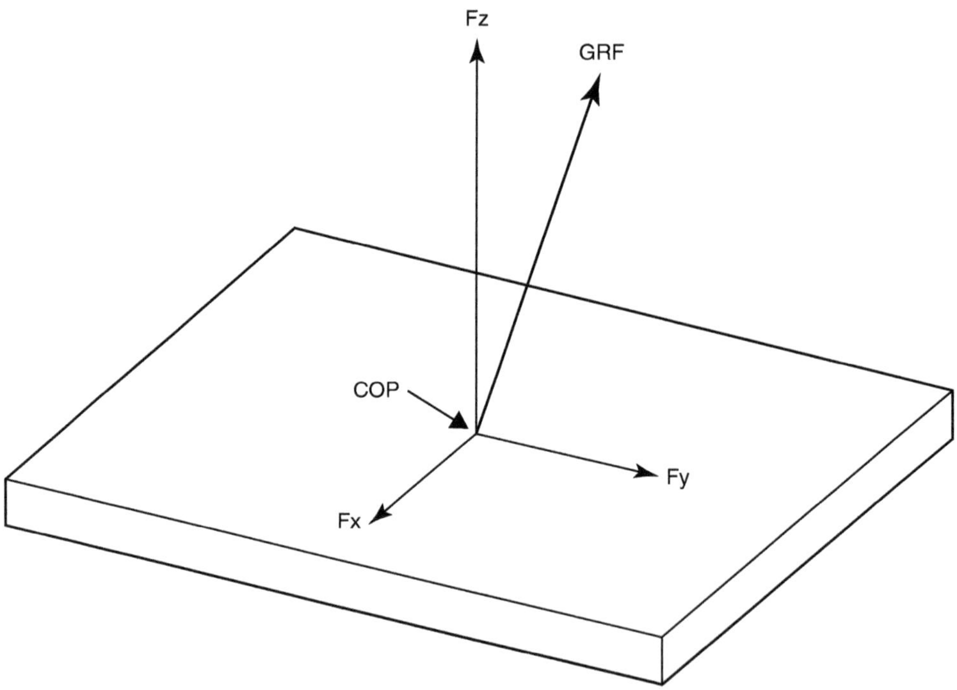


**Figure G3.4** Horizontal (Fx and Fy) and vertical (Fz) components of ground reaction force (GRF), including the centre of pressure (COP)

Fx occurs in the positive right (lateral) direction. It therefore follows that negative Fz, Fy, and Fx act downwards, backwards and in the right medial direction. The convention adopted by the International Society of Biomechanics (ISB) replaces Fz with Fy, Fy with Fx, and Fx with Fz.

To **understand and interpret GRF–time curves,** it is imperative to have knowledge of Newton's laws of motion. Forces acting on a runner during the stance phase include the weight of the athlete ($F_w$), air resistance ($F_a$), and the components of GRF (*Figure G3.5*). In accordance with Newton's second law of motion, the sum of all the forces acting in each (X, Y, Z) direction is proportional to the acceleration experienced by the athlete in that direction (i.e. $\Sigma F = m \times a$); as shown in Equations G3.1 to G3.3. Dividing both sides of each equation by the mass of the runner yields the acceleration of the runner's COM. Assuming that both $F_w$ and $F_a$ are constant, the shape of the resulting acceleration–time curves would be identical to the force–time curves.

$$F_z - F_w = m \times a_z \qquad \textbf{(Equation G3.1)}$$

$$F_y - F_a = m \times a_y \qquad \textbf{(Equation G3.2)}$$

$$F_x = m \times a_x \qquad \textbf{(Equation G3.3)}$$

Hypothetical Fz– and Fy–time curves are shown in *Figure G3.6*, together with free body diagrams that coincide approximately with three points during the stance phase. The Fx–time curve has been omitted, as it shows far greater inter-individual variability and is of a magnitude that is much smaller than the other two; meaning that it has less effect on the acceleration of the body.

In the vertical direction, as the athlete's foot first contacts the ground, the magnitude of the Fz component of GRF is smaller than that of the body weight (i.e. Fz < Fw). In accordance with Equation G3.1, this results in a negative (i.e. downwards) force and, therefore, acceleration acting on the runner. As the runner is already moving downwards at this time, acceleration in

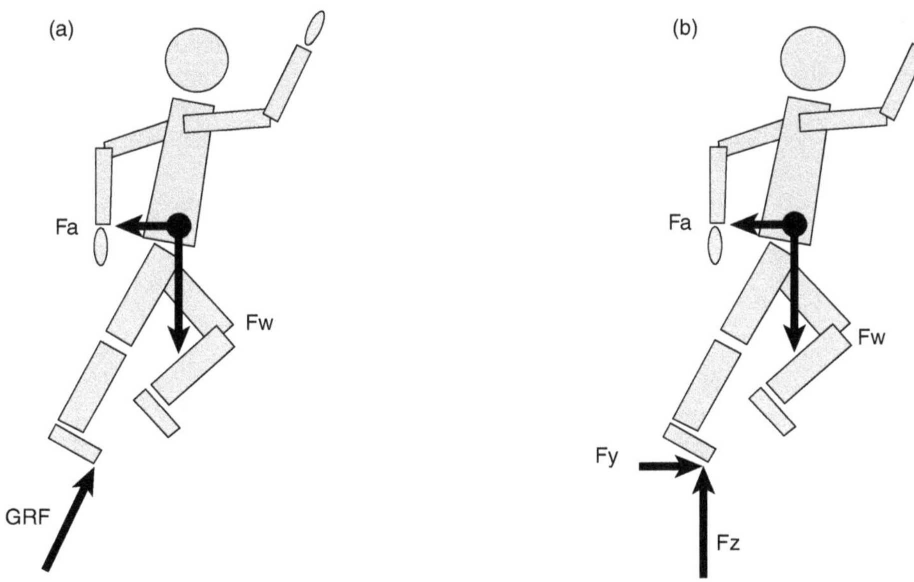


**Figure G3.5** Free body diagram of a runner showing a) resultant ground reaction force and b) Fy and Fz components of the ground reaction force

the same direction results in an increase in the downwards velocity of the COM. This pattern is quickly reversed as Fz exceeds Fw, resulting in a positive acceleration that acts in an upwards direction. Initially, this decreases the runner's downwards velocity until it reaches zero and their downwards motion is arrested. This occurs about halfway through the stance phase, after which the positive acceleration causes the runner's COM to move upward at an increasing velocity until just before toe-off. Here, the magnitude of Fz again drops below that of Fw, causing the acceleration to act in a downward direction. However, unlike at the start of the stance phase, the runner is moving upward at this time so the negative acceleration causes the velocity to decrease in this direction immediately before toe-off.

Interpretation of the Fy–time curve from the stance phase of running is generally simpler than the Fz curve as motion only occurs in one direction (i.e. forward along a line parallel to the y axis). Assuming air resistance to be so small as to be negligible, as the runner's foot contacts the plate in front of their COM, a braking force (i.e. negative Fy, prior to point b in *Figure G3.6*) is experienced that acts in a backwards direction. Again, in accordance with Newton's second law, this force acts to decelerate the forward motion of runner. This situation continues until the runner's COM passes over the point of support (i.e. point b in *Figure G3.6*) and the reaction force changes from negative to positive (i.e. forwards). Positive Fy (i.e. after point b in *Figure G3.6*) causes a positive horizontal acceleration that increases the forward horizontal velocity of the runner. Thus, when running at an approximately constant velocity, a braking force (negative Fy) acts during the first half of the stance phase that causes the horizontal velocity of the runner to decrease. Through the second part of the stance phase, a propulsive force (positive Fy) dominates, which causes the runner's COM to accelerate in the direction of motion.

Precisely how much the velocity of the runner's COM changes in any direction during the stance phase can be determined using the impulse–momentum relationship. Graphical integration of the force–time data, using Simpson's or the Trapezium rule, would yield the area bound by the curves and, hence, the impulse (*Figure G3.7*). In accordance with the impulse–momentum relationship, the change in velocity of the runner's COM can be obtained by dividing the net impulse by his/her mass. With regard to the forces acting in the direction of motion (i.e. Fy), if the braking impulse is greater than the propulsive impulse (*Figure G3.8a*) the runner will lose velocity during the stance phase. Conversely, if the braking impulse is less than the propulsive impulse the runner will gain velocity as he/she passes over the plate (*Figure G3.8b*). If the two impulses are equal (i.e. zero net impulse), then the runner will complete the

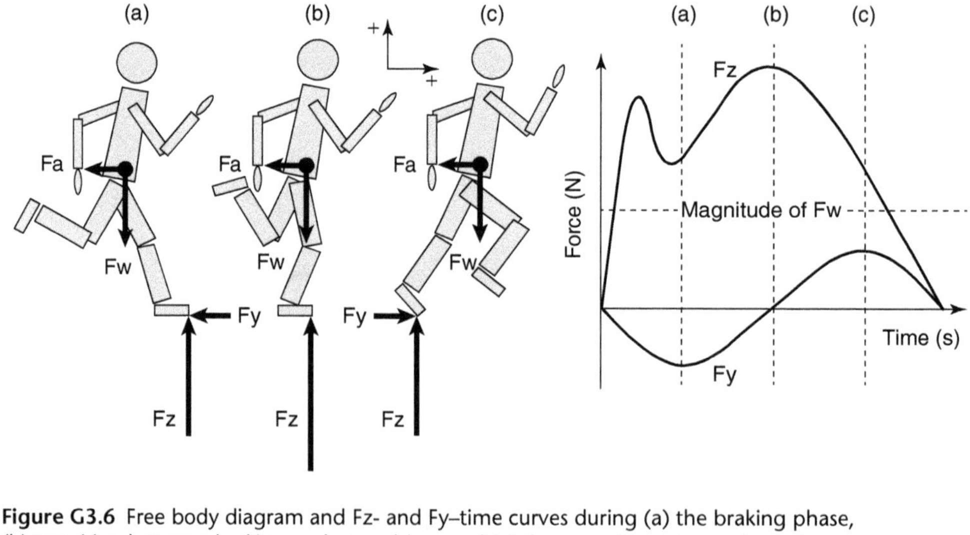


**Figure G3.6** Free body diagram and Fz- and Fy–time curves during (a) the braking phase, (b) transition between braking and propulsion, and (c) the propulsive phase of running

stance phase with the same velocity with which they started it (*Figure G3.8c*). In the unlikely situation of a runner experiencing zero net impulse over successive strides, their overall velocity would decrease due to effect of air resistance during swing phases of each stride. Thus, in order to maintain a constant running velocity, the propulsive impulse should be slightly greater than the braking impulse during each stance phase.

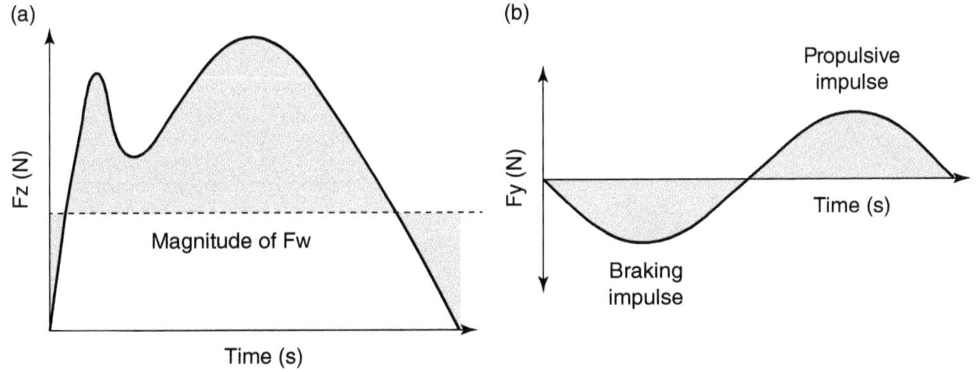


**Figure G3.7** Shaded areas depict (a) vertical and (b) horizontal impulse during the stance phase of running

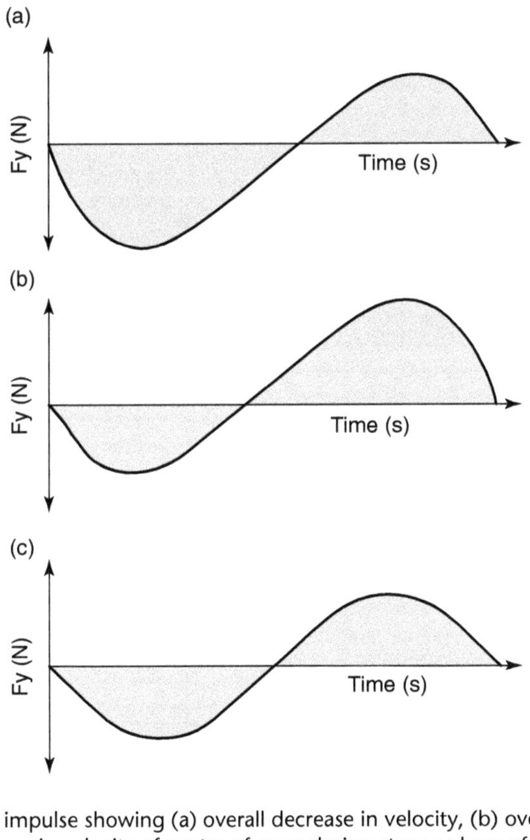


**Figure G3.8** Horizontal impulse showing (a) overall decrease in velocity, (b) overall increase in velocity, and (c) no change in velocity of centre of mass during stance phase of running

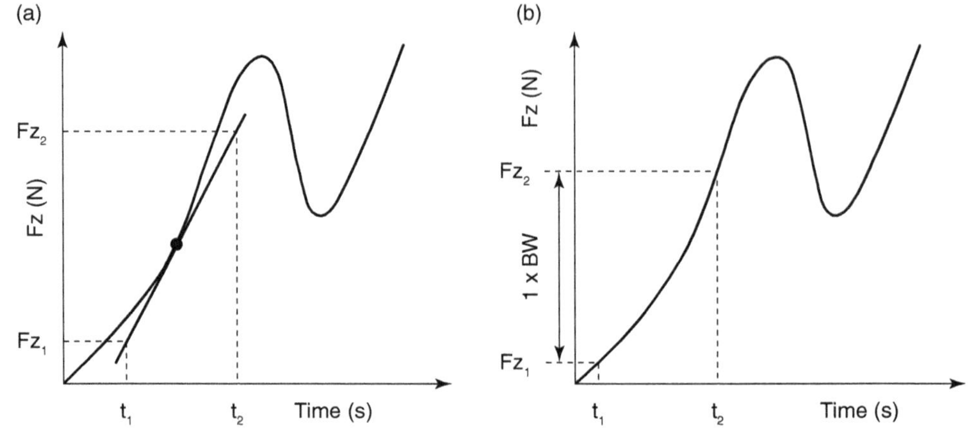


$$\text{Loading rate} = \frac{Fz_2 - Fz_1}{t_2 - t_1}$$

**Figure G3.9** Depiction of (a) instantaneous and (b) average loading rate from the initial phase of Fz–time curve during running

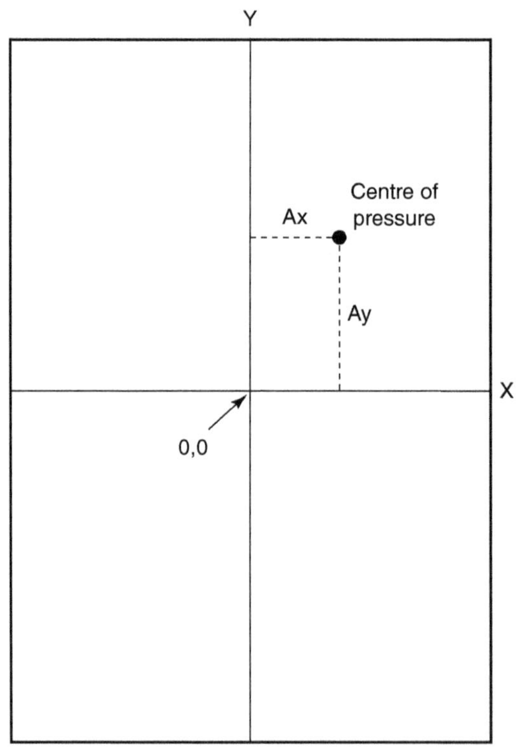


**Figure G3.10** Location of centre of pressure, at one point in time, using coordinates (Ax and Ay) in relation to the centre of the plate (0,0)

In biomechanics, there are a number of **GRF-related variables** that are of interest. For example, the rate at which the vertical GRF is applied to the body is often measured during running and landing, together with the peak force, as an indication of the risk of chronic injury due to such activities. The instantaneous loading rate can be obtained by measuring the gradient of a tangent that is drawn at the point on the Fz–time curve where the rate of change is considered to be greatest (*Figure G3.9a*). Alternatively, an average loading rate can be obtained by measuring the rate at which Fz rises by a force equal to the participant's body weight after an initial brief period during which the body is loaded to a predefined threshold (*Figure G3.9b*). For example, one method calculates loading rate between 50 N and a force that is equal to the participant's body weight plus 50 N. While this method masks the peak loading rate derived from the instantaneous method, its measure is more reliable and objective due to its systematic approach.

In addition to the GRF components and their impulses, the centre of pressure (COP) and free moment are often calculated from force plate data. The COP is the position of the GRF vector in relation to a plane parallel to and just below the surface of the plate. Two coordinates (Ax and Ay) give its location in relation to the origin, or centre, of the plate (*Figure G3.10*). Ax and Ay can be either positive (one side of the axis) or negative (the other side of the axis) and collectively designate the quadrant of the plate where the COP lies. If a person runs across or stands on the platform, the COP will lie somewhere beneath their foot (*Figure G3.11a*). Alternatively, during bipedal stance, the COP lies roughly between the two contact points (*Figure G3.11b*). In biomechanics, the pattern of movement of the COP beneath the foot during the stance phase of running is a common application. Motion of the COP also mirrors that of the COM during standing or during activities that require the body to be as stationary as possible (e.g. archery, shooting), so it is also used as a measure of stability during such activities.

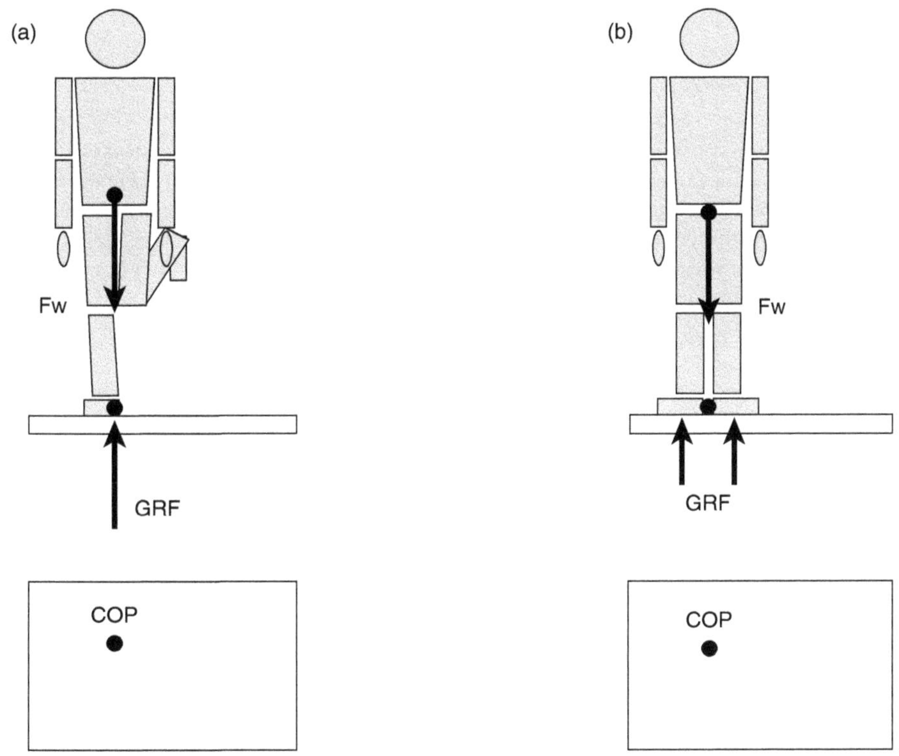


**Figure G3.11** Location of centre of pressure (COP) during (a) one-legged and (b) two-legged standing

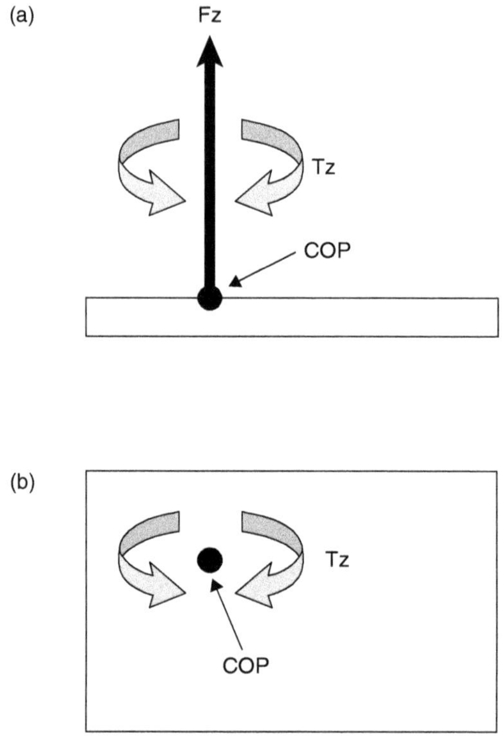


**Figure G3.12** Depiction of the free moment (Mz' or Tz) looking (a) along the plate's x-axis, and (b) along the z-axis

The free moment is commonly referred to as either Mz' or Tz and is the moment or torque about the vertical axis through the COP, which coincides with the Fz vector (*Figure G3.12*). It must be remembered that, like the components of force, Mz' is equal in magnitude but opposite in direction to the moment that is applied to the plate about the vertical axis. Mz' is therefore often used to measure the reaction to the moment exerted by individuals during activities that involve twisting around a vertical axis.

*Key notes*

| | |
|---|---|
| **The force plate** | Force plates (or force platforms) measure the ground reaction force (GRF) which, in accordance with Newton's third law of motion, is equal in magnitude and opposite in direction to the action force that is applied to the plate. In the United Kingdom, the vertical component of GRF is normally denoted as Fz, and the two horizontal components are denoted as Fy and Fx. Force plate transducers are usually piezoelectric or strain gauge types, which display high linearity, low hysteresis and minimal cross-talk between axes. |
| **Interpreting GRF–time curves** | In accordance with Newton's second law of motion, the sum of all of the forces acting on a body in a particular direction is proportional to the acceleration experienced by the body in the same direction. For example, during running the magnitude of Fz minus the athlete's body weight determines the magnitude of the vertical acceleration of their centre of mass (COM). Similarly, the magnitude of Fy minus the force of air resistance determines the runner's horizontal acceleration in the direction of running. |

| | |
|---|---|
| **GRF-related variables** | In addition to peak forces, impulse and loading rate, a number of other variables can also be obtained from force platform information. The centre of pressure (COP) is the position of the resultant GRF vector in a plane that is parallel to the surface of the plate. Two-dimensional coordinates (Ax and Ay) are used to locate the COP in relation to the centre, or origin, of the plate. The free moment (Mz′ or Tz) is the turning force or moment around a vertical axis through the COP. |

## FURTHER READING

The following resources provide further reading on the use of force platforms in sport and exercise biomechanics:

1 Munro, C. F., Miller, D. I., & Fuglevand (1987). Ground reaction forces in running: A re-examination. *Journal of Biomechanics, 20*(2), 147–155.
2 Hunter, J. P., Marshall, R. N., & McNair, P. J. (2005). Relationships between ground reaction force impulse and kinematics of sprint-running acceleration. *Journal of Applied Biomechanics, 21*(1), 31–43.

G4

# PRESSURE MEASUREMENT

*Michael Cole*

Pressure is a mechanical parameter that is relevant in many applications and represents the amount of force per unit of area and thus has units that are derived from these two components:

$$Pressure\ (N/m^2) = \frac{Force\ (N)}{Area\ (m^2)}$$

Another common unit used to describe pressure is the **pascal (Pa)**, which represents the pressure created by a 1 N force acting on a 1 $m^2$ area. As 1 pascal represents a relatively low force spread over a large area, it is more common to report **kilopascals (kPa)**, where 1 kPa is equivalent to 1000 pascals. An alternative to reporting pressure in pascals is to compare the measured pressure with the ambient pressure caused by the Earth's atmosphere. The pressure caused by the weight of the air in the Earth's atmosphere is referred to as **atmospheric** or **barometric pressure.** Imagine a column with a cross-sectional area of 1 $m^2$ that extends from the Earth's surface to the edge of the atmosphere. This column will contain a certain number of air particles that will collectively create a force due to their weight, their density and their distance from the Earth's surface. The **standard atmosphere** is a unit of pressure that represents one normal atmosphere and is defined as:

$$1\ \text{Atmosphere} = 101325\ \text{Pa or } 101.325\ \text{kPa}$$

or

$$1\ \text{Atmosphere} = 760\ \text{mmHg at } 0°\text{C } (32°\text{F})$$

The pressure experienced in any situation will be dependent upon both the magnitude of the applied force and area over which it acts. For example, if a 75 kg person was to stand on one foot and the area under the foot was 0.01 $m^2$, the pressure would be:

$$\text{Pressure} = \frac{\text{Force(N)}}{\text{Area}\,(\text{m}^2)} = \frac{75 \times 9.81}{0.01}$$

$$\text{Pressure} = 73575\ \mathbf{N/m^2}\ \text{or}\ 73.575\ \mathbf{kPa}\ \text{or}\ \frac{73.575}{101.325} = 0.73\,\mathbf{Atmospheres}$$

However, if the same person were to put on a pair of high-heel shoes and then also stood on one leg, the area in contact with the ground would decrease. If the new contact area was 0.002 $m^2$ the pressure would be:

$$\text{Pressure} = \frac{\text{Force (N)}}{\text{Area (m}^2\text{)}} = \frac{75 \times 9.81}{0.002}$$

$$\text{Pressure} = 367875\ \mathbf{N/m^2}\ \text{or}\ 367.875\ \mathbf{kPa}\ \text{or}\ \frac{367.875}{101.325} = 3.63\ \mathbf{Atmospheres}$$

In both these scenarios, the force was the same but the pressure differed significantly. This is important in human movement as the study of pressure gives an indication of the distribution of the load. For example, when a force is concentrated in a small area (i.e. a high pressure), the body is more likely to suffer damage than if the same load was applied over a larger area. Given this relationship, protective sports equipment such as helmets and shin pads are often designed to have a large contact area to reduce the pressure exerted on the underlying tissues and reduce the risk of injury. However, in some situations it may be desirable for some cutting or piercing tools (e.g. an arrow) to have a small contact area to maximize the pressure and allow penetration without the necessity for a large force.

There are many different approaches to measuring pressure, but the simplest method is using a **manometer**, which is an instrument that comprises two interconnected fluid-filled columns. When both columns are experiencing an equal pressure the fluid in each will rest at the same level (*Figure G4.1*). If a greater pressure is experienced on one side, the level of fluid on that side will go down and the level on the other side will rise (*Figure G4.2*). The difference in fluid height for the two columns will be related to the applied pressure, the cross-sectional area of the column and the density of the fluid contained within. Before calculating pressure, it is first necessary to calculate the weight of the fluid column using the following formula:

$$\text{Weight of fluid column} = \text{volume}_{\text{fluid}} \times \text{density}_{\text{fluid}}\ (\rho) \times \text{gravity (g)}$$

As $volume_{fluid}$ is the product of the fluid's height (h) and cross-sectional area (a), this formula can be expanded to:

$$\text{Weight of fluid column} = \text{h} \times \text{a} \times \rho \times \text{g}$$

Once we have determined the weight of the fluid column (i.e. its force), we are able to calculate pressure:

$$\text{Pressure} = \frac{\text{Force (N)}}{\text{Area }\left(\text{m}^2\right)}$$

Hence:

$$\text{Pressure} = \frac{\text{h} \times \text{a} \times \rho \times \text{g}}{\text{a}}$$

or

$$\textbf{Pressure} = \mathbf{h \times \rho \times g}$$

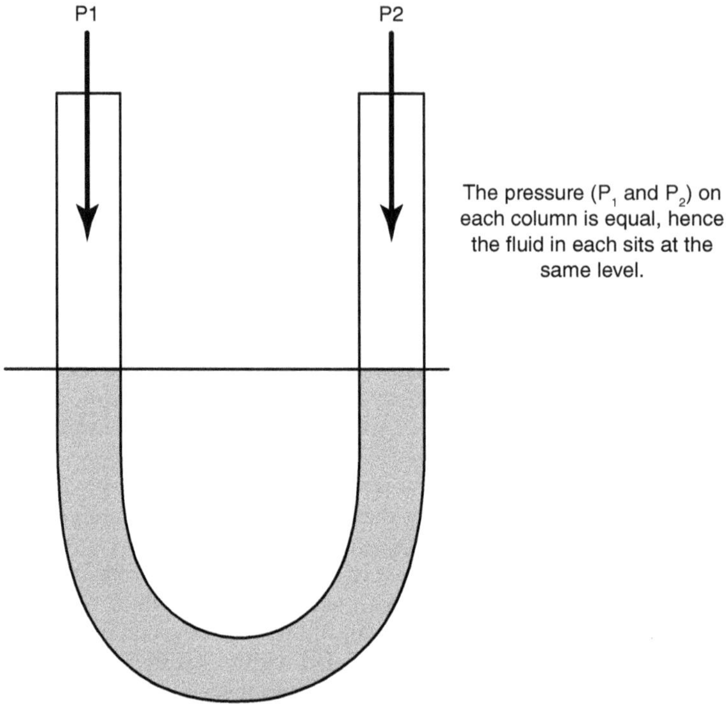


**Figure G4.1** A U-tube manometer under equilibrium conditions

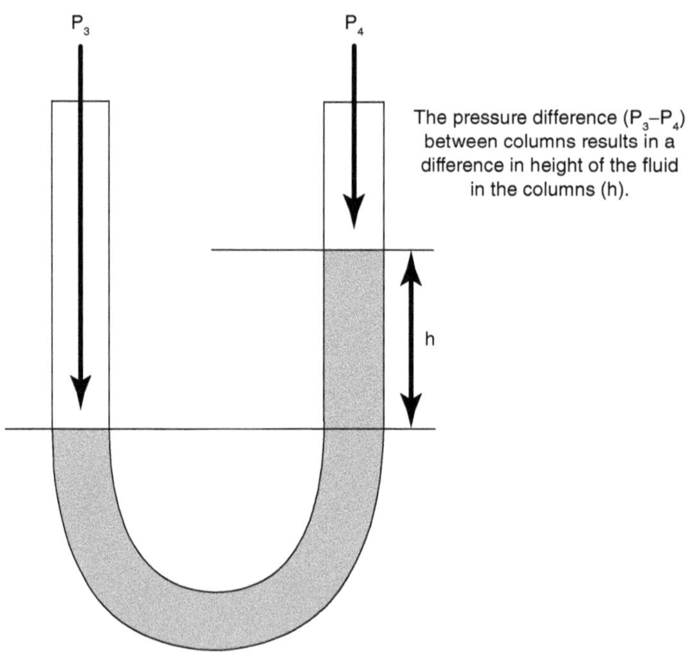


**Figure G4.2** A U-tube manometer under conditions of uneven pressure

In practice, pressure is often measured using a linear cylinder that has one end sealed, while the other is submerged in a dense or heavy liquid, such as mercury. For example, our ability to record a change in barometric pressure using a common mercury barometer is facilitated by the effect of this atmospheric pressure change on the equipment's mercury pool. Specifically, when atmospheric pressure increases, the pool of mercury is pushed down more, which leads to an increase in the height of the fluid in the glass cylinder. In contrast, when atmospheric pressure drops, the pressure placed on the mercury pool is reduced and the fluid level in the cylinder is lowered (*Figure G4.3*). When using this type of equipment, pressure is reported as the height (in millimetres) of the column of mercury (which has the chemical symbol of Hg) in the glass cylinder. At a pressure of 1 Atmosphere (101.325 kPa), the column height would be 760 mmHg at 0°C. Another example of a piece of equipment that uses this type of pressure measurement is a sphygmomanometer, which is used to measure blood pressure. Measures of blood pressure are normally reported as the systolic pressure (i.e. pressure created by the blood as it is pumped through the arteries by the heart) over the diastolic pressure (i.e. pressure in the arteries between heart beats). Hence, a blood pressure reading of 120 over 80 (120/80) actually reflects a systolic pressure of 120 mmHg and a diastolic pressure of 80 mmHg.

Thus far, only **static pressure** or **steady-state measurements** have been considered, but such techniques are useful only in applications where very slow changing conditions or equilibrium are experienced. If one were interested in the rate or pattern of change in pressure over short periods of time, **dynamic pressure measurements** that rely on **electromechanical pressure sensors** (e.g. strain gauges, variable capacitance, piezoelectric transducers) must be used. Electromechanical pressure sensors, or pressure transducers, convert motion generated by a pressure-sensitive device into an electrical signal that is proportional to the size of the applied pressure. A common application of dynamic pressure measurement in human movement is in the analysis of the pressure distribution beneath the foot during standing, walking or running. Such assessments often use thin pressure-sensitive insoles that comprise a large number of pressure sensors. These sensors allow detection of high- and low-pressure areas beneath the feet at any time point throughout the movement. This information is usually presented as a series of colours or shades that represent the different pressures (*Figure G4.4a*) or as a 3D graph that is characterised by bars of different heights (*Figure G4.4b*), each indicating the magnitude of pressure at each point. Foot pressure analysis has been widely used to investigate the effect of different types of footwear on pressure patterns and injury risk and research tends to suggest that higher pressures are indicative of a greater risk of injury.

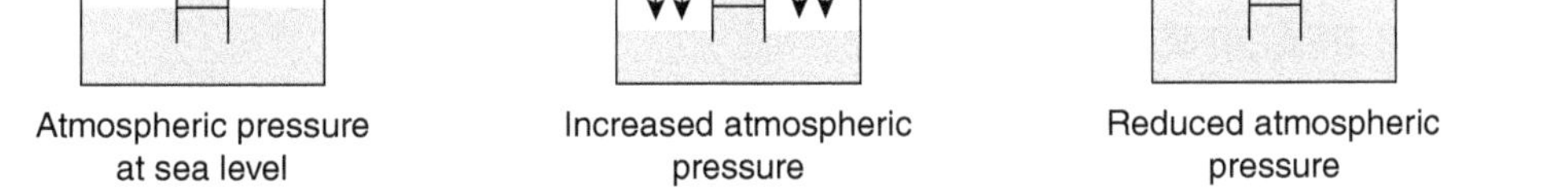


**Figure G4.3** The basic premise upon which common mercury barometers measure changes in atmospheric pressure

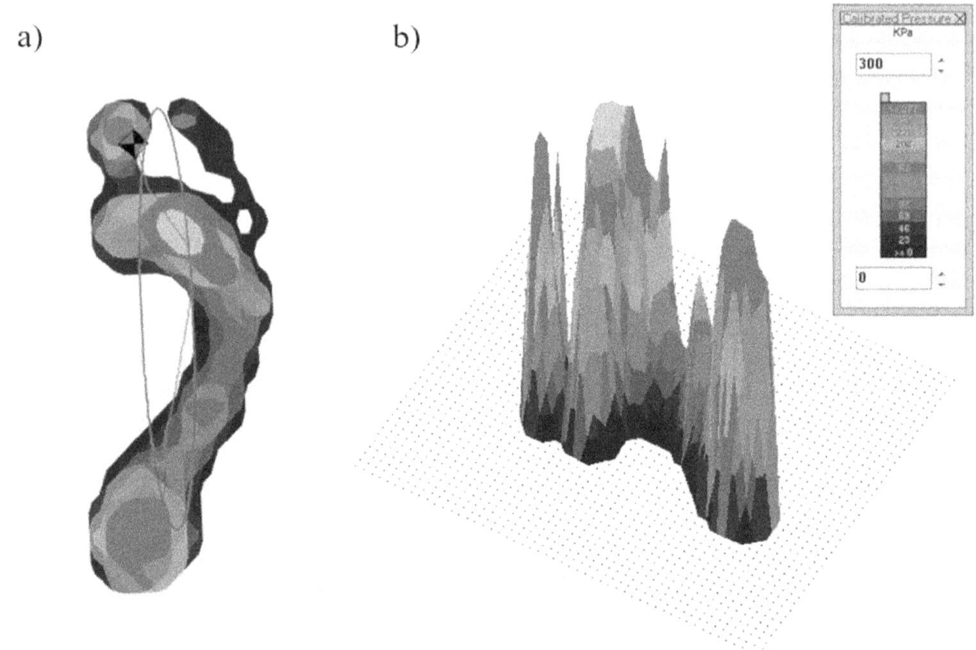


**Figure G4.4** A two-dimensional (a) and three-dimensional (b) pressure profile indicating the distribution of pressure between different areas of the foot during the stance phase of walking. In the three-dimensional profile, higher pressures are indicated by higher peaks

| *Key notes* | |
|---|---|
| **Pressure** | Pressure is defined as the amount of force per unit of area. |
| **Units of pressure** | While there are a number of common units used to report pressure, all are derived from the basic units of Newtons per metre squared (N/m$^2$). The **pascal (Pa)** is the pressure created by a 1 N force applied to a 1 m2$^2$ area. As a pascal represents a low force spread over a large area, it is more common to report **kilopascals (kPa)**, where 1 kPa is equal to 1000 pascals. **Atmospheric** or **barometric pressure** is the pressure caused by the weight of the air in the Earth's atmosphere. The **standard atmosphere** is the unit of pressure that is equivalent to 101325 Pa or 101.325 kPa. If measured using a mercury barometer, pressure is reported as the height (in mm) of a column of mercury (Hg), where 1 standard atmosphere equals 760 mmHg at 0°C. |

## FURTHER READING

The following article provides additional reading about the use of plantar foot pressure measurements to examine injury risk in military personnel.

1 Nunns, M., House, C., Rice, H., Mostazir, M., Davey, T., Stiles, V., Fallowfield, J., Allsopp, A., & Dixon, S. (2016). Four biomechanical and anthropometric measures predict tibial stress fracture: a prospective study of 1065 Royal Marines. *British Journal of Sports Medicine, 50*(19), 1206–1210.

# ELECTROMYOGRAPHY

*Adrian Burden*

Electromyography has been used since the 1960s to examine muscle function during sports and exercise to i) improve our understanding of how the body moves during complex actions; ii) evaluate the effects of muscle fatigue; and iii) predict the force in a muscle. Collectively, such studies have assisted with the design of more efficient and effective forms of training and rehabilitation. Electromyographical investigations involve synchronising the equipment involved with that used to record force and/or motion, to allow the timing of muscle activity to be known.

The fundamental unit of the neuromuscular system is the motor unit, which consists of the cell body and dendrites of a motor neuron, the multiple branches of its axon, and the muscle fibres that it innervates. Before tension is developed in a muscle, an action potential is generated by the motor neuron and propagates along the axon to the muscle fibres. At rest, muscle fibres have a potential difference of –60 to –90 mV with respect to the outside of the muscle. Propagation of the action potential along the muscle fibre reduces the potential difference (depolarisation) until it becomes positive (i.e. hyper-polarisation) before returning to resting levels once the action potential has passed (repolarisation). During a sustained muscle contraction, repeated depolarisation and repolarisation (i.e. the firing rate) can be in excess of 20 times per second.

Changes in the electrical potential of muscle fibres can be detected using electrodes placed either inside the muscle (fine-wire electrodes) or on the surface of the skin overlying the muscle (surface electrodes). The majority of sport and exercise science applications use surface electrodes which, depending on their size, can detect the signal from thousands of muscle fibres belonging to many (e.g. 20–50) different motor units. More recently, arrays containing many tiny surface electrodes have been developed that have the potential to be able to detect signals from fibres belonging to individual motor units. Regardless of the type of electrodes used, once the detected signal has been amplified and recorded, it is known as the **electromyogram or EMG.** A typical raw EMG recorded using surface electrodes is shown in *Figure G5.1*.

To accurately record muscle activation, it is necessary to use specific **electromyography equipment** that meets a number of very specific technological standards. It is generally accepted that the peak amplitude of the raw EMG recorded using surface electromyography does not exceed 5 mV and that its frequency spectrum is between 0 and 1000 Hz, with most of the usable energy limited to below 500 Hz and the dominant energy between 50 and 150 Hz. When detecting and recording EMGs a major concern should be that the fidelity of the signal is maximised. This is partly achieved by maximising the signal-to-noise ratio (i.e. the ratio of the energy in the electromyographical signal to that in the noise). Noise can be considered as any signals that are not part of the electromyographical signal and can include movement artefacts, detection of the electrocardiogram, ambient noise from other machinery, and inherent noise in the recording equipment. Maximising the fidelity of the EMG is also achieved by minimising the distortion (i.e. alteration of the frequency components of the signal) that it receives during detection and recording. Collectively, this provides evidence that both the equipment and

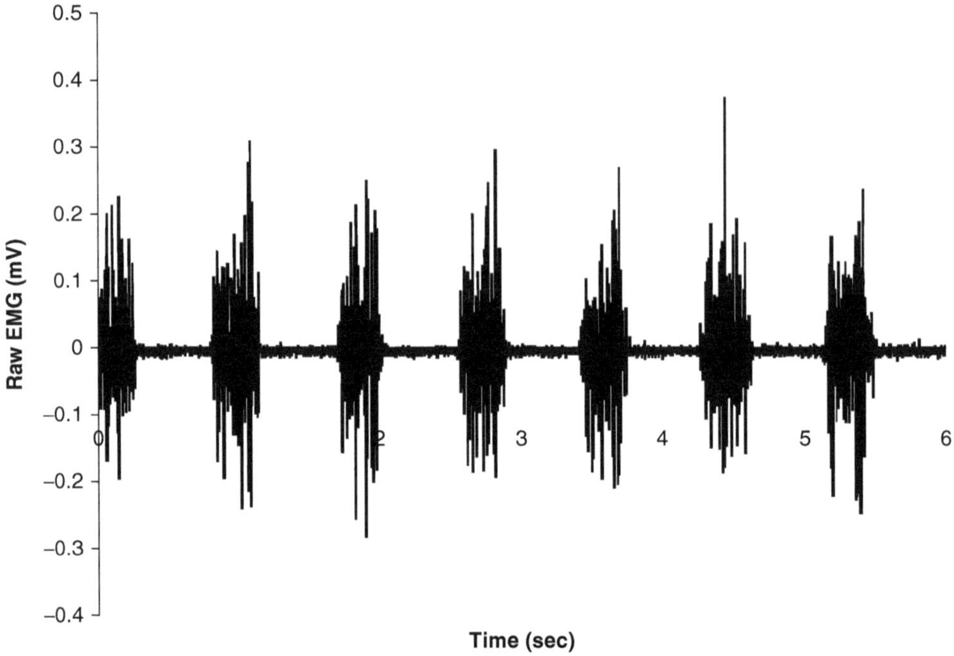


**Figure G5.1** Typical raw electromyogram (EMG)

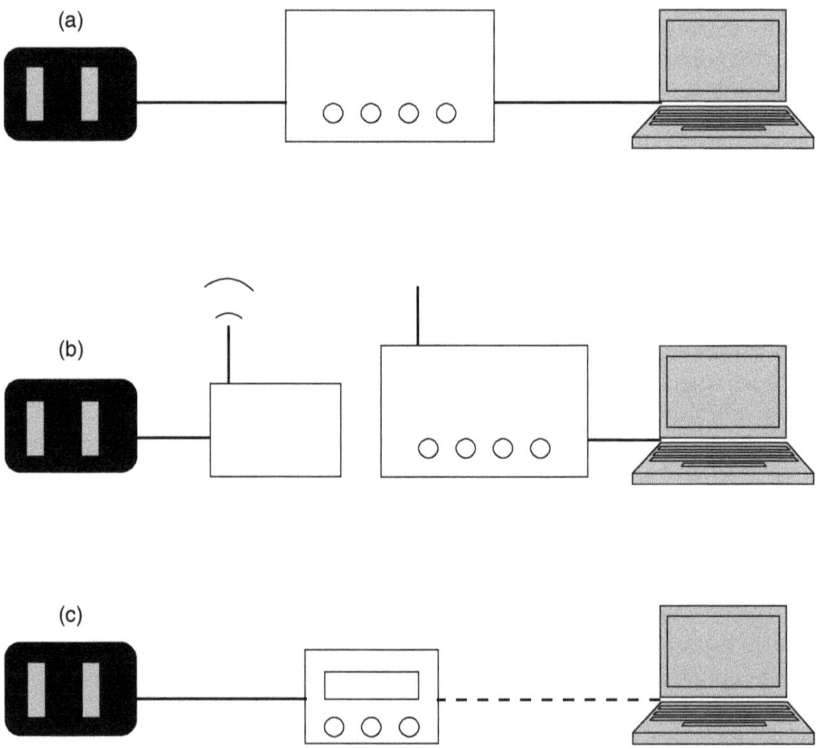


**Figure G5.2** Depiction of (a) hard-wired, (b) telemetry and (c) data logger electromyography systems

procedures used to detect and record EMGs have a major influence on their fidelity and should be given careful consideration.

Most commercially available electromyographical systems can be classified as either hard-wired, telemetry or data logger systems (*Figure G5.2*). A data logger or telemetry system is necessary if data is to be collected away from the main recording apparatus; however, data loggers typically do not allow online viewing of EMGs as they are being recorded and telemetry systems can be prone to ambient noise. Hard-wired systems do not suffer from these limitations, but obviously preclude data collection outside of the vicinity of the recording apparatus. The fidelity of the recorded EMG is dependent on the characteristics of the (differential) amplifier that is connected to the electrodes. These are listed below with recommended minimum specifications:

- Input impedance (>100 MΩ)
- Common mode rejection ratio (CMRR) (>80 dB [10,000])
- Input referred noise (<1–2 µV rms)
- Bandwidth (20–500 Hz)
- Gain (variable between 100 and 10,000)

While the requirements of amplifiers are generally agreed on by electromyographers, the configuration of electrodes and the material from which they are made are not. Some prefer pre-gelled silver/silver chloride (Ag/AgCl) electrodes that are circular with a diameter of 10 mm and a centre-to-centre distance of 20 mm. Others recommend silver bar electrodes that are 10 mm long, 1 mm wide, have a distance of 10 mm between them and are attached without the use of a gel.

When **recording EMGs**, the electrodes should be located between a motor point and a tendon to maximise the amplitude of the signal. If the motor point cannot be located using, for example, a stimulator, electrodes can be placed in the centre of the belly of the muscle while under contraction, although it should be recognised that this location could coincide with a motor point. Differential amplifiers subtract the signal detected by one recording electrode from that detected by the other. Thus, locating electrodes either side of a motor point will lead to the cancellation of symmetrical action potentials that are traveling in opposite directions from the neuromuscular junction and that reach the electrodes at approximately the same time. However, if both electrodes are placed to one side of a motor point, the signal is not cancelled to the same extent, as one electrode detects the signal slightly earlier than the other (*Figure G5.3*). Following the location of an appropriate site, the electrodes should be oriented along a line that is parallel to the direction of the underlying muscle fibres.

An improvement in the input impedance that is offered by many of today's amplifiers has diminished the need to reduce the skin–electrode impedance to, for example, below 10 kΩ. Skin preparation techniques that involve abrasion with fine sandpaper or scratching with a sterile lancet are, therefore, now largely redundant. Some preparation of the skin (to below 50 kΩ) is, however, still necessary in order to obtain a better electrode–skin contact and to improve the fidelity of the recorded signal. Typically, this involves cleansing the skin with soap and water and dry shaving it with a disposable razor. Additional rubbing with an alcohol-soaked pad and then allowing the alcohol to vaporise can be used to further reduce impedance in individuals with less sensitive skin. In addition to the recording electrodes, differential amplifiers require the use of a reference electrode that must be attached to electrically neutral tissue (e.g. a bony landmark). The degree of skin preparation given to the reference electrode site should be the same as that afforded to the muscle site. Most electromyographers also advise using an electrode gel or paste to facilitate detection of the underlying electromyographical signal. This can be accomplished either through the use of pre-gelled electrodes or by applying a gel or paste to the skin or electrode prior to attachment. Use of gel or paste is not always necessary when

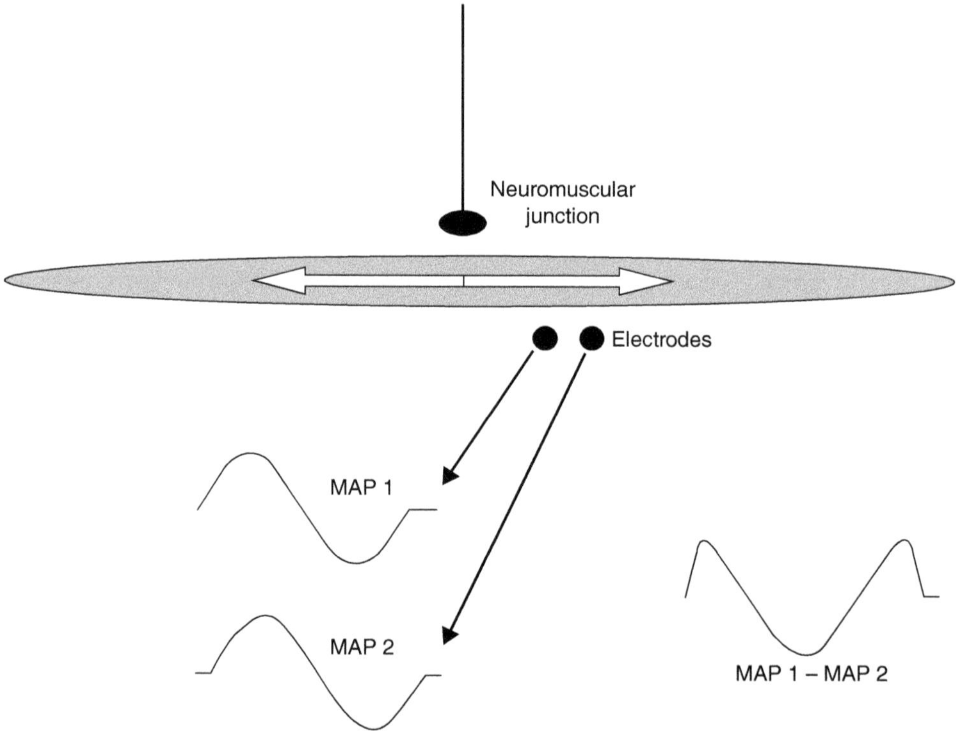


**Figure G5.3** Depiction of the recording of a single muscle fibre action potential (MAP) by a differential amplifier

using so-called "active electrodes" (i.e. those that are mounted onto a pre-amplifier). Here, the electrolytic medium is provided by the small amount of sweating that takes place when dry electrodes are applied to the skin.

Even if surface electrodes are placed close to the belly of the muscle it is possible that the detected signal may contain energy that emanates from other, more distant muscles. This is particularly pertinent when recording EMGs from muscles that are covered by thicker than normal amounts of subcutaneous fat, such as the gluteals and abdominals. The presence of cross-talk has traditionally been detected using functional tests that involve getting the participant to contract muscles that are adjacent to the one under investigation, without activating the one of interest. The detection of a signal from electrodes overlying the muscle of interest is, therefore, an indication of cross-talk. If possible, decreasing the size of the electrodes and/or the spacing between them reduces the chances of recording cross-talk. However, the most effective way of reducing it to almost negligible levels is to use a double differential, rather than a (single) differential amplifier. This has three, rather than two, detecting electrodes that are equally spaced apart, which calculates the difference between the signals detected by electrodes 1 and 2, and electrodes 2 and 3. These two (single differentiated) signals are then further differentiated (double differentiation) by the amplifier. This procedure works by significantly decreasing the detection volume of the three electrodes, and thereby filtering out signals from further away.

The Nyquist theorem (see the Sampling Theory section in this text) dictates that electromyographical signals, which are detected using surface electrodes, should be sampled at a minimum of 1000 Hz (ideally 2000 Hz) to avoid aliasing (i.e. loss of information from the signal). Sampling of the signal into a PC should also use an analogue-to-digital converter that has at least 12 bits (ideally 16 bits) to ensure that very small changes in muscle activity are able to be detected by the system.

If the electromyographer wishes to quantify the activity of a muscle or investigate how this activity changes over time, raw EMGs are processed in what is known as the **time domain**. This is usually achieved using either the Average Rectified EMG, Root Mean Square EMG or Linear Envelope, all of which provide an estimate of the amplitude of the raw EMG in µV or mV.

Calculation of the Average Rectified EMG value (ARV) involves first reversing all of the negative phases of the raw EMG (full-wave rectification). The integral of the rectified EMG is then calculated over a specific time period, or window (T), and the resulting **integrated EMG** is finally divided by T to form the ARV.

$$ARV = \frac{1}{T}\int_0^T |X(t)|\,dt$$

where:

$X(t)$ = the EMG signal

$T$ = the time over which the ARV is calculated

The Root Mean Square (RMS) EMG is the square root of the average power of the raw EMG calculated over a specific time period, or window (T).

$$RMS = \sqrt{\frac{1}{T}\int_0^T X^2(t)\,dt}$$

Both the ARV and RMS are recognised as appropriate processing methods and are commonly used by electromyographers, although the RMS yields a larger amplitude than the ARV (*Figure G5.4*).

As well as using a single calculation of the RMS or ARV, the raw EMG is often processed by making successive calculations throughout its duration, with the resulting series of values forming a type of moving average. For this, the duration (or width) of successive time windows (T) is selected by the investigator and can vary from 10 to 200 ms, depending on the duration and nature of the raw EMG. Selection of short duration window widths (e.g. 10–50 ms) may allow the detection of rapid alterations in activity, but the resulting curve will still resemble the rectified EMG (*Figure G5.5*). Thus, peak amplitudes from repetitions of the same task will remain highly variable. Adoption of longer widths (e.g. 100–200 ms) will reduce the variability of peak amplitudes, but the resulting curve will lose the trend of the underlying EMG (*Figure G5.6*). As such, rapid changes in muscle activity may go undetected. A possible solution is to use a moving average (either RMS or ARV) in which the time windows overlap instead of including discrete sections of data. Overlapping the windows by a progressively greater amount results in a curve that increasingly follows the trend of the underlying rectified EMG, but without the variable peaks that are evident in the rectified EMG (*Figure G5.7*).

The Linear Envelope is still also a popular processing method for use on EMGs from dynamic contractions. Similar to the moving average, this involves smoothing the rectified EMG with a low-pass filter and also results in a curve that follows the trend of the EMG. When using the Linear Envelope, the type, order and cut-off frequency need to be selected. Traditionally, a second-order Butterworth filter has been applied with a cut-off frequency between 3 and 80 Hz. Deciding on the cut-off frequency is similar to choosing the width and amount of overlap of the time window for the moving average. A low frequency will result in a very smooth curve, which will be unable to detect rapid changes in activation. Conversely, a higher frequency will closely follow rapid changes in activity and, hence, still bear the peaks that characterise the rectified EMG.

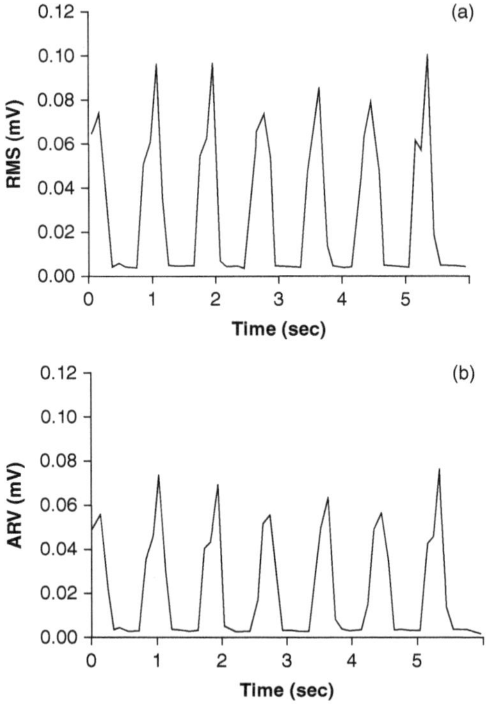


**Figure G5.4** The raw EMG from Figure G5.1, processed using (a) the Root Mean Square (RMS) and (b) the Average Rectified Value methods with a time window width of 100 ms

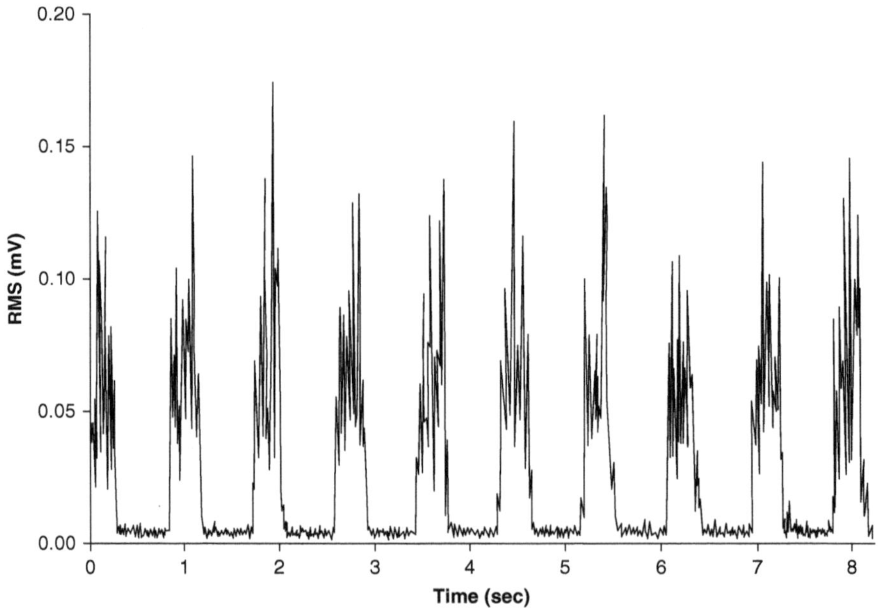


**Figure G5.5** The raw EMG from Figure G5.1, processed using the RMS method with a time window width of 10 ms

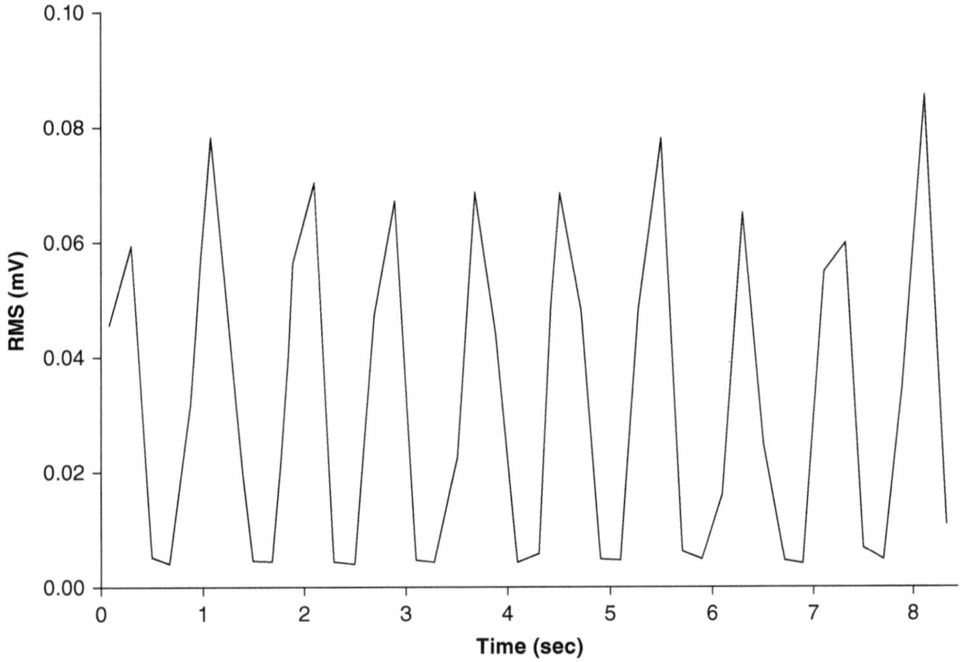


**Figure G5.6** The raw EMG from Figure G5.1, processed using the RMS method with a time window width of 200 ms

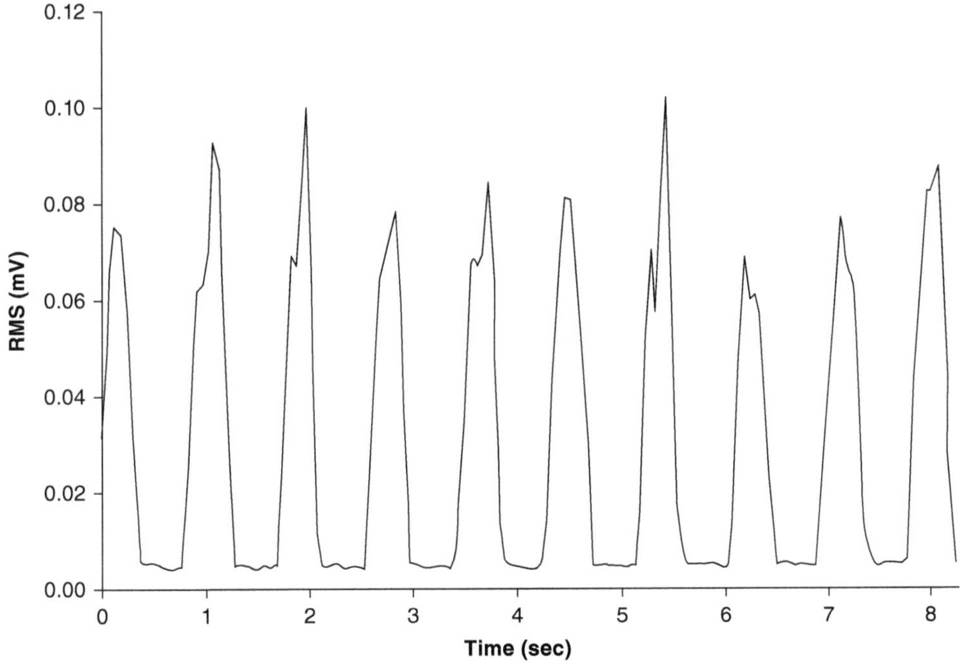


**Figure G5.7** The raw EMG from Figure G5.1, processed using the RMS method with a time window width of 100 ms overlapped at 50 ms

After processing, the EMG is often used to estimate when a muscle is active (i.e. on) or inactive (i.e. off). In order to determine the amplitude threshold at which the muscle can be considered to be active, the baseline EMG (or noise) is typically treated as a stochastic (or random) variable. The mean of this baseline is, for example, calculated over 50 ms and the muscle is deemed to be active when the EMG amplitude exceeds two standard deviations above the mean baseline activity for 20 ms or more.

The EMGs that are processed in the time domain can only be compared with those recorded from the same muscle at another time without the removal of electrodes (i.e. during the same testing session). Relocation of electrodes over the same muscle on subsequent occasions will invariably result in the detection of signals from different motor units. The skin–electrode impedance will also differ between sessions, regardless of how well skin preparation techniques are adhered to, which will affect the shape of the underlying signal. These and other factors will, therefore, affect the amplitude of the processed EMG. The amplitude of EMGs recorded from the same muscle on different occasions, as well as from different muscles and different individuals, cannot therefore be compared directly, even if they have been processed using the same method. This problem can be solved by **normalising EMGs** after they have been processed, which involves expressing each data point of the processed EMG from the specific task as a proportion or a percentage of the peak EMG from a reference contraction that has been processed in the same way. The reference contraction has traditionally been an isometric sub-maximal or maximal voluntary contraction (MVC) of the same muscle. More recently a dynamic MVC had been used from, for example, a maximal sprint or jump. Use of the EMG from an isometric or dynamic MVC has the added potential of revealing the activity of a muscle, in relation to its maximal activation capacity, during a specific task (*Figure G5.8*). However, in order to achieve reliable and valid EMGs from isometric or dynamic MVCs, individuals must practise them extensively. Previously unrehearsed MVCs will result in torque or force, and hence muscle activity, that is far from maximal.

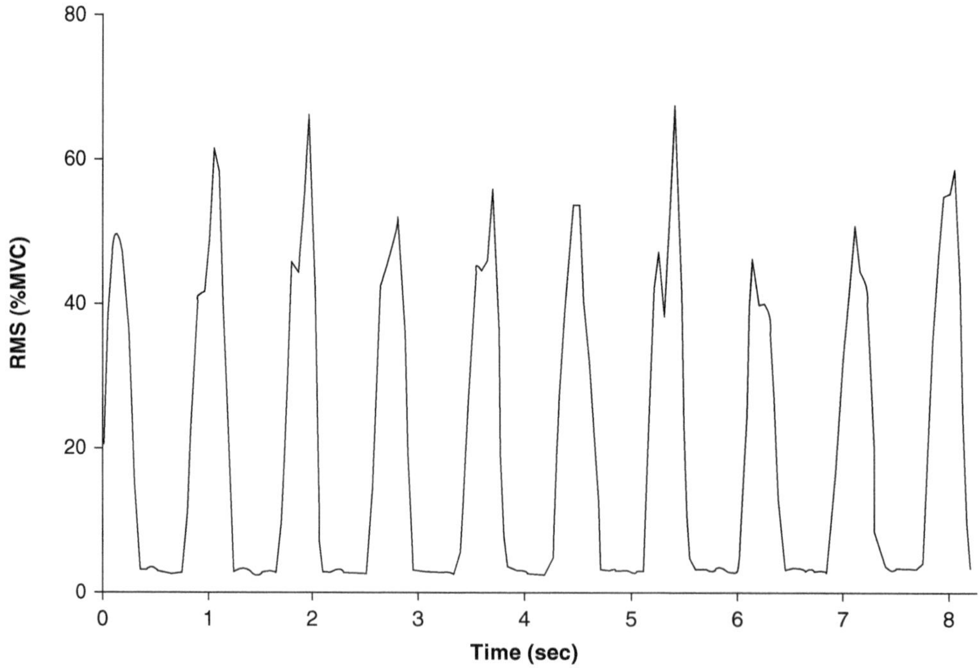


**Figure G5.8** The processed EMG from Figure G5.7 normalised to a maximal voluntary contraction (MVC) of the same muscle

Normalisation can also be used to reduce inter-individual variability of EMGs recorded from the same task. It is now well established that dividing each data point within the task EMG by either the mean or the peak EMG from the same task is the most effective way of improving group homogeneity. However, due to the nature of the denominator used in their normalisation equation, normalising EMGs in this manner cannot be used to compare EMG amplitudes between different muscles and individuals.

Processing EMGs in the time domain is useful for evaluating attributes, such as the amplitude or timing of muscle activity, but **frequency domain processing** is often used to investigate changes in the signal that accompany muscular fatigue. It is now well established that fatigue is associated with a compression of the frequency spectrum towards the lower frequencies (*Figure G5.9*) that occurs largely due to a decrease in the conduction velocity of action potentials.

Transformation of a raw EMG from the time domain to the frequency domain is typically achieved using a Fast Fourier Transform (FFT), which is usually performed between 0.5 and 1 s. The output of the FFT is typically represented as the power spectrum density (PSD), which shows the relative magnitudes of the range of frequencies present in the raw signal (*Figure G5.9*). To quantify the PSD, one of two parameters is commonly used; either the median frequency (MDF) or the mean frequency (MNF). The MDF is defined as the frequency that divides the PSD into equal halves, while the MNF is calculated as the sum of the product of the individual frequencies and their own power divided by the total power. The MDF is less sensitive to noise and more sensitive to spectral compression than the MNF and, as such, is more commonly used.

Regardless of which parameter is chosen, it is typically obtained from consecutive time windows, to enable changes in the signal that occur as a consequence of fatigue to be monitored. Successive values from the contraction period are then analysed using (linear) regression, with the intercept of the regression line being the initial frequency and the gradient representing the

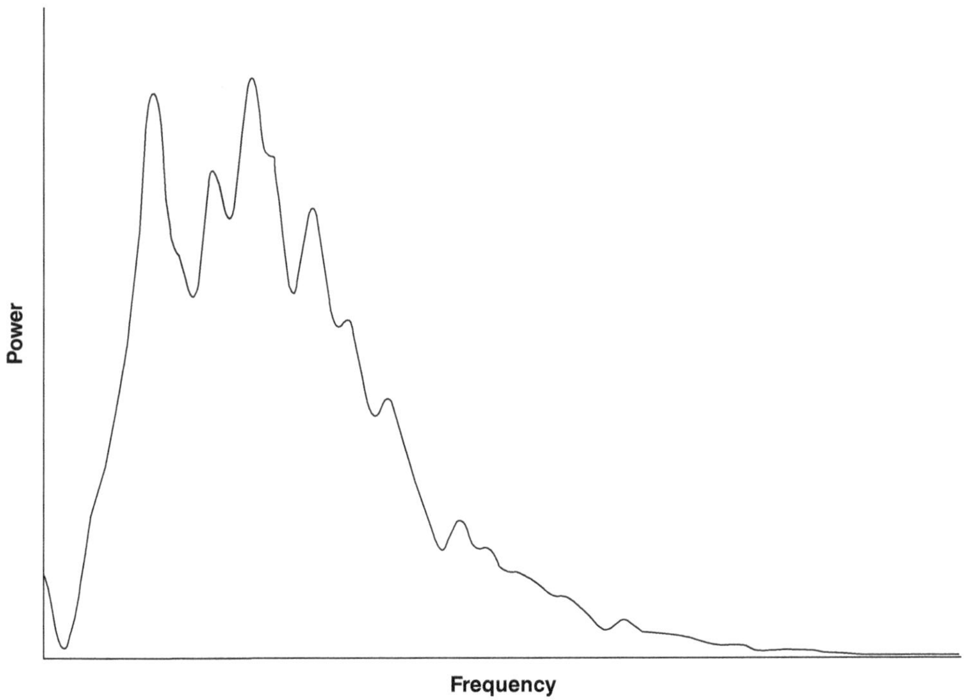


**Figure G5.9** Power spectrum density obtained from a section of the raw EMG in Figure G5.1

fatigue rate. In addition to fatigue, the frequency spectrum of the raw EMG is affected by a host of other factors. Similar to analysis in the time domain, specific frequencies (e.g. MDF) cannot, therefore, be compared directly when they are calculated from EMGs recorded from different muscles or individuals, or from the same muscle when the electrodes have been re-applied. However, comparisons can be made between the gradient of the regression line in order to investigate differences in fatigue rates between different muscles, occasions or individuals.

The FFT should only be used on EMGs that display high stability; that is typically those recorded from sustained force isometric contractions between 20 and 80 per cent MVC. Where EMGs are recorded from dynamic contractions they generally have less stable signals, largely as a consequence of recruitment and de-recruitment of different motor units. As such, the FFT should only be used in such circumstances when signal stability is reasonably high and parameters (i.e. MDF or MNF) should only be calculated at the same phase of repetitive cyclic events. The problem of obtaining spectral parameters from non-stationary signals has largely been overcome by using the joint time–frequency domain approach which estimates the change in frequency as a function of time. The simplest method that conforms to this approach is the short-time Fourier transform, which splits the EMG into small continuous or overlapping time windows, applies the FFT to each window and calculates the MDF or MNF as above. Recently, more sophisticated methods of time–frequency domain analysis have been applied to EMGs.

*Key notes*

| | |
|---|---|
| **The electromyogram (EMG)** | Electromyograms (EMGs) are recordings of electromyographical signals that emanate from muscle fibres prior to their contraction. Surface electromyography in sport and exercise biomechanics invariably involves recording signals from muscle fibres belonging to many different motor units. |
| **Electromyography equipment** | EMGs are usually detected using a pair of recording electrodes that are connected to a differential amplifier, which amplifies the difference between the signals detected from the two electrodes. Amplifiers can be connected directly to a PC for storage of EMGs (hard-wired system). Alternatively, amplified signals can be sent via radio waves to a receiver connected to a PC (telemetry system) or stored in a memory card before later being transferred to a PC (data logger system). |
| **Recording EMGs** | Surface electrodes should ideally be located between a motor point and a tendon, parallel to the direction of the underlying muscle fibres. Prior to this the skin should be shaved, washed, and ideally rubbed with an alcohol wipe to reduce skin–electrode impedance. Unwanted signals emanating from other adjacent muscles (cross-talk) should be minimised prior to recording EMGs. |
| **Time domain processing** | To quantify the magnitude of muscle activity over time, the raw EMG can be processed using the Average Rectified Value (ARV), Root Mean Square (RMS) or Linear Envelope. Both the ARV and the RMS are typically calculated over time windows that have a width of between 10 and 200 ms. The Linear Envelope typically constitutes a second order Butterworth filter with a cut-off frequency between 3 and 80 Hz. |

| | |
|---|---|
| **Normalising EMGs** | EMGs that have been processed in the time domain can only be compared with those from the same muscle at different times without removal of electrodes. To compare EMGs between different muscles and individuals, processed EMGs should be normalised by dividing them by the EMG, processed in exactly the same way, from a reference contraction. This can be an isometric sub-maximal or maximal voluntary contraction or a dynamic maximal voluntary contraction. Alternatively, if the aim is to improve the homogeneity of EMGs from a group of individuals, then each processed EMG should be normalised by dividing it by the mean or peak processed EMG from the same task. |
| **Frequency domain processing** | The frequency content of the raw EMG can be revealed by a Fast Fourier Transform (FFT). Typically, the FFT is calculated over intervals of 0.5–1 s, and the median frequency (MDF) is obtained from the resulting Power Density Spectrum. Changes in the MDF over time have traditionally been used as a measure of muscle fatigue. More recently, concerns over the use of the FFT on non-stationary signals has led to the development of more sophisticated joint time–frequency domain analysis techniques (e.g. wavelet analysis). |

## FURTHER READING

1 Reaz, M. B. I., Hussain, M. S., & Mohd-Yasin, F. (2006). Techniques of EMG signal analysis: Detection, processing, classification and applications. *Biological Procedures Online, 8*(1), 11–35.

# G6 ISOKINETICS

*Paul Grimshaw*

The term **isokinetic** is a word used to describe a muscle contraction when the rate of movement (velocity) is held constant. The term **dynamometer** is a word used to describe an apparatus for measuring force or power, especially during muscular effort. An **isokinetic dynamometer** is a device, which is usually electro-mechanical (both electrical and mechanical) in operation, that assesses isokinetic torque curves of muscles during different movement patterns.

Isokinetic dynamometers are used extensively within many forms of human movement and may even be used within sports settings, as exercise devices that develop specific muscles and/or muscle groups or within rehabilitation and medicine, to condition muscles following injury or muscle wasting.

There are many different commercial companies that produce isokinetic dynamometers and they can be used to test and assess almost any joint of the human body. Some examples of the more common commercial isokinetic dynamometer testing machines include Kin-Com, Ariel, Cybex, Biodex and Akron. *Figure G6.1* shows one of these commercial devices.

All these machines will incorporate a device that will control the rate of movement of the arm-crank of a machine that the limb or lever of the body is attached to. *Figure G6.2* helps to illustrate this in more detail.

Isokinetic devices can be set up to examine almost any joint within the human body. *Figure G6.2* shows an application on the shoulder during a flexion and extension movement. The machine, in this case, would assess the **agonist** and **antagonist** shoulder muscle function. The **agonist muscle** is defined as the muscle that contracts while another muscle resists or counteracts its motion. The **antagonist muscle** is defined as the muscle that offers a resistance during the action of the agonist muscle. This muscle contraction can be in the form of both a concentric and an eccentric type of contraction. **Concentric contraction** is defined as when muscle tension is developed to **accelerate** a lever arm or limb. In this case, the muscle contracts concentrically and the fibres of the muscle **shorten** (i.e. origin and insertion are drawn together). An **eccentric contraction** is when muscle tension is developed to **decelerate** a lever arm or limb. As the muscle contracts eccentrically, its fibres **lengthen** and the origin and insertion points are drawn apart. During the shoulder movement portrayed in *Figure G6.2*, the machine would assess the torque/strength possessed by both the flexor (pectoralis major and deltoid) and the extensor (latissimus dorsi and teres major) muscles of the shoulder joint.

Muscular contribution to joint stability is invaluable and it is helpful in the prevention of injuries. The knee joint is a typical example where muscular contribution to stability is essential. Knee joint muscular stability is provided by the muscles, which are termed the quadriceps (extensors) and the hamstrings (flexors). Following both injury and surgery to the knee joint (such as in the case of a ligament replacement), the muscles usually undergo severe muscle wasting. The isokinetic dynamometer is a device that can provide a measure of the amount of muscle wasting and, hence, a measure of the rehabilitation that is needed to regain the muscle balance (strength).

**Figure G6.1** The isokinetic dynamometer

Source: Serafino Mozzo / Shutterstock.com

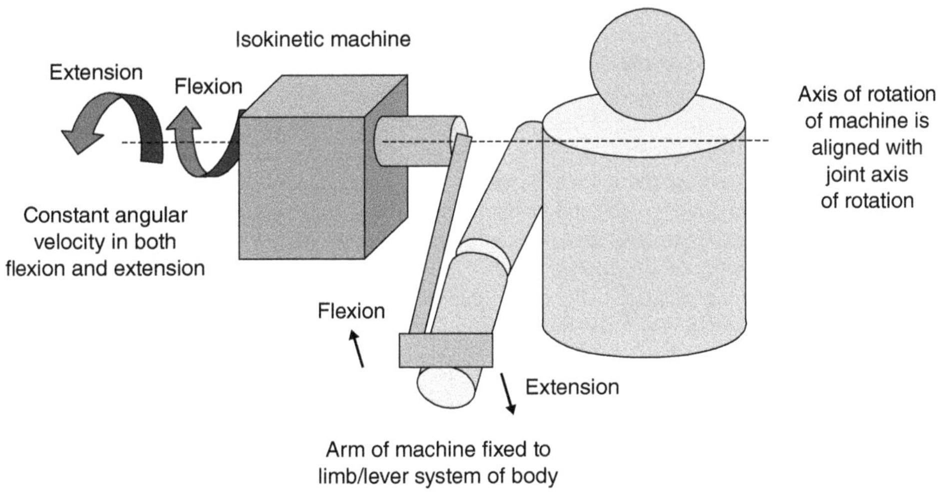


**Figure G6.2** Shoulder joint testing using an isokinetic dynamometer

The isokinetic device usually has the ability to test the limb in one of three testing modes: isokinetic, isometric and isotonic. **Isokinetic** assessment involves a fixed speed with a variable resistance, which accommodates the muscles' ability to generate force. It is characterised by a constant velocity at a pre-selected rate. The resistance offered by the machine varies to match the exact torque applied by the muscle. The unique application of this form of testing is that the joint is tested throughout its full range of movement (i.e. a situation that is similar to the actual condition within sport or exercise).

**Isometric** or static testing is a situation when the muscle develops tension and the fibres shorten, but no movement of the lever arm takes place. Since velocity is held constant (zero), the resistance automatically varies to match the force applied. An example of an isometric muscle contraction would be pushing against a wall (there is obviously force involved but basically once you have reached your maximum push capacity the arms and the wall do not move).

An **isotonic** form of testing utilises either a constant or pre-selected variable form of resistance. Due to changes in the musculoskeletal leverage system, the muscle tension needed to move the load changes as the joint moves to different positions (i.e. gaining and losing mechanical advantage). Hence, in this example, the muscle is only loaded at the joint's weakest mechanical point. An example of this type of muscle contraction would be in the case of moving the weights and bar during an arm curl exercise involving free weights.

Isokinetic dynamometers are used in medicine when the limb requires regular exercising after surgery to restore muscle power and prevent any seizing or limitation of movement. They can also be used to monitor the effectiveness of physiotherapy strength-training programmes. In sport, the device can be used to exercise and develop the power (strength) of certain muscle groups to an optimum level. Maximal exercise can take place throughout a full range of movement. The machine can be adjusted to simulate the exact movement used in the specific sport and even monitor the progress of an athlete in training and/or rehabilitation.

Within the modern gymnasium or training area, there are often many machines that are labelled with the term isokinetic. However, it is important to identify that these machines are often not the same devices as the sophisticated rehabilitation and training models produced by companies such as Kin-Com and Cybex. The machines seen in gymnasiums or fitness centres are modified exercise devices, which use the principle of cams (where they are able to load and unload at specific points) to create different mechanical loading patterns during the performance of an exercise. In this example, it would be unusual to see a true isokinetic dynamometer in this environment, primarily due to their extreme cost.

The true isokinetic dynamometer machine incorporates an electro-mechanical device, which keeps the limb at a constant pre-determined angular velocity. Any effort applied encounters an equal and opposite reaction force. The resistance developed is in proportion to the amount of force exerted. A maximal effort can be experienced as if a maximal load were being applied at all the points throughout the arc of motion. The anatomical axis of rotation of the joint (where the torque is created) is aligned with the machine's axis of rotation (where the torque is transmitted). Various arms and levers of adjustable length are available, so the device can accommodate most joints of the human body for a variety of subjects of different sizes. The results are presented as a measure of torque against angular displacement. Torque values can be plotted against the position of the limb at any instant. *Figure G6.3* depicts an example of a torque against angular displacement trace for a knee flexion – extension assessment using an isokinetic dynamometer.

Considering *Figure G6.3*, it is possible to see that the trace begins (on the left) at 90 degrees of knee flexion. As the quadriceps muscles extend the leg to full extension (180°), the device registers the torque generated by the quadriceps. Resistance is offered from the machine at every single point (angular position or displacement) throughout the full range of movement from 90 degrees to full extension (i.e. the 180° position). This resistance matches the torque generated by the muscles. During testing, the limb (technically the arm of the machine) is set to move at a constant pre-determined angular velocity. This angular velocity can be as low as

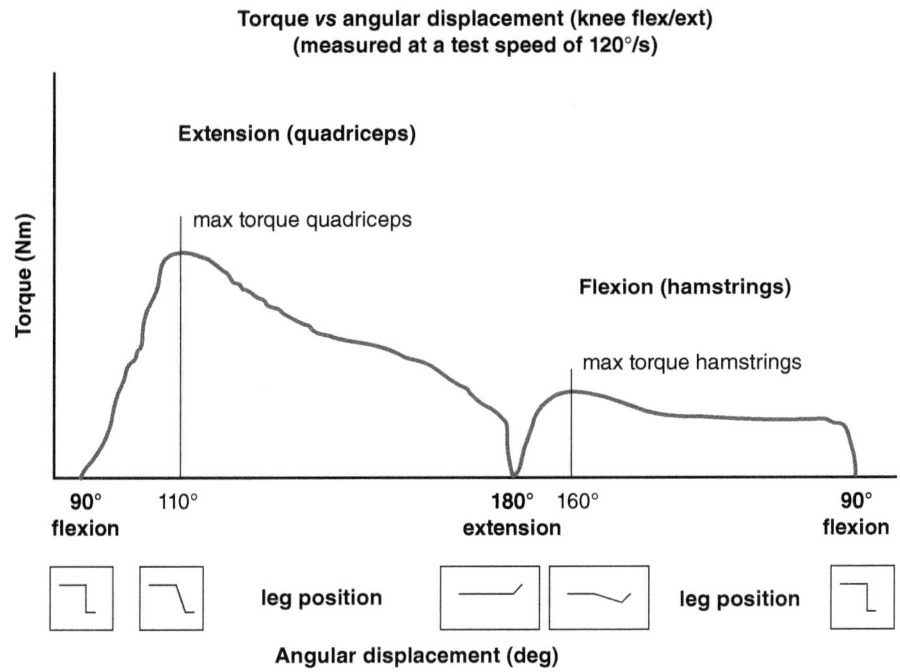


**Figure G6.3** Torque *vs* angular displacement graph from isokinetic testing

30°/s, but can be increased to in excess of 240°/s. Angular velocities that are low (30°/s) are said to measure the endurance capacity of the muscles, while higher angular velocities (e.g. >240°/s) measure the maximum strength capacity of the muscles. For example, it would be difficult for a muscle to maintain maximum force (torque) against a lever arm that is moving very slowly at only 30°/s (i.e. the contraction and force application would be required for a longer period of time). Conversely, at a faster speed of 240°/s, the muscle will reach its maximum contraction very quickly and it will not have to maintain this for too long a period of time. For this reason and more, the torque traces produced by the muscles at different speeds (angular velocities) will be different. *Figure G6.3* illustrates knee flexion and extension torque measured at 120°/s and it is possible to see that the maximum torque (force) generated by the quadriceps (left-hand trace) occurs at approximately 110° of knee flexion (or after only 20° of extension movement from the initial 90° flexion position). Considering the torque created by the hamstrings (right-hand side of the figure), it is possible to see that the maximum torque generated is less than that generated by the quadriceps and it occurs in a distinctly different pattern. For example, the quadriceps produce high acceleration and peak torque, whereas the hamstrings produce less acceleration and peak torque that is maintained for a longer time period. This indicates that the device is clearly able to show the differences between both the strength and function of the two muscle groups that support the knee joint.

Isokinetic dynamometers are able to measure several variables in relation to muscle function during both exercise and rehabilitation. For example, such variables as **peak torque, angle of peak torque, time to peak torque, agonist to antagonist muscle strength ratios, work done, power** and **torque decay rate** are some of the more common on most modern machines. In addition, it is also possible to have different test speeds for different functions of the muscle. For example, the quadriceps can be assessed at a speed of 120°/s during extension and, at the same time, the hamstrings can be tested at 30°/s during their flexion movement.

However, within isokinetic exercise and testing, two problems can occur if they are not accounted for. These are identified as follows:

1 During the test the subject must exert a maximum effort throughout the full range of movement. For example, for the machine to maintain the constant pre-determined test speed (angular velocity) the subject must maintain a maximum effort to the lever arm. In older type isokinetic machines, it was possible for the subject to exert an effort that was less than maximum and thus move the lever arm at a test speed that was lower than that of the pre-selected rate (i.e. the lever arm cannot move faster than the set speed but it can move slower). This, however, is now corrected in the modern machines by the data showing the exact speeds of the lever arm throughout testing and whether the testing speed is actually achieved.
2 In the context of testing using an isokinetic dynamometer, it is possible to observe that, sometimes, the movement under test will be acting against gravity (like in the case of knee extension) and sometimes the movement will be acting with gravity (like in the case of knee flexion). In this case it would be necessary to have to correct the torque generated by the muscles for the effects of resistance or assistance due to gravitational acceleration. Again, most modern isokinetic machines have this correction factor automatically built into the data collection.

Given that most modern isokinetic devices have such factors built into their software, it is now easier to correct a subject's slower movement patterns via encouragement and to account for gravitational acceleration by having the system weigh the subject's limb before testing begins. If encouragement is unsuccessful at improving the subject's movement patterns, the researcher will at least be aware of the potential errors that may be associated with not reaching the actual test speed values.

Isokinetic devices are a significant part of modern exercise testing and prescription and they are becoming essential for rehabilitation following injury. Unfortunately, however, the significant cost of these devices limits their availability to only technologically-advanced and financially-viable university, hospital or private industry laboratories.

*Key notes*

| | |
|---|---|
| **Isokinetics** | A movement that involves a fixed speed and variable resistance applied throughout the full range of movement of the limb/lever system. |
| **Isokinetic dynamometer** | This is a device that is used either to assess or to exercise the agonist and antagonist muscles that are located about the joints in the human body. The device can be used to test almost all the joints of the human body. The machine provides a constant pre-determined angular velocity of movement. The limb of the body moves the lever arm of the machine at this pre-determined angular velocity. Throughout the full range of movement, the limb will experience an equal and opposite resistance to the force it is applying to the lever arm in order to move it. There are various types of isokinetic dynamometers available and these are usually presented in the form of rehabilitation devices that are located in hospitals or universities. The device can examine the effectiveness of a strength-training programme that is prescribed following surgery or injury to a joint. Modern isokinetic dynamometers have the ability to test the limb at various speeds and in either an isokinetic, isometric, or isotonic mode of assessment. |

| | |
|---|---|
| **Isometric** | This is usually a muscle contraction during which the joint or position of the limb is held in a fixed angular position. As such, while the muscle develops tension, there is no change in the length of the muscle. |
| **Isotonic** | This type of exercise involves an equal amount of tension being developed in the muscle throughout the entire movement. As the muscle develops an equal amount of tension while its length continues to change, this situation is technically very difficult to achieve accurately in practice. |
| **Application** | The isokinetic dynamometer can be used in sport to exercise muscles to an optimum strength, or in medicine where it can be used to assess the degree of muscle wasting following injury or surgery to a limb/joint. The device can develop strength in a muscle throughout a full range of movement. The various modes of testing and various pre-determined angular velocities available allow a complex combination of both assessment and exercise. These devices are expensive and are rarely seen in gymnasiums (fitness centres) that are solely used for exercise purposes. |

## FURTHER READING

The following resources provide additional reading on the concept of isokinetics in sport and human movement.

1 Ellenbecker, T. S., & Davies, G. J. (2000). The application of isokinetics in testing and rehabilitation of the shoulder complex. *Journal of Athletic Training, 35*(3), 338–350.

2 Kellis, E., & Katis, A. (2007). Quantification of functional knee flexor to extensor moment ratio using isokinetics and electromyography. *Journal of Athletic Training,* 42(4), 477–485.

# G7 SPORTS EQUIPMENT

*Paul Grimshaw*

Sporting equipment can significantly affect the performance of athletes either detrimentally through injury or spectacularly through new world records. The future will see significant changes in sports surface construction, sport environment development and sport equipment design and technology that will prevent records being continually broken and yet sometimes assist in their achievement. Athletes are getting physically fitter, much stronger and significantly faster and their specific anthropometry (body shape and size) is rapidly changing. As a result, the technical demands of the event and the equipment will continue to dramatically change and it will be interesting to see how the future will respond to these changes. All this technology inevitably poses important questions, such as:

> How far will we go to enhance performance and when does the point arrive that we are measuring the equipment and surroundings rather than the individual athlete?

Or is it that this day has already arrived and it is the "fast" swimming pools like Sydney or Athens and the "sprint" track in Tokyo that will decide how medals are won and lost? This section shows some examples of how sporting equipment has developed to accommodate changes in athlete body shape and the technological demands of the sport. For clarity the section will look at examples from cycling and tennis.

## CYCLING

Technological advances in **cycling** have progressed primarily from the growth of the sport of triathlon (such as aerobars, steep seat angled frames (*Figure G7.1*) and forward riding positions). This has caused the traditional time trial racing posture to be subjected to much experimentation and change. The posture adopted on the bike is a direct function of the cyclist's body configuration and shape (hip, knee and ankle angles, body position (cyclist relative to pedals) and body orientation (trunk angle and horizontal)).

The importance of posture and anthropometry probably first emerged as a significant factor in 1989 when Greg LeMond cycled to a 57-second victory in the final time trial of the Tour de France. LeMond attributed this success and significant average speed of 54.545 km/h to his new aerodynamic body position.

In 1992, at the Barcelona Olympic Games, Chris Boardman shocked the cycling world with an astounding win in the 4000 m individual pursuit event. This win with its new "cycling position" and "technological bike" from Lotus (essentially a sports car company) created a revolution in bike design characteristics. The position adopted by Boardman allowed the rider to rest the arms on the tri-handlebar arrangement, which allowed an almost perfect time-trialling combination of athlete and bike, working together as one unit (*Figure G7.2*).

Advances in bike design and body position continued to evolve rapidly after this historic ride by Chris Boardman in 1992 and many new cycling positions and innovative bikes were

**Figure G7.1** The modern triathlon bike

Source: Radu Razvan / Shutterstock.com

suddenly seen to be developing. However, in 1993, while riding a homemade bike put together from what is often described as "parts of a washing machine", one rider, Graeme Obree, established a riding position that was to stay and change the shape of time trialling for some time. The "Obree"' position allowed the rider to rest the torso on the arms, which were tucked away underneath, thus totally eliminating them from the aerodynamic equation and reducing the drag by 15 per cent. This theoretically would give a speed gain of more than 2 km/h at speeds of 50 km/h. In addition, as a result of the particularly narrow bottom bracket he managed to reduce the drag even further by riding almost totally "knock-kneed" (*Figure G7.3*).

In 1993, Graeme Obree set the 4000 m single pursuit time at a new world record of 4 minutes 20.9 seconds, beating the record previously held by Chris Boardman. The following year saw the "Obree" position adopted by many other riders together with the same and also often modified bike design (with a chest pad added to aid comfort on the longer rides). Despite this interest and acceptance from many riders, this new bike design and anthropometric racing position could still not conquer the one-hour speed record set previously by Chris Boardman on the Lotus bike. Nevertheless, on 27 April 1994, Graeme Obree returned using the new "Obree" position, to set the one-hour speed record at an astounding 52.513 km/h.

Unfortunately, this success was short-lived as in May 1994 cycling's world governing body, the UCI (Union Cycliste Internationale), decided that the position would be banned from the forthcoming world championships. The position, however, was still used in the triathlete circuit where the UCI rules were not valid and many riders were seen taking from three to five seconds off each of their kilometre times.

Following the UCI ban, Graeme Obree returned to the cycling circuit in 1995 with a newly modified position and bike. This time, Obree used conventional equipment: a normal bike with

**Figure G7.2** Chris Boardman and the Lotus bike

Source: Allstar Picture Library / Alamy Stock Photo

a particularly long stem, with aerobars extended to their limits. The arms were now in an outstretched position with the hands about 30 cm in front of the front wheel hub (wheel spindle). This position was soon adopted and termed the "Superman position".

During the Atlanta Olympic games in 1996, many other riders used the "Superman Position" and it achieved great success in both the men and women's 4000 m individual pursuit titles. The world record for 4000 m had now been improved to a time of 4 minutes 19 seconds.

In the same year, at the world championships in Manchester, England, Chris Boardman returned with an adaptation to his Lotus bike of 1992, which now included a custom-built handlebar that allowed him also to adopt the new "Superman position". This combination of machine, athlete and aerodynamics was too much for the competition and the cycling world saw the 4000 m individual time trial record plummet to a staggering 4 minutes 11.114 seconds. Then, only one week after the world championships, Chris Boardman added to this spectacular performance by setting a new one-hour speed record of an outstanding 56.375 km/h.

To add to the controversy and confusion, the UCI subsequently decided to ban this "Superman position" and created rules that stated the handlebar must not exceed the front hub by more than 15 cm. In addition, they also added that the distance between the front hub and the bottom bracket (pedal spindle) could now be 75 cm. Considering this distance was only 60 cm for most conventional bikes, this new value of 75 cm appeared to put the specifications at a ridiculous level and only required the front part of the frame to be extended for a rider to adopt the banned "Superman position". However, despite this possibility, the UCI's specifications still made it particularly difficult to achieve precisely the position previously adopted by Graeme Obree, Chris Boardman and now many other riders.

**Figure G7.3** Graeme Obree and his modified bike, which has a very short reach that allowed him to tuck his elbows under his chest in a position that became known as the "Obree" riding position

Source: Anthony Collins Cycling / Alamy Stock Photo

The following years saw a considerable number of changes to the rules presented by the UCI and in 1999 they changed the specifications yet again. This time the maximum distance permitted between the front hub and the bottom bracket was to be 65 cm. Unfortunately, this had a dramatic effect and it meant that it was only possible for smaller riders to have any chance of achieving the previously successful "Superman position".

Cycle specifications and athlete positioning on the bike continue to this day to procure even more constraints and the rules around them are now substantial. For example, as from 2016 the rules state that "The distance between the vertical passing through the bottom bracket spindle and the front wheel spindle shall be between 54 cm minimum and 65 cm maximum." This means that handlebar extensions may now only project between 10 and 21 cm past a vertical line which passes through the front wheel spindle, hence many riders have no possibility of getting into any comfortable aerodynamic posture and the previous record-breaking "Superman position" is, for these riders, perhaps totally impossible to achieve (*Figure G7.4*). The current world record (at the time of print) for the individual pursuit track event is 4 min 07.251 seconds, which was achieved by Ashton Lambie (USA) in 2018 at the Pan American Championships in Mexico.

Such significant changes in the specifications by the UCI and the adoption of only "standard" frame designs will mean that the anthropometry of world-class riders will change. As a result, record breakers will be required to have a stature that suits the bike specifications and they will need to achieve some degree of aerodynamic control within these new specifications. Perhaps it may even be the case that the old records set with the "Superman position" will become, like many other records in numerous disciplines, set in conditions that can now not

**Figure G7.4** The modern time trial bike and rider position

Source: Radu Razvan / Shutterstock.com

be repeated. That is, they become a part of sport history, when athletes, coaches and spectators will talk of records that were achieved at a specific track, with a particular bike and by a certain type of athlete.

## TASK

Search the Internet to find the latest UCI rules for the different types of bikes used in competitive cycling.

## TENNIS

The **tennis** serve is probably the most important stroke in the game. The action involves a smooth, coordinated movement of different body parts delivered at an optimum height. The body adopts a link system of movement that is initiated from the legs and sequentially produces increased velocities at each subsequent segment in the kinetic chain (e.g. legs to pelvis, pelvis to torso). Finally, this velocity is transferred to the hand and racket, resulting in the maximum desired power of service.

According to empirical research, the tennis serve makes up to 20 to 30 per cent of all shots in both singles and doubles matches and accounts for 12 per cent of winning shots on clay, rising to 23 per cent of winning shots on grass. Furthermore, it is the only stroke that the opponent cannot directly affect and, therefore, it places the server at a distinct advantage.

The inter-relationships between different body positions, different types of service delivery (flat, topspin and slice), different racket trajectories and segment velocities have all been analysed. However, it is clear that there is limited published research that directly measures changes in these parameters due to contrasting anthropometry (*Figure G7.5*).

For an effective execution of the flat serve, it is necessary for the ball to be contacted as high above the ground as possible. This allows the ball to be struck at a flat or downward trajectory. According to researchers, this requires a contact height of approximately three metres to hit the service area with an adequate margin for error. However, few players can achieve this height, so the optimum serve becomes a compromise by varying downward projection angles by between 4 and 7 degrees. *Figure G7.5* illustrates the service action and identifies the height of ball contact.

In order to achieve maximum contact height during the serve, almost maximum extension of the joints such as the knee, hip and elbow (180°) is required. Research has suggested that contact heights in effective serves are approximately 150 per cent of the player's standing height, with joint extensions for the elbow and knee at approximately 173 and 165 degrees, respectively. Further research in the area has identified that a number of elite players are actually off the ground at the moment of ball and racket contact during the service action. This results from the rigorous leg drive towards the ball, which increases contact height.

It is suggested by some that the taller players will serve harder, with a higher velocity and with a much more controlled and "flatter" trajectory. The smaller players will therefore have to assume the "up" and "out" service method and, hence, hit the ball at a much-reduced velocity.

**Figure G7.5** The tennis serve

Source: Lev Radin / Shutterstock.com

Furthermore, it is proposed that, because of this clear anthropometric difference between players, smaller players will need to use the "foot up" technique of serving in order to try to increase the impact height relative to their standing posture. In contrast, taller players, like Roger Federer (1.85 m tall) from Switzerland and Rafael Nadal (1.85 m tall) from Spain, could use the "foot back" technique, which allows a faster progression towards the net. *Table G7.1* illustrates the differences between the heights of the top ten male and female professional tennis players in 2018.

In 1999, the average height of the top five male and female players at the US Open competition was 186.4 cm and 179.8 cm respectively (only the top five have been analysed for comparison due to limited data from 1999). This is 5 cm (191.4 cm average of top 5 men in 2018) shorter and 4.2 cm (175.6 cm average of top 5 female in 2018) taller than the male and female equivalents (top five in the world) in 2018.

With the top five male players of 2018 being some 5 cm (2 inches) taller than their equivalents in 1999, it is obvious that height is an important factor for the male professional players. Although these anthropometric qualities would facilitate long, fast and powerful strokes, the changing demands of the game also require each player to adopt a technique and, indeed, a racket that suits their own individual style and anthropometry to optimise performance.

The tennis racket has evolved dramatically since the wooden rackets of the 1970s (*Figure G7.6*), when the most popular choice was Canadian ashwood that was cut into long strips, then steam glued and pressed together. In addition, metal rackets moulded from aluminium (chosen for the high strength to weight ratio) were also a popular choice among players, for example the Wilson racket used by Jimmy Connors of the USA in the late 1970s. The typical racket of 1970 possessed a string area of 70 square inches (450 cm$^2$); it had a weight (mass) of 12.5 ounces and a racket frame of approximately 18 mm deep. Today, tennis rackets are complex, highly engineered components that are subjected to much research and development. Following such tennis racket progression over the years, one question seems apparent:

> Can these technological advances in racket design really make up for the significant differences in performance created by different anthropometries and, in particular, differences in serving power and speed?

**Table G7.1** The heights of the top ten male and female professional tennis players in 2018

| *Top 10 male 2018* | *Height (cm)* | *Top 10 female 2018* | *Height (cm)* |
|---|---|---|---|
| 1. Nadal (ESP) | 185 | 1. Halep (ROM) | 168 |
| 2. Federer (SUI) | 185 | 2. Wozniacki (DEN) | 177 |
| 3. Zverev (GER) | 198 | 3. Muguruza (VEN) | 182 |
| 4. Cilic (CRO) | 198 | 4. Svitolina (UKR) | 174 |
| 5. Dimitrov (BUL) | 191 | 5. Ostapenko (LAT) | 177 |
| 6. del Potro (ARG) | 198 | 6. Pliskova (CZECH REP) | 186 |
| 7. Anderson (RSA) | 203 | 7. Garcia (FRA) | 177 |
| 8. Thiem (AUT) | 185 | 8. Kvitova (CZECH REP) | 182 |
| 9. Goffin (BEL) | 180 | 9. V.Williams (USA) | 185 |
| 10. Isner (USA) | 208 | 10. Stephens (USA) | 170 |
| Mean | 193.1 | Mean | 177.8 |
| % over 182.9 cm (6 feet) | 90% | % over 175.5 cm (5 feet 9 inches) | 70% |

Data derived from the ATP (Association of Tennis Professionals) and WTA (Women's Tennis Association) websites (the governing bodies of the male and female professional tennis players): http://www.atpworldtour.com/ and http://www.wtatennis.com/

**Figure G7.6** The wooden tennis racket

Source: Wavebreak Media / Shutterstock.com

In 1976, the Prince racket emerged (which was originally developed by the aeronautical engineer Howard Head) and was to create a revolution in tennis racket design and construction. The aluminium Prince racket head was almost double the original size with a string area of 130 square inches (839 $cm^2$). The years that followed saw many copies of the Prince racket develop, some of which had even larger surface areas. This continued until 1980 when the ITF (International Tennis Federation) limited the string area to a maximum of 15.5 inches in length (approx. 40 cm) and 11.5 inches in width (approx. 30 cm). Hence, this ruling set a theoretical maximum available area of 178 square inches (1148 $cm^2$).

This standardisation had a dramatic effect on the industry, and the resulting years saw manufacturers moulding rackets of graphite and glass fibre, which produced a racket which was to be far more powerful than the older wooden rackets, and yet also significantly lighter.

At this time, research into racket design at Pennsylvania University in the United States of America showed that increasing the mass of the racket head by 33 per cent produced only a 5 per cent rise in the speed of the ball once it had been struck. However, a 33 per cent increase in racket head speed increased the ball speed by 31 per cent. This clearly showed the important link between the athlete and the racket (as the athlete is required to accelerate the racket to the ball) and the critical importance of lighter rackets (i.e. so the athlete can use the same force but move the racket much more quickly). In 1984, racket development then observed another significant change when an inventor from Southern Germany (Siegfried Kuebler) developed the wide-bodied frame. The theory was that the racket could be made even stiffer and lighter by increasing the frame depth or side cross-section, while narrowing the front profile. Basically, the stiffer the racket the more energy is returned to the ball. When the ball and racket connect, there is only a short time of contact and any flex in the racket is wasted energy.

**Figure G7.7** The large-framed Head tennis racket

Source: Radu Bercan / Shutterstock.com

Stiffer rackets generate more power and also have a significantly larger centre of percussion, which helps with directional control. The centre of percussion is considered to be the most responsive part of the racket's frame (usually in the string area). If a ball and racket make contact at the centre of percussion it is said that it produces the most powerful shot with no, or very little, vibration or shock. Current rackets will have a very large centre of percussion area, which is nowadays located at the top part of the racket where most top players in the modern game will hit the ball. For further information, see the Centre of Percussion section elsewhere in this text.

As with the Prince racket design, many manufacturers rushed into producing racket head frames up to 39 mm deep and only 10 mm wide. Materials such as Kevlar and complex thermoplastics were used in their construction, which helped allow the strings of the racket to act almost independently of the frame, leading to a reduced ball contact time, less lost energy and consequently more power.

In 1992, the modern racket had a surface area of 115 square inches (742 $cm^2$); it was 39 mm thick with an aerodynamic profile and all with a weight (mass) of only 10 ounces (283 grams). Compared to the wooden racket of the 1970s, it was 64 per cent bigger, 116 per cent thicker and yet 20 per cent lighter. The rackets of 2018 (and beyond) have an even larger centre of percussion and sometimes contain multiple points for optimum contacts with the ball, which will account for shots that are even mishit. This racket provides good ball speed without effort, is easy to manoeuvre, is aerodynamically efficient, and will not cause unnecessary fatigue to the athlete. They are regularly constructed from titanium or hyper-carbon and they will not be responsible for any injury to the body. To add to all this, some of them will interact with the anthropometry of the athlete, providing performance information or rebound efficiency, movability, precision, power and, at the same time, comfort.

Rafael Nadal (Spain), who was ranked number one in the world in 2018, uses a new Babolat Aero racket, which contains sensors (accelerometers and inertial measurement units (IMUs)) that are built into the racket frame and that record information from his matches (*Figure G7.8*). The racket incorporates an on–off switch in the handle, which he can operate. Information such as where the ball strikes the racket and what kind of shot he uses is immediately recorded. In 2014, the International Tennis Federation (ITF) relaxed the rules for the use of such a racket and they allowed the racket to be used in formal matches. Therefore, the information collected during a match can be downloaded to his smartphone after completion for analysis when he is off-court (i.e. not during the match). From this, it is clear that players are turning to sports technology to get every edge that they can to win. Sport is a huge global business that is worth billions of pounds (and dollars) every year and winning at the elite level is almost everything to the athlete.

Today's elite tennis players will select a racket that is carefully balanced to their particular needs and indeed, to some extent, their individual anthropometry. For example, if they are a powerful serve-and-volley player, it is likely that this player will use a different racket to a player whose strength is to play from the baseline. This customisation and precise selectivity will continue to play a critical role in the development and progression of the sport of tennis into and perhaps even beyond the twenty-first century.

**Figure G7.8** Rafael Nadal and the Babolat racket with sensor technology

Source: Leonard Zhukovsky / Shutterstock.com

# TASK

Search the internet to find more information on the latest sensor technology Babolat Aero tennis racket used by Rafael Nadal.

| *Key notes* | |
|---|---|
| **Anthropometry** | Anthropometry is the measurement and study of the human body, its parts and capacities. Athletes' body shapes are changing, such that they are becoming stronger, fitter and faster. These changes occur as a result of the need to improve and the need to accommodate changes in equipment design. |
| **Sports equipment design** | The technical demands of sporting events are also changing and manufacturers of sports equipment continue to look for ways to improve athletic performances while adhering to the strict design rules imposed by the governing bodies of the sport. History shows that sports equipment design can have a profound effect on athletic performances, whether this is improved aerodynamics in cycling or lighter, stronger and more balanced rackets in tennis. Nevertheless, just as specific alterations to sports equipment can improve performance and/or reduce injury risk, other alterations may impede performance and/or promote the development of acute or overuse injuries. |
| **Sports equipment in cycling** | Technological advances in cycling essentially developed from the sport of triathlon. The posture and body orientations maintained on the bike are critical to successful performance and both the rider and the bike can significantly affect this performance. However, the rules and restrictions presented by governing bodies such as the UCI can cause riders to have to continually modify their riding positions. |
| **Sports equipment in tennis** | Obvious differences in athlete anthropometry, such as the height of male tennis players, can create different requirements in skill and the sports equipment required to perform the skill. Taller, stronger players may have an advantage with regards to the tennis service action. Tennis equipment is developing rapidly which now allows elite athletes of different anthropometry to all be potential world champions. |

# FURTHER READING

The following four resources provide additional reading on sports technology developments in cycling and tennis.

1 Haake, S. (2009). The impact of technology on sporting performance in Olympic sports. *Journal of Sports Sciences, 27*(13), 1421–1431.
2 Elliott, B. (2006). Biomechanics and tennis. *British Journal of Sports Medicine, 40*(5), 392–396.
3 International Tennis Federation (2017). *The 2018 rules of tennis*. Nassau, Bahamas: International Tennis Federation Limited. Retrieved from https://www.itftennis.com/media/277603/277603.pdf.
4 Union Cycliste Internationale (2018). *Clarification guide of the UCI technical regulation*. Aigle, Switzerland: Union Cycliste Internationale. Retrieved from http://en.static.uci.deltatre.net/inside-uci/rules-and-regulations/equipment-165067/.

# DATA SMOOTHING, FILTERING AND PROCESSING

*Michael Cole*

When experimentally collecting kinematic and kinetic data in sport and exercise biomechanics, it is common for the measures to include varying degrees of error (commonly referred to as noise). While there are many different sources of noise in biomechanics, some of the most common include: i) digitisation error; ii) skin movement artefact; iii) other electrophysiological signals; and/or iv) environmental factors. **Digitisation errors** commonly occur due to: i) difficulties with consistently and/or accurately identifying reference points and joint centres; or ii) limitations in the resolution of the digitising system (e.g. the size of the image relative to the actual field of view). Errors associated with **skin movement artefact** are common in biomechanics experiments that involve the placement of measuring equipment on the skin's surface (e.g. joint markers, wearable sensors, surface electromyography). Given that the skin and skeletal structures can move independently of each other and that they are separated by varying amounts of subcutaneous and adipose tissue, it is widely recognised that measures taken with skin-based equipment are subject to some degree of error.

The introduction of noise via **other electrophysiological signals** and **environmental factors** is most common during the assessment of muscle activation patterns (via surface electromyography) and ground reaction forces (via a force platform), but may be evident in other datasets also. In surface electromyography, recording electrodes are positioned on the surface of the skin overlying the belly of the muscle of interest. However, when examining muscle activity or ground reaction forces in the frequency domain, it is possible that one may observe a spike in the data around the fundamental frequency of the country's alternating current (AC) power supply (e.g. 50 Hz in the UK or 60 Hz in the USA). Furthermore, when assessing the trunk muscles (e.g. the erector spinae), it is not uncommon for the activation patterns of these muscles to be contaminated by the phasic activations of the cardiac muscle (i.e. the electrocardiogram (ECG) signal). While it is impossible to account for all potential sources of error at the time of data collection, it is sensible to try to reduce the error by adopting appropriate procedures.

For many biomechanics data sets, the presence of noise can be qualitatively recognised by its "rough" or "ragged" appearance. However, the attenuation of noise in collected data is not guided by aesthetics, but rather by the need to be exact with estimates of movement parameters. For example, the data presented in *Figure G8.1* depicts the raw or "untouched" horizontal velocity of the body's centre of mass during walking and was collected using a conventional three-dimensional motion analysis system. Each heel contact made by the participant during the walking trial is represented by the vertical lines drawn on the figure. If the task was to report the horizontal velocity of the body's centre of mass at each of these time points, it is clear that there would be less confidence in the measures due to the noise in the data set.

A further problem with using noisy data, such as that depicted in *Figure G8.1*, for an analysis is that the error that is present in this measure will ultimately influence the accuracy of other measures that are derived from it. For example, just as velocity is the rate of change of displacement, acceleration is the rate of change of velocity. As such, when acceleration data are derived from erroneous velocity data, these errors are further magnified in the accelerations.

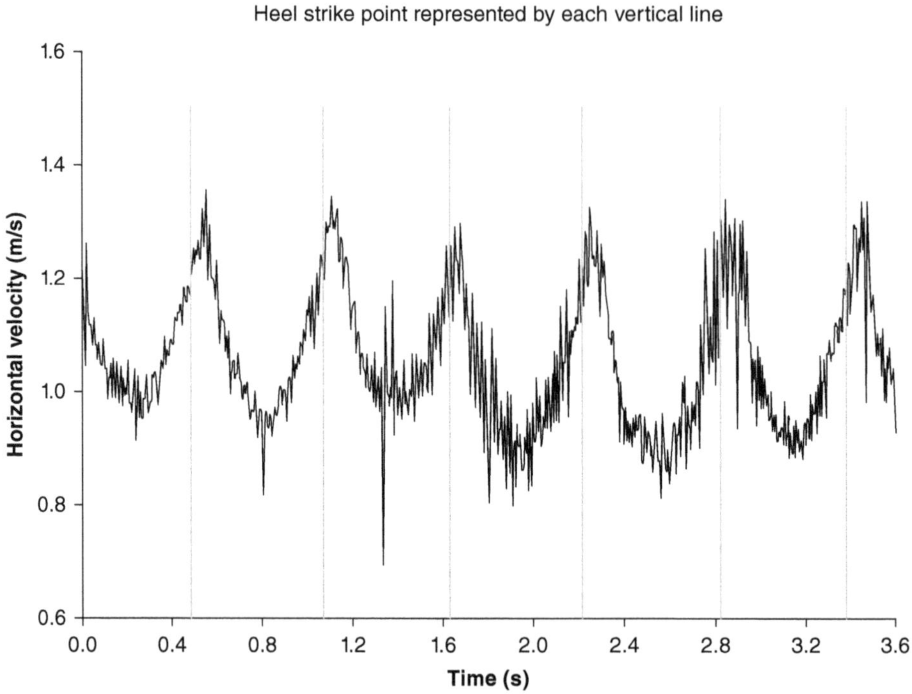


**Figure G8.1** Typical "raw" horizontal velocity data for the body's centre of mass during walking gait, with the timing of successive heel strikes indicated by the vertical lines

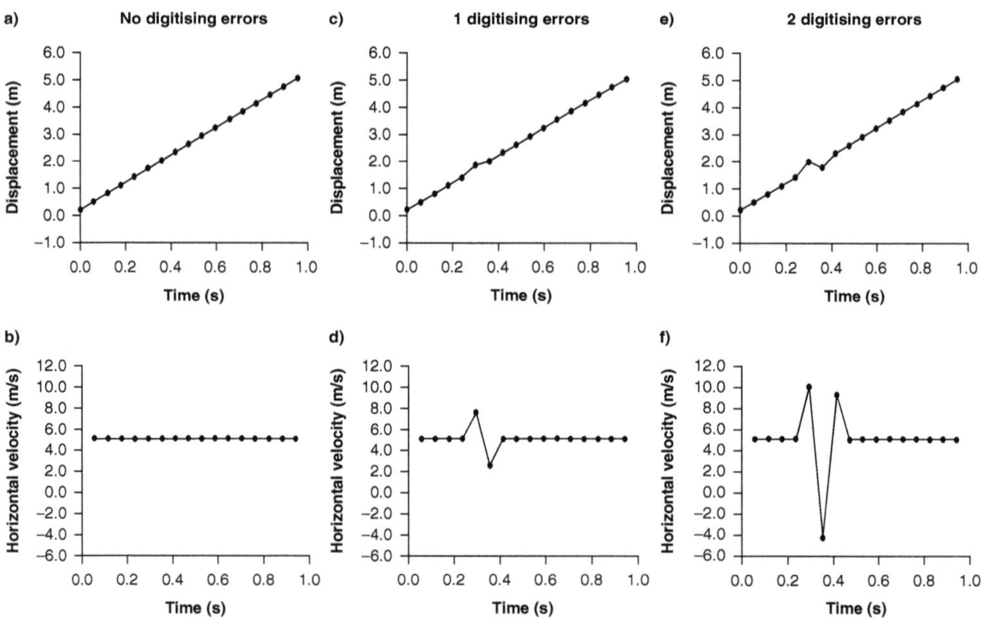


**Figure G8.2** The influence of small displacement errors that are introduced during the digitisation process on the subsequent calculation of linear velocity. The plotted data depict digitised displacements and subsequent velocities in scenarios where there are (a–b) no errors, (c–d) one error and (e–f) two errors

To provide a clearer example of the influence of small errors in displacement on the subsequent calculation of velocity, consider a projectile that is moving at a constant velocity; that is, for each unit of time, its displacement changes by the same amount (*Figure G8.2a*). When the position of the projectile is perfectly digitised at each time point, the subsequent velocity is shown by a flat line; that is, the projectile's constant horizontal velocity is correctly recorded (*Figure G8.2b*). However, if one (*Figure G8.2c*) or two (*Figure G8.2e*) small errors were made during the digitisation process, these errors would be magnified in the subsequent calculation of horizontal velocity (*Figure G8.2d* and *Figure G8.2f*). When the data has one error (middle column) the effect is to first overestimate one of the velocity calculations, before subsequently underestimating the next. This produces spikes in the velocity data that may be erroneously interpreted to suggest that the projectile is not travelling at a constant velocity, but rather is accelerating or decelerating. This problem is exacerbated in the third column, where two small errors have been made during the digitisation process and the subsequent error in the horizontal velocity measure is markedly worse. In general cases, where all motion capture data (for example) inherently has some degree of error, it can be difficult to obtain a meaningful velocity measure without first attenuating the noise in the displacement data.

Measurement errors, such as those described thus far, can be reduced by applying an algorithm to **smooth** (or **filter**) the data. The algorithm has a similar effect to that achieved by hand when a smooth curve is drawn through some ragged data. A common smoothing algorithm is the **Hanning** algorithm, which is a moving average procedure that is described below:

$$y_i = 0.25x_{i-1} + 0.5x_i + 0.25x_{i+1}$$

where:

x = the original (unfiltered/unsmoothed) data

y = the new (filtered/smoothed) data

n = the number of data samples

i = sample number 1 to n–1

This formula is applied to all of the data as a moving average. It can be applied more than once for increased smoothing. To illustrate its application, consider the horizontal velocity data in *Figure G8.1*: the first 0.2 seconds (collected at 150 Hz) of which is tabulated in *Table G8.1*.

Using the Hanning algorithm, the smoothed data (y) for Sample #2 is calculated as:

$$y_2 = (0.25 \times 1.186) + (0.5 \times 1.216) + (0.25 \times 1.234)$$

$$y_2 = 0.297 + 0.608 + 0.308$$

$$y_2 = 1.213\,\mathrm{m/s}$$

Similarly, for Sample #3:

$$y_3 = (0.25 \times 1.216) + (0.5 \times 1.234) + (0.25 \times 1.252)$$

$$y_3 = 1.234\ \mathrm{m/s}$$

Of course, doing this sample by sample is very time-consuming, so the process is often automated using formulae in a spreadsheet or specially written computer programs (*Figure G8.3*).

**Table G8.1** Typical raw horizontal velocity data for the body's centre of mass during walking gait (150 Hz) and the same data smoothed based on a Hanning algorithm

| *Sample #* | *Raw velocity data* | *Smoothed velocity data* |
|---|---|---|
| 1 | 1.186 | 1.189 |
| **2** | **1.216** | **1.213** |
| **3** | **1.234** | **1.234** |
| 4 | 1.252 | 1.241 |
| 5 | 1.224 | 1.237 |
| 6 | 1.250 | 1.241 |
| 7 | 1.241 | 1.250 |
| 8 | 1.266 | 1.253 |
| 9 | 1.237 | 1.243 |
| 10 | 1.232 | 1.245 |
| 11 | 1.279 | 1.278 |
| 12 | 1.322 | 1.295 |
| 13 | 1.256 | 1.276 |
| 14 | 1.272 | 1.288 |
| 15 | 1.354 | 1.307 |
| 16 | 1.248 | 1.262 |
| 17 | 1.198 | 1.234 |
| 18 | 1.293 | 1.250 |
| 19 | 1.217 | 1.232 |
| 20 | 1.200 | 1.203 |
| 21 | 1.196 | 1.195 |
| 22 | 1.189 | 1.185 |
| 23 | 1.164 | 1.187 |
| 24 | 1.231 | 1.201 |
| 25 | 1.179 | 1.174 |
| 26 | 1.107 | 1.119 |
| 27 | 1.085 | 1.104 |
| 28 | 1.140 | 1.122 |
| 29 | 1.125 | 1.128 |
| 30 | 1.124 | 1.120 |

It should be noted that the moving average algorithm cannot calculate data for the first or last data points in the array as this method needs to have a data point to represent i–1 and i+1. For the first point, there is no i–1, while the last point has no i+1. To overcome this issue, the Hanning algorithm uses the following end-point routines:

$$y_1 = 0.5 \times (x_1 + x_2)$$

$$y_n = 0.5 \times (x_{n-1} + x_n)$$

An advantage of the Hanning algorithm is that it can easily be programmed into a spreadsheet or a customised computer program, but a shortcoming of this method is that it lacks some of the flexibility that is needed to attenuate the noise that is inherent in biomechanics data. By comparing the data presented in *Figure G8.1* and *Figure G8.3*, it is clear that the Hanning algorithm has improved the smoothness of the velocity data, but it still contains some of the features that led to data smoothing in the first place. Frequently in sport and exercise

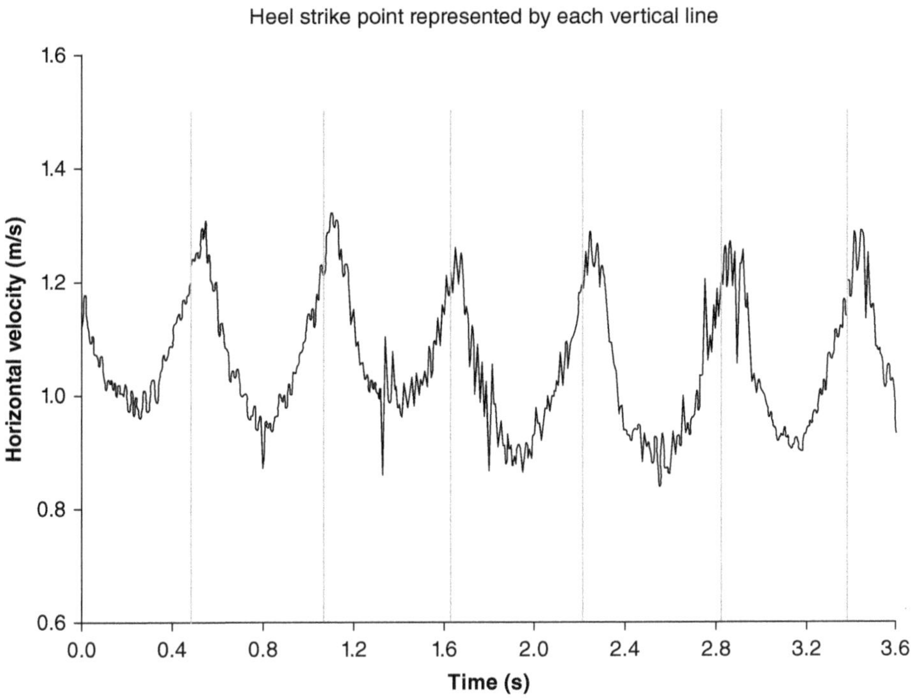


**Figure G8.3** The horizontal velocity data for the body's centre of mass from Figure G8.1 smoothed using the Hanning algorithm. The vertical lines represent the timing of successive heel strikes throughout the walking gait pattern

biomechanics, it is necessary to have **greater smoothing** than the Hanning algorithm can supply and, for this reason, other algorithms have been developed.

The **second-order Butterworth** smoothing algorithm uses similar principles to the Hanning algorithm, but is more flexible and, hence, is potentially better suited to biomechanics data. Within the biomechanics literature, the second-order Butterworth algorithm is frequently referred to as the **fourth-order Butterworth algorithm**, as it is usually applied twice. Although it is only necessary to apply the algorithm once to smooth the data, a single application of the algorithm also leads to a temporal shift that is problematic for the biomechanist. To correct this temporal shift, the algorithm is applied a second time, but in the reverse direction (i.e. starting at the last data point and working backwards to the first data point). The fourth-order Butterworth algorithm is very effective for smoothing a wide variety of biomechanical data (*Figure G8.4*) and is generally available in motion analysis software used for producing kinematic data.

**Cubic** and **quintic splines** are also two common alternatives to the Hanning and fourth-order Butterworth algorithms that are routinely used in biomechanics to smooth noisy kinematic data. The essential principle of these methods is that the algorithm uses a third-order (cubic) or fifth-order (quintic) polynomial to sequentially fit a smoothed curve over adjacent collections of data points, repeating this process until the end of the data structure is reached. To achieve the best outcomes, the model of the original (noisy) data that is produced using the cubic or quintic spline algorithms is systematically varied until the internal consistency of the original data set is maximised and the validity of the algorithm is optimal. Advocates for spline-based smoothing techniques suggest that they are more suited for processing kinematic data, as they better replicate the smooth nature of human movement and provide more precise results

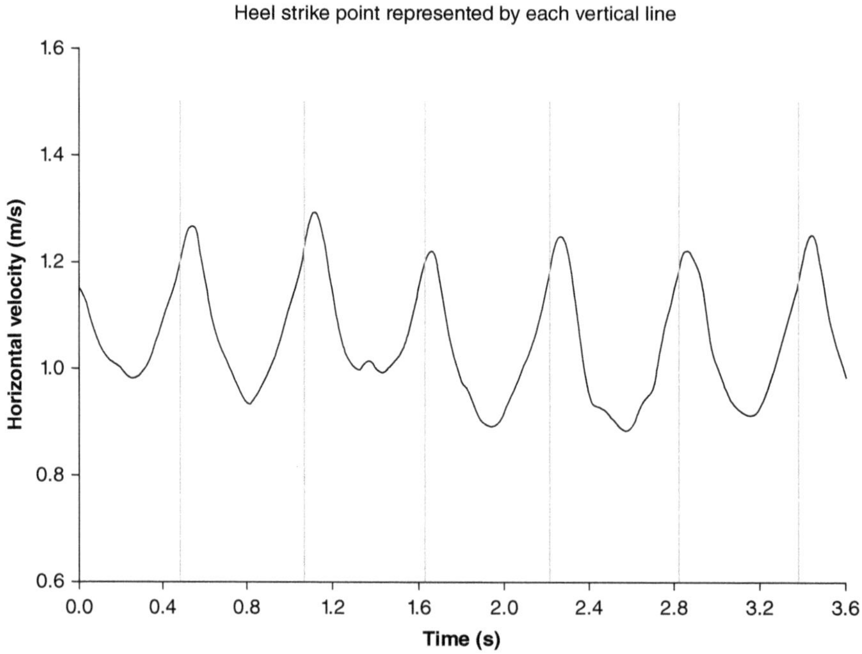


**Figure G8.4** The horizontal velocity data for the body's centre of mass from Figure G8.1 smoothed using the fourth-order Butterworth algorithm with a low-pass cut-off frequency of 8 Hz. The vertical lines represent the timing of successive heel strikes throughout the walking gait pattern

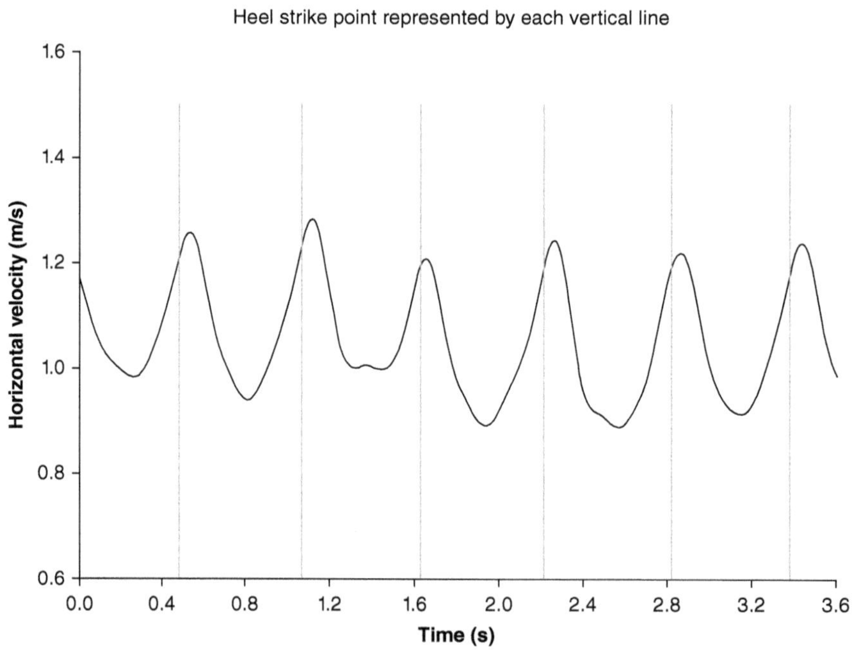


**Figure G8.5** The horizontal velocity data for the body's centre of mass from Figure G8.1 smoothed using the quintic spline algorithm. The vertical lines represent the timing of successive heel strikes throughout the walking gait pattern

for the first and second derivatives (*Figure G8.5*). Like the fourth-order Butterworth algorithm, the algorithms underpinning splines are also complex, but are often available in motion analysis software used for producing kinematic data.

*Key notes*

| | |
|---|---|
| **Data smoothing** | Data smoothing is required to reduce the effects of errors related to digitisation, skin movement artefacts, environmental factors and other electrophysiological signals. Collectively, these types of error are called noise and, without appropriate smoothing, they can introduce significant problems when first and second derivatives are computed (i.e. velocities and accelerations). To facilitate data smoothing, several algorithms are available in the computer software used for biomechanical analyses. |
| **Hanning algorithm** | The Hanning algorithm uses a moving average technique to smooth the data and reduce the effect of errors. This smoothing process improves the estimates of velocity and acceleration that are derived from displacement data using numerical differentiation. As a moving average technique, the Hanning algorithm uses three consecutive data points (e.g. points 1, 2, 3) to produce each smoothed coordinate, and repeats this process for the length of the data set. However, despite its simplicity, the Hanning algorithm lacks the flexibility that is often needed to attenuate noise in biomechanical data, which necessitates the use of more complex smoothing algorithms. |
| **Second-order Butterworth algorithm** | The second-order Butterworth algorithm is a more flexible and often more effective alternative to the Hanning algorithm. However, when this algorithm is applied to the data once, it introduces an unwanted time shift. To ameliorate this side effect, the algorithm is applied a second time, but in the opposite direction (i.e. from last to first data point). For this reason, this algorithm is often referred to as a fourth-order Butterworth algorithm in the biomechanics literature. |
| **Cubic and quintic splines** | Cubic and quintic splines are both widely used as alternatives to the fourth-order Butterworth algorithm to smooth biomechanical data; however, quintic splines tend to be better suited for complex human motion. Both algorithms try to fit a smoothed curve over three (cubic) or five (quintic) adjacent data points. This process is then systematically repeated for the next collection of data points until the end of the data set is reached. |

## FURTHER READING

The following resources provide additional reading about the smoothing algorithms used to attenuate noise in biomechanical data:

1 Sinclair, J., Taylor, P. J., & Hobb, S. J. (2013). Digital filtering of three-dimensional lower extremity kinematics: An assessment. *Journal of Human Kinetics, 39*, 25–36.
2 Giakas, G., & Baltzopoulos, V. (1997). A comparison of automatic filtering techniques applied to biomechanical walking data. *Journal of Biomechanics, 30*(8), 847–850.
3 Challis, J. H., & Kerwin, D. G. (1988). An evaluation of splines in biomechanical data analysis. In G. D. Groot, A. P. Hollander, P. A. Huijing & G. J. van Ingen Schenau (Eds.), *Biomechanics XI-B* (pp. 1057–1061). Basel: Karger.
4 Woltring, H. J. (1985). On optimal smoothing and derivative estimation from noisy displacement data in biomechanics. *Human Movement Science, 4*(3), 229–245.

# SECTION H

# Appendices

# PLANES AND AXES OF MOTION



*Paul Grimshaw*

In a two-dimensional (2D) example (like the page of this text that you are reading) we have two dimensions of space. Vertical and horizontal (or height and width) are terms that are used to express two-dimensional space. The pages of this text will have a vertical distance (height) and a horizontal distance (width). In this context, we often use X and Y axes to represent the two-dimensional space we are considering. The X axis would be drawn in the horizontal direction and the Y axis would be drawn in the vertical direction. The X axis is often termed the **abscissa** and the Y axis as the **ordinate.** The point at which the two axes **intersect** (cross) is called the **origin.** In three-dimensional (3D) space a third axis is needed to describe the movement and this is usually denoted as the Z axis. This axis also acts through the origin but is perpendicular to both the X and Y axes described previously.

The three planes of motion that have been created from the three axes of motion can also be translated to the human body. In this case the origin of these planes and axes is usually expressed at the centre of mass of the body. Using the example of the body that is shown in *Figure H1.1*, this is at approximately the mid-point between the two hips. It is important to note that in this figure the body is presented in what is called the **anatomical position** (facing forwards, arms by the side, feet forwards and parallel, and palms forwards with fingers extended). However, in many cases, within human movement the body will be in different positions and it will be important to be able to describe the movement in these positions relative to the three **cardinal planes** (a plane that passes through the centre of mass of the body), and axes of motion described.

The **sagittal plane** runs from a front-to-back and superior-to-inferior (top-to-bottom) orientation. The plane divides the body into equal left and right portions. The **transverse axis** of rotation is **perpendicular** (at 90°) to the sagittal plane (*Figure H1.1*). A typical movement in this plane and about this axis of rotation would be a **somersault.**

The **frontal plane** runs from a side-to-side and superior-to-inferior orientation. The plane in this example divides the body into equal front (anterior) and rear (posterior) portions. The **anterior–posterior axis** of rotation (sagittal axis) is perpendicular to this plane (*Figure H1.1*). A typical movement in this plane and about this axis of rotation would be a **cartwheel.** A cartwheel is where you perform a lateral handspring with the arms and legs extended (i.e. in the same plane/manner as rolling a coin from left to right and looking at it from the front).

The **transverse plane** runs from a side-to-side and anterior-to-posterior orientation. The plane in this example divides the body into equal upper and lower portions (superior and inferior parts). The **longitudinal axis** of rotation is perpendicular to the transverse plane (*Figure H1.1*). A typical movement in this plane and about this axis of rotation would be a **pirouette** (as in the sport of ice skating).

Although the three cardinal planes are usually expressed as originating from the body's centre of mass, it is important to realise that all movements of the body or segments of the body that occur parallel to these cardinal planes are occurring within that plane. In this case we can see that these planes can be applied to any location within the body, but the sagittal plane is always perpendicular (at 90°) to the frontal plane, and each of these is always perpendicular to

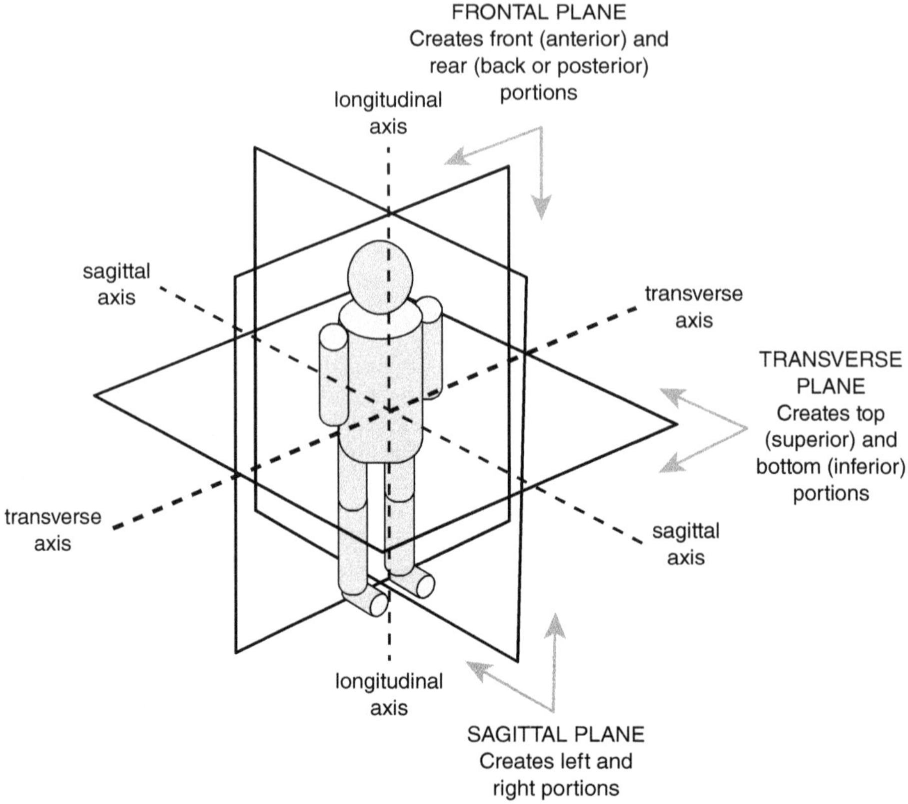


**Figure H1.1** Three-dimensional planes and axes of motion

the transverse plane. The rotational movement of a body or a body part in any given plane will occur around an **anatomical axis** that runs perpendicular to that plane of motion. Therefore, an anterior–posterior axis will always be perpendicular to a transverse axis and these will always be perpendicular to a longitudinal axis.

Some anatomical movements can be described with reference to one plane and one axis of rotation; these are known as **single planar movements.** For example, flexion or extension of the knee, elbow or shoulder would occur in the **sagittal plane** and about the **transverse axis** of rotation; abduction and adduction movement at most joints and valgus and varus rotation at the knee would occur in the **frontal plane** and about the **anterior–posterior axis** of rotation; internal and external rotation of the knee joint (or more precisely internal and external rotation of the tibia/fibula or femur) and horizontal abduction and adduction would occur in the **transverse plane** and about the **longitudinal axis** of rotation.

However, many movements of the human body occur across several planes and about several axes of rotation; these are termed **multi-planar movements.** A good example of this is the movement of the foot in **pronation** and **supination. Pronation** and **supination** are complex movements that involve motion in three planes and about three axes of rotation simultaneously. These movements are often described at the subtalar joint, which is in the foot. However, pronation and supination can also occur at the wrist joint. Pronation and supination at the subtalar joint in the foot involve the following. Pronation: calcaneal **eversion**, ankle **dorsiflexion** and forefoot **abduction.** Supination: calcaneal **inversion**, forefoot **adduction** and ankle **plantarflexion.**

# FREE BODY DIAGRAMS

*Paul Grimshaw*

Free body diagrams are pictures (diagrams) of forces acting on a body. They allow us to be able to analyse the effect of all the external forces acting on a body more easily (i.e. the effect of the net force). As we have seen within human movement, there are a number of different types of forces that can act on a body: i) gravitational force (weight); ii) frictional force; iii) normal reaction forces; iv) applied contact forces; v) tensile, shear and compressive forces; vi) muscle and joint forces; and vii) centripetal, tangential and centrifugal forces. In human movement, it is often the case that several forces will act on the body simultaneously. As we have seen earlier, force is a vector quantity and thus a force can be expressed or represented by lines with both magnitude and direction. The net effect of these forces (the resultant) acting on a body can be determined through representing all the forces acting on a body using a **free body diagram.**

In drawing free body diagrams, there are a number of steps that we should go through in order to assist us in the accurate representation of all the forces acting on a system or body. These can be outlined as follows:

**First:** isolate the body from its surroundings. Then draw the body upon which the forces are considered to act. For example, if we are interested in the forces on the lower arm (the ulna and radius), then **only draw the ulna and radius.** Do not draw any other aspect that the body of interest may or may not be in contact with.

**Second:** take time to identify all the external forces that are acting on the body or system. This is usually the most difficult part. As a guide, it is useful to systematically go through the different forces that could be acting. For example, if the body has mass, then there will be a weight force acting and this will be through the centre of mass of the body. Similarly, if the body is in contact with any other body there must be a normal force (acting perpendicular to the surfaces in contact) and a frictional force (acting along/parallel to the surface of contact) acting between the two bodies. When there is no rotational component of force (torque), the position and location of our force vectors is not so important, provided we maintain consistent lines and directions of force (orientation) application. When expressing forces on a free body diagram, the line of action of the force is located through the point of application. If two forces are equal and opposite and they lie on the same line of action, the resultant effect of the forces will be zero and we could represent these forces anywhere along the line of action. Often it is useful to break the forces down into their component parts (i.e. horizontal and vertical components of a resultant force).

**Third:** once all the components in the second stage have been classified, the next step is to draw all these forces on the diagram and include the magnitude and direction (angles and orientations) of each known force. In addition represent all the unknown forces on the diagram.

**Finally:** select a coordinate system of conventional representation in order to identify the positive and negative components of the force. It should be noted that if we are concerned with torques and moments, the positioning of these components on the free body diagram is critical.

## EXAMPLE

A mass (i.e. a body) is being pushed up an inclined plane by a force (i.e. parallel to the inclined plane). Draw the free body diagram of the external forces acting on the body. *Figure H2.1* shows the solution of this in more detail.

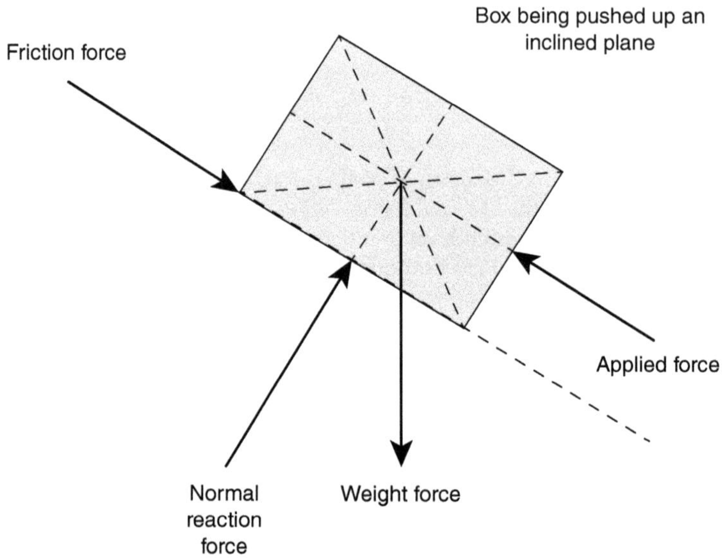


**Figure H2.1** Free body diagram

# SAMPLING THEORY

*Neil Fowler (edited by Paul Grimshaw)*

During the analysis of movement, it is often important to collect data at more than one single moment. Indeed, in most biomechanical analyses, it is critical to be able to record changes in the key variables over a period of time. Most measurement devices work by sampling and recording data at regular intervals during the measurement period. The number of samples in any given period is known as the **sample frequency** and is usually recorded in **hertz** (**Hz**), that is, the number of measures per second. For example most digital video cameras (like the one on your phone) now operate at rates of above 100 frames per second (100 Hz), while it is common to sample force platform data at around 1000 Hz (1000 measurements every second).

Consider the simple example shown in *Figure H3.1*, which is used to illustrate the effect of sample rate on the data recorded. If the changes in the knee angle (broken line and the right axis) and the vertical ground reaction (solid line and the left axis) force during a drop-jump take-off are recorded, the input signal being recorded is the knee angle and this is changing in a continuous

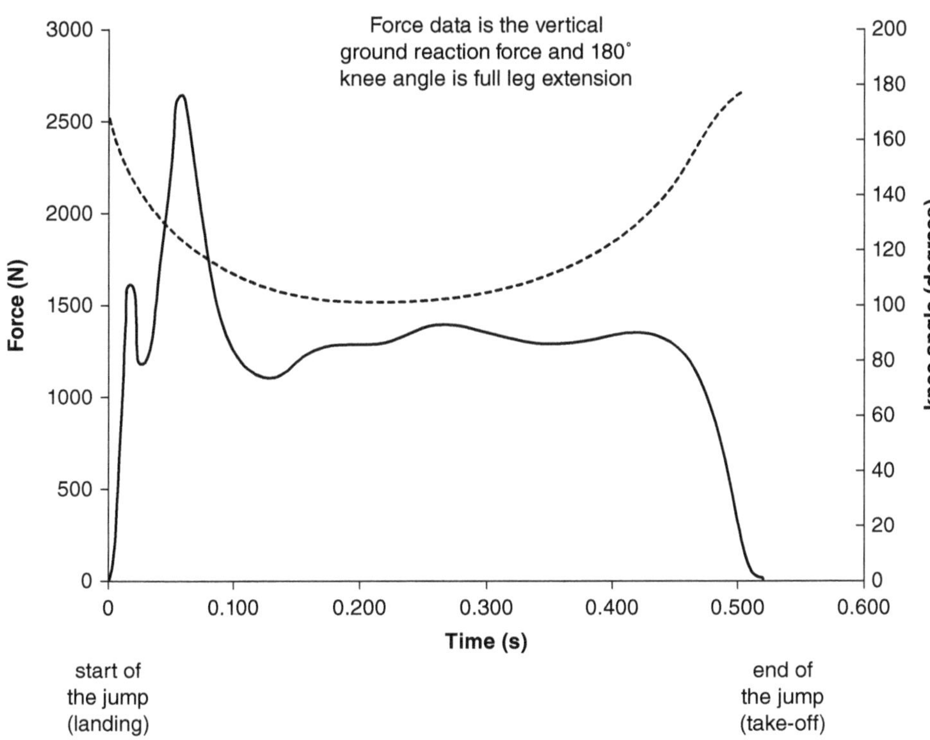


**Figure H3.1** Knee angle and vertical ground reaction force during a drop-jump

fashion; in other words there is always a knee angle present that could be recorded. When data are sampled at a set frequency, a series of "snapshots" are recorded creating a discontinuous record of the magnitude of the knee angle.

By looking at these two graphs it is possible to see that while it may be acceptable to sample the knee angle data at say 100 Hz it would not be acceptable to sample the force data at this rate as key moments in the input signal are missed. This is because the knee angle and vertical ground reaction force change at different rates and are said to have a different **frequency content.**

The frequency content describes the make-up of the signal and reflects the rate at which changes in the magnitude of the variable happen. In the case of *Figure H3.1*, it is clear that the knee angle undergoes a gradual change from flexion (decreasing value) to extension (increasing value) with only one major direction change in the movement. On the other hand the force data is more complex with more rapid changes in the magnitude.

To better define **frequency content**, it is often helpful to consider how a signal may be composed. *Figure H3.2* shows how three separate signals (i.e. three sets of data) can be added together to form a composite signal. If the three input signals are sampled separately, these would create the three data sets shown on the left in the figure; each with its own unique frequency of oscillation. If these three signals were part of a single composite signal, then the data set produced would be the summation of the three independent signals at each moment in time (i.e. at each sample moment the three input signals would be added together). The final waveform would look different from each of the three initial graphs. The frequency content of the final waveform would have a lower value of 1 Hz and an upper value of 8 Hz (shown on the right of *Figure H3.2*).

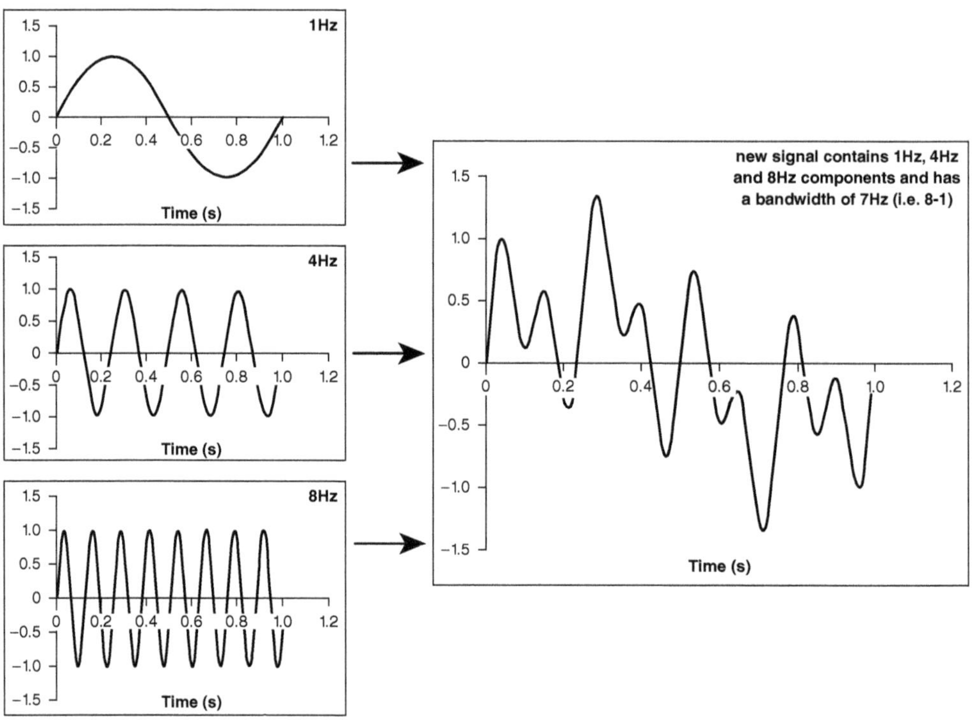


**Figure H3.2** Three separate waveforms (left) that, when added together, form one composite waveform (right)

Before starting an experiment it is therefore important that thought is given to selecting the most appropriate sample frequency. This can be done by applying the **sampling theorem.** The sampling theorem is sometimes called the **Nyquist–Shannon sampling theorem** or **Whittaker–Nyquist–Kotelnikov–Shannon sampling theorem** after the scientists credited with its development. The theorem was first formulated by Harry Nyquist in 1928 but was only formally proved by Claude E. Shannon in 1949. The theorem states that:

> The sampling frequency must be greater than twice the bandwidth of the input signal in order to be able to reconstruct the original perfectly from the sampled version.

It is important to note that the theorem refers to the **bandwidth** of the signal and not simply to the greatest frequency within the signal. The bandwidth is considered to be the range captured between the highest and lowest frequencies in the signal. In *Figure H3.2*, this would refer to the range between 1 Hz and 8 Hz; that is, the bandwidth is 7 Hz, whereas the upper frequency would be 8 Hz. While in human movement the law is often simplified by only considering the upper frequency value, there are other sampling conditions where consideration of the actual bandwidth is critical to ensure that the appropriate sample frequency is used.

If the sampling condition is not satisfied, then frequencies will overlap and the nature of the recorded signal will be different from the input signal. This overlap is called **aliasing.** To prevent aliasing, you can either 1) increase the sampling frequency or 2) introduce an anti-aliasing filter (or make the anti-aliasing filter more stringent). The anti-aliasing filter is used to restrict the bandwidth of the signal to satisfy the sampling condition. This holds in theory, but cannot be satisfied in practice as there may be some elements of the real signal that fall outside of the sampled range, and thus the recorded signal will not include all of the real signal. However, in most situations, the amount of information lost may be small enough that the aliasing effects are negligible.

If the sampling frequency is exactly twice the highest frequency of the input signal, then phase mismatches between the sampler and the signal will distort the signal. For example, sampling $\cos(\pi * t)$ at $t = 0,1,2...$ will give a discrete signal $\cos(\pi * n)$, as desired. However, sampling the same signal at $t = 0.5,1.5,2.5...$ will generate a constant zero signal, because the cosine of 90°, 270°, 450° ($0.5\,\pi$, $1.5\,\pi$, $2.5\,\pi$) and so on, will be zero. These two sets of samples, which differ only in phase and not in frequency, give dramatically different results because they sample at exactly the critical frequency. It is thus important that the selected sample frequency is **more** than twice the signal bandwidth and not exactly twice this value.

H4

# BIOMECHANICS APPS (APPLICATIONS)

*Adrian Burden*

Many people now have smartphones and tablets, which they use to run apps as part of their daily lives. In recent years, a number of apps have been developed to aid people's understanding of biomechanics, particularly in the field of sport and exercise science. This section reviews some of these apps and highlights their features and limitations. Included are apps that take advantage of the accelerometer in the device to display real-time acceleration and velocity, and those that allow images and videos from the camera to be analysed biomechanically. The apps were available at the time of this text going to press and are free or available for a minimal cost. For convenience, these have been sub-divided into: i) general biomechanics apps; ii) video analysis apps; and iii) modelling apps.

## GENERAL BIOMECHANICS APPS

### 1) Biomech

Developer: Iain Hunter

Cost: Free

Devices: iPad/iPhone

Description: Includes the following apps:

**Projectile motion:** Displays the parabolic flight path (height *vs* distance), and allows the user to manipulate the launch, including mass and gravity, and view output variables.

**Angular motion:** Displays a rotating rod and its angular velocity, and allows the user to manipulate its angular momentum, mass and radius.

**Acceleration:** Displays the motion of a dot, and allows the user to manipulate the initial velocity and acceleration.

**Friction:** Turns the device into an inclined plane, onto which objects can be placed to determine their coefficient of static friction. As well as the angle and coefficient, the normal (perpendicular to plane) and frictional (parallel to plane) components of the weight of the device are displays.

**Terminal velocity:** Displays a skier on a slope and a graph of resultant force (drag, friction and downhill) from the start of motion until terminal velocity is reached. The user can manipulate input variables, including drag coefficient, and view output variables.

#### *Key features*

- Allows user to easily manipulate many input variables.

### *Limitations*

- Apart from terminal velocity (skier), the objects are basic (i.e. dots and lines).
- Output variables are display only and not downloadable (e.g. to a spreadsheet).

## 2) Wind Tunnel

Developer: Algorizk

Cost: Free and £3.99 versions

Devices: iPad/iPhone

Description: Turns the device into a two-dimensional (side) view of a wind tunnel which allows you to observe the fluid flow around a variety of shapes, including an airfoil (aerofoil). The fluid flow around the object can be represented by particles or smoke, and colour changes can also be added to show differences in fluid pressure and speed around the object.

### *Key features*

- Variety of shapes to upload (11 for free and 15 for cost version), and shapes can also be drawn by the user.
- Lift and drag forces displayed (Lift/Drag ratio in cost version).
- Friction, speed and viscosity of the fluid can be manipulated (cost version only).

### *Limitations*

- Only the cost version allows you to save and share free drawn shapes.
- Shapes (e.g. aerofoil) cannot be rotated.
- Display only and not downloadable (e.g. to a spreadsheet).
- More detailed aerodynamic parameters are available in the more expensive CFD version.

## 3) Vernier Graphical Analysis

Developer: Vernier Software and Technology

Cost: Free

Devices: iPad/iPhone

Description: Allows the user to produce graphs and tables, and perform basic numerical functions. Data can be imported from the Graphical Analysis app (see above), input by hand, or collected from a built in accelerometer.

### *Key features*

- Provides basic statistical analysis between two points (mean, max, min, etc.).
- Provides some curve-fitting options (linear, quadratic, etc.).
- Data can be exported in various formats to other packages.

### *Limitations*

- No options to "smooth" coordinate data (e.g. from the Video Physics app).

## 4) Newton's Laws of Motion

Developer: Vidyatech Solutions Private Limited

Cost: Free

Devices: iPad/iPhone

Description: Detailed explanations of Newton's three laws of motion.

### *Key features*

- Includes text, images (some interactive) and interactive self-test questions.

### *Limitations*

- Most of the examples are not sport or exercise related.

## 5) ForceData

Developer: Carlos Balsalobre-Fernández

Cost: Free

Devices: iPad/iPhone

Description: Provides a real-time graphical output of the velocity, acceleration and force of the device.

### *Key features*

- Vertical and horizontal axes can be stretched.
- Sampling frequency can be manipulated between 25 and 200 Hz.
- Mass can be manipulated.
- Data can be exported as a .csv file.

### *Limitations*

- Size of graph is limited to only half of the screen.

# VIDEO ANALYSIS APPS

## 1) iAnalyze

Developer: Winning Edge Apps, Inc

Cost: £0.79

Devices: iPad/iPhone

Description: Allows the user to obtain basic biomechanical data from an uploaded video.

### *Key features*

- Stopwatch, drawing objects/lines, three-point angle calculator and speed between two points.
- Can compare two videos and measurements on a horizontal split screen.

### *Limitations*

- Display only and not downloadable (e.g. to a spreadsheet).

### 2) Video Physics

Developer: Vernier Software and Technology

Cost: £3.99

Devices: iPad/iPhone

Description: Allows the user to digitise a point from an uploaded video and obtain its position and velocity.

### *Key features*

- Scaling
- Can set position of origin and angle of axes (e.g. make the x-axis level with the ground).
- Manual digitising and automatic tracking
- Displays graphs of both horizontal and vertical position and velocity *vs* time.
- Can export video, video with points, and data file.

### *Limitations*

- Horizontal or vertical, although not both.
- Single point only on same "experiment". No option to alter digitising (sampling) frequency.
- Data file can only be opened in Graphical Analysis.

## MODELLING APPS

### 1) 3DSSPP

Developer: University of Michigan

Cost: Free

Devices: iPad/iPhone

Description: Calculates the lower back compression force during lifts, pulls and pushes in a variety of postures.

### *Key features*

- Allows user to manipulate the force at the hand.
- Provides an indication of what percentage of the population would have the trunk and shoulder strength to perform the exercise.
- Stick figure can be rotated and zoomed in/out.

### *Limitations*

- Only compression force presented, and in numerical form only.
- Only hand force vectors shown on stick figure, and these do not change when hand force is altered.

## 2) The Kinetic Spine

Developer: Brigham Young University

Cost: Free

Devices: iPad/iPhone

Description: Allows the user to animate a three-dimensional model of the lumbar spine.

### *Key features*

- Model can be rotated and viewed from any angle.
- Flexion–extension, lateral flexion and rotation are all modelled.
- Speed of motion can be altered.
- Motion of healthy, fused and degenerated spines can be observed and compared (side by side or overlaid).

### *Limitations*

- Only bones (vertebrae) are displayed, with no ligaments, muscles or discs.

## 3) Projectile Motion Calculator

Developer: Andre Yonadam

Cost: Free

Devices: iPad/iPhone

Description: Provides solutions for projectiles that take off and land from the same height, fall from a height, and are projected from a height.

### *Key features*

- Multiple input options for each of the three launch situations.
- Answers are provided to many decimal places, to allow subtle changes to input data to be compared.
- Acceleration due to gravity can be customised.

### *Limitations*

- Data only, with no animation of the projectile flight path.

# FOUNDATIONS OF MOTION ANALYSIS

*Adrian Burden*

Video analysis (either original or digital) can be **qualitative** or **quantitative**, depending on the aims of the investigation. Qualitative analysis involves observation of video and diagnosis of particular aspects of technique. It is a subjective process that seldom requires the camera to be located in a specific or stationary position and often does not require any additional equipment. Quantitative analysis requires kinematic information (e.g. linear position, velocity and acceleration of body segment endpoints and/or angular position, velocity and acceleration of body segments) to be obtained from video. Qualitative and quantitative analyses are often combined, particularly by coaches, and there are numerous software packages that are commercially available that allow the user to display a number of images on the same screen.

Quantitative analysis generally requires the video camera(s) to be stationary and located in a specific position(s), and the images subsequently to be stored in and displayed on a computer. Each image is then digitised to provide horizontal (x) and vertical (y) coordinates of selected points on the body; usually segment end points. These coordinates are then scaled (2D) or reconstructed (3D) to provide real-life coordinates and smoothed to reduce errors that inevitably occur during their collection. They are also often combined with temporal information to obtain velocities and accelerations.

## VIDEO SYSTEMS, FORMATS AND CAMERA SETTINGS

High-definition (HD) video now offers 720 vertical x 1280 horizontal pixels (921,600 pixels per image) or 1080 vertical x 1920 horizontal pixels (2,073,600 pixels), which are known as 720 or 1080 formats. The "i" or "p" that follows the 1080 format (i.e. 1080i or 1080p) denotes whether the image is formed from interlaced or progressive scanning. In interlaced scanning each set of odd- and even-numbered scanned lines constitutes a separate image or video field that belongs to the same video frame. Thus, the image resolution is effectively halved when using interlaced scanning. When projected or displayed on a large screen, low-resolution images can become pixelated (see *Figure H5.1*) and are unsuitable for further analysis.

In the Phase Alternating Line or PAL system, which is used in the United Kingdom, video fields can be displayed sequentially to provide 50 fields per second or 50 Hz, with a time between fields (or temporal resolution) of 0.02 s. The National Television System Committee (NTSC) standard used in the United States of America has a greater number of fields per second (almost 60 Hz). Commercially available high-speed video cameras are able to record more than 1000 images per second. Such cameras are particularly important for recording detail during rapid movements that would be missed if conventional cameras were used.

The high-speed shutter facility on a video camera should be used to prevent blurred or smeared images of moving bodies. Most cameras have variable electronic shutters that open for 1/120th, 1/250th, 1/500th, and 1/1000th of a second or less (0.008, 0.004, 0.002, or 0.001 s respectively), which allows light to pass into the camera for progressively shorter periods of

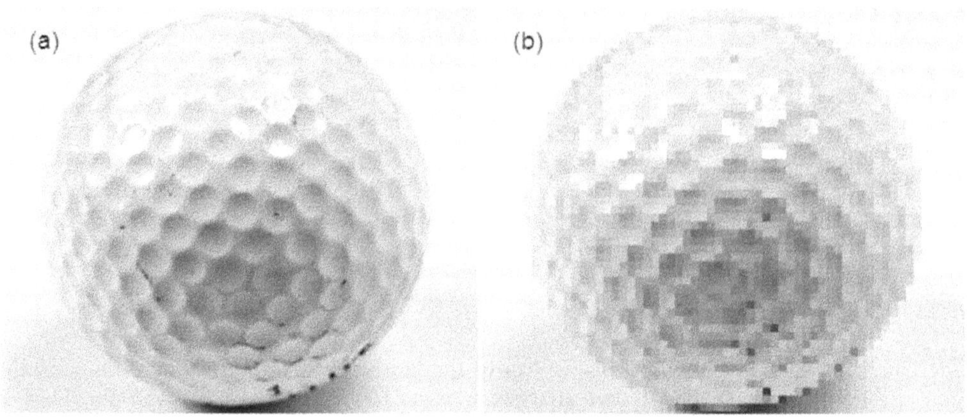


**Figure H5.1** Pixelisation of an image recorded with (a) higher and (b) lower resolution

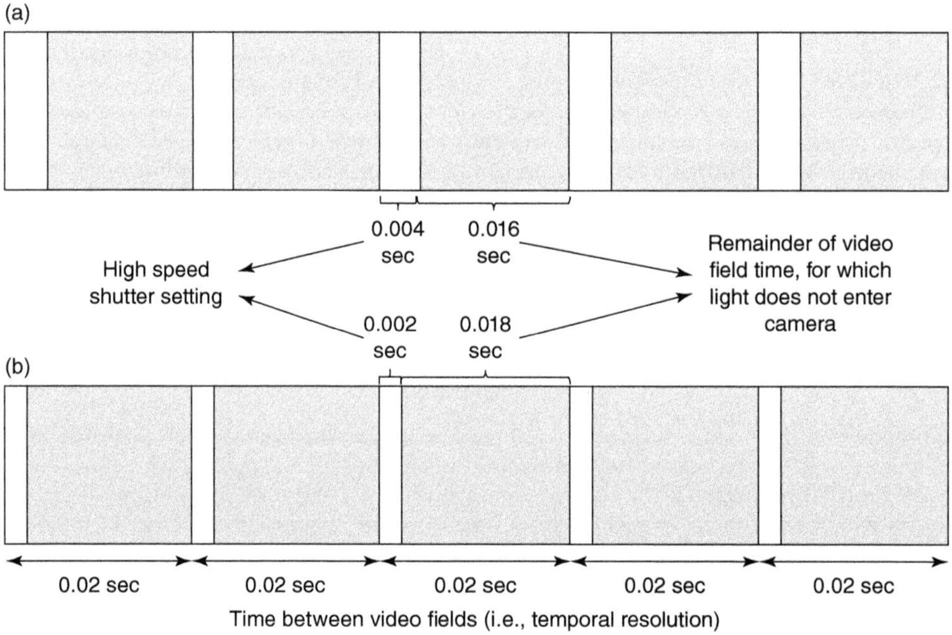


**Figure H5.2** Depiction of high-speed shutter settings of (a) 1/250th and (b) 1/500th of a second

time. Rapid movements, such as a golf club's motion around ball strike, will only be clear if shutter speeds in excess of 1/1000th of a second are used. However, for slower movements, such as walking, a shutter speed of 1/250th of a second would be adequate. The disadvantage of using higher shutter speeds (e.g. 1/500th or 1/1000th of a second) is that, depending on the light available, the image can appear dark due to the lack of light entering the lens of the camera. This is not usually a problem when filming outdoors, but in conditions of poorer light, that typically occur indoors, additional floodlighting may need to be used to improve the quality of the image.

High-speed shutter settings should not be confused with the temporal resolution of the video system/camera. *Figure H5.2* demonstrates the relationship between two different high-speed shutter settings and temporal resolution for a conventional video camera operating under the PAL System. Much larger high-speed shutter settings (i.e. smaller fraction of a second) are required for high-speed video cameras and are often as high as 1/100,000th of a second (i.e. 0.00001 s or 0.01 ms or 10 μs).

The manual focus setting on a video camera should also be used when conducting a quantitative analysis. In automatic focus mode, to which cameras often default, the lens is focused on the object nearest to it, which may not be the subject of interest. Thus, with the camera positioned appropriately (see the video recording procedures that follow) and the camera set to manual focus, the telephoto lens should be used to zoom in as close as possible to the participant standing in the centre of the activity area. The camera is then focused until the image of the participant is sharp, and the telephoto lens is then used to zoom out to the required image size. This will ensure that the lens remains focused in the plane of motion.

The high-speed shutter and manual focus settings are the most important when using video cameras for quantitative analysis; however, image quality will also be affected by the white balance setting and other features (e.g. filters for recording in different environments) that are now commonly available on most digital video cameras.

## VIDEO RECORDING PROCEDURES

The same video formats and camera settings discussed previously can be used regardless of whether a two-dimensional (2D) or three-dimensional (3D) video analysis is to be undertaken. A 2D analysis typically uses only one camera to record activities that are essentially planar in nature, and in which the plane of motion coincides with the photographic plane (i.e. at 90° to the optical axis of the camera). Alternatively, a 3D analysis uses two or more cameras and should always be used to investigate activities that do not occur in a single plane. The procedures used to record images for 2D and 3D analyses are generally different and are therefore largely dealt with separately.

Irrespective of whether a 2D or 3D analysis is to be used, the camera(s) should be mounted on a rigid tripod and, normally, once in the required position should not be moved during recording. Once data collection is complete, the locations of body segment end points (often joint centres) are digitised using appropriate software. To facilitate this process, individuals should ideally wear minimal and tight-fitting clothing that contrasts with the colour of the background during recording. The background should also be uncluttered and non-reflective. Joint centres can also be marked on the skin, either directly with a soluble pen or using stickers that contrast with the skin's colour. While such markers are useful in identifying joint centres, they should not be relied upon to accurately represent the underlying segment end points; particularly when segments rotate out of the photographic plane.

For 2D analyses, **the camera should be positioned as far from the plane of motion as possible** to reduce the effects of perspective error. The telephoto zoom lens can then be used to bring the image of the participant to the required size in the field of view. This should not be too large, so as to cut some of the activity, or too small so that the individual cannot be digitised accurately. Perspective error occurs when objects or parts of objects that are closer to the lens appear larger than those that are further away. One of the effects of perspective error is the apparent shortening of body segments when they move out of the plane of motion (*Figure H5.3*), which inevitably occurs during even the most apparently planar of activities (e.g. running). In addition, perspective error results in angles between segments becoming more obtuse when they are moved out of the plane (*Figure H5.4*). As stated above, 3D analysis should be used when the motion of body segments does not occur in a single plane and information gained from a 2D analysis would largely be inaccurate due to perspective error.

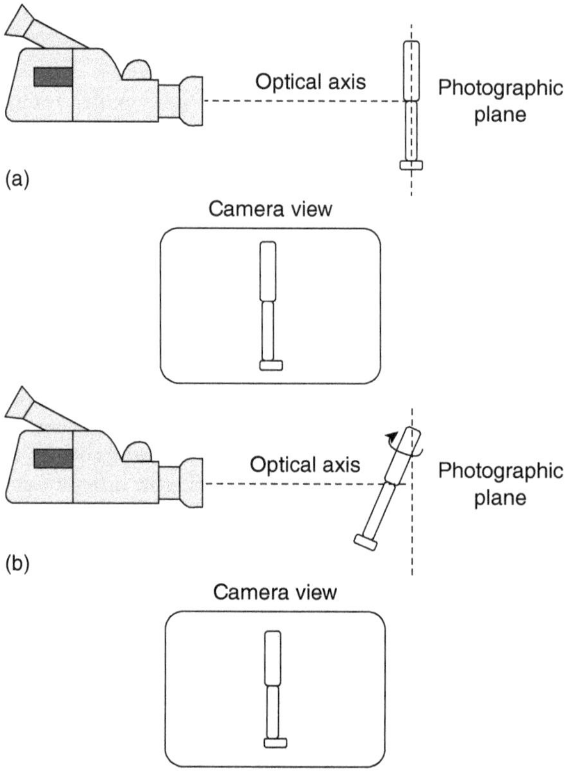


**Figure H5.3** (a) Plane of body segments coincides with photographic plane; and (b) body segments moved out of photographic plane, showing shorter segment length in camera view

To limit the effect of perspective error, the optical axis of the camera (i.e. an imaginary line passing through the middle of the lens) must be oriented at 90 degrees to the intended plane of motion (*Figure H5.5a*). Assuming that the plane of motion is vertical (e.g. during running), this can be partly achieved by placing a spirit level on top of the camera and positioned both parallel and perpendicular to its optical axis. Of course, this assumes that the top surface of the camera is both horizontal and level. If not, the height of the centre of the lens can be measured and a marker placed in the plane of motion at the same height. The lens can then be zoomed in on the marker, which should remain in the middle of the image. When the required field of view is confirmed, the bottom of the image should also be parallel with a line that is known to be horizontal (e.g. the ground). A 3–4–5 triangle (or multiples thereof) can be used to ensure that the optical axis is aligned at 90 degrees to the plane of motion, in the horizontal plane (*Figure H5.5b*). A plumb line should be used to ensure that the apex of the triangle, or a line extending from this point, is positioned directly below the centre of the lens (i.e. the optical axis).

Once the camera is appropriately oriented, vertical and horizontal scaling objects (e.g. one-metre rulers) need to be placed in the photographic plane/plane of motion and included in the field of view during recording. These objects are required to facilitate the later conversion of measures made in the software's digital environment (e.g. pixels/second) to more meaningful real-world outcomes (e.g. metres/second). A vertical reference (e.g. a plumb line) should also be recorded.

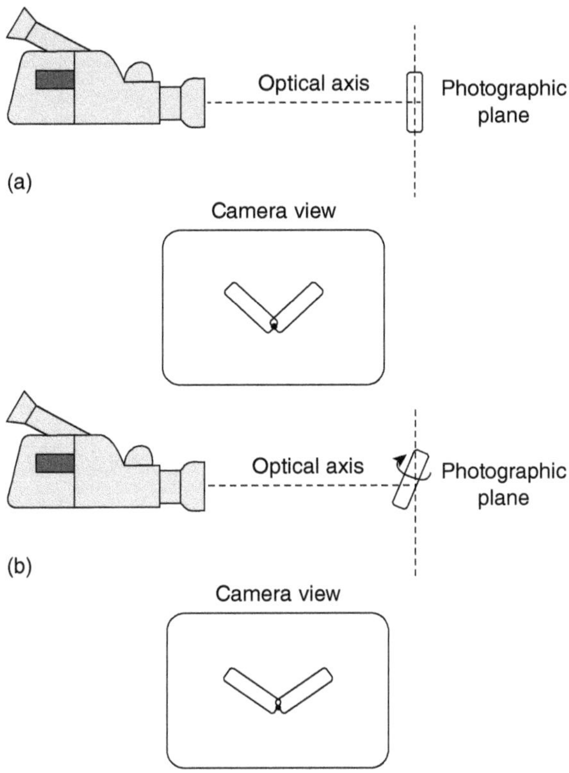


**Figure H5.4** (a) Plane of body segments coincides with photographic plane; and (b) body segments moved out of photographic plane, showing more obtuse segment angle in camera view

If the action occurs over a relatively long path (e.g. long jump or cricket bowler's run-up), use of a single camera will result in a field of view where the individual is too small to be digitised accurately. In such conditions, a number of synchronised cameras can be used with their fields of view overlapping slightly. Furthermore, in conditions of poor lighting (e.g. indoors with a shutter setting in excess of 1/250th of a second), it is likely that floodlights positioned at about 30 degrees to the plane of performance will be needed to illuminate the action. A short video series concerning 2D motion analysis can be accessed at: http://www.youtube.com/playlist?list=PLE90993667582FD1D (last accessed January 2019)

## 3D analysis

Most 3D assessments of human movement are now performed using the more sophisticated high-speed optoelectronic motion analysis systems that are described elsewhere in this text. However, as more traditional video-based 3D motion analysis systems are still used in some laboratories around the world, the following introduces the techniques underpinning this approach.

To perform a video-based 3D analysis of human movement, two or more cameras are required to film the activity. Ideally the optical axes of these cameras should intersect at

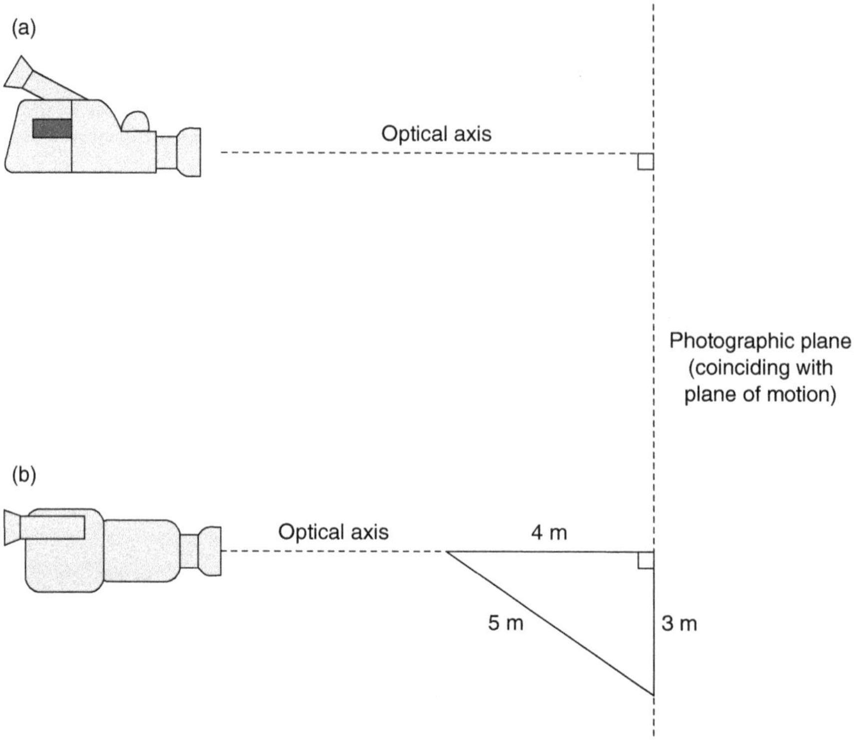


**Figure H5.5** Optical axis of the camera aligned at 90° to the photographic plane (a) when viewed from the side, and (b) when viewed from above

approximately 90 degrees, but this angle can range between 60 and 120 degrees (*Figure H5.6*). To facilitate accurate reconstruction of the data, the cameras should also be gen-locked, so that their shutters open at exactly the same time and their respective video fields remain perfectly synchronised. However, if this is not possible, due to the cabling required (*Figure H5.6*), a timing device should be included within the field of view of all cameras.

In replacement of the scaling object used in 2D analysis, at least six control points should be distributed throughout the volume in which the activity takes place. Use of more than six points provides more accurate reconstruction of real-life positions of body markers from digitised coordinates. For activities that occur in a relatively small volume the control points are usually contained within a calibration frame that can be dismantled after use (*Figure H5.7*). The exact location of each control point must be known and is usually expressed in relation to one of the points on the frame that forms the origin of three orthogonal (X, Y, and Z) axes (*Figure H5.6* and *Figure H5.7*). Where the activity takes place in a larger volume, the frame can be repositioned throughout it, or alternatively a series of poles containing control points can be used. Each control point must be visible by all cameras, and the structure containing them obviously must be removed prior to the activity being recorded. Similar to 2D analysis, in conditions of poor lighting (e.g. indoors with a shutter setting in excess of 1/250th of a second), the action should be illuminated with floodlights positioned beside each camera.

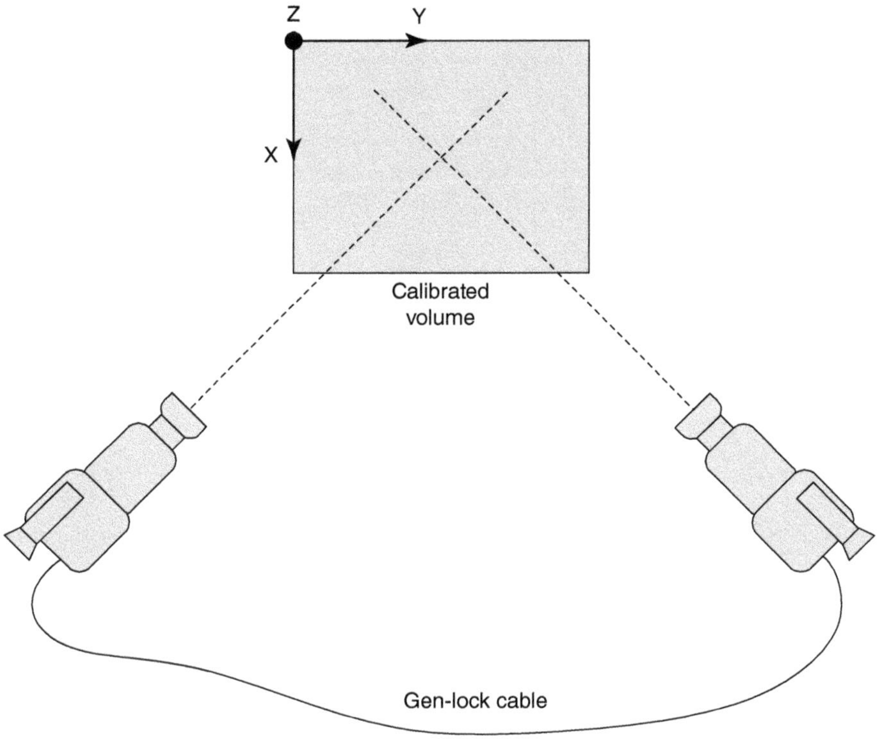


**Figure H5.6** Plan view of typical camera position used in a 3D video analysis

## DIGITISING

For video-based motion analysis techniques, the process of digitising essentially places an imaginary grid over the video field, or image, with the same number of coordinates as pixels. The bottom left-hand corner of the grid coincides with that of the image and shows the x and y coordinates as 0,0 (*Figure H5.8*). From this point each of the grid's horizontal lines represents a new vertical or y coordinate, which increases from the bottom to the top of the image. Similarly, each vertical line constitutes a different horizontal or x coordinate that increases from left to right. The original process of digitising involved using the computer mouse to move a cursor over the image to locate points of interest, which are usually body segment end points. Clicking a mouse button would record the x and y coordinates of the point (*Figure H5.9*). If the kinematics of the whole body centre of mass are required then 18 points on the body are typically digitised, although this depends on the anthropometric model used. Digitised points are joined together by the computer to form a stick figure (*Figures H5.8* and *H5.9*) or often, in many cases, a more humanoid representation. Modern motion analysis systems are now able to perform the digitisation process automatically.

The number of x and y coordinates that form the digitising grid is known as the resolution of the digitising system, which represents the smallest change in position that it can detect (typically 768 × 576). Recent software developments that enable the image to be zoomed have

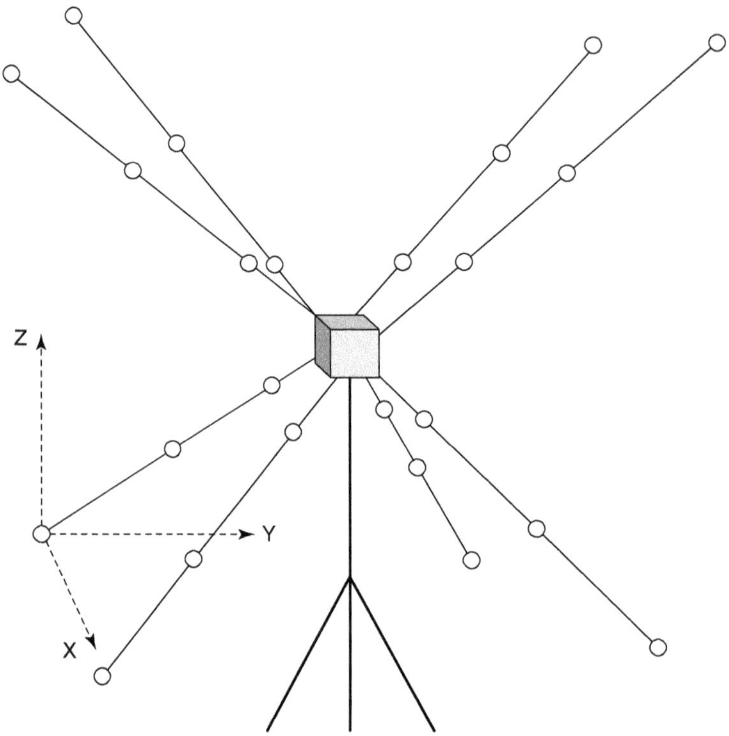


**Figure H5.7** Example of a calibration frame used in 3D analysis

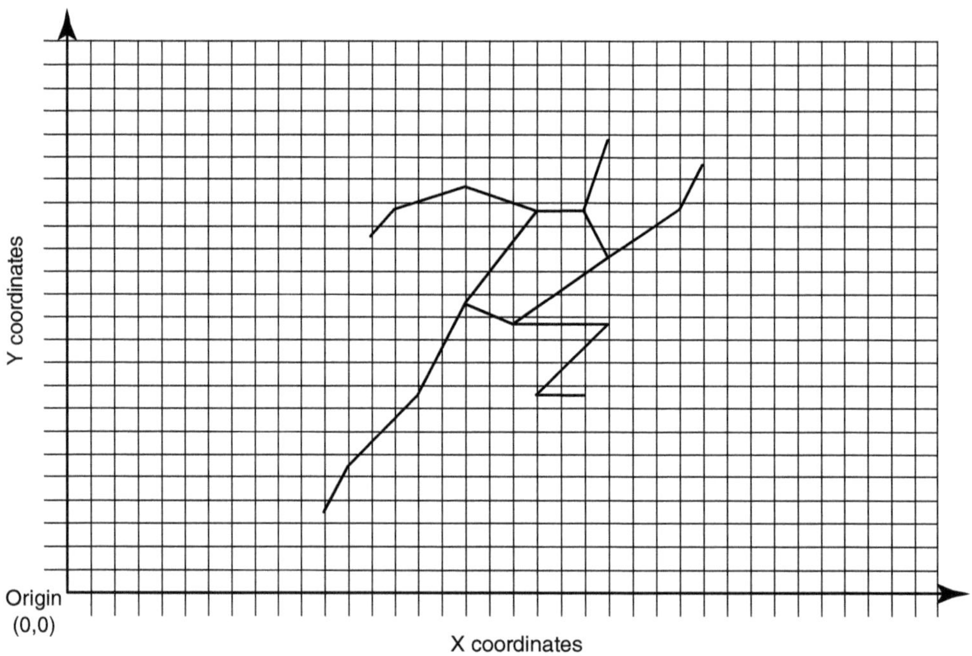


**Figure H5.8** Depiction of a video digitising system

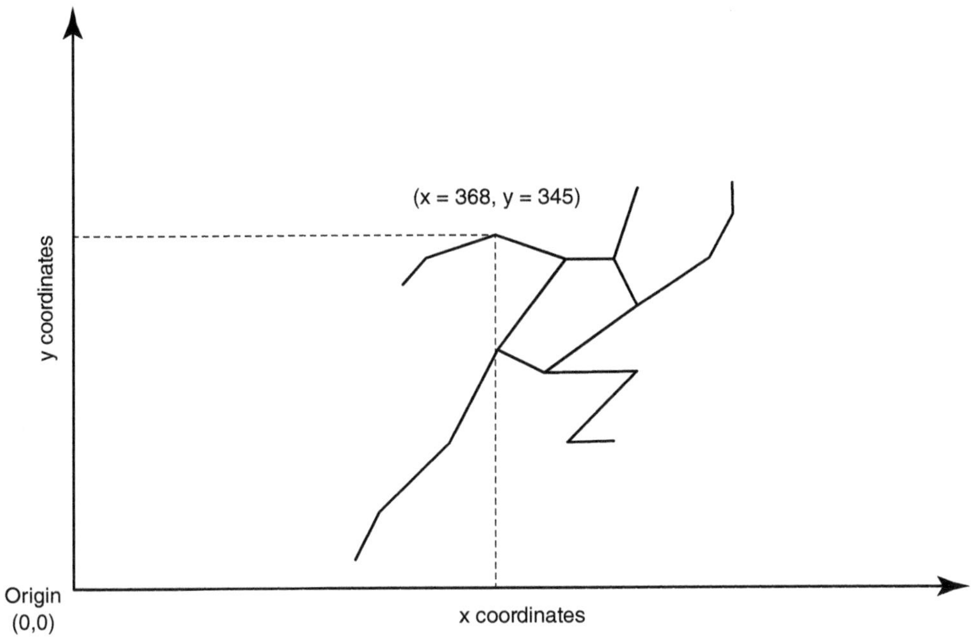


**Figure H5.9** Depiction of digitising of the elbow joint

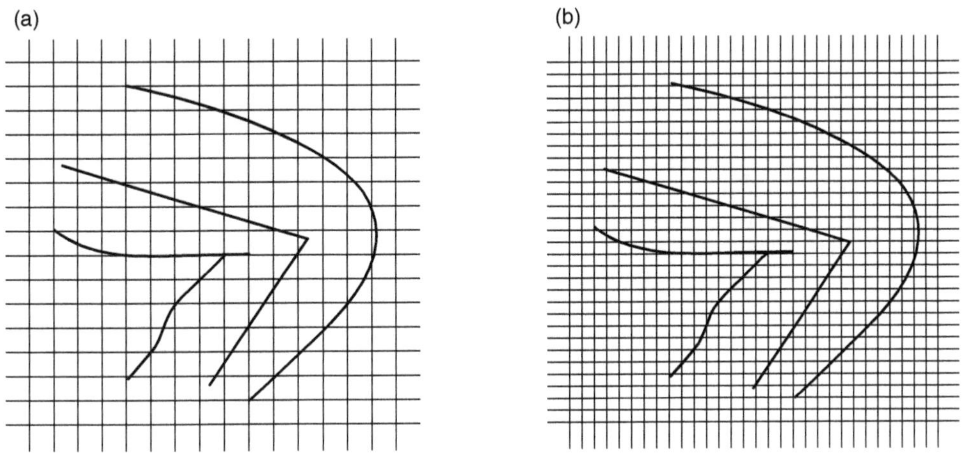


**Figure H5.10** Effects of improved resolution of the digitising system from (a) to (b)

improved the resolution of video systems. Higher digitiser resolution (*Figure H5.10*), now used in most modern motion analysis systems, improves the accuracy of the digitised coordinates.

## SCALING AND RECONSTRUCTION

Digitised coordinates need to be converted into real-life horizontal and vertical positions of body landmarks. This is commonly referred to as "scaling" in 2D video analysis and, assuming that the plane of motion coincides with the photographic plane, is achieved by firstly digitising the horizontal and vertical scaling objects that were included in the field of view. The length of the scaling object (in metres) is then divided by the number of coordinates that it represents to produce a scaling factor for the horizontal and vertical directions. Coordinates of body landmarks are multiplied by the respective scaling factor to give their true position (in metres) in relation to the origin, which usually coincides with the bottom left corner of the image.

Following digitisation in a video-based 3D analysis, the 2D coordinates of body landmarks are obtained from the images recorded by each camera. Before these sets of x–y coordinates can be reconstructed into real-life horizontal (X and Y) and vertical (Z) positions, it is imperative that the images from each camera are synchronised. If the cameras are not gen-locked, then the timing device included in the fields of view can be used for synchronisation. This can be done by selecting one of the cameras to be the time base. A curve consisting of a series of third-degree polynomials, known as a cubic spline, can then be fitted to the coordinate time data from the other camera(s) and the data interpolated to the same time base.

Now that two (or more) sets of synchronised x-y coordinates have been obtained, they can be reconstructed into a single set of X, Y, Z real-life coordinates. Most video-based motion analysis software uses the Direct Linear Transformation (DLT), which has been described elsewhere in this text and is represented in the following equations:

$$x = \frac{\ell_1 X + \ell_2 Y + \ell_3 Z + \ell_4}{\ell_9 X + \ell_{10} Y + \ell_{11} Z + 1}$$

$$y = \frac{\ell_5 X + \ell_6 Y + \ell_7 Z + \ell_8}{\ell_9 X + \ell_{10} Y + \ell_{11} Z + 1}$$

The eleven calibration coefficients, or parameters ($\ell 1$ to $\ell 11$), included in the DLT equations above represent the location and orientation of the cameras and replace the scaling factor used in 2D analysis. In order to calculate these parameters, the six (or more) control points that were included within the field of view of each camera must first be digitised. Assuming only six control points were used, this results in twelve new equations for each camera; one for each digitised x and y coordinate from each control point. As the X, Y, and Z coordinates of each control point are known, the twelve equations can be solved using a least squares approach to obtain the eleven DLT parameters.

With the eleven DLT parameters known, the DLT equations can now be used to find the X, Y, and Z coordinates of the digitised body landmarks. Again, assuming only two cameras are used, each body landmark will have two pairs of digitised x–y coordinates. These are inserted into the DLT equations to form four new equations, which can be rearranged and solved to find the X, Y, and Z real-life coordinates of the landmarks.

# MATHEMATICS

## Algebra and trigonometry

*Paul Grimshaw*

The following examples of algebra and trigonometry are commonly used in biomechanics and are representative of some of those that have been used within this text. **Algebra** refers to the branch of mathematics that generalises arithmetic by using variables for numbers (i.e. x + y = a + b).

### ALGEBRA

#### The rules of signs

| | | | | | |
|---|---|---|---|---|---|
| + | × | + | = | + | PLUS |
| + | × | – | = | – | MINUS |
| – | × | + | = | – | MINUS |
| – | × | – | = | + | PLUS |

Any number multiplied by zero (0) equals zero (0)

#### Example of multiplication of different signs

| | | | | | |
|---|---|---|---|---|---|
| –8 | × | 3 | × | –6 | = |

Carry out parts of the calculation first and introduce brackets.

| | | | | | | | |
|---|---|---|---|---|---|---|---|
| (–8 | × | 3) | × | –6 | = | | |
| (–8 | × | 3) | | | = | –24 | (Part 1) |
| –24 | × | –6 | | | = | +144 | (Part 2) |

(Note: it is not normally necessary to put the plus sign before a number)

Summary of rules for division of positive and negative integers. An integer is classified as a number that may be expressed as the sum or difference of two natural numbers. A natural number is any positive integer (i.e. 1, 2, 3, 4 etc.).

| | | | | | |
|---|---|---|---|---|---|
| +a | ÷ | +b | = | + | (a/b) |
| +a | ÷ | –b | = | – | (a/b) |
| –a | ÷ | +b | = | – | (a/b) |
| –a | ÷ | –b | = | + | (a/b) |

Example

| | | | | | | | | |
|---|---|---|---|---|---|---|---|---|
| 64 | ÷ | 8 | ÷ | −2 | ÷ | 2 | = |
| (64/8) | ÷ | −2 | ÷ | 2 | = | | |
| (8/−2) | ÷ | 2 | = | | | | |
| −4 | ÷ | 2 | = | −2 | | | |

## Solution of problems involving two or more arithmetic operations

Rules of precedence:

1 Evaluate terms in brackets
2 Multiplication and division
3 Addition and subtraction

Order of working can be remembered by using the **BODMAS** rule

| B | O | D | M | A | S |
|---|---|---|---|---|---|
| ( ) | of | ÷ | × | + | − |
| First | | | | | Last |

**BODMAS** tells you the order in which to perform calculations if you have a choice. Brackets (B) first, then of (O) (such as square root **of** 4 or 3 to the power **of** 5 (i.e., operations)), then divide (D), multiply (M), add (A) and subtract (S).

| Example | | | | | | | | | | |
|---|---|---|---|---|---|---|---|---|---|---|
| 3 × | (2 | + | 5) | + | 6 × | (7 | − | 4) | = | |
| 3 × | 7 | | | + | 6 × | 3 | | | = | |
| 21 | | | | + | | 18 | | | = | 39 |

## PERCENTAGES

32% of 69

$$= \frac{32 \times 69}{100}$$

= **22.08**

What % of 79 is 37?

$$\frac{37}{79} \times 100$$

= **46.84%**

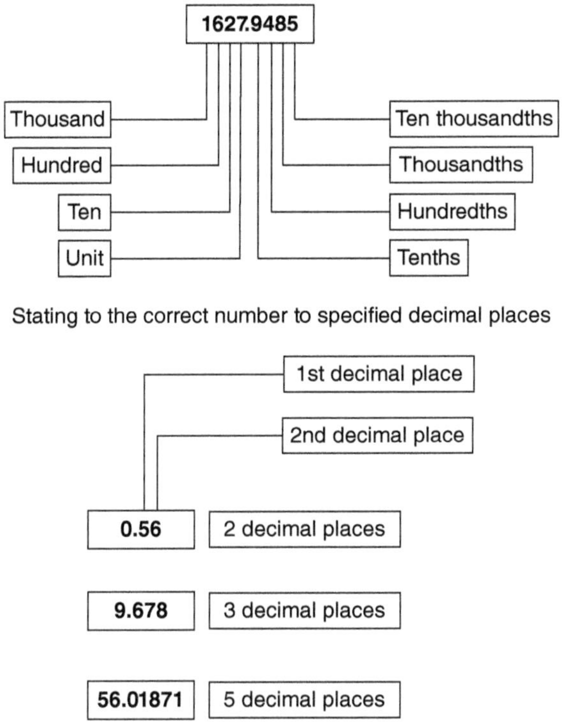


**Figure H6.1** Presenting a number to the correct decimal place

# DECIMALS

## Rules

The last decimal place that is specified is **unchanged** if the digit that follows it is 4 or less. The last decimal place specified is **increased** by 1 if the digit that follows it is 5 or more.

## Example

| | |
|---|---|
| 19.3864 | Stated as 19.386 when specified to 3 decimal places |
| 0.01337 | Stated at 0.0134 when specified to 4 decimal places |

### *Powers*

| | |
|---|---|
| 4 × 4 | Raised to the power of 2 or squared |
| 4 × 4 × 4 | Raised to the power of 3 or cubed |
| 6 × 6 × 6 × 6 | Raised to the power of 4 |

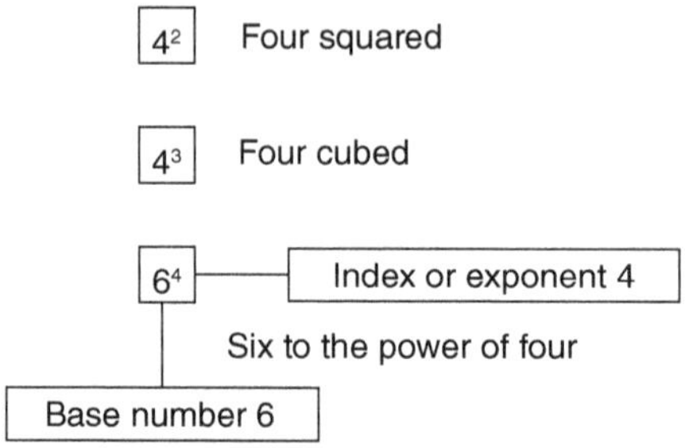


**Figure H6.2** Showing the written expression

# WRITTEN EXPRESSION

## Rules of indices

### *Multiplication rule*

| | |
|---|---|
| $10^2 \times 10^4 = 10^6$ | Add the indices |

This works when the base numbers are the same

$2^2 \times 2^4 = 2^6$

$4 \times 16 = 64$

However when the base numbers are not the same

$3^4 \times 5^5 = 81 \times 3125$

$81 \times 3125 = 253125$

Any number expressed to the power of zero equals 1.

$6.65^0 = 1$

$3^0 = 1$

### *Division rule*

| | |
|---|---|
| $10^5 \div 10^2 = 10^3$ | Subtract the indices |

This works when the base numbers are the same.

$2^8 \div 2^4 = 2^4$

$256 \div 16 = 16$

However when the base numbers are not the same

$8^4 \div 3^6 = 4096 \div 729$

$4096 \div 729 = 5.62$ (2 decimal places)

### *Raising indices to a power rule*

MULTIPLY INDICES

$(10^3)^2 = 10^{3 \times 2} = 10^6$

$(8^4)^3 = 8^{4 \times 3} = 8^{12}$

### *Summary*

| | | |
|---|---|---|
| MULTIPLICATION | $a^m \times a^n = a^{m+n}$ | add indices |
| DIVISION | $a^m \div a^n = a^{m-n}$ | subtract indices |

Applies when base numbers are the same

| | | |
|---|---|---|
| RAISING TO A POWER | $(a^m)^n = a^{m \times n}$ | multiply indices |
| $a^0 = 1$ | | For all numbers |

## TRANSPOSITION OF FORMULA

$$6x - 10 = \frac{3x}{3}$$

Solve for x Multiply both sides by 3 to remove the fractional component on the right

$$3(6x - 10) = 3\frac{3x}{3}$$

Cancel out

$$3(6x - 10) = \cancel{3}\frac{3x}{\cancel{3}}$$

$$3(6x - 10) = 3x$$

| | |
|---|---|
| 3 (6x – 10) | = 3x |
| 18x – 30 | = 3x |
| 18x – 3x – 30 | = 0 |
| 18x – 3x | = 30 |
| 15x | = 30 |
| x | $= \frac{30}{15}$ |
| x | = 2 |

$$3x + 4 = 2$$

$$3x = 2 - 4$$

$$x = \frac{2-4}{3}$$

$$x = -\frac{2}{3}$$

*Example*

$$9x + 8 - 4 = 6x$$

$$9x - 6x + 8 - 4 = 0$$

$$9x - 6x = -8 + 4$$

$$3x = -4$$

$$x = -\frac{4}{3}$$

$$x = -1.33$$

## General rules regarding transposition of formula

Negative quantity on one side of the equation becomes a positive quantity when it is transferred (transposed) to the other side of the equation. Similarly a divisor on one side of the equation becomes a multiplier when it is transferred to the other side of the equation (i.e. either side of the equal sign).

Solve for x:

$$8(x + 2) = 3(x - 3) + 45$$

Multiply out the brackets:

$$8x + 16 = 3x - 9 + 45$$

Transpose formula:

$$8x - 3x = -9 + 45 - 16$$

$$5x = 20$$

$$x = \frac{20}{5}$$

$$x = 4$$

Solve for x when:

$$\frac{4x+9}{3}+\frac{3x+7}{4}=11$$

Determine the lowest common denominator = 3 × 4

Reduce the fractions by multiplying both sides 3 × 4

$$3\times4\left(\frac{4x+9}{3}+\frac{3x+7}{4}\right)=3\times4\times11$$

Multiply out brackets:

$$3\times4\left(\frac{4x+9}{3}\right)+3\times4\left(\frac{3x+7}{4}\right)=3\times4\times11$$

Cancel out where possible:

$$\cancel{3}\times4\times\left(\frac{4x+9}{\cancel{3}}\right)+3\times\cancel{4}\left(\frac{3x+7}{\cancel{4}}\right)=3\times4\times11$$

Multiply out brackets:

| | |
|---|---|
| 4(4x + 9) + 3(3x + 7) | = 3 × 4 × 11 |
| 16x + 36 + 9x + 21 | = 3 × 4 × 11 |
| 16x + 9x | = (12 × 11) – 36 – 21 |
| 25x | = 132 – 57 |
| 25x | = 75 |
| x | $= \frac{75}{25}$ |
| x | = 3 |

## POWERS

$ay^2$ means a × y × y

2ay means 2 × a × y

## Solution of an expression

Example 1

Find the value of: $xy + 2yz + 3zx$

When $x = 3$, $y = 2$ and $z = 1$

$= (3 \times 2) + (2 \times 2 \times 1) + (3 \times 1 \times 3)$

$= 6 + 4 + 9$

$= 19$

Example 2

Find the value of: $\dfrac{2a + 4ab + 3c}{a + 2b + 3c}$

When $a = 6$, $b = 3$ and $c = 2$

$$= \frac{(2 \times 6) + (4 \times 6 \times 3) + (3 \times 2)}{6 + (2 \times 3) + (3 \times 2)}$$

$$= \frac{12 + 72 + 6}{6 + 6 + 6}$$

$$= \frac{90}{18}$$

$= 5$

### *Multiplication and division of algebraic functions*

$(-a) \times (-b) = +ab$

$(+a) \times (+b) = +ab$

$(-c) \times (+d) = -cd$

$(+c) \times (-d) = -cd$

To multiply two or more expressions, the rule is: find the product of the coefficients and prefix this product by the sign obtained from applying the rule of signs

### *Examples*

$4b \times 2b = 8b^2$

$-3a \times 4a = -12a^2$

$-4x \times -6x = 24x^2$

$+5y \times y = 5y^2$

Power (indices)

$4b^3 \times 3b^2 = 12b^5$

$6bc^2 \times 5b^4c^3 = 30b^5c^5$

$2xy^5 \times 8x = 16x^2y^5$

$3y^2 \times 4x^3y^4 = 12x^3y^6$

### *Expressions containing two or more terms*

$(2a + 2)(4a + 3)$

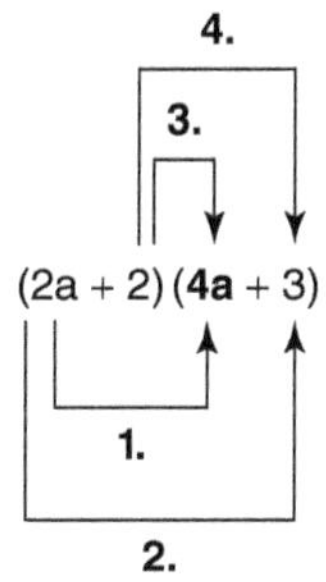


**Figure H6.3** The order of operation of terms in brackets

## ORDER OF OPERATION

$= 8a^2 + 6a + 8a + 6$

$= \mathbf{8a^2 + 14a + 6}$

### Example 1

$(3x^2 + 6)(4 - 2x)$

$12x^2 - 6x^3 + 24 - 12x$

### Example 2

$(2x^2 + 4y - 2)(2 + 3x - 4y)$

$4x^2 + 6x^3 - 8x^2y + 8y + 12xy - 16y^2 - 4 - 6x + 8y$

Combine like terms:

$4x^2 + 6x^3 - 8x^2y + 16y + 12xy - 16y^2 - 4 - 6x$

## TRIGONOMETRY

In addition to algebra, **trigonometry** is used a great deal in biomechanics and, hence, it is important to have a good understanding of the more common relationships. Essentially, this will be a revision of the trigonometry used within mathematics studied at school level, however, it is important in the context of this text. Many of the principles used within biomechanics are based on the right-angled triangle which is shown in the following examples.

Sum of the angles of a triangle = 180°

Right-angled triangle

| | |
|---|---|
| Pythagoras Rule | a = opposite |
| $b^2 = a^2 + c^2$ | b = hypotenuse |
| $b = \sqrt{a^2 + c^2}$ | c = adjacent |

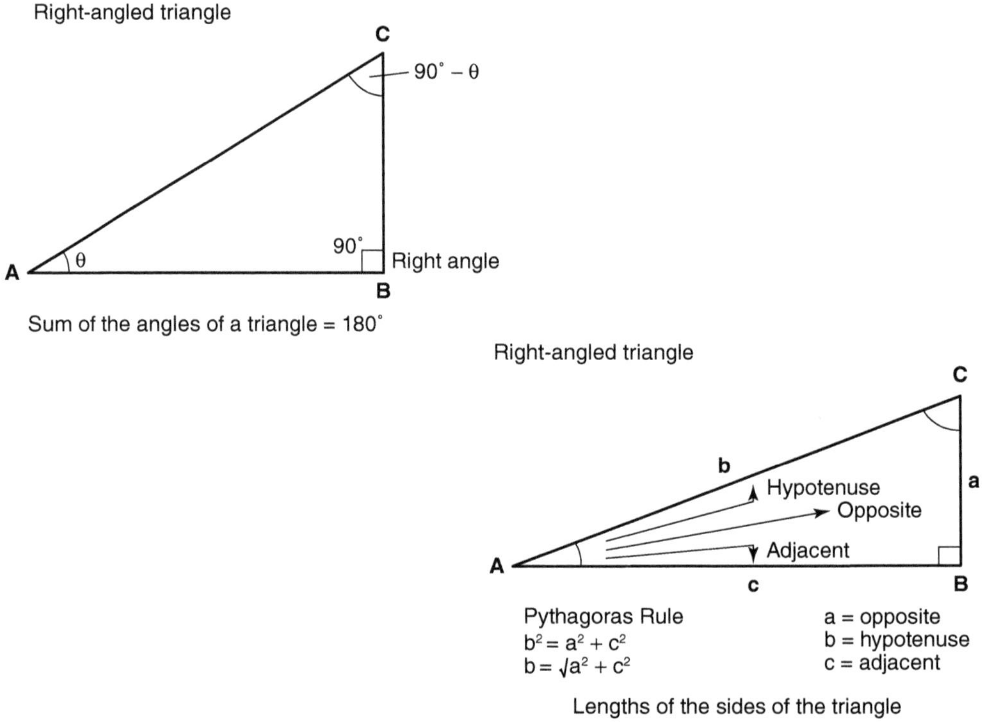


**Figure H6.4** The right-angled triangle

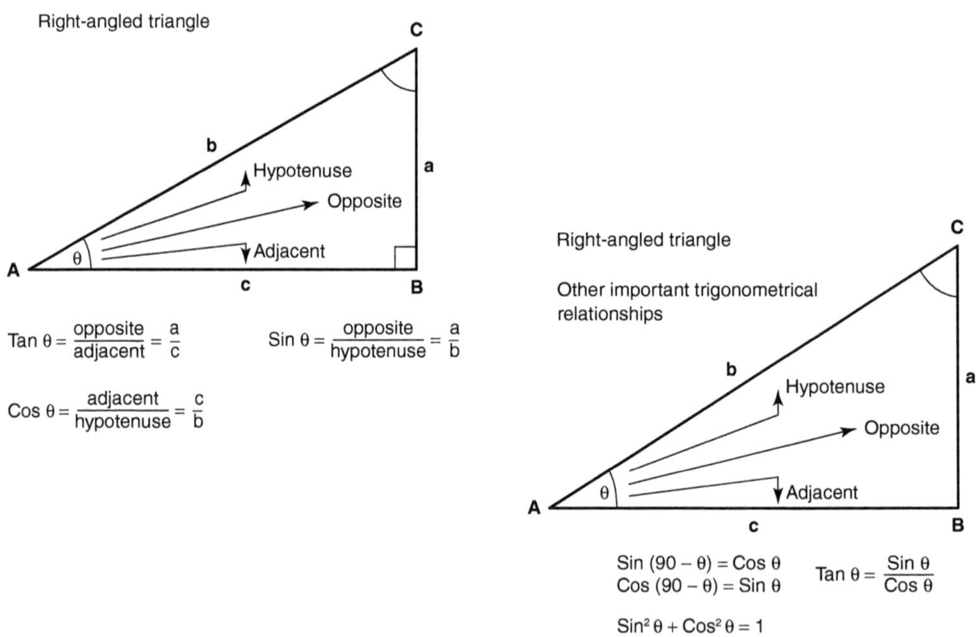


**Figure H6.5** The right-angled triangle: important relationships

Lengths of the sides of the triangle:

$$\text{Tan } \theta = \frac{\text{opposite}}{\text{adjacent}} = \frac{a}{c}$$

$$\text{Sin } \theta = \frac{\text{opposite}}{\text{hypotenuse}} = \frac{a}{b}$$

$$\text{Cos } \theta = \frac{\text{adjacent}}{\text{hypotenuse}} = \frac{c}{b}$$

$$\text{Sin } (90 - \theta) = \text{Cos } \theta$$

$$\text{Cos } (90 - \theta) = \text{Sin } \theta$$

$$\text{Tan } \theta = \frac{\text{Sin } \theta}{\text{Cos } \theta}$$

$$\text{Sin}^2 \theta + \text{Cos}^2 \theta = 1$$

Within biomechanics, there are also many examples where the application of trigonometry is required in triangles that are not right-angled. The following formulae are useful in this context.

## Application of trigonometry – all triangles

| Sine rule | Cosine rule |
|---|---|
| $\frac{a}{\text{Sin A}} = \frac{b}{\text{Sin B}} = \frac{c}{\text{Sin C}}$ | $a^2 = b^2 + c^2 - 2bc \text{ Cos A}$<br>$b^2 = a^2 + c^2 - 2ac \text{ Cos B}$<br>$c^2 = a^2 + b^2 - 2ab \text{ Cos C}$ |

## Example 1

Using the sine rule in the example shown in *Figure H6.6* to solve the problem for the length of the sides **a** and **b**.

### *Solution*

$$A = 28°$$

$$C = 32°$$

$$c = 23 \text{ cm}$$

Calculate the lengths of the sides **a** and **b**:

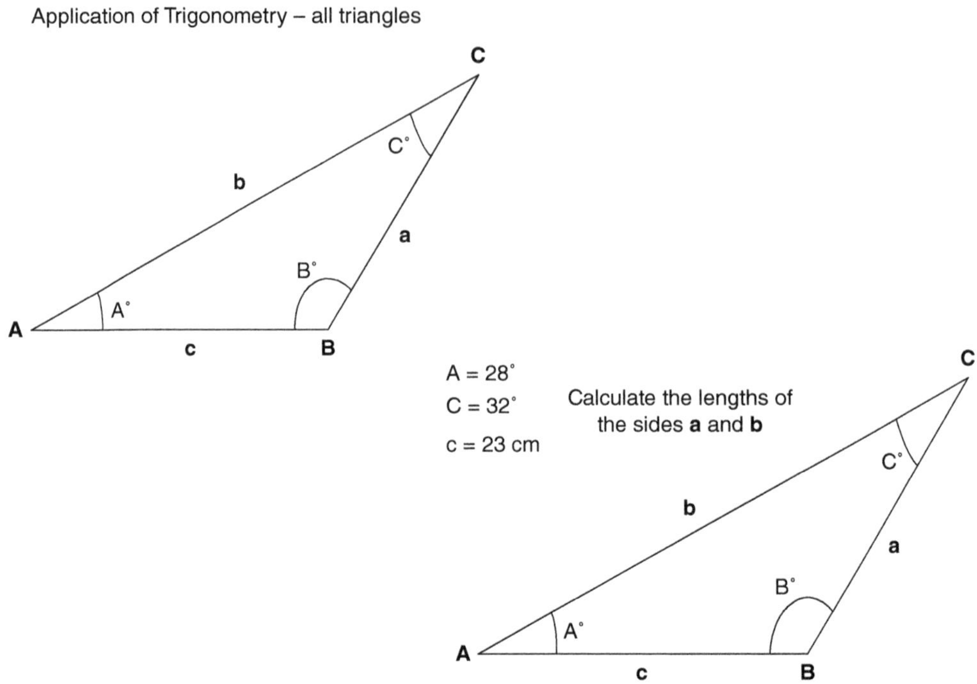


**Figure H6.6** Application of trigonometry for all triangles

---

Sine rule

$$\frac{a}{\text{Sin A}} = \frac{b}{\text{Sin B}} = \frac{c}{\text{Sin C}}$$

Using

$$\frac{a}{\text{Sin A}} = \frac{c}{\text{Sin C}}$$

Solution

$$\frac{a}{\text{Sin } 28°} = \frac{23 \text{ cm}}{\text{Sin } 32°}$$

$$\frac{a}{0.47} = \frac{23}{0.53}$$

$$a = 0.47 \times \frac{23}{0.53}$$

$$a = 20.40 \textbf{ cm}$$

---

## *Solution – distance b*

### SOLUTION USING THE SINE RULE

$$\text{Angle B} = 180° - (28 + 32)$$

$$\text{Angle B} = 120°$$

$$\frac{b}{\text{Sin } 120°} = \frac{23 \text{ cm}}{\text{Sin } 32°}$$

$$\frac{b}{0.87} = \frac{23}{0.53}$$

$$b = 0.87 \times \frac{23}{0.53}$$

$$b = 37.75 \textbf{ cm}$$

# INDEX

For Product Safety Concerns and Information please contact our EU
representative GPSR@taylorandfrancis.com
Taylor & Francis Verlag GmbH, Kaufingerstraße 24, 80331 München, Germany

www.ingramcontent.com/pod-product-compliance
Ingram Content Group UK Ltd.
Pitfield, Milton Keynes, MK11 3LW, UK
UKHW052224190926
475702UK00007B/1191
* 9 7 8 1 1 3 8 6 4 0 2 3 8 *